CO-ARD-662

THIRD EDITION

INTRODUCTORY ALGEBRA

IGNACIO BELLO

Hillsborough Community College

with
JACK R. BRITTON

University of South Florida
Emeritus

DELLEN PUBLISHING COMPANY

San Francisco

COLLIER MACMILLAN PUBLISHERS

London

divisions of Macmillan, Inc.

IN MEMORY OF MY FATHER

On the cover: The painting on the cover, executed by Ronald Davis in 1988, is vinyl-acrylic copolymer and dry pigments on canvas. Ronald Davis is involved in expressing the illusion of a three-dimensional object rendered on a flat surface. His work can be seen in leading museums throughout the world, including the San Francisco Museum of Modern Art, the Los Angeles County Museum of Art, and the Whitney Museum of American Art. In Los Angeles, Davis is represented by the Asher–Faure gallery.

© Copyright 1990 by Dellen Publishing Company,
a division of Macmillan, Inc.

Printed in the United States of America

Permissions: Dellen Publishing Company
               400 Pacific Avenue
               San Francisco, California 94133

Orders: Dellen Publishing Company
               c/o Macmillan Publishing Company
               Front and Brown Streets
               Riverside, New Jersey 08075

Collier Macmillan Canada, Inc.

LIBRARY OF CONGRESS CATALOGING IN PUBLICATION DATA

Bello, Ignacio.
   Introductory algebra/Ignacio Bello with Jack R. Britton.—3rd ed.

  ISBN 0-02-307961-4
   1. Algebra.   I. Britton, Jack Rolf. 1908–   .   II. Title.
QA152.2.B452   1990   90–2830
512.9—dc20   CIP

Printing: 1 2 3 4 5 6 7 8 9   Year: 0 1 2 3 4

ISBN 0-02-307961-4

CONTENTS

CHAPTER 3   SOLVING EQUATIONS AND INEQUALITIES   188

CHAPTER 4   POLYNOMIALS   282

HOW TO USE THIS BOOK

1. Take the Pretest. If you score 80% or more, go to the next chapter.

2. If you score less than 80% on the Pretest, do the following:

    a. Read the Reviews at the beginning of each section.

    b. Read the Objectives in each section.

    c. Read and study the explanations and Examples.

    d. Do the Problems in the margin.

    e. Do the odd-numbered problems at the end of each section.

    f. Read the Summary at the end of the chapter.

3. Do part (a) of each problem in the Review Exercises. If you have the correct answer for 80% or more of the Review Exercises, take the Practice Test. Study the answers you missed and go on to the next chapter.

    If fewer than 80% of your answers are correct, consult your instructor or a qualified tutor, then do part (b) of the Review Exercises. If 80% or more of your answers are correct, take the Practice Test, study the answers you missed, and go on to the next chapter. If not, consult your instructor or a qualified tutor. You can also look at the complete solutions for the Practice Test in the *Student's Solutions Manual* or watch the videotape corresponding to the chapter you are working on. Keep trying the Review Exercises until 80% or more of your answers are correct, then take the Practice Test, study the answers you missed, and go on to the next chapter. The flowchart on the next page may simplify things for you.

We always welcome students' comments and suggestions. You may send them to us at the following address:

Ignacio Bello
Hillsborough Community College
P.O. Box 32127
Tampa, Florida 33622

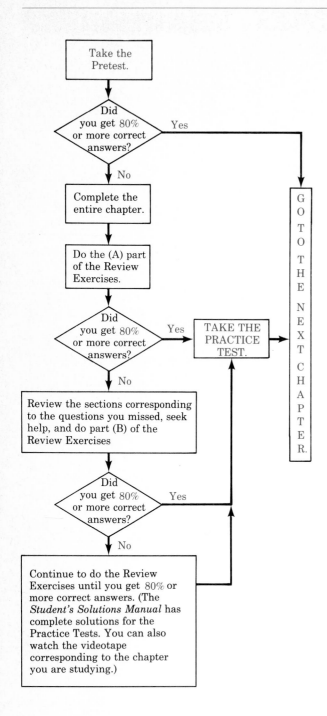

Take the
Pretest.

Did
you get 80%
or more correct
answers? — Yes

No

Complete the
entire chapter.

Do the (A) part
of the Review
Exercises.

Did
you get 80%
or more correct
answers? — Yes

No

TAKE THE
PRACTICE
TEST.

Review the sections corresponding
to the questions you missed, seek
help, and do part (B) of the
Review Exercises

Did
you get 80%
or more correct
answers? — Yes

No

Continue to do the Review
Exercises until you get 80% or
more correct answers. (The
Student's Solutions Manual has
complete solutions for the
Practice Tests. You can also
watch the videotape
corresponding to the chapter
you are studying.)

GO TO THE NEXT CHAPTER.

This book is intended for students who need help with their elementary or introductory algebra. To help these students, the following features have been incorporated in the book.

Pretests	At the beginning of each chapter. Pinpoint student's strengths and weaknesses. With diagnostic answers.
Reviews	At the beginning of each section. Tell students what they need to know in order to complete the section.
Objectives	At the beginning of each section. Tell students what they must learn. Coordinated with the Examples and Exercises.
Examples	Illustrate the topics under discussion. Coordinated with the Objectives and Exercises (Objective A goes with Example 1).
Margin Problems	Correlated with the corresponding Examples and designed to check the student's facility in applying the concepts developed. The answers are given in the margin on the next spread.
Exercises	To practice the principles being learned. Coordinated with the Objectives and Examples (Group A goes with Objective A).
Applications	Given at the end of most Exercise sets. Show students how the material is applied to other areas.
Skill Checkers	Test previous material needed in the succeeding section. For example: Review factorization before reducing fractional expressions.
Using Your Knowledge	Follows the Exercises and applies the ideas discussed in each section to different situations.
Calculators	The **Calculator Corner** section shows the students how the material under discussion can be handled using calculators.
Summary	At the end of each chapter. Highlights important concepts and their meaning. Examples are included when possible.
Review Exercises	At the end of each chapter. Review all objectives covered. Tell students in which section the question under consideration appears.
Practice Tests	With diagnostic answers. Tell students in which section, Example, and page the question under consideration appears.

Ancillary Materials

Student's Solutions Manual (odd answers and solutions to Pretests and Practice Tests)
Instructor's Manual (tests and all answers)
IPS Testing System (IBM and Apple)
Videos (lectures; solutions to Practice Tests)

SUGGESTED COURSES USING THIS BOOK

We have class-tested the entire book with both standard and special classes at Hillsborough Community College. From this testing and the many suggestions received by the reviewers, it is evident that the book can be used in various ways:

1. As a textbook in a traditional lecture course. Simply skip the margin problems and lecture as usual.
2. As a textbook in an individualized study course. Assign the reading portion and the exercises one day; answer questions and collect homework the next day.
3. As a lab text. Use as any other textbook in the mathematics lab.
4. As a combination of these methods.

WHAT IS NEW IN THE THIRD EDITION?

We have followed the suggestions of many users and reviewers of the first two editions to clarify the exposition, expand coverage, and, in general, improve the book. Specifically, we have:

• Added the **Reviews** at the beginning of each section.
• Introduced subsections A, B, C, and so on for ease in identifying the topic under discussion.
• Added the **Skill Checkers** at the end of the Exercises.
• Added two new chapters (**Systems of Linear Equations** and **Applied Geometry**)
• Provided many new worked examples, Using Your Knowledge problems, and hundreds of new Applications and problems.

ACKNOWLEDGMENTS

The author would like to express his appreciation to the following persons:

First Edition:

Stephen I. Gendler, Clarion Community College

Virginia M. Gies, Lincoln Land Community College

Leland R. Halberg, Lane Community College

Calvin A. Lathan, Monroe Community College

John Lester, Miami Dade Community College

Second Edition:

Wayne Andrepont, University of Louisiana

Charles Blanchard, University of Louisiana

Rosanne Donohoe, Northern Virginia Community College

Margaret Esser, Hillsborough Community College

George Grisham, Illinois Central College

Paul N. Hutchens, Florissant Valley Community College
Donald R. Johnson, Scottsdale Community College
George Kosan, Hillsborough Community College
Robert Kossanovich, Ferris State College
Peter Lindstrom, North Lake College
Susan C. Meyers, Sinclair Community College
Leonard M. Wapner, El Camino College

Cathy Sagendorf (typist)
Robert Balman, Joseph Clemente, and John Hunt (answer checkers)

REVIEWERS (THIRD EDITION)

Barbara Burrows, Santa Fe Community College
Diana B. Fernandez, Hillsborough Community College
Virginia Harris, Hillsborough Community College
Rosemary M. Karr, Eastern Kentucky University
Shirley Markus, University of Louisville
Sharon Ross, DeKalb College
Patricia Stanley, Ball State University
Ara B. Sullenberger, Tarrant County Jr. College District

Special thanks go to Liana Fox, who gave many suggestions and teaching insights; Josephine Rinaldo, who proofread the manuscript during several stages of production and finally decided to do it right by authoring the solutions manual; and Dr. Jack Britton, a gentleman, scholar, and bon vivant who edited the complete book several times while trying to retire. Last but not least, thanks to the wonderful production department provided by our editor Don Dellen. Without their help, this book would not have been possible.

CREDITS FOR CHAPTER OPENING SCULPTURES

Chapter 1: Ernest Trova, *Gox No. 4*, 1975. Stainless steel, height 9 feet. Storm King Art Center, Mountainville, New York. Purchased with the aid of funds from the National Endowment for the Arts.

Chapter 2: Isaac Witkin, *Kumo*, 1971. Cor-ten steel, height 16 feet 3 inches. Storm King Art Center, Mountainville, New York. Purchase.

Chapter 3: Menashe Kadishman, *Suspended*, 1977. Cor-ten steel, height 23 feet. Storm King Art Center, Mountainville, New York. Gift of Muriel and Philip I. Berman.

Chapter 4: Jerome Kirk, *Orbit*, 1972. Stainless steel, height 12 feet. Storm King Art Center, Mountainville, New York. Purchase.

Chapter 5: David Smith, *XI Books III Apples*, 1959. Stainless steel, height 94 inches. Storm King Art Center, Mountainville, New York. Gift of the Ralph E. Ogden Foundation.

Chapter 6: Tal Streeter, *Endless Column*, 1968. Steel painted red, height 62 feet 7 inches. Storm King Art Center, Mountainville, New York. Purchased with the aid of funds from the National Endowment for the Arts.

Chapter 7: Joseph Pillhofer, *Untitled*, circa 1950–60. Stone, height 91 inches. Storm King Art Center, Mountainville, New York. Gift of Joan O. Stern.

Chapter 8: Barnett Newman, *Broken Obelisk*. Modern Museum of Art.

Chapter 9: Joseph Konzal, *Simplex*, 1970. Cor-ten steel, height 91 inches. Storm King Art Center, Mountainville, New York. Purchase.

Chapter 10: Charles Ginnever, *Fayette: For Charles and Medgar Evers*, 1971. Cor-ten steel, height 8 feet 4 inches. Storm King Art Center, Mountainville, New York. Purchase.

THIRD EDITION

INTRODUCTORY ALGEBRA

CHAPTER 1

NUMBERS AND THEIR PROPERTIES

In this chapter we introduce the numbers we use in algebra (fractions and decimals), the operations we perform with these numbers, and the properties of these operations. We close the chapter with a section dealing with formulas.

(Answers on pages 6–7)

1. Write -13 as a fraction with a denominator of 1.

   1. _____

2. Find a fraction equivalent to $\dfrac{2}{7}$ with a denominator of 21.

   2. _____

3. Find a fraction equivalent to $\dfrac{6}{15}$ with a denominator of 5.

   3. _____

4. Reduce $\dfrac{27}{36}$ to lowest terms.

   4. _____

5. Find $5\dfrac{1}{4} \cdot \dfrac{20}{21}$.

   5. _____

6. Find $2\dfrac{1}{4} \div \dfrac{5}{8}$.

   6. _____

7. Find $1\dfrac{3}{10} + \dfrac{5}{12}$.

   7. _____

8. Find $4\dfrac{1}{6} - 1\dfrac{7}{10}$.

   8. _____

9. Write 48.123 in expanded form.

9. _____

10. Write $\dfrac{19}{1000}$ as a decimal.

10. _____

11. Write $\dfrac{3}{11}$ as a decimal.

11. _____

12. Write 0.025 as a reduced fraction.

12. _____

13. Write 2.12 as a reduced fraction.

13. _____

14. Write 43.5% as a fraction.

14. _____

15. Write 34.8% as a decimal.

15. _____

16. Write 0.49 as a percent.

16. _____

17. Write $\dfrac{3}{8}$ as a percent.

17. _____

18. Which law is illustrated in each of the following statements?
    a. $b \cdot (c \cdot a) = b \cdot (a \cdot c)$ **b.** $(5 + 9) + 2 = 5 + (9 + 2)$

18. a. _____ b. _____

19. Find.
    a. $(9 \times 8) + 4$ **b.** $9 \times (8 + 4)$

19. a. _____ b. _____

20. Find.
    a. $4(x + 7)$ **b.** $8(x + y + z)$

20. a. _____ b. _____

21. Remove parentheses (simplify).
    a. $(8 + 2a) + 4$ **b.** $(8a) \cdot 9$

21. a. _____ b. _____

22. Evaluate.

    $°C = \dfrac{5}{9}(°F - 32)$ when $°F = 50$

22. _____

23. A rectangle is 8.1 in. long and 1.9 in. wide. Find **(a)** its area and **(b)** its perimeter.

23. a. _____ b. _____

24. A circle has a 9-in. radius. Find **(a)** its area and **(b)** its circumference. (Use 3.14 for the value of π.)

24. a. _____ b. _____

25. Evaluate.

    $y \div 3x - z + 5$ for $x = 2$, $y = 12$, $z = 1$

25. _____

IF YOU MISSED QUESTION	SECTION	EXAMPLES	PAGE	ANSWERS
1	1.1	1	10	1. $\dfrac{-13}{1}$
2	1.1	2	11	2. $\dfrac{6}{21}$
3	1.1	3	11	3. $\dfrac{2}{5}$
4	1.1	4	12	4. $\dfrac{3}{4}$
5	1.2	1, 2, 3	17, 18	5. 5
6	1.2	4, 5	19, 20	6. $\dfrac{18}{5}$
7	1.2	6	22	7. $1\dfrac{43}{60}$, or $\dfrac{103}{60}$
8	1.2	8	24	8. $\dfrac{37}{15}$ or $2\dfrac{7}{15}$
9	1.3	1	32	9. $40 + 8 + \dfrac{1}{10} + \dfrac{2}{100} + \dfrac{3}{1000}$
10	1.3	2	32	10. 0.019
11	1.3	3	33	11. $0.\overline{27}$
12	1.3	4	34	12. $\dfrac{1}{40}$
13	1.3	5	35	13. $\dfrac{53}{25}$
14	1.4	1	41	14. $\dfrac{87}{200}$
15	1.4	2	42	15. 0.348
16	1.4	3	43	16. 49%
17	1.4	4	44	17. 37.5%
18a	1.5	2	52	18. a. Commutative $\times$
b	1.5	2	52	b. Associative $+$
19a	1.5	1	51	19. a. 76
b	1.5	1	51	b. 108
20a	1.5	3	53	20. a. $4x + 28$
20b	1.5	3	53	b. $8x + 8y + 8z$
21a	1.5	4	54	21. a. $2a + 12$
21b	1.5	4	54	b. $72a$
22	1.6	1, 2	61, 62	22. $10°$

IF YOU MISSED QUESTION	SECTION	EXAMPLES	PAGE	ANSWERS
23a	1.6	4	64	**23. a.** 15.39 in.^{2}
23b	1.6	4	64	**b.** 20 in.
24a	1.6	6	66	**24. a.** 254.34 in.^{2}
24b	1.6	6	66	**b.** 56.52 in.
25	1.6	7	67	**25.** 12

1.1 NUMBERS

OBJECTIVES

REVIEW

Before starting this section you should know how to add, subtract, multiply, and divide natural numbers.

OBJECTIVES

You should be able to:

A. Write an integer as a fraction.
B. Find a fraction equivalent to a given one with a specified denominator.
C. Reduce a fraction to lowest terms.

The symbols on the sundial and the Roman clock have something in common. They both use numerals to name the numbers from 1 to 12. Algebra and arithmetic also have something in common. They use the same numbers and the same rules.

A. NUMBERS OF ARITHMETIC

In arithmetic you learned about the counting numbers. The numbers used for counting are the **natural numbers:**

1, 2, 3, 4, 5, and so on

These numbers are used in algebra. We also use the **whole numbers: 0, 1, 2, 3, 4,** and so on. Later on, you probably learned about the **integers.** The integers include the **positive integers,**

$+1, +2, +3, +4, +5$, and so on

(read "positive one, positive two," and so on), the **negative integers,**

$-1, -2, -3, -4, -5$, and so on

and the number **0,** which is neither positive nor negative. Thus the integers are

$\ldots, -2, -1, 0, 1, 2, \ldots$

where the dots ($\ldots$) indicate that the enumeration continues without end. Note that $+1 = 1$, $+2 = 2$, $+3 = 3$, and so on. Thus, the positive integers are the natural numbers.

All the preceding numbers can be written as **common fractions** of the form $\frac{a}{b}$ (or a/b), in which the numerator a and the denominator b are both integers and **the denominator is not 0.** If a number can be written in this form, it is called a **rational number.** Thus, $\frac{1}{3}$, $\frac{-5}{2}$, and $\frac{0}{7}$ are rational numbers. Of course, any integer can be written as a fraction by writing it with a denominator of **1.** For example,

$$4 = \frac{4}{1}, \quad 8 = \frac{8}{1}, \quad 0 = \frac{0}{1}, \quad \text{and} \quad -3 = \frac{-3}{1}$$

EXAMPLE 1 Write the following numbers as fractions with a denominator of 1.

a. 10   **b.** -15

Solution

a. $10 = \dfrac{10}{1}$

b. $-15 = \dfrac{-15}{1}$ ▲

Problem 1 Write as fractions with a denominator of 1.*
a. 47   **b.** -13

The rational numbers we have discussed are part of a larger set of numbers, the set of real numbers. The **real numbers** include the *rational numbers* and the *irrational numbers*. Thus each real number is either rational or irrational.

The **irrational numbers** are numbers that cannot be written as the ratio of two integers. For example, $\sqrt{2}$, π, $-\sqrt[3]{10}$, and $\frac{\sqrt{3}}{2}$ are irrational numbers.

We will say more about the irrational numbers later in the book.

B. EQUIVALENT FRACTIONS

Can you find other ways of writing 10 as a fraction? Here are some:

$$10 = \frac{10}{1} = \frac{10 \times 2}{1 \times 2} = \frac{20}{2}$$

$$10 = \frac{10}{1} = \frac{10 \times 3}{1 \times 3} = \frac{30}{3}$$

and

$$10 = \frac{10}{1} = \frac{10 \times 4}{1 \times 4} = \frac{40}{4}$$

Note that $\frac{2}{2} = 1$, $\frac{3}{3} = 1$, and $\frac{4}{4} = 1$. As you can see, the fraction $\frac{10}{1}$ is **equivalent** to (has the same value as) many other fractions. We can always obtain other fractions equivalent to any given fraction by *multiplying* the numerator and denominator of the original fraction by the *same* nonzero number. This is the same as multiplying the fraction by 1. For example,

$$\frac{3}{5} = \frac{3 \times 2}{5 \times 2} = \frac{6}{10}$$

$$\frac{3}{5} = \frac{3 \times 3}{5 \times 3} = \frac{9}{15}$$

and

$$\frac{3}{5} = \frac{3 \times 4}{5 \times 4} = \frac{12}{20}$$

(In Section 1.2 you will learn that $\frac{3}{5} \times 1 = \frac{3}{5} \times \frac{2}{2} = \frac{6}{10}$, $\frac{3}{5} \times 1 = \frac{3}{5} \times \frac{3}{3} = \frac{9}{15}$, and $\frac{3}{5} \times 1 = \frac{3}{5} \times \frac{4}{4} = \frac{12}{20}$).

*The answers to the problems in the margin always appear at the bottom of the margin on the next spread.

EXAMPLE 2 Find a fraction equivalent to $\frac{3}{5}$ with a denominator of 20.

Solution We must solve the problem

$$\frac{3}{5} = \frac{?}{20}$$

Now, the denominator, 5, was multiplied by 4 to get 20.

$$\frac{3}{5} = \frac{?}{20} \quad \text{If you multiply the denominator by 4,}$$
$$\underline{\hspace{0.5cm}\text{Multiply by 4}\hspace{0.5cm}}$$

$$\overline{\hspace{0.5cm}\text{Multiply by 4}\hspace{0.5cm}}$$
$$\frac{3}{5} = \frac{12}{20} \quad \text{you have to multiply the numerator by 4.}$$

So we also multiplied the numerator, 3, by 4 to get 12. ▲

Here is a slightly different problem. Can we find a fraction equivalent to $\frac{15}{20}$ with a denominator of 4? We shall do this in the next example.

EXAMPLE 3 Find a fraction equivalent to $\frac{15}{20}$ with a denominator of 4.

Solution

$$\frac{15}{20} = \frac{?}{4} \quad \text{20 was divided by 5 to get 4.}$$
$$\underline{\hspace{0.5cm}\text{Divide by 5}\hspace{0.5cm}}$$

$$\overline{\hspace{0.5cm}\text{Divide by 5}\hspace{0.5cm}}$$
$$\frac{15}{20} = \frac{3}{4} \quad \text{15 was divided by 5 to get 3.} \quad ▲$$

We can summarize our work with equivalent fractions in the following rule:

> To obtain an **equivalent fraction,** *multiply* or *divide* both numerator and denominator of the fraction by the *same* nonzero number.

C. REDUCING FRACTIONS TO LOWEST TERMS

The preceding rule can be used to *reduce* fractions to lowest terms. A fraction is *reduced* to lowest terms when there is no number (except 1) that will divide the numerator and the denominator. This is done as follows:

> To **reduce** a fraction to *lowest* terms, divide the numerator and denominator by the *largest* whole number that will divide them exactly.*

---

*"Divide them exactly" means that the quotients are whole numbers.

Problem 2 Find a fraction equivalent to $\frac{7}{6}$ with a denominator of 18.

Problem 3 Find a fraction equivalent to $\frac{18}{30}$ with a denominator of 5.

Thus, to reduce $\frac{12}{30}$ to lowest terms, divide the numerator and denominator by 6, the *largest* whole number that divides 12 and 30 exactly. (6 is sometimes called the *greatest common divisor* (GCD) of 12 and 30) Thus,

$$\frac{12}{30} = \frac{12 \div 6}{30 \div 6} = \frac{2}{5}$$

This reduction is sometimes shown like this:

$$\frac{\overset{2}{\cancel{12}}}{\underset{5}{\cancel{30}}} = \frac{2}{5}$$

EXAMPLE 4   Reduce to lowest terms.

a. $\dfrac{15}{20}$ **b.** $\dfrac{30}{45}$ **c.** $\dfrac{60}{48}$

Problem 4   Reduce to lowest terms.

a. $\dfrac{10}{15}$ **b.** $\dfrac{18}{24}$

Solution

a. The *largest* whole number exactly dividing 15 and 20 is **5**. Thus,

$$\frac{15}{20} = \frac{15 \div 5}{20 \div 5} = \frac{3}{4}$$

b. The *largest* whole number exactly dividing 30 and 45 is **15**. Hence,

$$\frac{30}{45} = \frac{30 \div 15}{45 \div 15} = \frac{2}{3}$$

c. The *largest* whole number dividing 60 and 48 is **12**. We then write:

$$\frac{60}{48} = \frac{60 \div 12}{48 \div 12} = \frac{5}{4} \qquad\qquad ▲$$

What if you are unable to find the *largest* number dividing the numerator and denominator in $\frac{30}{45}$? There is no problem, but it takes a little longer. Say you notice that 30 and 45 are both divisible by **5**. You then write:

$$\frac{30}{45} = \frac{30 \div 5}{45 \div 5} = \frac{6}{9}$$

But 6 and 9 are both divisible by 3. Thus,

$$\frac{6}{9} = \frac{6 \div 3}{9 \div 3} = \frac{2}{3}$$

This is the same answer! The whole procedure can be written as

$$\frac{\overset{2}{\cancel{\underset{\cancel{30}}{6}}}}{\underset{\underset{3}{\cancel{9}}}{\cancel{45}}} = \frac{2}{3}$$

_____
_____
_____
_____
_____

NAME _____

CLASS _____

SECTION _____

ANSWERS

A.   Write as a fraction with a denominator of 1.

1. 28

2. 93

3. -42

4. -86

5. 0

6. 1

7. -1

8. -17

1. _____

2. _____

3. _____

4. _____

5. _____

6. _____

7. _____

8. _____

B.   Find the missing number.

9. $\dfrac{1}{8} = \dfrac{?}{24}$

10. $\dfrac{7}{9} = \dfrac{?}{27}$

11. $\dfrac{7}{12} = \dfrac{?}{60}$

12. $\dfrac{5}{6} = \dfrac{?}{48}$

13. $\dfrac{5}{3} = \dfrac{?}{15}$

14. $\dfrac{9}{8} = \dfrac{?}{32}$

15. $\dfrac{7}{11} = \dfrac{?}{33}$

16. $\dfrac{11}{7} = \dfrac{?}{35}$

17. $\dfrac{1}{8} = \dfrac{4}{?}$

18. $\dfrac{3}{5} = \dfrac{27}{?}$

19. $\dfrac{5}{6} = \dfrac{5}{?}$

20. $\dfrac{9}{10} = \dfrac{9}{?}$

21. $\dfrac{8}{7} = \dfrac{16}{?}$

22. $\dfrac{9}{5} = \dfrac{36}{?}$

23. $\dfrac{6}{5} = \dfrac{36}{?}$

24. $\dfrac{5}{3} = \dfrac{45}{?}$

25. $\dfrac{21}{56} = \dfrac{?}{8}$

26. $\dfrac{12}{18} = \dfrac{?}{3}$

27. $\dfrac{36}{180} = \dfrac{?}{5}$

28. $\dfrac{8}{24} = \dfrac{4}{?}$

9. _____

10. _____

11. _____

12. _____

13. _____

14. _____

15. _____

16. _____

17. _____

18. _____

19. _____

20. _____

21. _____

22. _____

23. _____

24. _____

25. _____

26. _____

27. _____

28. _____

29. $\dfrac{18}{12} = \dfrac{3}{?}$

30. $\dfrac{56}{21} = \dfrac{8}{?}$

29. _____

30. _____

C. Reduce to lowest terms.

31. $\dfrac{15}{12}$

32. $\dfrac{30}{28}$

33. $\dfrac{13}{52}$

34. $\dfrac{27}{54}$

35. $\dfrac{56}{24}$

36. $\dfrac{56}{21}$

37. $\dfrac{22}{33}$

38. $\dfrac{26}{39}$

39. $\dfrac{100}{25}$

40. $\dfrac{0}{3}$

31. _____

32. _____

33. _____

34. _____

35. _____

36. _____

37. _____

38. _____

39. _____

40. _____

D. Applications

In problems 41 through 49, *reduced* means written in lowest terms.

41. A baseball player collected 220 hits in 660 times at bat. What reduced fraction of the time did he get a hit?

41. _____

42. Of every 100 Americans who die, 35 die from heart disease, 21 die from cancer, and 44 die from something else.
   a. What reduced fraction of these 100 Americans die from heart disease?
   b. What reduced fraction of these 100 Americans die from cancer?
   c. What reduced fraction of these 100 Americans die from something else?

42. **a.** _____
    b. _____
    c. _____

43. Four out of 10 hospital patients in this country have operations. What reduced fraction of the patients have operations?

43. _____

44. Of every 100 Americans:
   a. 15 say they never had a headache. What reduced fraction of the 100 is that?
   b. 18 have high blood pressure. What reduced fraction of the 100 is that?

44. **a.** _____
    b. _____
    c. _____

45. Of every 98 households in America, 21 consist of a single person. What reduced fraction is that?

45. _____

ANSWER (to problem on page 12)

4. **a.** $\dfrac{2}{3}$ **b.** $\dfrac{3}{4}$

In problems 46 through 50, use the formula

$$\text{Gear ratio} = \frac{\text{number of teeth in the driven gear}}{\text{number of teeth in the driver gear}}$$

46. The reduced gear ratio in a machine is $\frac{11}{22}$. Rewrite this ratio so that the driven gear has 44 teeth.

46. _____

47. The reduced gear ratio in a machine is $\frac{7}{9}$. Rewrite this ratio so that the driver gear has 72 teeth.

47. _____

48. The gear ratio in a certain machine is $\frac{24}{48}$.
  a. What is the reduced gear ratio?
  b. Write an equivalent ratio so that the driven gear has 32 teeth.

48. a. _____
  b. _____

49. The gear ratio in a certain machine is $\frac{32}{64}$.
  a. What is the reduced gear ratio?
  b. Write an equivalent ratio so that the driver gear has 48 teeth.

49. a. _____
  b. _____

50. The gear ratio in a machine is $\frac{7}{9}$. If the driven gear has 21 teeth, how many teeth does the driver gear have?

50. _____

1.1 USING YOUR KNOWLEDGE

A **ratio** is a way of comparing two or more numbers by using division. For example, if in a group of 10 people there are 3 women and 7 men, the ratio of *women* to *men* is

$3 \leftarrow$ Number of women
$\overline{7} \leftarrow$ Number of men

On the other hand, if there are 6 women and 4 men in the group, the *reduced ratio* of *men* to *women* is

$\dfrac{4}{6} = \dfrac{2}{3}$ $\leftarrow$ Number of men
$\phantom{\dfrac{4}{6} = \dfrac{2}{3}}$ $\leftarrow$ Number of women

1. A class is composed of 25 girls and 30 boys. Find the reduced ratio of girls to boys.

1. _____

2. Stock analysts consider the price to earnings (PE) ratio when buying or selling stock. If the price of a certain stock is $20 and its earnings are $5, what is the reduced PE ratio of the stock?

2. _____

3. Management measures the solvency of a business by using the current ratio (CR). This ratio is defined to be the ratio of current assets (CA) to current liabilities (CL); that is,

$$CR = \frac{CA}{CL}$$

  a. If the current assets of a company amount to $22,080 and the current liabilities are $3680, find the reduced CR.
  b. It is recommended that the CR of a business be about 2 to 1. If a business has $3500 in liabilities, how much should its assets be to maintain the desired 2 to 1 ratio?

3. a. _____
  b. _____

4. In economics, the average cost AC is defined to be the total cost TC divided by the output O. Find the reduced average cost of a certain item if it is known that:
   a. It costs $160 to produce 4 of the items.
   b. It costs $340 to produce 5 of the items.

   4. a. _____
      b. _____

5. The power gain (G_p) (read "G sub p") of an amplifier is defined as the ratio of output power (P_o) to input power (P_i). Find the reduced power gain G_p written as a fraction if
   a. $P_o = 50,000$ watts (W) and $P_i = 400$ W
   b. $P_o = 18$ milliwatts (mW) and $P_i = 60$ mW

   5. a. _____
      b. _____

1.2 OPERATIONS WITH FRACTIONS

Courtesy Fotoart.

How much sugar do the cups contain? To find the answer, we can multiply 3 by $\frac{1}{4}$, that is, find

$3 \cdot \dfrac{1}{4}$ (The dot, ·, indicates multiplication.)

Since we have 3 one-quarter cups of sugar, which makes $\frac{3}{4}$ cup,

$$3 \cdot \frac{1}{4} = \frac{3}{4}$$

Note that

$$3 \cdot \frac{1}{4} = \frac{3}{1} \cdot \frac{1}{4} = \frac{3 \cdot 1}{1 \cdot 4} = \frac{3}{4}$$

Similarly,

$$\frac{4}{9} \cdot \frac{2}{5} = \frac{4 \cdot 2}{9 \cdot 5} = \frac{8}{45}$$

$4 \cdot 2$ means 4×2 and $9 \cdot 5$ means 9×5. The numbers to be multiplied are called **factors.**

A.   MULTIPLICATION OF FRACTIONS

Here is the general rule for multiplying fractions.

$$\frac{a}{b} \cdot \frac{c}{d} = \frac{a \cdot c}{b \cdot d}$$

EXAMPLE 1   Find $\dfrac{9}{5} \cdot \dfrac{3}{4}$.

Solution

$$\frac{9}{5} \cdot \frac{3}{4} = \frac{9 \cdot 3}{5 \cdot 4} = \frac{27}{20}$$

Problem 1   Find $\dfrac{8}{3} \cdot \dfrac{4}{7}$.

▲

When multiplying fractions, it saves time if common factors are divided out *before* you multiply. For example,

$$\frac{2}{5} \cdot \frac{5}{7} = \frac{2 \cdot 5}{5 \cdot 7} = \frac{10}{35} = \frac{2}{7}$$

We can save time by writing

$$\frac{2}{\overset{1}{\cancel{5}}} \cdot \frac{\cancel{5}}{7} = \frac{2 \cdot 1}{1 \cdot 7} = \frac{2}{7} \quad \text{This can be done because } \frac{5}{5} = 1.$$

EXAMPLE 2 Find.

a. $\dfrac{3}{7} \cdot \dfrac{7}{8}$ **b.** $\dfrac{5}{8} \cdot \dfrac{4}{15}$

Solution

a. $\dfrac{3}{7} \cdot \dfrac{\overset{1}{\cancel{7}}}{8} = \dfrac{3 \cdot 1}{1 \cdot 8} = \dfrac{3}{8}$

b. $\dfrac{\cancel{5}}{\underset{2}{\cancel{8}}} \cdot \dfrac{\overset{1}{\cancel{4}}}{\underset{3}{\cancel{15}}} = \dfrac{1 \cdot 1}{2 \cdot 3} = \dfrac{1}{6}$

▲

If we wish to multiply a fraction by a **mixed number** such as $3\frac{1}{4}$, we must convert the mixed number to a fraction first. The number $3\frac{1}{4}$ (read "3 and $\frac{1}{4}$") means $3 + \frac{1}{4} = \frac{12}{4} + \frac{1}{4} = \frac{13}{4}$. (This addition will be clearer to you after studying the addition of fractions.) For now, we can shorten the procedure by using the following diagram:

$$3\frac{1}{4} \longrightarrow \frac{13}{4}$$

Work clockwise. *First* multiply the denominator by the whole number part; add the numerator. This is the new numerator. Use the same denominator.

EXAMPLE 3 $5\dfrac{1}{3} \cdot \dfrac{9}{16}$.

Solution

$$5\frac{1}{3} = \frac{3 \cdot 5 + 1}{3} = \frac{16}{3}$$

Thus,

$$5\frac{1}{3} \cdot \frac{9}{16} = \frac{\overset{1}{\cancel{16}}}{\underset{1}{\cancel{3}}} \cdot \frac{\overset{3}{\cancel{9}}}{\underset{1}{\cancel{16}}} = \frac{1 \cdot 3}{1 \cdot 1} = 3$$

Problem 2 Find.

a. $\dfrac{2}{5} \cdot \dfrac{5}{7}$ **b.** $\dfrac{3}{14} \cdot \dfrac{7}{6}$

Problem 3 Find $4\dfrac{1}{2} \cdot \dfrac{8}{9}$.

B. DIVISION OF FRACTIONS

If we wish to divide one number by another, we can indicate the division by a fraction. Thus, to divide 2 by 5 we can write

$$2 \div 5 = \frac{2}{5} = 2 \cdot \frac{1}{5}$$

with "Multiply" pointing to the product, and "The divisor is inverted" noted below.

Note that to divide 2 by 5 we multiplied 2 by $\frac{1}{5}$, where the fraction $\frac{1}{5}$ was obtained by *inverting* $\frac{5}{1}$ to obtain the $\frac{1}{5}$. (In mathematics $\frac{5}{1}$ and $\frac{1}{5}$ are called **reciprocals**.)

Now let us try the problem $5 \div \frac{5}{7}$. If we try to do it like the preceding problem, we write

$$5 \div \frac{5}{7} = 5 \cdot \frac{7}{5} = \frac{\overset{1}{\cancel{5}}}{1} \cdot \frac{7}{\underset{1}{\cancel{5}}} = 7$$

with "Multiply" and "Invert" noted.

Thus, to divide $\frac{a}{b}$ by $\frac{c}{d}$, we multiply $\frac{a}{b}$ by the **reciprocal** of $\frac{c}{d}$—that is, $\frac{d}{c}$. Here is the rule:

$$\frac{a}{b} \div \frac{c}{d} = \frac{a}{b} \cdot \frac{d}{c}$$

EXAMPLE 4 Find.

a. $\dfrac{3}{5} \div \dfrac{2}{7}$ **b.** $\dfrac{4}{9} \div 5$

Solution

a. $\dfrac{3}{5} \div \dfrac{2}{7} = \dfrac{3}{5} \cdot \dfrac{7}{2} = \dfrac{21}{10}$

b. $\dfrac{4}{9} \div 5 = \dfrac{4}{9} \cdot \dfrac{1}{5} = \dfrac{4}{45}$ ▲

Problem 4 Find.

a. $\dfrac{5}{7} \div \dfrac{5}{6}$

b. $\dfrac{3}{4} \div 7$

As in the case of multiplication, if mixed numbers are involved, they are changed to fractions first, like this:

$$2\frac{1}{4} \div \frac{3}{5} = \frac{9}{4} \div \frac{3}{5} = \frac{\overset{3}{\cancel{9}}}{4} \cdot \frac{5}{\underset{1}{\cancel{3}}} = \frac{15}{4}$$

with "Change" and "Invert" noted.

EXAMPLE 5 Find.

a. $3\frac{1}{4} \div \frac{7}{8}$ **b.** $\frac{11}{12} \div 7\frac{1}{3}$

Solution

a. $3\frac{1}{4} \div \frac{7}{8} = \frac{13}{4} \div \frac{7}{8} = \frac{13}{\underset{1}{4}} \cdot \frac{\overset{2}{8}}{7} = \frac{26}{7}$

b. $\frac{11}{12} \div 7\frac{1}{3} = \frac{11}{12} \div \frac{22}{3} = \frac{\overset{1}{11}}{\underset{4}{12}} \cdot \frac{\overset{1}{3}}{\underset{2}{22}} = \frac{1}{8}$ ▲

Now we are ready to add fractions.

C. ADDITION OF FRACTIONS

The photo shows that 1 quarter plus 2 quarters equals 3 quarters. In symbols,

$$\frac{1}{4} + \frac{2}{4} = \frac{1+2}{4} = \frac{3}{4}$$

To add fractions with the *same* denominator, add the numerators and keep the denominator, as shown.

$$\frac{a}{b} + \frac{c}{b} = \frac{a+c}{b}$$

Thus,

$$\frac{1}{5} + \frac{2}{5} = \frac{1+2}{5} = \frac{3}{5}$$

and

$$\frac{3}{8} + \frac{1}{8} = \frac{3+1}{8} = \frac{4}{8} = \frac{1}{2}$$

Note that in the last addition we reduced $\frac{4}{8}$ to $\frac{1}{2}$ by dividing the numerator and denominator by 4.

Now suppose you wish to add $\frac{5}{12}$ and $\frac{1}{18}$. Since these two fractions do not have the same denominators, our rule does not work. In order to add them, we must learn how to write $\frac{5}{12}$ and $\frac{1}{18}$ as equivalent fractions with the same denominator. To keep things simple, we should also try to make this denominator as small as possible; that is, we should first find the **least common denominator (LCD)** of the fractions. The LCD of two fractions is the **least** (*smallest*) **common mul-**

tiple (LCM) of their denominators. Thus, to find the LCD of $\frac{5}{12}$ and $\frac{1}{18}$, we find the LCM of 12 and 18. There are several ways of doing this. One way is to select the larger number (18) and find its multiples. The first multiple of 18 is $2 \cdot 18 = 36$, and 12 divides into 36. Thus, 36 is the LCD of 12 and 18. Unfortunately, this method is not practical in algebra. A method that is consists of writing 12 and 18 in *factored* form. We can start by writing 12 as $2 \cdot 6$. In turn, $6 = 2 \cdot 3$; thus

$$12 = 2 \cdot 2 \cdot 3$$

Similarly,

$$18 = 2 \cdot 9, \quad \text{or} \quad 2 \cdot 3 \cdot 3$$

Now, the smallest number that is a multiple of 12 and 18 must have at least two 2's (there are two 2's in 12) and two 3's (there are two 3's in 18). Thus, the LCD of $\frac{5}{12}$ and $\frac{1}{18}$ is

$$\underbrace{2 \cdot 2}_{\text{two 2's}} \cdot \underbrace{3 \cdot 3}_{\text{two 3's}} = 36$$

Fortunately, there is a shorter way of finding the LCM of 12 and 18. Note that what we want to do is find the *common* factors in 12 and 18, so we can divide 12 and 18 by the smallest divisor common to both numbers (2) and then the next (3), and so on. If we multiply these divisors by the final quotient, the result is the LCM. Here is the shortened version.

$$\begin{array}{ll} \text{Divide by 2.} & 2\,\overline{)\,12 \quad 18} \\ \text{Divide by 3.} & 3\,\overline{)\,\;6 \quad\; 9} \\ & \qquad\; 2 \text{—} 3 \rightarrow \text{Multiply } 2 \cdot 3 \cdot 2 \cdot 3. \end{array}$$

The LCM is $2 \cdot 3 \cdot 2 \cdot 3 = 36$.

In general, we use the following rule to find the LCD of two fractions.

TO FIND THE LCD OF TWO FRACTIONS

1. Write the denominators in a horizontal row and divide each number by a *divisor* common to both numbers.

2. Continue the process until the resulting quotients have no common divisor.

3. The product of the *divisors* and the *final quotients* is the LCD.

Now that we know that the LCD of $\frac{5}{12}$ and $\frac{1}{18}$ is 36, we can add the two fractions by writing each as an equivalent fraction with a denominator of 36. This can be done by multiplying numerator and denominator of $\frac{5}{12}$ by 3 and of $\frac{1}{18}$ by 2. Thus, we get

$$\frac{5}{12} = \frac{5 \cdot 3}{12 \cdot 3} = \frac{15}{36}$$

$$\frac{1}{18} = \frac{1 \cdot 2}{18 \cdot 2} = \frac{2}{36}$$

We then have

$$\frac{5}{12} + \frac{1}{18} = \frac{15}{36} + \frac{2}{36} = \frac{15 + 2}{36} = \frac{17}{36}$$

EXAMPLE 6   Find $\dfrac{1}{20} + 1\dfrac{1}{18}$.

Solution   We first find the LCM of 20 and 18.

STEP 1. Divide by 2. $2 \overline{)\;20\quad 18}$
$\phantom{2 \overline{)}}\quad 10 \text{---} 9 \rightarrow$ Multiply $2 \cdot 10 \cdot 9$.

Since there is no other divisor for 10 and 9, the LCM is

$2 \cdot 10 \cdot 9 = 180$

(You can also find the LCM by writing $20 = 2 \cdot 2 \cdot 5$ and $18 = 2 \cdot 3 \cdot 3$; the LCM is $2 \cdot 2 \cdot 3 \cdot 3 \cdot 5 = 180$.)

STEP 2. Now

$$\frac{1}{20} = \frac{1 \cdot 9}{20 \cdot 9} = \frac{9}{180}$$

and

$$1\frac{1}{18} = \frac{19}{18} = \frac{19 \cdot 10}{18 \cdot 10} = \frac{190}{180}$$

thus

$$\frac{1}{20} + 1\frac{1}{18} = \frac{9}{180} + \frac{190}{180} = \frac{199}{180}, \quad \text{or} \quad 1\frac{19}{180} \qquad \blacktriangle$$

Can we use the same procedure to add three or more fractions? Almost. Except we need to know how to find the LCD for three or more fractions. The procedure to do this is very similar to that used for finding the LCM of two numbers. We can write 15, 21, and 28 in factored form.

$15 = 3 \cdot 5$
$21 = 3 \cdot 7$
$28 = 2 \cdot 2 \cdot 7$

The LCM must contain at least $2 \cdot 2$, 3, 5, and 7. (Note that we select each factor the *greatest* number of times it appears in any factorization.) Thus, the LCM of 15, 21, and 28 is $2 \cdot 2 \cdot 3 \cdot 5 \cdot 7 = 420$. We can also use this shortened procedure.

FINDING THE LCD OF THREE OR MORE FRACTIONS

1. Write the denominators in a horizontal row (we are finding the LCM of 15, 21, and 28 in the margin), and divide the numbers by a divisor common to two or more of the numbers. If any of the other numbers is not divisible by this divisor, *circle* the number and carry it to the next line.

2. Repeat Step 1 with the quotients and carrydowns until *no two numbers* have a common divisor (except 1).

3. The LCD is the *product* of the divisors from the preceding steps and the numbers in the final row.

Problem 6   Find $1\dfrac{1}{10} + \dfrac{3}{14}$.

1. Divide by 3: $3 \overline{)\;15\quad 21\quad \textcircled{28}}$
2. Divide by 7: $7 \overline{)\;5\quad 7\quad 28}$
$\phantom{2. \text{Divide by 7: } 7 \overline{)}}\; 5\quad 1\quad 4$
3. The LCD is

$3 \cdot 7 \cdot 5 \cdot 1 \cdot 4 = 420$.

ANSWER   (to problem on page 20)

5. **a.** $\dfrac{22}{3}$ **b.** $\dfrac{3}{11}$

Thus to add

$$\frac{1}{16} + \frac{3}{10} + \frac{1}{28}$$

we first find the LCD of 16, 10, and 18. (Use either method.)

METHOD 1

You can find the LCD by writing

$16 = 2 \cdot 2 \cdot 2 \cdot 2$

$10 = 2 \cdot 5$

$28 = 2 \cdot 2 \cdot 7$

and selecting each factor the *greatest* number of times it appears in any factorization. Since 2 appears four times and 5 and 7 appear once, the LCD is

$2 \cdot 2 \cdot 2 \cdot 2 \cdot 5 \cdot 7 = 560.$

METHOD 2

Use the 3-step procedure given in the text.

STEP 1. Divide by 2. $2 \,)\, \overline{16 \ 10 \ 28}$

STEP 2. Divide by 2. $2 \,)\, \overline{8 \ \ ⑤ \ 14}$
$$ 4 \ \ 5 \ \ 7$$

The LCD is $2 \cdot 2 \cdot 4 \cdot 5 \cdot 7 = 560.$

We then write $\frac{1}{16}$, $\frac{3}{10}$, and $\frac{1}{28}$ with a denominator of 560.

$$\frac{1}{16} = \frac{1 \cdot 35}{16 \cdot 35} = \frac{35}{560}$$

$$\frac{3}{10} = \frac{3 \cdot 56}{10 \cdot 56} = \frac{168}{560}$$

and

$$\frac{1}{28} = \frac{1 \cdot 20}{28 \cdot 20} = \frac{20}{560}$$

Thus,

$$\frac{1}{16} + \frac{3}{10} + \frac{1}{28} = \frac{35}{560} + \frac{168}{560} + \frac{20}{560} = \frac{223}{560}$$

D. SUBTRACTION OF FRACTIONS

Now that you know how to add, *subtraction* is no problem. All the rules we have mentioned still apply! Thus,

$$\frac{5}{8} - \frac{2}{8} = \frac{5 - 2}{8} = \frac{3}{8}$$

$$\frac{7}{9} - \frac{1}{9} = \frac{7 - 1}{9} = \frac{6}{9} = \frac{2}{3}$$

The next example shows how to subtract fractions involving *different* denominators.

EXAMPLE 7   Find $\dfrac{7}{12} - \dfrac{1}{18}$.

Solution   We first get the LCD of the fractions.

Problem 7   Find $\dfrac{7}{12} - \dfrac{1}{10}$.

METHOD 1	**METHOD 2**

Write

$$12 = 2 \cdot 2 \cdot 3$$
$$18 = 2 \cdot 3 \cdot 3$$

STEP 1. $2 \overline{)\,12 \quad 18}$

STEP 2. $3 \overline{)\,6 \quad 9}$
$$\qquad\qquad 2 \quad 3$$

The LCD is $2 \cdot 2 \cdot 3 \cdot 3 = 36$.   **STEP 3.** The LCD is $2 \cdot 3 \cdot 2 \cdot 3 = 36$.

We then write each fraction with 36 as the denominator.

$$\frac{7}{12} = \frac{7 \cdot 3}{12 \cdot 3} = \frac{21}{36} \quad \text{and} \quad \frac{1}{18} = \frac{1 \cdot 2}{18 \cdot 2} = \frac{2}{36}$$

Thus,

$$\frac{7}{12} - \frac{1}{18} = \frac{21}{36} - \frac{2}{36} = \frac{21 - 2}{36} = \frac{19}{36}$$

▲

The rules we mentioned for adding fractions also apply to subtraction. If mixed numbers are involved, we have two methods for subtraction, as illustrated next.

EXAMPLE 8 Subtract.

$$3\frac{1}{6} - 2\frac{5}{8}$$

Solution   The LCM of 6 and 8 is 24.

METHOD 1.   First convert to improper fractions.

$$3\frac{1}{6} = \frac{19}{6} \quad \text{and} \quad 2\frac{5}{8} = \frac{21}{8}$$

Thus,

$$3\frac{1}{6} - 2\frac{5}{8} = \frac{19}{6} - \frac{21}{8}$$

$$= \frac{19 \cdot 4}{6 \cdot 4} - \frac{21 \cdot 3}{8 \cdot 3}$$

$$= \frac{76}{24} - \frac{63}{24}$$

$$= \frac{13}{24}$$

METHOD 2.   Vertical subtraction

Some students prefer to subtract mixed numbers by setting the problem in a column, as shown:

$$3\frac{1}{6}$$
$$-2\frac{5}{8}$$
$$\overline{\phantom{-2\frac{5}{8}}}$$

The fractional part is subtracted first, followed by the whole number part. Unfortunately, $\frac{5}{8}$ can not be subtracted from $\frac{1}{6}$ at this time, so we rename $1 = \frac{6}{6}$ and rewrite the problem like this:

Problem 8   Subtract.

$$4\frac{1}{6} - 3\frac{2}{9}$$

$$2\frac{7}{6} \quad \left(\text{Note that } 3\frac{1}{6} = 2 + \frac{6}{6} + \frac{1}{6} = 2\frac{7}{6}.\right)$$

$$-2\frac{5}{8}$$

Since the LCD is 24, rewrite $\frac{7}{6}$ and $\frac{5}{8}$ with 24 as denominators.

$$2\frac{7 \cdot 4}{6 \cdot 4} = 2\frac{28}{24}$$

$$-2\frac{5 \cdot 3}{8 \cdot 3} = 2\frac{15}{24}$$

$$\frac{13}{24}$$

The complete procedure can be written as follows:

$$3\frac{1}{6} \quad\to\quad 2\frac{7}{6} \quad\to\quad 2\frac{28}{24}$$

$$-2\frac{5}{8} \qquad\quad -2\frac{5}{8} \qquad\quad -2\frac{15}{24}$$

$$\frac{13}{24}$$

Of course, the answer is the same as before. ▲

Finally, we do a problem involving addition and subtraction of fractions.

EXAMPLE 9 Perform the indicated operations.

$$1\frac{1}{10} + \frac{5}{21} - \frac{3}{28}$$

Solution We first write $1\frac{1}{10}$ as $\frac{11}{10}$ and then find the LCM of 10, 21, and 28. (Use either method.)

METHOD 1.

$$10 = 2 \cdot 5$$
$$21 = 3 \cdot 7$$
$$28 = 2 \cdot 2 \cdot 7$$

The LCD is $2 \cdot 2 \cdot 3 \cdot 5 \cdot 7 = 420$.

METHOD 2.

STEP 1. Divide by 2: $2\,)\,10 \quad \textcircled{21} \quad 28$

STEP 2. Divide by 7: $7\,)\,\textcircled{5} \quad 21 \quad 14$

$$\qquad\qquad\qquad\qquad 5 \quad\ 3 \quad\ 2$$

The LCD is $2 \cdot 7 \cdot 5 \cdot 3 \cdot 2 = 420$.

Now

$$1\frac{1}{10} = \frac{11}{10} = \frac{11 \cdot 42}{10 \cdot 42} = \frac{462}{420}$$

$$\frac{5}{21} = \frac{5 \cdot 20}{21 \cdot 20} = \frac{100}{420}$$

Problem 9 Find.

$$1\frac{1}{15} + \frac{1}{21} - \frac{1}{28}$$

and

$$\frac{3}{28} = \frac{3 \cdot 15}{28 \cdot 15} = \frac{45}{420}$$

Thus,

$$1\frac{1}{10} + \frac{5}{21} - \frac{3}{28} = \frac{462}{420} + \frac{100}{420} - \frac{45}{420}$$

$$= \frac{462 + 100 - 45}{420} = \frac{517}{420}$$

NAME

CLASS

SECTION

ANSWERS

A. Find (reduce answers).

1. $\dfrac{2}{3} \cdot \dfrac{7}{3}$

2. $\dfrac{3}{4} \cdot \dfrac{7}{8}$

3. $\dfrac{6}{5} \cdot \dfrac{7}{6}$

4. $\dfrac{2}{5} \cdot \dfrac{5}{3}$

5. $\dfrac{7}{3} \cdot \dfrac{6}{7}$

6. $\dfrac{5}{6} \cdot \dfrac{3}{5}$

7. $7 \cdot \dfrac{8}{7}$

8. $10 \cdot 1\dfrac{1}{5}$

9. $2\dfrac{3}{5} \cdot 2\dfrac{1}{7}$

10. $2\dfrac{1}{3} \cdot 4\dfrac{1}{2}$

B. Find (reduce answers).

11. $7 \div \dfrac{3}{5}$

12. $5 \div \dfrac{2}{3}$

13. $\dfrac{3}{5} \div \dfrac{9}{10}$

14. $\dfrac{2}{3} \div \dfrac{6}{7}$

15. $\dfrac{9}{10} \div \dfrac{3}{5}$

16. $\dfrac{3}{4} \div \dfrac{3}{4}$

17. $1\dfrac{1}{5} \div \dfrac{3}{8}$

18. $3\dfrac{3}{4} \div 3$

19. $2\dfrac{1}{2} \div 6\dfrac{1}{4}$

20. $3\dfrac{1}{8} \div 1\dfrac{1}{3}$

C. Find (reduce answers).

21. $\dfrac{1}{5} + \dfrac{2}{5}$

22. $\dfrac{1}{3} + \dfrac{1}{3}$

23. $\dfrac{3}{8} + \dfrac{5}{8}$

24. $\dfrac{2}{9} + \dfrac{4}{9}$

25. $\dfrac{7}{8} + \dfrac{3}{4}$

26. $\dfrac{1}{2} + \dfrac{1}{6}$

27. $\dfrac{5}{6} + \dfrac{3}{10}$

28. $3 + \dfrac{2}{5}$

1. _____

2. _____

3. _____

4. _____

5. _____

6. _____

7. _____

8. _____

9. _____

10. _____

11. _____

12. _____

13. _____

14. _____

15. _____

16. _____

17. _____

18. _____

19. _____

20. _____

21. _____

22. _____

23. _____

24. _____

25. _____

26. _____

27. _____

28. _____

29. $2\frac{1}{3} + 1\frac{1}{2}$

30. $1\frac{3}{4} + 2\frac{1}{6}$

31. $\frac{1}{5} + 2 + \frac{9}{10}$

32. $\frac{1}{6} + \frac{1}{8} + \frac{1}{3}$

33. $3\frac{1}{2} + 1\frac{1}{7} + 2\frac{1}{4}$

34. $1\frac{1}{3} + 2\frac{1}{4} + 1\frac{1}{5}$

D. Find (reduce answers).

35. $\frac{5}{8} - \frac{2}{8}$

36. $\frac{3}{7} - \frac{1}{7}$

37. $\frac{1}{3} - \frac{1}{6}$

38. $\frac{5}{12} - \frac{1}{4}$

39. $\frac{7}{10} - \frac{3}{20}$

40. $\frac{5}{20} - \frac{7}{40}$

41. $\frac{8}{15} - \frac{2}{25}$

42. $\frac{7}{8} - \frac{5}{12}$

43. $2\frac{1}{5} - 1\frac{3}{4}$

44. $4\frac{1}{2} - 2\frac{1}{3}$

45. $3 - 1\frac{3}{4}$

46. $2 - 1\frac{1}{3}$

Perform the indicated operations.

47. $\frac{5}{6} + \frac{1}{9} - \frac{1}{3}$

48. $\frac{3}{4} + \frac{1}{12} - \frac{1}{6}$

49. $1\frac{1}{3} + 2\frac{1}{3} - 1\frac{1}{5}$

50. $3\frac{1}{2} + 1\frac{1}{7} - 2\frac{1}{4}$

E. Applications

51. The weight of an object on the moon is $\frac{1}{6}$ of its weight on earth. How much did the Lunar Rover, weighing 450 lb on earth, weigh on the moon?

52. Do you want to meet a millionaire? Your best bet is to go to Idaho. In this state, $\frac{1}{38}$ of the population happens to be millionaires. If there are about 912,000 persons in Idaho, about how many millionaires are there in the state?

53. The Actors Equity has 28,000 members, but only $\frac{1}{5}$ of the membership is working at any given moment. How many working actors is that?

54. The voltage of a standard C battery is $1\frac{1}{2}$ volts (V). What is the total voltage of 6 such batteries?

29. _____
30. _____
31. _____
32. _____
33. _____
34. _____

35. _____
36. _____
37. _____
38. _____
39. _____
40. _____
41. _____
42. _____
43. _____
44. _____
45. _____
46. _____

47. _____
48. _____
49. _____
50. _____

51. _____

52. _____

53. _____

54. _____

55. An earthmover removed $66\frac{1}{2}$ yd^3 of sand in 4 hr. How many cubic yards did it remove per hour?

55. _____

56. A stock is selling for $\$3\frac{1}{2}$. How many shares can be bought with \$98?

56. _____

57. In a recent survey, it was found that $\frac{9}{50}$ of all American households make more than \$25,000 annually and that $\frac{3}{25}$ make between \$20,000 and \$25,000. What total fraction of the American households make more than \$20,000?

57. _____

58. Between 1971 and 1977, $\frac{1}{5}$ of the immigrants coming to America were from Europe and $\frac{9}{20}$ came from this hemisphere. What total fraction of the immigrants were in these two groups?

58. _____

59. U.S. workers work an average of $46\frac{3}{5}$ hours per week, while Canadians work $38\frac{9}{10}$. How many more hours per week do U.S. workers work?

59. _____

60. Human bones are $\frac{1}{4}$ water, $\frac{3}{10}$ living tissue, and the rest minerals. Thus, the fraction of the bone that is minerals is $1 - \frac{1}{4} - \frac{3}{10}$. Find this fraction.

60. _____

✓ **SKILL CHECKER**

The **Skill Checker** exercises review skills previously studied. They will help you maintain the skills you have mastered. They will appear periodically in the exercise sets.

Divide.

61. 3 by 4

61. _____

62. 5 by 8

62. _____

63. 2 by 3   (Give the answer to two decimal places.)

63. _____

64. 2 by 11   (Give the answer to two decimal places.)

64. _____

1.2 USING YOUR KNOWLEDGE

In this section we learned that the LCM of two numbers is the *smallest* multiple of the two numbers. Can you ever apply this idea to anything else except adding and subtracting fractions? You bet! Hot dogs and buns. Have you noticed that hot dogs come in packages of 10 but buns come in packages of 8 (or 12)?

1. What is the smallest number of packages of hot dogs (10 to a package) and buns (8 to a package) you must buy so that you have as many hot dogs as you have buns? *Hint:* Think of multiples.

1. _____

2. If buns are sold in packages of 12, what is the smallest number of package of hot dogs (10 to a package) and buns you must buy so that you have as many hot dogs as you have buns?

2. _____

An Odd-lot Problem When selling less than 100 shares of stock (odd lots), odd-lot brokers charge $\frac{1}{8}$ of a point (dollar) per share to their customers. Thus, if your stock has a market value of $\$5\frac{1}{4}$, the broker would pay you, per share,

$$5\frac{1}{4} - \frac{1}{8} = 5 + \frac{1}{4} - \frac{1}{8} = 5 + \frac{2}{8} - \frac{1}{8} = 5\frac{1}{8}$$

3. Find the odd-lot price of a stock whose market value is $\$3\frac{1}{4}$.

4. Find the odd-lot price of a stock whose market value is $\$2\frac{1}{4}$.

3. _____

4. _____

1.3 DECIMALS

REVIEW

Before starting this section you should know how to divide one natural number by another and give the answer to the indicated number of decimal places.

OBJECTIVES

You should be able to:

A. Write a decimal in expanded form.
B. Write a fraction as a decimal.
C. Write a decimal as a fraction.

The price of gas is $1.029 (read "one point zero two nine"). This number is called a **decimal.** The word *decimal* means that we count by *tens,* that is, we use *base* ten. The dot in 1.029 is called the *decimal point.* The decimal 1.029 consists of two parts: the whole number part (i.e., 1, the number to the *left* of the decimal point) and the decimal part, .029, as shown.

$$\overset{\text{Whole number}}{1.\underset{\text{Decimal}}{029}}$$

The decimal part .029 has three decimal digits, 0, 2, and 9. When writing whole numbers the decimal point can be omitted. Thus,

012.00 = 12   We omit the 0 to the *left* of the 1.
   13. = 13

and

197.00 = 197

Moreover, a decimal such as .21 is usually written as **0**.21 to avoid confusing 21 with .21. Similarly,

.1   is written as 0.1
.14   is written as 0.14

and

.123 is written as 0.123

A. EXPANDED FORM OF A DECIMAL

The number 11.664 can be written in a diagram, as shown.

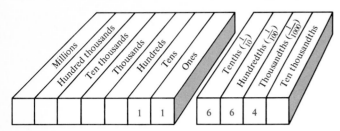

With the help of this diagram, we can write the number in *expanded form* like this:

$$\boxed{1}\,\boxed{1}\,.\,\boxed{6}\,\boxed{6}\,\boxed{4}$$

$$10 + 1 + \frac{6}{10} + \frac{6}{100} + \frac{4}{1000}$$

The number 11.664 is read by reading the number to the left of the decimal (*eleven*), using the word *and* for the decimal point, and reading the number to the right of the decimal point followed by the place value of the *last* digit (*six hundred sixty-four thousandths*).

EXAMPLE 1 Write 35.216 in expanded form.

Solution

$$\boxed{3}\,\boxed{5}\,.\,\boxed{2}\,\boxed{1}\,\boxed{6}$$

$$30 + 5 + \frac{2}{10} + \frac{1}{100} + \frac{6}{1000}$$

Problem 1 Write 48.846 in expanded form.

B. FROM FRACTIONS TO DECIMALS

Fractions can be written as decimals. For example,

$$\frac{3}{10} = \text{three tenths} = 0.3$$

One zero   One decimal
(tenths)   digit

$$\frac{7}{100} = \text{seven hundredths} = 0.07$$

Two zeros   Two decimal
(hundredths)   digits

and

$$\frac{29}{1000} = \text{twenty-nine thousandths} = 0.029$$

Three zeros   Three decimal
(thousandths)   digits

EXAMPLE 2 Write as decimals.

a. $\dfrac{9}{10}$ **b.** $\dfrac{38}{100}$ **c.** $\dfrac{9}{1000}$

Problem 2 Write as decimals.

a. $\dfrac{7}{10}$ **b.** $\dfrac{48}{100}$ **c.** $\dfrac{3}{1000}$

Solution

a. $\dfrac{9}{10} = 0.9$

b. $\dfrac{38}{100} = 0.38$

c. $\dfrac{9}{1000} = 0.009$ ▲

Since a fraction is an indicated division, we can write a fraction as a decimal by dividing. For example, $\frac{3}{4}$ means $3 \div 4$:

See Calculator Corner at the end of this section to learn how to do this division with a calculator.

$$\begin{array}{r} 0.75 \\ 4\,\overline{)\,3.00} \quad \text{Note that } 3 = 3.00 \\ \underline{2\ 8} \\ 20 \\ \underline{20} \\ 0 \end{array}$$

Here the division **terminates** (has a 0 remainder). Thus, $\frac{3}{4} = 0.75$ and 0.75 is called a **terminating decimal.**

Since $\frac{2}{3}$ means $2 \div 3$,

$$\begin{array}{r} 0.666\ldots \\ 3\,\overline{)\,2.00} \\ \underline{1\ 8} \\ 20 \\ \underline{18} \\ 20 \\ \underline{18} \\ 2 \quad \text{The division continues, because the remainder repeats.} \end{array}$$

Hence $\frac{2}{3} = 0.666\ldots$. Such decimals (called **repeating decimals**) may be written by placing a bar over the repeating digits; that is, $\frac{2}{3} = 0.\overline{6}$.

EXAMPLE 3 Write as decimals.

a. $\dfrac{5}{8}$ **b.** $\dfrac{2}{11}$

Problem 3 Write as decimals.

a. $\dfrac{3}{8}$ **b.** $\dfrac{4}{11}$

Solution

a. $\frac{5}{8}$ means $5 \div 8$.

$$\begin{array}{r} 0.625 \\ 8\,\overline{)\,5.00} \\ \underline{4\ 8} \\ 20 \\ \underline{16} \\ 40 \\ \underline{40} \\ 0 \end{array}$$

← The remainder is 0. Thus, the answer is a *terminating* decimal.

$\dfrac{5}{8} = 0.625$ ←

b.
$$11\overline{)2.0000}$$

$$\begin{array}{r} 0.1818\ldots \\ \hline 1\,1 \\ \hline 90 \\ 88 \\ \hline 20 \\ 11 \\ \hline 90 \\ 88 \\ \hline 20 \end{array}$$

The remainders 90 and 20 alternately repeat. Thus, the answer is a **repeating decimal**.

$$\frac{2}{11} = 0.\overline{18}$$

C. FROM DECIMALS TO FRACTIONS

Terminating decimals can be converted to fractions. For example,

$$0.3 = \frac{3}{10}$$

$$0.18 = \frac{18}{100} = \frac{9}{50}$$

and

$$0.150 = \frac{150}{1000} = \frac{3}{20}$$

Here is the rule for changing a terminating decimal to a fraction.

RULE FOR CHANGING A TERMINATING DECIMAL TO A FRACTION

1. Write the digits of the number, omitting the decimal point, as the *numerator* of the fraction.

2. The *denominator* is a 1 *followed by as many zeros as there are decimal digits in the decimal.*

EXAMPLE 4 Write as a reduced fraction.

a. 0.035   **b.** 0.0275

Solution

a. $0.035 = \dfrac{35}{1000} = \dfrac{7}{200}$

   3 digits 3 zeros

b. $0.0275 = \dfrac{275}{10,000} = \dfrac{11}{400}$

   4 digits 4 zeros

In case the decimal has a whole-number part, write the digits of the number omitting the decimal point as the numerator of the fraction.

Problem 4 Write as a reduced fraction.
a. 0.045   **b.** 0.0175

For example, to write 4.23 as a fraction, write

$$4.\underset{\text{2 digits}}{\underbrace{23}} = \frac{423}{\underset{\text{2 zeros}}{\underbrace{100}}}$$

EXAMPLE 5 Write as a reduced fraction.

a. 3.11   **b.** 5.15

Solution

a. $3.11 = \dfrac{311}{100}$

b. $5.15 = \dfrac{515}{100} = \dfrac{103}{20}$ ▲

We have now studied terminating and repeating decimals. As it turns out, every decimal that terminates or repeats can be written as a ratio and is called a **rational number.** There are other numbers that *cannot* be written as ratios. When written in decimal form they never terminate and never repeat. They are called **irrational numbers.** As we have mentioned, the rational and the irrational numbers together form the set of *real numbers*.

> The **real numbers** are all the *rational* and all the *irrational* numbers together.

We give some properties of the real numbers in Section 1.5.

Problem 5 Write as a reduced fraction.
a. 4.19   **b.** 6.25

EXERCISE 1.3

A. Write in expanded form.

1. 4.7

2. 3.9

3. 5.62

4. 9.28

5. 16.123

6. 18.845

7. 49.012

8. 93.038

9. 57.104

10. 85.305

1. _____

2. _____

3. _____

4. _____

5. _____

6. _____

7. _____

8. _____

9. _____

10. _____

B. Write as decimals.

11. $\dfrac{9}{10}$

12. $\dfrac{7}{10}$

13. $\dfrac{9}{100}$

14. $\dfrac{8}{100}$

15. $\dfrac{11}{1000}$

16. $\dfrac{17}{1000}$

17. $\dfrac{2}{1000}$

18. $\dfrac{7}{1000}$

19. $\dfrac{187}{1000}$

20. $\dfrac{472}{1000}$

21. $\dfrac{1}{5}$

22. $\dfrac{1}{2}$

23. $\dfrac{7}{8}$

24. $\dfrac{1}{8}$

25. $\dfrac{3}{16}$

26. $\dfrac{5}{16}$

27. $\dfrac{2}{9}$

28. $\dfrac{4}{9}$

11. _____

12. _____

13. _____

14. _____

15. _____

16. _____

17. _____

18. _____

19. _____

20. _____

21. _____

22. _____

23. _____

24. _____

25. _____

26. _____

27. _____

28. _____

29. $\dfrac{6}{11}$ 30. $\dfrac{7}{11}$

31. $\dfrac{3}{11}$ 32. $\dfrac{5}{11}$

33. $\dfrac{1}{6}$ 34. $\dfrac{5}{6}$

35. $\dfrac{10}{9}$ 36. $\dfrac{11}{9}$

C. Write as reduced fractions.

37. 0.6   38. 0.4

39. 0.9   40. 0.7

41. 0.06   42. 0.08

43. 0.12   44. 0.18

45. 0.054   46. 0.062

47. 2.13   48. 3.41

49. 19.18   50. 27.34

51. 40.25   52. 50.64

53. 38.175   54. 97.275

D. Applications

55. The odds for getting a busy signal when dialing long distance are 1 to 8.9. Write 8.9 in expanded form.

56. The average daily consumption of thiamin (a vitamin) is 2.09 milligrams. Write 2.09 in expanded form.

57. In this country, $\frac{3}{4}$ of the people finish high school and $\frac{9}{20}$ go to to college.
    a. Write $\frac{3}{4}$ as a decimal.
    b. Write $\frac{9}{20}$ as a decimal.

58. 1 in 6 Americans wears a full set of false teeth. Write $\frac{1}{6}$ as a repeating decimal.

Answer lines:

29. _____
30. _____
31. _____
32. _____
33. _____
34. _____
35. _____
36. _____

37. _____
38. _____
39. _____
40. _____
41. _____
42. _____
43. _____
44. _____
45. _____
46. _____
47. _____
48. _____
49. _____
50. _____
51. _____
52. _____
53. _____
54. _____

55. _____

56. _____

57. a. _____
     b. _____

58. _____

59. Here are some figures regarding financial help to families in the United States.

    a. 0.155 of the families receive Social Security. Write 0.155 as a reduced fraction.

    b. 0.78 of the families receive other types of help. Write 0.78 as a reduced fraction.

60. In a recent year, the top three killers (per 1000 people) in the United States were:

Heart disease (337.2 deaths)

Cancer (175.8 deaths)

Cerebrovascular disease (87.9 deaths)

    a. Write 337.2 as a reduced fraction.

    b. Write 175.8 as a reduced fraction.

    c. Write 87.9 as a reduced fraction.

✓ **SKILL CHECKER**

Reduce.

61. $\dfrac{36}{100}$

62. $\dfrac{480}{1000}$

63. Write $\dfrac{42.1}{100}$ as a fraction with a denominator of 1000.

59. a. _____
    b. _____

60. a. _____
    b. _____
    c. _____

61. _____

62. _____

63. _____

1.3 USING YOUR KNOWLEDGE

In this section we have seen that fractions can be changed to decimals by dividing the numerator by the denominator. The result may be a terminating decimal or a repeating decimal.

In business transactions we would like to work with terminating decimals. Look at the accompanying group of the 10 most active stocks. The fractional parts appearing are $\frac{7}{8}$, $\frac{3}{4}$, $\frac{5}{8}$, $\frac{1}{2}$, $\frac{3}{8}$, and $\frac{1}{4}$.

MOST ACTIVE STOCKS

	OPEN	HIGH	LOW	CLOSE	CHG.	VOLUME
Am Airlin	14⅜	14⅜	13⅞	13⅞	−⅝	216,100
GenTel&El	29	29	28¾	28¾	−⅛	212,900
Exxon...........	53	53	52½	52½	−½	160,300
Richmnd Cp	20½	20⅝	20	20⅛	−⅞	152,200
Am Tel&Tel......	60½	60⅝	59⅞	60⅛	−¼	144,200
Watkins Jhn	14	15⅞	14	15⅞	−3¼	137,100
Norton Sim	20¾	21⅛	20¾	20⅞	+⅜	133,700
Weyerhsr.........	41	41⅛	40⅝	40¾	−¼	131,000
Chrysler..........	21⅜	21⅜	20½	20⅝	−⅞	130,700
IntTelTel	31⅞	32	31¼	31⅜	−⅜	114,400

1. Write as a decimal.

    a. $\dfrac{7}{8}$

    b. $\dfrac{3}{4}$

1. a. _____
    b. _____

EXERCISE 1.3

2. Write as a decimal.

   a. $\dfrac{5}{8}$

   b. $\dfrac{1}{2}$

3. Write as a decimal.

   a. $\dfrac{3}{8}$

   b. $\dfrac{1}{4}$

4. What kind of fractions (terminating or repeating) are $\frac{7}{8}, \frac{3}{4}, \frac{5}{8}, \frac{1}{2}$, and $\frac{3}{8}$, and why do you think this is?

5. Here are some fractions and their decimal equivalents.

TERMINATING	REPEATING
$\dfrac{7}{8} = \dfrac{7}{2 \times 2 \times 2} = 0.875$	$\dfrac{1}{3} = \dfrac{1}{3} = 0.\overline{3}$
$\dfrac{3}{4} = \dfrac{3}{2 \times 2} = 0.75$	$\dfrac{2}{3} = \dfrac{2}{3} = 0.\overline{6}$
$\dfrac{5}{8} = \dfrac{5}{2 \times 2 \times 2} = 0.625$	$\dfrac{1}{6} = \dfrac{1}{2 \times 3} = 0.1\overline{6}$
$\dfrac{1}{2} = \dfrac{1}{2} = 0.5$	$\dfrac{1}{9} = \dfrac{1}{3 \times 3} = 0.\overline{1}$
$\dfrac{3}{8} = \dfrac{3}{2 \times 2 \times 2} = 0.375$	$\dfrac{1}{12} = \dfrac{1}{2 \times 2 \times 3} = 0.08\overline{3}$
$\dfrac{1}{5} = \dfrac{1}{5} = 0.2$	$\dfrac{1}{7} = \dfrac{1}{7} = 0.\overline{142857}$
$\dfrac{1}{10} = \dfrac{1}{2 \times 5} = 0.1$	$\dfrac{1}{11} = \dfrac{1}{11} = 0.\overline{09}$

Look at the denominators of the terminating fractions. Now look at the denominators of the repeating fractions. Can you make a conjecture (guess) about the denominators of the fractions that terminate?

CALCULATOR CORNER

To convert a fraction to a decimal using a calculator is very simple. Thus, to write $\frac{3}{4}$ as a decimal, we simply recall that $\frac{3}{4}$ means $3 \div 4$. Pressing $\boxed{3}\ \boxed{\div}\ \boxed{4}\ \boxed{=}$ will give us the correct answer, 0.75. (The calculator is nice enough to write the zero to the left of the decimal in the final answer.) You can also do Example 3 using division. Moreover, if your instructor permits it, you can do Problems 21 through 36 in Exercise 1.3.

1.4 PERCENTS

COUPON

20% OFF
ANY OTHER PURCHASE

WITH THIS COUPON
LIMIT ONE COUPON
PER PURCHASE
Expires 2/28/95

In the preceding sections we studied fractions and decimals. We are now ready to study **percents.** For example, the ad says you can get 20% (read "20 percent") off. The percent symbol % means *per hundred*, and so a percent can be expressed as a fraction whose denominator is 100. Thus,

$$20\% = \frac{20}{100}$$

$$9.7\% = \frac{9.7}{100}$$

$$187\% = \frac{187}{100}$$

A. CONVERTING PERCENTS TO FRACTIONS

As you can see from the pattern, percents can be written as fractions using the following rule:

$$n\% \text{ means } \frac{n}{100}$$

EXAMPLE 1 Write as a fraction.

a. 36%   **b.** 42.1%   **c.** $12\frac{1}{2}\%$

Solution

a. $36\% = \frac{36}{100} = \frac{9}{25}$

b. $42.1\% = \frac{42.1}{100} = \frac{42.1 \cdot 10}{100 \cdot 10} = \frac{421}{1000}$

Here we multiplied by $\frac{10}{10}$ to make the numerator a whole number. In general, the numerator and denominator of a common fraction are whole numbers.

REVIEW

Before starting this section, you should know:

1. How to reduce fractions.
2. How to divide one natural number by another.

OBJECTIVES

You should be able to:

A. Convert percents to fractions.
B. Convert percents to decimals.
C. Convert decimals to percents.
D. Convert fractions to percents.
E. Solve problems involving percents.

Problem 1 Write as a fraction.

a. 32%   **b.** 31.7%   **c.** $10\frac{1}{2}\%$

c. Since $\frac{1}{2} = 0.5$,

$$12\frac{1}{2}\% = 12.5\% = \frac{12.5}{100} = \frac{12.5 \cdot 10}{100 \cdot 10} = \frac{125}{1000} = \frac{1}{8}$$

B. CONVERTING PERCENTS TO DECIMALS

Since we know how to convert fractions to decimals, we can now convert percents to decimals. For example,

$$43\% = \frac{43}{100} = 0.43$$

$$87\% = \frac{87}{100} = 0.87$$

$$4.5\% = \frac{4.5}{100} = \frac{45}{1000} = 0.045$$

Here is the general rule:

CONVERTING PERCENTS TO DECIMALS
Move the decimal point in the number *two* places to the *left* and omit the % symbol.

EXAMPLE 2   Write as a decimal.

a. 98%   **b.** 34.7%   **c.** 7.2%

Solution

a. $98\% = {}_{\curvearrowleft}98. = 0.98$ Recall that a decimal point follows every whole number.

b. $34.7\% = {}_{\curvearrowleft}34.7 = 0.347$

c. $7.2\% = {}_{\curvearrowleft}07.2 = 0.072$ (Note that a 0 was inserted so we could move the decimal point two places to the left.)

C. CONVERTING DECIMALS TO PERCENTS

As you can see from Example 2,

$$0.98 = 98\%$$

and

$$0.347 = 34.7\%$$

Thus, we can reverse the previous rule to convert decimals to percents. Here is the way we do it:

CONVERTING DECIMALS TO PERCENTS
Move the decimal point *two* places to the *right* and attach the % symbol.

Problem 2   Write as a decimal.
a. 73%   **b.** 81.4%   **c.** 8.3%

EXAMPLE 3   Write as a percent.

a. 0.53   **b.** 3.19   **c.** 64.7

Solution

a. $0.53 = 0.53\% = 53\%$

b. $3.19 = 3.19\% = 319\%$

c. $64.7 = 64.70\% = 6470\%$ (Note that a 0 was inserted so we could move the decimal point two places to the right.)

Problem 3   Write as a percent.
a. 0.47   **b.** 8.17   **c.** 71.8

D.   CONVERTING FRACTIONS TO PERCENTS

Do you remember how to convert $\frac{3}{4}$ to a decimal? It is done by dividing 3 by 4, obtaining

$$
\begin{array}{r}
0.75 \\
4\overline{)3.00} \\
2\,8 \\
\hline
20 \\
20 \\
\hline
0
\end{array}
$$

Hence,

$$\frac{3}{4} = 0.75$$

Since $0.75 = 75\%$, we have $\frac{3}{4} = 0.75 = 75\%$. Similarly, to change $\frac{3}{8}$ to a percent, we divide 3 by 8, carrying the division to three places, as shown:

$$
\begin{array}{r}
0.375 \\
8\overline{)3.000} \\
2\,4 \\
\hline
60 \\
56 \\
\hline
40 \\
40 \\
\hline
0
\end{array}
$$

$$0.375 = 37.5\% = 37\frac{1}{2}\%$$

Thus,

$$\frac{3}{8} = 37\frac{1}{2}\%$$

Here is the rule we used:

CHANGING A FRACTION TO A PERCENT

Divide the *numerator* by the *denominator* (carry the division to three decimal places) and change the resulting decimal to a percent by *moving the decimal point two places to the right.*

EXAMPLE 4   Write as a percent.

a. $\dfrac{7}{8}$ **b.** $\dfrac{2}{3}$

Solution

a. Divide 7 by 8.

```
      0.875
  8 ) 7.00
      6 4
      ───
       60
       56
       ──
        40
        40
        ──
         0
```

Thus,

$$\frac{7}{8} = 0.875 = 87.5\% = 87\frac{1}{2}\%$$

b. Divide 2 by 3.

```
      0.666
  3 ) 2.000
      1 8
      ───
       20
       18
       ──
        20
        18
        ──
         2
```

Since the quotient is 0.666 . . . , or $0.66\bar{6}$, $\frac{2}{3} = 0.66\bar{6} = 66.\bar{6}\%$.

E.   APPLICATIONS

Many problems involve the idea of percents. For example, the ad at the beginning of the section says that you can have 20% off a purchase. Suppose an item costs $70. How much off do you get? We need to find:

$$20\% \text{ of } 70$$

This is translated as 0.20×70 Note that $20\% = 0.20$ and "of" is translated as "multiply."

Thus,

$$0.20 \times 70 = 14 \quad 0.20 \times 70 = 14\underset{\curvearrowleft}{00}$$

that is, we get $14 off.

Note: In this problem we have used the multiplication sign $\times$ instead of the raised dot because this raised dot can be confused with the decimal point.

Problem 4   Write as a percent.

a. $\dfrac{5}{8}$ **b.** $\dfrac{1}{3}$

EXAMPLE 5   For the year 1984, the combined Social Security and Medicare tax was 6.7% of your salary. Carlos Perez earned $20,000 in 1984. What was his tax?

Solution   We need to find:

$$\underbrace{6.7\%}_{\downarrow} \text{ of } \underbrace{20,000}_{\downarrow}$$

Translating $0.067 \times 20,000 = \$1340$

Thus, his tax amounted to $1340.

Problem 5   Find the combined tax, if Perez' salary is $30,000.

NAME

CLASS

SECTION

ANSWERS

A. Write as a reduced fraction.

1. 12%

2. 28%

3. 18.5%

4. 14.5%

5. 6%

6. 2%

7. 8.1%

8. 9.3%

9. 10.5%

10. 18.5%

1. _____
2. _____
3. _____
4. _____
5. _____
6. _____
7. _____
8. _____
9. _____
10. _____

B. Write as a decimal.

11. 33%

12. 52%

13. 5%

14. 9%

15. 300%

16. 500%

17. 11.8%

18. 89.1%

19. 0.5%

20. 0.7%

11. _____
12. _____
13. _____
14. _____
15. _____
16. _____
17. _____
18. _____
19. _____
20. _____

C. Write as a percent.

21. 0.05

22. 0.07

23. 0.39

24. 0.74

25. 0.416

26. 0.829

27. 0.003

28. 0.008

21. _____
22. _____
23. _____
24. _____
25. _____
26. _____
27. _____
28. _____

NAME _____

29. 1.00

30. 2.1

29. _____
30. _____

D. Write as a percent.

31. $\dfrac{1}{8}$

32. $\dfrac{1}{2}$

33. $\dfrac{3}{5}$

34. $\dfrac{2}{5}$

35. $\dfrac{1}{6}$

36. $\dfrac{5}{6}$

37. $\dfrac{3}{20}$

38. $\dfrac{11}{20}$

39. $\dfrac{3}{40}$

40. $\dfrac{7}{40}$

31. _____
32. _____
33. _____
34. _____
35. _____
36. _____
37. _____
38. _____
39. _____
40. _____

E. Applications

41. In 1989 the combined Social Security and Medicare tax was 7.51% of your earnings. If you made $20,000 in 1989, what would your tax be?

41. _____

42. In 1990 the combined Social Security and Medicare tax will be 7.65% of your earnings. If you make $30,000 in 1990, what will your tax be?

42. _____

43. An item is advertised at 30% off. If the item sells regularly for $50
  a. How much off do you get?
  b. How much do you have to pay for the item?

43. a. _____
    b. _____

44. A certain store offers a 25% discount. If an item is marked $60
  a. How much is the discount?
  b. How much do you have to pay for the item?

44. a. _____
    b. _____

45. Do you have any credit cards? About 64% of Americans do. Write 64% as a reduced fraction.

45. _____

46. Today, women account for 42% of professional technical workers and 20.8% of the managers.
  a. Write 42% as a reduced fraction.
  b. Write 20.8% as a reduced fraction.

46. a. _____
    b. _____

47. In a recent year, about 95% of all homes in the United States had a phone and about 90% of the customers made at least one long-distance call during the year.
  a. Write 95% as a decimal.
  b. Write 90% as a decimal.

47. a. _____
    b. _____

48. A recent survey indicated that 11.1% of all calls result in a busy signal, while 12.7% go unanswered.
  a. Write 11.1% as a decimal.
  b. Write 12.7% as a decimal.

48. a. _____
    b. _____

49. A recent Gallup poll indicated that 47.7% of the population exercised daily compared to only 24.1% that did so 17 years earlier.
   a. Write 47.7% as a decimal.
   b. Write 24.1% as a decimal.

49. a. _____
    b. _____

50. An orange provides 0.02 of the recommended daily allowance for proteins. Write 0.02 as a percent.

50. _____

51. Coffee is brewed in 0.84 of the homes in America but 0.30 of the population drink decaffeinated coffee.
   a. Write 0.84 as a percent.
   b. Write 0.30 as a percent.

51. a. _____
    b. _____

52. The fraction of American adults who cook for fun is 0.26. Of these, 0.065 ride a motorcycle.
   a. Write 0.26 as a percent.
   b. Write 0.065 as a percent.

52. a. _____
    b. _____

53. One out of every 50 Americans is manic depressive, and 1 in 27 is neurotic.
   a. Write $\frac{1}{50}$ as a percent.
   b. Write $\frac{1}{27}$ as a percent.

53. a. _____
    b. _____

54. In 1965 the divorce rate was 2.5 per 1000. Now it is about 6.1 per 1000.
   a. Write $\frac{2.5}{1000}$ as a percent.
   b. Write $\frac{6.1}{1000}$ as a percent.

54. a. _____
    b. _____

55. Have you bought a car lately? Here are some figures on the subject: $\frac{9}{20}$ of buyers are installment buyers. Of these, 1.1 out of 4 finance through the dealer.
   a. Write $\frac{9}{20}$ as a percent.
   b. Write $\frac{1.1}{4}$ as a percent.

55. a. _____
    b. _____

✓ **SKILL CHECKER**

Multiply.

56. $10 \cdot \frac{1}{10}$ **57.** $12 \cdot \frac{5}{6}$

56. _____

57. _____

Divide.

58. $10 \div \frac{1}{10}$ **59.** $7 \div \frac{1}{7}$

58. _____

59. _____

60. $\frac{1}{8} \div 8$

60. _____

1.4 USING YOUR KNOWLEDGE

The Pursuit of Happiness *Psychology Today* conducted a survey about happiness. Here are some conclusions from that report.

1. 7 out of 10 people said they had been happy over the last 6 months. What percent of the people is that?

1. _____

2. 70% expected to be happier in the future than now. What fraction of the people is that?

2. _____

3. 0.40 of the people felt lonely.
   a. What percent of the people is that?
   b. What fraction of the people is that?

3. **a.** _____
   b. _____

4. Only 4% of the men were ready to cry. Write 4% as a decimal.

4. _____

5. Of the people surveyed, 49% were single. Write 49% as
   a. A fraction.
   b. A decimal.

5. **a.** _____
   b. _____

Do you wonder how they came up with some of these percents? They used their knowledge. You do the same and fill in the spaces in the table below which refers to the marital status of the 52,000 people surveyed. For example, in the first line 25,480 persons out of 52,000 were single. This is

$$\frac{25,480}{52,000} = 49\%$$

Fill in the percents in the last column.

MARITAL STATUS	NUMBER	PERCENT
Single	25,480	49%
6. Married (first time)	15,600	____
7. Remarried	2600	____
8. Divorced, separated	5720	____
9. Widowed	520	____
10. Cohabitating	2080	____

6. _____

7. _____

8. _____

9. _____

10. _____

CALCULATOR CORNER

If your calculator has a percent key $\boxed{\%}$ it may automatically change percents to decimals. Thus, to do Example 2: Change 98% to a decimal by pressing $\boxed{9}\,\boxed{8}\,\boxed{\%}$. The display shows the decimal representation, 0.98. Similarly, to change 34.7% to a decimal, key in $\boxed{3}\,\boxed{4}\,\boxed{.}\,\boxed{7}\,\boxed{\%}$ and the answer, 0.347, will be displayed.

Finally, if you wish to find the *actual* price of an article that is marked down by a certain percent, the calculator can do it for you automatically. For example, if a $150 stereo set is marked 25% off, the actual price of the stereo can be obtained by pressing

$\boxed{1}\,\boxed{5}\,\boxed{0}\,\boxed{-}\,\boxed{2}\,\boxed{5}\,\boxed{\%}\,\boxed{=}$

giving an answer of $112.50. If your instructor permits it, you can check the answers in Problems 41 through 45.

1.5 PROPERTIES OF NUMBERS

OBJECTIVES

REVIEW

Before starting this section you should know how to add, subtract, multiply, and divide natural numbers.

OBJECTIVES

You should be able to:

A. Do calculations involving parentheses.
B. Tell which laws are illustrated in a given statement.
C. Use the distributive law to do calculations.
D. Remove parentheses in a given expression.
E. State which property is illustrated in a given statement.

In this section we shall discuss some properties that apply to the numbers we have studied.

A. GROUPING SYMBOLS

What does the sign in the picture mean? It is *intended* to mean (small-engine) mechanic—that is, a mechanic that works on small engines. What does $2 + 3 \cdot 4$ mean? If we add 2 and 3 *first* and then multiply by 4, we get 20. If we multiply 3 by 4 and then add 2, the result is 14. To indicate which operation we must do *first*, we use *grouping symbols* such as parentheses ( ). Thus,

$(2 + 3) \cdot 4$ means $5 \cdot 4 = 20$

whereas

$2 + (3 \cdot 4)$ means $2 + 12 = 14$

EXAMPLE 1 Find.

a. $(4 \cdot 5) + 6$ **b.** $4 \cdot (5 + 6)$

Solution

a. $(4 \cdot 5) + 6$ means $20 + 6 = 26$
b. $4 \cdot (5 + 6)$ means $4 \cdot 11 = 44$

Problem 1 Find.
a. $(3 \cdot 4) + 7$ **b.** $3 \cdot (4 + 7)$

B. GROUPING AND ORDER OF OPERATIONS

Now, what about $2 + 3 + 4$? Does it mean $(2 + 3) + 4$ or $2 + (3 + 4)$? It does not matter! The result is the same. This fact can be stated as follows:

ASSOCIATIVE LAW OF ADDITION
$a + (b + c) = (a + b) + c$ for any numbers a, b, and c

The associative law of addition tells us that the *grouping* does not matter in addition.

What about multiplication? Does the grouping matter? For example, is $2 \cdot (3 \cdot 4)$ the same as $(2 \cdot 3) \cdot 4$? Since both calculations give 24, the manner in which we group the numbers does not matter. This fact is stated in the **associative law of multiplication.**

ASSOCIATIVE LAW OF MULTIPLICATION

$a \cdot (b \cdot c) = (a \cdot b) \cdot c$ for any numbers a, b, and c

---

The associative law of multiplication tells us that the grouping does not matter in multiplication.

We now have seen that grouping numbers differently in addition and multiplication yields the same answer. What about order? As it turns out, the order in which we do additions or multiplications does not matter. For example, $2 + 3 = 3 + 2$ and $5 \cdot 4 = 4 \cdot 5$. In general, we have the following laws:

COMMUTATIVE LAW OF ADDITION

$a + b = b + a$ for any numbers a and b

COMMUTATIVE LAW OF MULTIPLICATION

$a \cdot b = b \cdot a$ for any numbers a and b

The commutative laws of addition and multiplication tell us that *order* does not matter in addition or multiplication.

EXAMPLE 2 Which law is illustrated in each of the statements?

a. $a \cdot (b \cdot c) = (b \cdot c) \cdot a$
b. $(5 + 9) + 2 = 5 + (9 + 2)$
c. $(5 + 9) + 2 = (9 + 5) + 2$
d. $(6 \cdot 2) \cdot 3 = 6 \cdot (2 \cdot 3)$

Solution

a. We have changed the *order* of multiplication. The commutative law of multiplication was used.

b. We have changed the *grouping* of the numbers. The associative law of addition was used.

c. We changed the *order* of the 5 and the 9 within the parentheses. The commutative law of addition was used.

d. Here we changed the *grouping* of the numbers. We used the associative law of multiplication.

C. THE DISTRIBUTIVE LAW

All of the properties discussed contain a *single* operation. Now, suppose you wish to multiply a number, say 7, by the sum of 4 and 5. As

Problem 2 Which law is illustrated in each of the statements?
a. $a \cdot (b \cdot c) = a \cdot (c \cdot b)$
b. $(4 + 3) + 5 = (3 + 4) + 5$
c. $(2 + 8) + 1 = 2 + (8 + 1)$
d. $(a \cdot b) \cdot c = a \cdot (b \cdot c)$

ANSWER (to problem on page 51)
1. a. 19   **b.** 33

it turns out $7 \cdot (4 + 5)$ can be obtained in two ways:

$$7 \cdot (4 + 5)$$
$$7 \cdot \quad 9 \quad\Bigg\}\quad \text{Adding within the parentheses first}$$
$$63$$

$$(7 \cdot 4) + (7 \cdot 5)$$
$$\underset{28}{\;} \; + \; \underset{35}{\;} \quad\Bigg\}\quad \text{Multiplying and then adding}$$
$$63$$

Thus,

$$7 \cdot (4 + 5) = (7 \cdot 4) + (7 \cdot 5) \quad 7(4 + 5) = 7 \cdot 4 + 7 \cdot 5$$

means to multiply 4 by 7 and then 5 by 7.

The parentheses in $(7 \cdot 4) + (7 \cdot 5)$ can be *omitted* as long as we agree that multiplications must be done *first*. With this agreement, the distributive law is stated as follows:

DISTRIBUTIVE LAW

$$a(b + c) = ab + ac \qquad \text{for any numbers } a, b, \text{ and } c$$

Note that $a(b + c)$ means $a \cdot (b + c)$, ab means $a \cdot b$ and ac means $a \cdot c$. The distributive law can be extended to more than two numbers inside the parentheses. Thus, $a(b + c + d) = ab + ac + ad$. Moreover, $(b + c)a = ba + ca$.

EXAMPLE 3 Use the distributive law to multiply.
a. $8(2 + 4)$
b. $3(x + 5)$, where x is a real number
c. $3(x + y + z)$
d. $(x + y)z$

Problem 3 Multiply.
a. $7(6 + 3)$
b. $4(6 + x)$, where x is a real number.
c. $5(a + b + c)$
d. $(c + d)e$

Solution

a. $8(2 + 4) = 8 \cdot 2 + 8 \cdot 4$
$\qquad\qquad = 16 + 32 \qquad$ Multiply first
$\qquad\qquad = 48 \qquad\qquad$ Add next
b. $3(x + 5) = 3x + 3 \cdot 5 \quad$ This expression cannot be simplified further,
$\qquad\qquad = 3x + 15 \qquad$ since we do not know the value of x.
c. $3(x + y + z) = 3x + 3y + 3z$
d. $(x + y)z = xz + yz$

D. REMOVING PARENTHESES

All of these laws can be used to simplify expressions such as $(5 + a) + 4$ and $(4a) \cdot 5$, where a represents a real number, by removing the parentheses. Here is the way we do it:

$$(5 + a) + 4 = (a + 5) + 4 \quad \text{By the commutative law}$$
$$= a + (5 + 4) \quad \text{By the associative law}$$
$$= a + 9$$

Similarly,

$$(4a) \cdot 5 = 5 \cdot (4a) \quad \text{By the commutative law}$$
$$= (5 \cdot 4)a \quad \text{By the associative law}$$
$$= 20a$$

EXAMPLE 4   Remove parentheses (simplify).

a. $(7 + 3a) + 8$ **b.** $(3a)6$

Solution

a. $(7 + 3a) + 8 = (3a + 7) + 8$
$$= 3a + (7 + 8)$$
$$= 3a + 15$$

b. $(3a)6 = 6(3a)$
$$= (6 \cdot 3)a$$
$$= 18a$$

Problem 4   Remove parentheses (simplify).
a. $(2 + 5a) + 11$ **b.** $(6a)7$

E.   IDENTITIES AND INVERSES

The properties we have mentioned are applicable to any real numbers. We now want to discuss two special numbers that have unique properties, the numbers zero and one. If we add **0** to a number, the number is unchanged, that is, the number 0 preserves the identity of all numbers under addition. Thus, **0** is called the **identity element for addition.** This fact is stated next.

0 IS THE IDENTITY ELEMENT FOR ADDITION
$a + 0 = 0 + a = a$ for any number a

The number 1 preserves the identity of all numbers under multiplication. Thus, if we multiply a number by 1 the number remains unchanged. Thus, **1** is called the **identity element for multiplication,** as stated.

1 IS THE IDENTITY ELEMENT FOR MULTIPLICATION
$a \cdot 1 = 1 \cdot a = a$ for any number a

We complete our list of properties by stating two ideas that will be more fully discussed in the next chapter.

ADDITIVE INVERSE (OPPOSITE)
For every number a there exists another number $-a$ called its **additive inverse** (or **opposite**) such that $a + (-a) = 0$.

ANSWERS   (to problems on pages 52–53)
2. a. Commutative ×
  b. Commutative +   **c.** Associative +
  d. Associative ×
3. a. $7 \cdot 6 + 7 \cdot 3 = 42 + 21 = 63$
  b. $24 + 4x$ **c.** $5a + 5b + 5c$ **d.** $ce + de$

When dealing with multiplication, the *multiplicative inverse* is called the *reciprocal*.

MULTIPLICATIVE INVERSE (RECIPROCAL)

For every number **a** (except 0) there exists another number $\dfrac{1}{a}$ called the **multiplicative inverse** (or **reciprocal**) such that

$$a \cdot \frac{1}{a} = 1$$

EXAMPLE 5 Which property is illustrated in each of the following statements?

a. $5 + (-5) = 0$ **b.** $8 \cdot \dfrac{1}{8} = 1$

c. $0 + 9 = 9$ **d.** $7 \cdot 1 = 7$

Solution

a. $5 + (-5) = 0$ Additive inverse

b. $8 \cdot \dfrac{1}{8} = 1$ Multiplicative inverse

c. $0 + 9 = 9$ Identity element for addition

d. $7 \cdot 1 = 7$ Identity element for multiplication   ▲

Finally, for easy reference we list all the properties we have studied.

Problem 5 Which property is illustrated in each of the following statements?

a. $2 \cdot \dfrac{1}{2} = 1$ **b.** $1 + (-1) = 0$

c. $1 + 0 = 1$ **d.** $1 \cdot 1 = 1$

PROPERTIES OF THE REAL NUMBERS

If a, b, and c are real numbers

$a + b = b + a$	$a \cdot b = b \cdot a$	Commutative laws
$a + (b + c) = (a + b) + c$	$a \cdot (b \cdot c) = (a \cdot b) \cdot c$	Associative laws
$a + 0 = 0 + a = a$	$a \cdot 1 = 1 \cdot a = a$	Identity laws
$a + (-a) = 0$	$a \cdot \dfrac{1}{a} = 1 \ (a \neq 0)$	Inverse laws
$a(b + c) = ab + ac$		Distributive law

ANSWERS (to problems on pages 54–55)
4. **a.** $5a + 13$ **b.** $42a$
5. **a.** Multiplicative inverse
   b. Additive inverse
   c. Additive identity
   d. Multiplicative identity

NAME

CLASS

SECTION

ANSWERS

A. Find.

1. $(3 \cdot 5) + 8$

2. $3 \cdot (5 + 8)$

3. $6 \times (4 + 2)$

4. $(6 \times 4) + 2$

5. $(7 \cdot 8) + 1$

6. $7 \cdot (8 + 1)$

7. $9 \times (4 + 0)$

8. $(9 \times 4) + 0$

9. $(10 \cdot 7) + 8$

10. $10 \cdot (7 + 8)$

1. _____
2. _____
3. _____
4. _____
5. _____
6. _____
7. _____
8. _____
9. _____
10. _____

B. Which law is illustrated in each statement?

11. $9 + 8 = 8 + 9$

12. $b \cdot a = a \cdot b$

13. $4 \cdot 3 = 3 \cdot 4$

14. $(a + 4) + b = a + (4 + b)$

15. $3(x + 6) = 3x + 3 \cdot 6$

16. $8(2 + x) = 8 \cdot 2 + 8x$

17. $a \cdot (b \cdot c) = a \cdot (c \cdot b)$

18. $a \cdot (b \cdot c) = (a \cdot b) \cdot c$

19. $a + (b + 3) = (a + b) + 3$

20. $(a + 3) + b = (3 + a) + b$

11. _____
12. _____
13. _____
14. _____
15. _____
16. _____
17. _____
18. _____
19. _____
20. _____

C. Use the distributive law to multiply.

21. $6(4 + x)$

22. $3(2 + x)$

23. $8(x + y + z)$

24. $5(x + y + z)$

25. $6(x + 7)$

26. $7(x + 2)$

27. $(a + 5)b$

28. $(a + 2)c$

21. _____
22. _____
23. _____
24. _____
25. _____
26. _____
27. _____
28. _____

29. $6(5 + b)$

30. $3(7 + b)$

29. _____

30. _____

D. Remove parentheses (simplify).

31. $(4 + 2a) + 7$

32. $(2a + 4) + 7$

31. _____

32. _____

33. $(7a + 9) + 8$

34. $(2a + 5) + 11$

33. _____

34. _____

35. $9(4a)$

36. $8(7a)$

35. _____

36. _____

37. $(2a)10$

38. $(5a)7$

37. _____

38. _____

39. $(6a)7$

40. $(9a)3$

39. _____

40. _____

Use the distributive law to fill in the blank.

41. $8(9 + \underline{\hspace{1cm}}) = 8 \cdot 9 + 8 \cdot 3$

41. _____

42. $7(\underline{\hspace{1cm}} + 10) = 7 \cdot 4 + 7 \cdot 10$

42. _____

43. $\underline{\hspace{1cm}}(3 + b) = 15 + 5b$

43. _____

44. $3(\underline{\hspace{1cm}} + b) = 3 \cdot 8 + 3b$

44. _____

45. $\underline{\hspace{1cm}}(b + c) = ab + ac$

45. _____

E. Which property is illustrated in each of the following statements?

46. $9 \cdot \dfrac{1}{9} = 1$

46. _____

47. $10 \cdot 1 = 10$

47. _____

48. $9[3 + (-3)] = 9(0)$

48. _____

49. $0 + 15 = 15$

49. _____

50. $1 \cdot 27 = 27$

50. _____

✓ **SKILL CHECKER**

Write as a decimal.

51. $\dfrac{6}{10,000}$

52. $\dfrac{8}{1000}$

51. _____

52. _____

Find.

53. $2.3 \cdot 1.4$

54. $\dfrac{1}{2} \cdot 20 \cdot 10$

53. _____

54. _____

EXERCISE 1.5

1.5 USING YOUR KNOWLEDGE

The distributive law can be used to simplify certain multiplications. For example, to multiply 8 by 43 we can write

$$8(43) = 8(40 + 3)$$
$$= 320 + 24$$
$$= 344$$

Use this idea to multiply:

1. 7(38)

2. 8(23)

3. 6(46)

4. 9(52)

1. _____

2. _____

3. _____

4. _____

CALCULATOR CORNER

Most modern calculators contain a set of parentheses on the keyboard. These keys will allow you to specify the exact order in which you wish operations to be performed. Thus, to perform the operations in Example 1, you key in

$$\boxed{(}\ \boxed{4}\ \boxed{\times}\ \boxed{5}\ \boxed{)}\ \boxed{+}\ \boxed{6}\ \boxed{=}$$

to obtain 26. Similarly, if you key in

$$\boxed{4}\ \boxed{\times}\ \boxed{(}\ \boxed{5}\ \boxed{+}\ \boxed{6}\ \boxed{)}\ \boxed{=}$$

you obtain 44. However, in most calculators you will not be able to find the answer to the problem 4(5 + 6) unless you key in the multiplication sign $\boxed{\times}$ between the 4 and the parentheses. With this in mind (and if your instructor is willing), do Problems 21 through 24.

1.6 FORMULAS

OBJECTIVES

REVIEW

Before starting this section, you should know:

1. How to reduce fractions.
2. How to add and multiply decimals.

OBJECTIVES

You should be able to:

A. Use a given formula to solve a problem.
B. Use symbols to write a formula given in words.
C. Use formulas from geometry to find areas and perimeters.
D. Use the correct order of operations to evaluate a given expression.
E. Apply the concepts studied to practical problems.

Many problems in algebra can be solved if the proper formula is used. For example, why doesn't the man in the picture get electrocuted when holding the two wires? You may know the answer. It is because the voltage V in the battery (12 volts) is too small. How do we know that? Well, there is a formula in physics that tells us that the current I going through the man's body is

$I = \dfrac{V}{R}$ This equation was discovered by George Simon Ohm and is called Ohm's Law.

where V is the voltage and R the resistance of his body. A car battery carries $V = 12$ volts (V) and $R = 20{,}000$ ohms (Ω); so the current is

$$I = \frac{12}{20{,}000} = \frac{6}{10{,}000} = 0.0006 \text{ amperes (A)}$$

It actually takes about 0.001 A to give you a slight shock.

A. USING FORMULAS

EXAMPLE 1 Anthropologists know how to estimate the height of a man (in centimeters, cm) by using a bone as a clue. To do this, they use the formula

$$H = 2.89h + 70.64$$

Height of the man   Length of the humerus

Estimate the height of a man whose humerus bone is 30 cm long.

Solution We write 30 in place of h (in parentheses to indicate multiplication), obtaining

$H = 2.89(30) + 70.64$

$= 86.70 + 70.64$

$= 157.34 \text{ cm}$ ▲

Problem 1 The formula to estimate the height of a woman using the humerus as a clue is

$$H = 2.75h + 71.48$$

Estimate the height of a woman whose humerus is 20 cm long.

In the preceding examples, we have **evaluated** (found the value of) simple expressions. To evaluate more complicated ones, you need to remember some of the rules we have studied.

EXAMPLE 2 The formula for converting degrees Fahrenheit (°F) to degrees Celsius (°C) is

$$°C = \frac{5}{9}(°F - 32)$$

On a hot summer day, the temperature is 95°F. How many degrees Celsius is that?

Solution In this case, °F = 95. So,

$$°C = \frac{5}{9}(°F - 32) = \frac{5}{9}(95 - 32)$$

$$= \frac{5}{9}(63) \quad \text{Subtracting inside the parentheses first}$$

$$= \frac{5 \cdot 63}{9}$$

$$= 35$$

Thus the temperature is 35°C.

Problem 2 Find the temperature (in degrees Celsius) on a winter day in which the thermometer reads 41°F.

B. WRITING FORMULAS

In science, business, and mathematics, the formulas we have studied are sometimes written in words. When this happens, we must first re-write them in the language of algebra. For example, you are probably aware that the interest I is found by multiplying the principal P by the rate r by the time t. This is translated as:

$I = Prt$

(Note that when we write the letters P, r, and t together we agree that they are to be multiplied; that is, $I = Prt$ means $I = P \cdot r \cdot t$, where P, r, and t are called **factors**.) Thus, if you deposit \$100 in a savings account that pays 6% interest per year, we know that the principal (the amount you deposit) $P = 100$, the rate $r = 6\% = 0.06$, and the time $t = 1$. Hence,

$I = Prt$

$I = 100 \times 0.06 \times 1$

$\quad = \$6$

That is, the interest is \$6 for 1 year.

EXAMPLE 3 The retail selling price R of an item is obtained by adding the original cost C and the markup M on the item.

a. Write a formula for the retail selling price.

b. Find the retail selling price R of an item that originally cost \$50 and has a \$10 markup.

Problem 3 The total cost T of an item is obtained by adding its cost C and the tax t on the item.
a. Write a formula for the total cost T.
b. Find the total cost of an item that costs \$60 and has a \$3 tax.

Solution

a. Write the problem in words and then translate it.

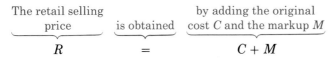

The retail selling price $\underbrace{}$ is obtained $\underbrace{}$ by adding the original cost C and the markup M $\underbrace{}$

$$R \qquad = \qquad C + M$$

b. Here $C = \$50$ and $M = \$10$. Substituting in

$$R = C + M$$

we have

$$R = 50 + 10 = 60$$

C. GEOMETRIC FORMULAS

Many of the formulas we encounter in algebra come from geometry. For example, to find the *area* of a figure, we must find the number of square units contained in the figure. Unit squares look like these:

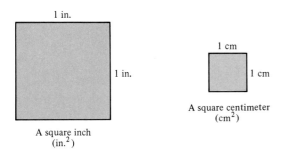

1 in.

1 in.

A square inch (in.^{2})

1 cm

1 cm

A square centimeter (cm^2)

Now, to find the **area** of a figure, say a rectangle, we must find the number of square units it contains. For example, the area of a rectangle 3 cm by 4 cm is 3 cm × 4 cm = 12 cm^2 (read "12 square centimeters"), as shown in the diagram.

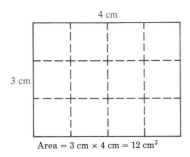

4 cm

3 cm

Area = 3 cm × 4 cm = 12 cm^2

In general, we can find the area A of a rectangle by multiplying its length L by its width W, as given in the next formula.

The area A of a rectangle	is found	By multiplying its length L by its width W
A	$=$	LW

ANSWER (to problem on page 61)
1. 126.48 cm

What about a rectangle's **perimeter** (distance around)? We can find the perimeter by adding the lengths of the four sides. Since we have two sides of length L and two sides of length W, the perimeter of the rectangle is

$$W + L + W + L = 2L + 2W$$

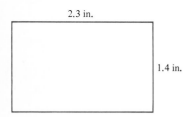

In general, we have the following formula:

> The perimeter P of a rectangle of length L and width W is
>
> $$P = 2L + 2W$$

EXAMPLE 4 Find.

a. The area of the rectangle.
b. The perimeter of the rectangle.

2.3 in.

1.4 in.

Solution

a. The area:

$$A = LW$$
$$= (2.3 \text{ in.}) \cdot (1.4 \text{ in.})$$
$$= 3.22 \text{ in.}^2$$

b. The perimeter:

$$P = 2L + 2W$$
$$= 2(2.3 \text{ in.}) + 2(1.4 \text{ in.})$$
$$= 4.6 \text{ in.} + 2.8 \text{ in.}$$
$$= 7.4 \text{ in.}$$
▲

Note that the area is given in square units, whereas the perimeter is a length and is given in linear units.

If we know the area of a rectangle, we can always calculate the area of the triangle in the figure (shaded). The area of the triangle (shaded) is $\frac{1}{2}$ the area of the rectangle, which is bh. Thus, the area of the triangle is

$$\frac{1}{2}bh$$

Problem 4 Find **a.** the area and **b.** the perimeter of a 3.2 cm by 1.8 cm rectangle.

ANSWERS (to problems on page 62)
2. 5° C   **3. a.** $T = C + t$ **b.** $T = \$63$

Note that this time we used b and h instead of L and W.

If a triangle has base b and perpendicular height h, its area A is

$$A = \frac{1}{2}bh$$

This formula holds true for any type of triangle.

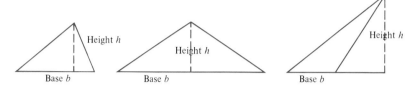

EXAMPLE 5 Find the area of a triangular piece of cloth 20 cm long and 10 cm high.

Solution

$$A = \frac{1}{2}bh$$

$$= \frac{1}{2}(20 \text{ cm}) \cdot (10 \text{ cm})$$

$$= 100 \text{ cm}^2 \qquad \blacktriangle$$

The approximate area of a circle can easily be found if we know the distance from the center of the circle to its edge, the **radius** r, of the circle. Here is the formula:

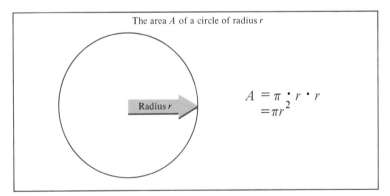

The area A of a circle of radius r

Radius r

$$A = \pi \cdot r \cdot r$$
$$= \pi r^2$$

As you can see, the formula to find the area of a circle involves the number π (read "pie"). The number π is irrational; it cannot be written as a terminating or repeating decimal or a fraction but can be *approximated*. In most of our work we shall say that π is about 3.14 or $\frac{22}{7}$; that is, $\pi \approx 3.14$ or $\pi \approx \frac{22}{7}$. The number π is also used in finding the perimeter (distance around) a circle. This perimeter is called the **circumference** C of the circle and is found by using the following formula:

The circumference C of a circle of radius r is

$$C = 2\pi r$$

Problem 5 Find the area of a triangular piece of material 10 in. long by 5 in. high.

EXAMPLE 6 A circle has a radius of 5 in. Find.

a. The area of the circle.
b. The circumference of the circle.

Solution

a. $A = \pi \cdot r \cdot r = 3.14 \times (5\text{ in.}) \times (5\text{ in.})$
$$= 78.5\text{ in.}^2$$

b. $C = 2 \times 3.14 \times 5\text{ in.} = 31.4\text{ in.}$

D. EVALUATING EXPRESSIONS

In the examples above we have *evaluated* (found the value of) simple formulas. To evaluate more complicated ones, we need some rules to determine what operations we must do first. For example, consider the expression

$$x + y \div z$$

If $x = 8$, $y = 4$, and $z = 2$, we can evaluate the expression in two ways:

$x + y \div z$			$x + y \div z$	
$8 + 4 \div 2$			$8 + 4 \div 2$	
$12 \div 2$	Add first		$8 + 2$	Divide first
6	Divide next		10	Add next

Since the answers are different, we must agree to do operations in a certain order. Here is the rule.

EVALUATING EXPRESSIONS

1. Do the operations *inside* the grouping symbols (such as parentheses) *first*, starting with the innermost grouping symbols.

2. Perform *multiplications* and *divisions* as they occur from left to right.

3. Perform *additions* and *subtractions* as they occur from left to right.

You can recall the order of operations by remembering the following:

Parentheses	Please
Multiplications	My
Divisions	Dear
Additions	Aunt
Subtractions	Sally

With this convention, the expression

$$x + y \div z$$

where $x = 8$, $y = 4$, and $z = 2$, is evaluated as follows:

$x + y \div z = 8 + 4 \div 2$	There are no parentheses involved.
$= 8 + 2$	Multiplications and divisions are done first.
$= 10$	Additions and subtractions next.

Problem 6 A circle has a radius of 3 cm. Find.
a. Its area.
b. Its circumference.

EXAMPLE 7 Evaluate

$$y \div 2x - (z + 4)$$

when $x = 3$, $y = 6$, $z = 1$.

Problem 7 Evaluate $y \div 3x - (z + 2)$ when $x = 2$, $y = 9$, and $z = 2$.

Solution

$$
\begin{aligned}
y \div 2x - (z + 4) &= 6 \div 2 \cdot 3 - (1 + 4) && \text{Substituting } x = 3, y = 6, z = 1 \\
&= 6 \div 2 \cdot 3 - 5 && \text{Adding 1 and 4 inside the parentheses} \\
&= 3 \cdot 3 - 5 && \text{Dividing } 6 \div 2 = 3 \\
&= 9 - 5 && \text{Multiplying } 3 \cdot 3 = 9 \\
&= 4 && \text{Subtracting } 9 - 5 = 4 \quad \blacktriangle
\end{aligned}
$$

Note that according to our rules of order, $y \div 2x$ *does not* mean $\dfrac{y}{2x}$ but rather $\dfrac{y}{2} \cdot x$. To avoid confusion, if you wish to have $\dfrac{y}{2x}$, write $y \div (2x)$.

E. APPLICATIONS

Have you ever tried administering medicine to a child but felt in doubt about the proper amount (dose)? There are several formulas that tell us the corresponding dose of medication for children when the adult dosage is known. However, before you use any of them make sure you consult your health-care professional.

1. Friend's rule (for children under 2 yr)

   (Age in months · adult dose) ÷ 150 = child's dose

2. Clark's rule (for children over 2 yr)

   (Weight of child · adult dose) ÷ 150 = child's dose

3. Young's rule (for children between 3 and 12)

   (Age · adult dose) ÷ (Age + 12) = child's dose

EXAMPLE 8

a. Suppose a child is 10 months old and the adult dose of aspirin is a 300-milligram (mg) tablet. What is the child's dose?

b. If a 7-yr-old child weighs 75 lb and the adult dose is 4 tablets a day, what is the child's dose?

c. Suppose a child is 6 yr old and the adult dose of an antibiotic is 6 tablets every 12 hr. What is the child's dose?

Problem 8

a. If a child is 20 months old and the adult dose of aspirin is a 300-mg tablet, what is the child's dose?

b. If a 9-yr-old child weighs 75 lb and the adult dose is 8 tablets a day, what is the child's dose?

c. Suppose a child is 4 yr old and the adult dose is 4 tablets every 12 hr. What is the child's dose?

Solution

a. We substitute 10 and 300 in the first formula, obtaining:

$$
\begin{aligned}
(10 \cdot 300) \div 150 &= 3000 \div 150 && \text{Multiplying inside the parentheses} \\
&= 20 && \text{Dividing 3000 by 150}
\end{aligned}
$$

Thus, the child's dose is 20 mg of aspirin.

b. Substitute 75 and 4 in the formula, obtaining

$$(75 \cdot 4) \div 150 = 300 \div 150 \quad \text{Multiplying inside the parentheses}$$
$$= 2 \quad \text{Dividing 300 by 150}$$

Thus, the child's dose is 2 tablets a day.

c. This time we substitute 6 and 4 in the formula. We have

$$(6 \cdot 6) \div (6 + 12) = 36 \div 18 \quad \text{Doing the operations inside the parentheses}$$
$$= 2 \quad \text{Dividing 36 by 18}$$

Thus, the child's dose is 2 tablets every 12 hr.

ANSWERS (to problems on pages 66–67)
6. **a.** 28.26 cm² **b.** 18.84 cm
7. 2
8. **a.** A 40-mg tablet
   b. 4 tablets a day
   c. 1 tablet every 12 hr

NAME

CLASS

SECTION

A.

ANSWERS

1. The number of miles D traveled in T hours by an object moving at a rate R (in miles per hour) is given by $D = RT$.
   a. Find D when $R = 30$ and $T = 4$.
   b. Find the distance traveled by a car going 55 mi/h for 5 hr.

   1. a. _____
      b. _____

2. The rate of travel R of an object moving a distance D in time T is given by

   $$R = \frac{D}{T}$$

   a. Find R when $D = 240$ mi and $T = 4$ hr.
   b. Find the rate of travel of a train that traveled 140 mi in 4 hr.

   2. a. _____
      b. _____

3. The height H of a man (in inches) is related to his weight W (in pounds) by the formula $W = 5H - 190$. If a man is 60 in. high, what should his weight be?

   3. _____

4. The number of hours H a growing child should sleep is

   $$H = 17 - \frac{A}{2}$$

   where A is the age of the child in years. How many hours should a 6-yr-old sleep?

   4. _____

5. The formula for converting degrees Celsius to degrees Fahrenheit is

   $$°F = \frac{9°C}{5} + 32$$

   If the temperature is 15°C, what is the corresponding Fahrenheit temperature?

   5. _____

6. The profit P a business makes is obtained by subtracting the expenses E from the income I.
   a. Write a formula for the profit P.
   b. Find the profit of a business with $2700 in income and $347 in expenses.

   6. a. _____
      b. _____

B.

7. The energy efficiency ratio (EER) for an air conditioner is obtained by dividing the British thermal units (Btu) by the watts (W).
   a. Write a formula that will give the EER of an air conditioner.
   b. Find the EER of an air conditioner with a 9000-Btu capacity (per hour) and a rating of 1000 W.

   7. a. _____
      b. _____

8. The capital C of a business is the difference between the assets A and the liabilities L.
   a. Write a formula that will give the capital of a business.
   b. If a business has $4800 in assets and $2300 in liabilities, what is the capital of the business?

8. a. _____
   b. _____

9. The selling price S of an item is the sum of the cost C and the margin (markup) m.
   a. Write a formula for the selling price of a given item.
   b. A merchant wishes to have a $15 margin (markup) on an item costing $52. What should be the selling price of this item?

9. a. _____
   b. _____

10. The tip speed TS of a propeller is equal to π times the diameter d of the propeller times the number N of revolutions per second. ($\pi \approx 3.14$.)
    a. Write a formula for TS.
    b. If a propeller has a 2-m diameter and it is turning at 1000 revolutions per second, find TS.

10. a. _____
    b. _____

C. Find the area and perimeter of the following.

11. A 10-cm by 20-cm rectangle.

11. _____

12. A 7-in. by 8-in. rectangle.

12. _____

13. A 4.2-m by 3.1-m rectangle.

13. _____

14. A 10.4-ft by 5.2-ft rectangle.

14. _____

15. A 14.1-yd by 7.9-yd rectangle.

15. _____

Find the area of the following.

16. A triangle whose base is 8 in. long and whose height is 10 in.

16. _____

17. A triangle of base 10 cm and height 7 cm.

17. _____

18. A triangle with a 50-ft base and a 20-ft height.

18. _____

19. A triangular piece of land with a 30-yd base and 20-yd height.

19. _____

20. A triangle of base 20 in. and height 5 in.

20. _____

21. A circle with a 10-in. radius.

21. _____

22. A circle with a 5-cm radius.

22. _____

23. A circle with a 20-ft radius.

23. _____

24. A circle with a 2-yd radius.

24. _____

25. A circle with a 7-m radius.

25. _____

Find the circumference of the circle with the given radius.

26. 3 cm

26. _____

27. 15 in.

27. _____

28. 20 yd

28. _____

29. 10 ft

29. _____

30. 15 m

30. _____

D. Evaluate the following expressions when $x = 3$, $y = 2$, $z = 4$.

31. $(x + y)(x + z)$ **32.** $(x + z)(y + z)$

33. $(x + y)(z - y)$ **34.** $(x + z)(z - x)$

35. $xz \div y - z$ **36.** $xz \div y - x$

37. $(x - y + z \div y) \div x$ **38.** $(z - x + z \div y) \div x$

39. $8(z \div y + x - z)$ **40.** $9(z \div y + y - x)$

41. $(z - x)[6y \div z - (x - y)]$ **42.** $(z - y)[4x \div z - (z - y)]$

43. $(x - y)(z - y)(z - x)$ **44.** $(x - y)(z - 2y)(z - x)$

45. $9(z - y) \div x$ **46.** $12(z - y) \div x$

47. $6z \div (x - y + z \div y)$ **48.** $3y \div (z - x + z \div y)$

49. $(x - y + z \div y) \div (z - x + z \div y)$

50. $9(z - x + z \div y) \div (x - y + z \div y)$

E. Applications

51. A rectangular roof is 20 ft by 15 ft. How many square feet does the roof cover?

52. The largest pizza ever made (in Glens Falls, New York) had a 40-ft radius. What area was covered by this circular pizza? ($\pi \approx 3.14$.)

53. The largest proven circular crater, located in northern Arizona, has a 2600-ft radius. What is the circumference of this crater?

54. The largest stained glass window in the world is at Kennedy International Airport and measures 300 ft by 23 ft. What is the area of this rectangular window?

55. The largest omelet was cooked at Conestoga College, in Ontario, Canada, and measured 30 ft by 10 ft. What was the area of this rectangular omelet?

✓ **SKILL CHECKER**

56. Find a fraction equivalent to $\frac{3}{8}$ with a denominator of 32.

31. _____

32. _____

33. _____

34. _____

35. _____

36. _____

37. _____

38. _____

39. _____

40. _____

41. _____

42. _____

43. _____

44. _____

45. _____

46. _____

47. _____

48. _____

49. _____

50. _____

51. _____

52. _____

53. _____

54. _____

55. _____

56. _____

57. Find a fraction equivalent to $\dfrac{28}{52}$ with a denominator of 13.

57. _____

58. Find $4\dfrac{3}{8} - 3\dfrac{5}{7}$.

58. _____

59. Find $6\dfrac{1}{3} - 5\dfrac{7}{8}$.

59. _____

60. Write 37.4% as a decimal.

60. _____

1.6 USING YOUR KNOWLEDGE

Many practical problems around the house require some knowledge of the formulas we have studied. For example, let us say that you wish to carpet your living room. You need to use the formula for the area A of a rectangle of length L and width W. The formula is $A = L \cdot W$.

1. How many square yards of carpet are needed to carpet a living room measuring 7 yd by 5 yd?

1. _____

2. If you wish to plant new grass in your yard, you can buy sod-squares of grass that can be simply laid on the ground. Each sod-square is approximately 1 ft². How many squares do you need to plant a lawn that is 90 ft by 60 ft?

2. _____

3. Do you know how efficient your air conditioner is? The formula for the efficiency is

$$\text{EER} = \frac{\text{Btu}}{W}$$

where EER means energy efficiency ratio, Btu means British thermal unit (a unit that measures how much cool air is produced each hour), and watts (W) is a measurement of the electric power consumed per hour. Find the EER for an air conditioner that puts out 8000 Btu per hour while consuming 1000 W of power.

3. _____

4. If you wish to fence a yard you need to know its perimeter (the distance around the yard). If the yard is W ft by L ft, the perimeter P is given by

$$P = 2W + 2L$$

Find the amount of fence needed for a yard that is 70 ft by 50 ft.

4. _____

5. Do you know how to figure how much it costs to operate your appliances and lights? To find the answer you should first know that the amount of energy E you consume can be obtained by multiplying the power P (in watts) by the time T; that is, $E = PT$.
  a. How much energy is used by a 100-W lamp that is on for 15 hr?
  b. If an electric company charges 5¢ per each 1000 W used in an hour (a kilowatt-hour), how much does it cost to operate the 100-W lamp for the 15 hr?

5. **a.** _____

  b. _____

SUMMARY

SECTION	ITEM	MEANING	EXAMPLE
1.1A	Natural numbers	1, 2, 3, 4, 5, and so on	4, 13, and 497 are natural numbers.
1.1A	Whole numbers	0, 1, 2, 3, 4, and so on	0, 92, and 384 are whole numbers.
1.1A	Positive integers	1, 2, 3, 4, 5, and so on	5, 10, and 999 are positive integers.
1.1A	Negative integers	$-1, -2, -3, -4,$ and so on	$-8, -50, -459$ are negative integers.
1.1A	Integers	$\ldots -2, -1, 0, 1, 2, \ldots$	$-98, 0,$ and 459 are integers.
1.1A	Rational number	A number that can be written in the form $\frac{a}{b}$, a and b integers and b not 0	$\frac{1}{5}, \frac{-3}{4}$, and $\frac{0}{5}$ are rational numbers.
1.1A	Irrational numbers	Numbers that cannot be written as the ratio of two integers	$\sqrt{2}, -\sqrt[3]{2}$, and π are irrational numbers.
1.1A	Real numbers	The rationals and the irrationals	$3, 0, -9, \sqrt{2}$, and $\sqrt[3]{5}$ are real numbers.
1.1B	Equivalent fractions	Two fractions of equal value	$\frac{1}{2}$ and $\frac{3}{6}$ are equivalent fractions.
1.1B	Fraction reduced to lowest terms	A fraction is reduced to **lowest terms** when there is no number (except 1) that will divide the numerator and the denominator.	$\frac{4}{9}$ and $\frac{5}{16}$ are reduced to lowest terms but $\frac{7}{28}$ is not. $\left(\frac{7}{28}\right.$ can be reduced to $\left.\frac{1}{4}.\right)$
1.2A	Multiplication of fractions	$\frac{a}{b} \cdot \frac{c}{d} = \frac{a \cdot c}{b \cdot d}$	$\frac{3}{4} \cdot \frac{5}{7} = \frac{3 \cdot 5}{4 \cdot 7} = \frac{15}{28}$
1.2B	Division of fractions	$\frac{a}{b} \div \frac{c}{d} = \frac{a}{b} \cdot \frac{d}{c}$	$\frac{3}{4} \div \frac{2}{7} = \frac{3}{4} \cdot \frac{7}{2} = \frac{21}{8}$
1.2C	Addition of fractions	$\frac{a}{b} + \frac{c}{b} = \frac{a + c}{b}$	$\frac{3}{5} + \frac{1}{5} = \frac{3 + 1}{5} = \frac{4}{5}$
1.2C	Least common denominator (LCD)	The smallest multiple of the denominators	The LCD of $\frac{3}{8}$ and $\frac{5}{14}$ is 56.
1.2D	Subtraction of fractions	$\frac{a}{b} - \frac{c}{b} = \frac{a - c}{b}$	$\frac{3}{5} - \frac{1}{5} = \frac{3 - 1}{5} = \frac{2}{5}$
1.3A	Expanded form of a decimal	A decimal written as a sum of tens, ones, tenths, and so on	$10 + 2 + \frac{3}{10} + \frac{7}{100}$ is written in expanded form.
1.3B	Terminating decimal	A number whose decimal representation terminates	$\frac{2}{5} = 0.4$ is terminating.
1.3B	Repeating decimal	A number whose decimal representation repeats	$\frac{2}{3} = 0.666 \ldots$

(Continued)

SECTION	ITEM	MEANING	EXAMPLE
1.5B	Associative law of addition	For any numbers a, b, c, $a + (b + c) = (a + b) + c$.	$3 + (4 + 9) = (3 + 4) + 9$
1.5B	Associative law of multiplication	For any numbers a, b, c, $a \cdot (b \cdot c) = (a \cdot b) \cdot c$.	$5 \cdot (2 \cdot 8) = (5 \cdot 2) \cdot 8$
1.5B	Commutative law of addition	For any numbers a and b, $a + b = b + a$.	$2 + 9 = 9 + 2$
1.5B	Commutative law of multiplication	For any numbers a and b, $a \cdot b = b \cdot a$.	$18 \cdot 5 = 5 \cdot 18$
1.5C	Distributive law	For any numbers a, b, c, $a(b + c) = ab + ac$.	$3(4 + x) = 3 \cdot 4 + 3 \cdot x$
1.5E	Identity for addition	0 is the identity for addition.	$3 + 0 = 0 + 3 = 3$
1.5E	Identity for multiplication	1 is the identity for multiplication.	$1 \cdot 7 = 7 \cdot 1 = 7$
1.5E	Additive inverse (opposite)	For any number a, its additive inverse is $-a$.	The additive inverse of 5 is -5 and that of -8 is 8.
1.5E	Multiplicative inverse (reciprocal)	The inverse of a is $\dfrac{1}{a}$ if a is not 0 (0 has no reciprocal.)	The multiplicative inverse of $\dfrac{3}{4}$ is $\dfrac{4}{3}$.
1.6C	Area of a rectangle	$A = L \cdot W$, where L is the length and W is the width.	The area of a rectangle 6 in. long and 4 in. wide is $6 \cdot 4 = 24$ in.^{2}
1.6C	Perimeter of a rectangle	$P = 2L + 2W$, where L is the length and W is the width.	The perimeter of a rectangle 6 in. long and 4 in. wide is $P = 2 \cdot 6 + 2 \cdot 4 = 20$ in.
1.6C	Area of a triangle	The area A of a triangle with base b and perpendicular height h is $A = \dfrac{1}{2}bh$.	The area A of a triangle 6 in. high and with an 8-in. base is $A = \dfrac{1}{2} \cdot 8 \cdot 6 = 24$ in.^{2}
1.6C	Area of a circle	The area A of a circle of radius r is $A = \pi r^2$.	The area A of a circle of radius 6 in. is $A = \pi \cdot 6^2 = 36\pi$.
1.6C	Circumference of a circle	The circumference C of a circle of radius r is $C = 2\pi r$.	The circumference C of a circle of radius 6 in. is $C = 2 \cdot \pi \cdot 6 = 12\pi$.

NAME

CLASS

SECTION

ANSWERS

(*If you need help with these exercises, look in the section indicated in brackets.*)

1. [1.1A] Write as a fraction with a denominator of 1.
   a. -16
   b. -14
   c. -97

2. [1.1B] Write the fraction $\frac{3}{8}$ with a denominator of:
   a. 16
   b. 24
   c. 40

3. [1.1B] Write the fraction $\frac{12}{48}$ with a denominator of:
   a. 4
   b. 8
   c. 12

4. [1.1C] Reduce to lowest terms.
   a. $\frac{8}{12}$
   b. $\frac{8}{16}$
   c. $\frac{27}{45}$

5. [1.2A] Find.
   a. $3\frac{1}{8} \cdot \frac{16}{25}$
   b. $2\frac{1}{4} \cdot \frac{16}{9}$
   c. $5\frac{1}{6} \cdot \frac{12}{31}$

6. [1.2B] Find.
   a. $3\frac{2}{3} \div \frac{11}{6}$
   b. $2\frac{1}{2} \div \frac{5}{6}$
   c. $4\frac{3}{4} \div \frac{19}{8}$

1. a. _____
   b. _____
   c. _____

2. a. _____
   b. _____
   c. _____

3. a. _____
   b. _____
   c. _____

4. a. _____
   b. _____
   c. _____

5. a. _____
   b. _____
   c. _____

6. a. _____
   b. _____
   c. _____

7. [1.2C] Find.

a. $1\dfrac{1}{6} + \dfrac{1}{8}$

b. $2\dfrac{3}{4} + \dfrac{5}{6}$

c. $3\dfrac{1}{10} + \dfrac{1}{8}$

8. [1.2D] Find.

a. $3\dfrac{1}{6} - 1\dfrac{9}{10}$

b. $4\dfrac{1}{6} - 1\dfrac{9}{10}$

c. $5\dfrac{1}{6} - 1\dfrac{9}{10}$

9. [1.3A] Write in expanded form.
   a. 68.421
   b. 68.422
   c. 68.423

10. [1.3B] Write as a decimal.

a. $\dfrac{21}{1000}$

b. $\dfrac{23}{1000}$

c. $\dfrac{27}{1000}$

11. [1.3B] Write as a decimal.

a. $\dfrac{2}{11}$

b. $\dfrac{3}{11}$

c. $\dfrac{4}{11}$

12. [1.3C] Write as a reduced fraction.
   a. 0.015
   b. 0.025
   c. 0.045

13. [1.3C] Write as a reduced fraction.
   a. 3.14
   b. 3.16
   c. 3.18

7. a. _____
   b. _____
   c. _____

8. a. _____
   b. _____
   c. _____

9. a. _____
   b. _____
   c. _____

10. a. _____
    b. _____
    c. _____

11. a. _____
    b. _____
    c. _____

12. a. _____
    b. _____
    c. _____

13. a. _____
    b. _____
    c. _____

14. [1.4A] Write as a fraction.
   a. 58.5%
   b. 68.5%
   c. 78.5%

14. a. _____
    b. _____
    c. _____

15. [1.4B] Write as a decimal.
   a. 54.8%
   b. 64.8%
   c. 74.8%

15. a. _____
    b. _____
    c. _____

16. [1.4C] Write as a percent.
   a. 0.19
   b. 0.29
   c. 0.39

16. a. _____
    b. _____
    c. _____

17. [1.4D] Write as a percent.

   a. $\dfrac{1}{8}$

   b. $\dfrac{5}{8}$

   c. $\dfrac{7}{8}$

17. a. _____
    b. _____
    c. _____

18. [1.5B] Which law is illustrated in each of the following statements?
   a. $x + (y + z) = x + (z + y)$
   b. $6 \cdot (8 \cdot 7) = (6 \cdot 8) \cdot 7$
   c. $(x + y) + z = x + (y + z)$

18. a. _____
    b. _____
    c. _____

19. [1.5C] Find.
   a. $(8 \times 3) + 4$
   b. $8 \times (3 + 4)$

19. a. _____
    b. _____

20. [1.5C] Find.
   a. $8(x + 5)$
   b. $7(a + b + c)$

20. a. _____
    b. _____

21. [1.5D] Remove parentheses (simplify).
   a. $(5 + 4x) + 2$
   b. $(3 + 5x) + 6$
   c. $(2 + 7x) + 5$

21. a. _____
    b. _____
    c. _____

22. [1.5D] Remove parentheses (simplify).
   a. $(5x) \cdot 2$
   b. $(3x) \cdot 6$
   c. $(2x) \cdot 5$

22. a. _____
    b. _____
    c. _____

23. [1.6C] Find the area and perimeter of the rectangle with the given dimensions.
   a. 3.1 in. long and 1.9 in. wide
   b. 3.2 in. long and 1.8 in. wide
   c. 3.3 in. long and 1.7 in. wide

23. a. _____
    b. _____
    c. _____

REVIEW EXERCISES CHAPTER 1

NAME ______

24. [1.6C] Find the area and circumference of the circle with the given radius. (Use $\pi \approx 3.14$.)
    a. 1 in.
    b. 2 in.
    c. 3 in.

25. [1.6D] Evaluate for $x = 2$, $y = 3$, and $z = 4$.
    a. $2y \div x + 5 - z$
    b. $4y \div x + 4 - z$
    c. $6y \div x + 1 - z$

24. a. ______
b. ______
c. ______

25. a. ______
b. ______
c. ______

NAME

CLASS

SECTION

ANSWERS

(*Answers on pages 82–83*)

1. Write -18 as a fraction with a denominator of 1.

   1. _____

2. Find a fraction equivalent to $\dfrac{3}{7}$ with a denominator of 21.

   2. _____

3. Find a fraction equivalent to $\dfrac{9}{15}$ with a denominator of 5.

   3. _____

4. Reduce $\dfrac{27}{54}$ to lowest terms.

   4. _____

5. Find $5\dfrac{1}{4} \cdot \dfrac{32}{21}$.

   5. _____

6. Find $2\dfrac{1}{4} \div \dfrac{3}{8}$.

   6. _____

7. Find $1\dfrac{1}{10} + \dfrac{5}{12}$.

   7. _____

8. Find $4\dfrac{1}{6} - 1\dfrac{9}{10}$.

   8. _____

9. Write 68.428 in expanded form.

9. _____

10. Write $\dfrac{29}{1000}$ as a decimal.

10. _____

11. Write $\dfrac{8}{11}$ as a decimal.

11. _____

12. Write 0.045 as a reduced fraction.

12. _____

13. Write 3.12 as a reduced fraction.

13. _____

14. Write 48.5% as a fraction.

14. _____

15. Write 84.8% as a decimal.

15. _____

16. Write 0.69 as a percent.

16. _____

17. Write $\dfrac{5}{8}$ as a percent.

17. _____

18. Which law is illustrated in each of the following statements?
    a. $b \cdot (c \cdot a) = b \cdot (a \cdot c)$
    b. $(6 + 8) + 2 = 6 + (8 + 2)$

18. **a.** _____
    b. _____

19. Find.
    a. $(9 \times 2) + 4$
    b. $9 \times (2 + 4)$

19. **a.** _____
    b. _____

20. Find.
    a. $8(x + 2)$
    b. $5(x + y + z)$

20. **a.** _____
    b. _____

21. Remove parentheses (simplify).
    a. $(8 + 3a) + 6$
    b. $(8a) \cdot 6$

21. **a.** _____
    b. _____

22. Evaluate.

$$^{\circ}\text{C} = \frac{5}{9}(^{\circ}\text{F} - 32) \text{ when } ^{\circ}\text{F} = 59$$

22. _____

23. A rectangle is 6.1 in. long and 1.9 in. wide. Find.
    a. Its area
    b. Its perimeter

23. **a.** _____
    b. _____

24. A circle has a 4-in. radius. (Use $\pi \approx 3.14$.) Find.
    a. Its area
    b. Its circumference

24. **a.** _____
    b. _____

25. Evaluate.

$$y \div 2x - z + 8 \text{ for } x = 3, y = 18, z = 2$$

25. _____

IF YOU MISSED QUESTION	SECTION	EXAMPLES	PAGE	ANSWERS
1	1.1	1	10	1. $\dfrac{-18}{1}$
2	1.1	2	11	2. $\dfrac{9}{21}$
3	1.1	3	11	3. $\dfrac{3}{5}$
4	1.1	4	12	4. $\dfrac{1}{2}$
5	1.2	1, 2, 3	17, 18	5. 8
6	1.2	4, 5	19, 20	6. 6
7	1.2	6	22	7. $\dfrac{91}{60}$
8	1.2	8	24	8. $\dfrac{34}{15}$
9	1.3	1	32	9. $60 + 8 + \dfrac{4}{10} + \dfrac{2}{100} + \dfrac{8}{1000}$
10	1.3	2	32	10. 0.029
11	1.3	3	33	11. $0.\overline{72}$
12	1.3	4	34	12. $\dfrac{9}{200}$
13	1.3	5	35	13. $\dfrac{78}{25}$
14	1.4	1	41	14. $\dfrac{97}{200}$
15	1.4	2	42	15. 0.848
16	1.4	3	43	16. 69%
17	1.4	4	44	17. 62.5%
18a	1.5	2	52	18. a. Commutative $\times$
18b	1.5	2	52	b. Associative $+$
19a	1.5	1	51	19. a. 22
19b	1.5	1	51	b. 54
20a	1.5	3	53	20. a. $8x + 16$
20b	1.5	3	53	b. $5x + 5y + 5z$
21a	1.5	4	54	21. a. $3a + 14$
21b	1.5	4	54	b. $48a$
22	1.6	1, 2	61, 62	22. $15°$

IF YOU MISSED QUESTION	SECTION	EXAMPLES	PAGE	ANSWERS
23a	1.6	4	64	**23.** **a.** 11.59 in.^{2}
23b	1.6	4	64	**b.** 16 in.
24a	1.6	6	66	**24.** **a.** 50.24 in.^{2}
24b	1.6	6	66	**b.** 25.12 in.
25	1.6	7	67	**25.** 33

CHAPTER 2

INTEGERS, RATIONALS, AND EXPRESSIONS

In this chapter we expand the system of natural numbers and study integers and rational numbers. As you will see, algebra turns out to be a generalized type of arithmetic. You will become familiar with the language used in algebra and learn to work with algebraic expressions. We end the chapter with the study of negative exponents and an important application: scientific notation.

_____

_____

_____

_____

_____

NAME _____

CLASS _____

SECTION _____

(Answers on pages 88–89)

ANSWERS

1. Find the additive inverse (opposite) of -4.

1. _____

2. Find $|-3|$.

2. _____

3. Find.
   a. $7 + (-4)$
   b. $(-3) + (-5)$
   c. $(-15) + 7$
   d. $10 + (-3)$

3. a. _____
   b. _____
   c. _____
   d. _____

4. Find.
   a. $-20 - 3$
   b. $-12 - (-3)$
   c. $-5 - (-7)$

4. a. _____
   b. _____
   c. _____

5. Find.
   a. $3 \cdot 6$
   b. $-7 \cdot 5$
   c. $8 \cdot (-4)$
   d. $-8 \cdot (-6)$

5. a. _____
   b. _____
   c. _____
   d. _____

6. Find.
   a. $(-5)^2$
   b. -5^2

6. a. _____
   b. _____

7. Find.

   a. $\dfrac{-48}{6}$

   b. $\dfrac{-36}{-9}$

   c. $-40 \div (-10)$

   d. $\dfrac{8}{0}$

7. a. _____
   b. _____
   c. _____
   d. _____

8. Find the additive inverse (opposite) of -5.6.

8. _____

9. Find $\left|-4\dfrac{1}{5}\right|$.

9. _____

10. Find.
   a. $-7.9 + 5.2$

   b. $-\dfrac{3}{5} + \left(-\dfrac{1}{8}\right)$

10. a. _____
   b. _____

11. Find.
    a. $-5.9 - (-4.7)$

    b. $\dfrac{5}{6} - \dfrac{9}{4}$

12. Find.
    a. $-4.1(3.2)$

    b. $-\dfrac{3}{5}\left(-\dfrac{3}{7}\right)$

13. Find.

    a. $\dfrac{3}{5} \div \dfrac{-4}{15}$

    b. $\dfrac{-5}{6} \div \left(-\dfrac{9}{2}\right)$

14. Simplify.
    a. $6(x - 4)$
    b. $-(5x - 3y)$
    c. $-2(x + 2y - 4)$

15. Write in symbols.
    a. The quotient of $(x - y)$ and z
    b. The markup M of an item is obtained by subtracting the cost C from the selling price S.

16. Write the indicated products using juxtaposition.
    a. $-a$ times b times c
    b. $5 \cdot 5 \cdot b \cdot a \cdot a \cdot b \cdot b$

17. Combine like terms.
    a. $(-3x) + (-7x)$
    b. $-6ab^2 - (-3ab^2)$

18. Simplify.
    a. $9x - 2(x - 3) - (x + 1)$
    b. $[x - (x + 5)] + [2 + 3(x - 1)]$

19. Find.
    a. $(3a^2b)(-5ab^2)$
    b. $(-3x^2yz)(-4xy^2z^3)$

20. Find.

    a. $\dfrac{18x^3y^7}{-6xy^3}$

    b. $\dfrac{-9xy^5}{-27x^5y}$

21. Multiply and simplify.
    a. $2^7 \cdot 2^{-5}$
    b. $y^{-3} \cdot y^{-4}$

22. Find.

    a. $\dfrac{x}{x^6}$

    b. $\dfrac{x^{-3}}{x^{-3}}$

    c. $\dfrac{x^{-5}}{x^{-6}}$

23. Simplify.
    a. $(x^3)^{-4}$
    b. $(2x^3y^{-4})^3$
    c. $(3x^{-3}y^2)^{-2}$

24. Write in scientific notation.
    a. 57,000,000
    b. 0.0000047

25. Perform the indicated operations.
    a. $(4 \times 10^4) \times (8.1 \times 10^5)$

    b. $\dfrac{1.24 \times 10^{-2}}{3.1 \times 10^{-3}}$

22. a. _____
    b. _____
    c. _____

23. a. _____
    b. _____
    c. _____

24. a. _____
    b. _____

25. a. _____
    b. _____

IF YOU MISSED QUESTION	SECTION	EXAMPLES	PAGE	ANSWERS
1	2.1	1	92	**1.** 4
2	2.1	2	93	**2.** 3
3a	2.1	3	93	**3. a.** 3
3b	2.1	4	94	**b.** -8
3c	2.1	5a	95	**c.** -8
3d	2.1	5b	95	**d.** 7
4a	2.1	6b	96	**4. a.** -23
4b	2.1	6c	96	**b.** -9
4c	2.1	6d	96	**c.** 2
5a	2.2	1a	102	**5. a.** 18
5b	2.2	1b	102	**b.** -35
5c	2.2	1c	102	**c.** -32
5d	2.2	1d	102	**d.** 48
6a	2.2	2a	103	**6. a.** 25
6b	2.2	2b	103	**b.** -25
7a	2.2	4b	104	**7. a.** -8
7b	2.2	4c	104	**b.** 4
7c	2.2	4d	104	**c.** 4
7d	2.2	4e	104	**d.** Not defined
8	2.3	1	110	**8.** 5.6
9	2.3	2	110	**9.** $4\frac{1}{5}$
10a	2.3	3	110	**10. a.** -2.7
10b	2.3	4	111	**b.** $-\dfrac{29}{40}$
11a	2.3	5a, b	111	**11. a.** -1.2
11b	2.3	5c, d	111–112	**b.** $-\dfrac{17}{12}$
12a	2.3	6a	112	**12. a.** -13.12
12b	2.3	6a	112	**b.** $\dfrac{9}{35}$
13a	2.3	8a	114	**13. a.** $-\dfrac{9}{4}$
13b	2.3	8b	114	**b.** $\dfrac{5}{27}$

IF YOU MISSED QUESTION	SECTION	EXAMPLES	PAGE	ANSWERS
14a	2.4	1	120	**14. a.** $6x - 24$
14b	2.4	2	121	**b.** $-5x + 3y$
14c	2.4	3	122	**c.** $-2x - 4y + 8$
15a	2.5	1, 7	128, 131	**15. a.** $\dfrac{x - y}{z}$
15b	2.5	3	129	**b.** $M = S - C$
16a	2.5	4	129	**16. a.** $-abc$
16b	2.5	5	130	**b.** $5^2 a^2 b^3 = 25 a^2 b^3$
17a	2.6	1	138	**17. a.** $-10x$
17b	2.6	2	139	**b.** $-3ab^2$
18a	2.6	4	141	**18. a.** $6x + 5$
18b	2.6	5	142	**b.** $3x - 6$
19a	2.7	1, 2	148, 149	**19. a.** $-15 a^3 b^3$
19b	2.7	2	149	**b.** $12 x^3 y^3 z^4$
20a	2.7	3a, b	151	**20. a.** $-3 x^2 y^4$
20b	2.7	3c	151	**b.** $\dfrac{y^4}{3x^4}$
21a	2.8	3a, b	159	**21. a.** 4
21b	2.8	3c, d	159	**b.** $\dfrac{1}{y^7}$
22a	2.8	4b	160	**22. a.** $\dfrac{1}{x^5}$
22b	2.8	4c	160	**b.** 1
22c	2.8	4d	160	**c.** x
23a	2.8	5	161	**23. a.** $\dfrac{1}{x^{12}}$
23b	2.8	6b	161	**b.** $\dfrac{8x^9}{y^{12}}$
23c	2.8	6c	161	**c.** $\dfrac{x^6}{9y^4}$
24a	2.9	1, 2	168	**24. a.** 5.7×10^7
24b	2.9	1, 2	168	**b.** 4.7×10^{-6}
25a	2.9	3	169	**25. a.** 3.24×10^{10}
25b	2.9	4	169	**b.** 4

2.1 ADDITION AND SUBTRACTION OF INTEGERS

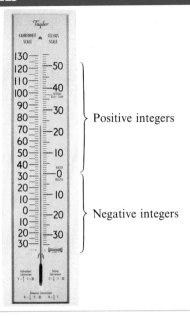

Positive integers

Negative integers

OBJECTIVES

REVIEW

Before starting this section, you should know how to add, subtract, multiply, and divide natural numbers.

OBJECTIVES

You should be able to:

A. Find the additive inverse (opposite) of a given integer.
B. Find the absolute value of an integer.
C. Add two integers.
D. Subtract one integer from another.
E. Add several integers.
F. Solve applications using the concepts studied.

The thermometer uses **integers** to measure temperature. The integers are . . . −3, −2, −1, 0, 1, 2, 3, ... These numbers are used in many everyday situations. When you earn $20, you have 20 dollars (or +20 dollars, if you prefer). When you spend $5, you write −5 dollars. The number 20 is a **positive integer,** and the number −5 (read "negative 5") is a **negative integer.** Here are some other quantities that use integers in their measure:

A loss of $25	−25	A $25 gain	25
10 ft below sea level	−10	10 ft above sea level	10
15° below zero	−15	15° above zero	15

A. ADDITIVE INVERSES (Opposites)

The temperature 15 degrees below zero is shown on the thermometer in the margin. If we take the scale on this thermometer and turn it sideways so that the positive numbers are on the right, the resulting scale is called a **number line** (see the following figure). Clearly, on a number line the positive integers are to the right of 0, the negative integers are to the left of 0, and 0 itself is called the **origin.**

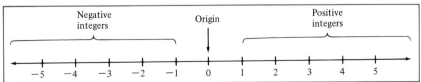

Note that our number line was drawn 5 units long on each side, but the arrows at either end indicate that the line could be drawn to any desired length. Moreover, for every *positive* integer, there is a corre-

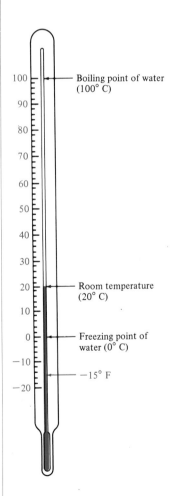

sponding *negative* integer. Thus, associated with the positive integer 4, we have the negative integer −4. Since 4 and −4 are the same distance from the origin but in opposite directions, 4 and −4 are called **opposites.** Moreover, 4 + (−4) = 0. Hence, 4 and −4 are also called **additive inverses.** Similarly, the additive inverse (opposite) of −3 is 3 and the additive inverse (opposite) of 2 is −2. Note that −3 + 3 = 0 and 2 + (−2) = 0. In general, we have

$$a + (-a) = (-a) + a = 0 \qquad \text{for any integer } a$$

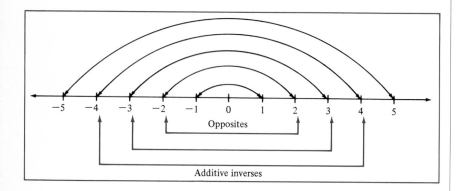

The additive inverse (opposite) of any number a is $-a$.

You can read $-a$ as "the opposite of a" or "the additive inverse of a." Note that a and $-a$ are inverses of each other. Thus, 10 and −10 are inverses of each other. Similarly, −7 and 7 are inverses of each other.

EXAMPLE 1 Find the additive inverse (opposite).

a. 5   **b.** −4   **c.** 0

Solution

a. The additive inverse of 5 is −5 (see the following figure).
b. The additive inverse of −4 is −(−4) = 4 (see the following figure).
c. The additive inverse of 0 is 0.

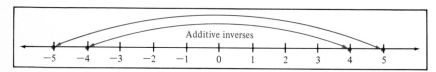

Problem 1 Find the additive inverse (opposite).
a. 17   **b.** −10   **c.** −0

B. ABSOLUTE VALUE

Now, let us go back to the number line. What is the distance between 3 and 0? The answer is 3 units. What about between −3 and 0? The answer is *still* 3 units. (See the following figure.) The distance between any number n and 0 is called the *absolute value* of the number and is denoted by $|n|$. Thus, $|-3| = 3$ and $|3| = 3$.

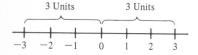

The **absolute value** of a number n is its distance from 0 and is denoted by $|n|$.

You can think of the absolute value of a number as the number of units disregarding the sign.

EXAMPLE 2   Find.

a. $|-8|$ **b.** $|7|$ **c.** $|0|$

Solution

a. $|-8| = 8$ -8 is 8 units from 0
b. $|7|$ $= 7$ 7 is 7 units from 0
c. $|0|$ $= 0$ 0 is 0 units from 0

Problem 2   Find.
a. $|-19|$ **b.** $|6|$ **c.** $|-0|$

C. ADDITION OF INTEGERS

We are now ready to do additions on the number line. Here is how to do it.

> **RULE FOR ADDING INTEGERS**
>
> To add $a + b$ on the number line
>
> 1. Start at 0 and move to a (to the *right* if a is *positive*, to the *left* if a is *negative*).
>
> 2. **a.** If b is *positive,* move *right* b units.
>   **b.** If b is *negative,* move *left* $|b|$ units.
>   **c.** If b is 0, stay at a.

For example, the sum $2 + 4$, or $(+2) + (+4)$, is found by starting at 0, moving 2 units to the right, followed by 4 more units to the right. Thus, $2 + 4 = 6$, as shown in the following figure.

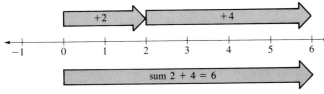

EXAMPLE 3   Find $5 + (-3)$.

Solution   Start at 0. Move 5 units to the right and then 3 units to the left. The result is 2. Thus, $5 + (-3) = 2$, as shown in the following figure.

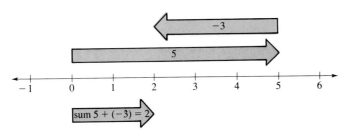

Problem 3   Find $4 + (-1)$.

▲

This same procedure can be used to add two negative numbers. However, we have to be careful when writing such problems. For example, to add -3 and -2, we should write

$(-3) + (-2)$

Why the parentheses? Because writing

$-3 + -2$

is confusing. Never write two signs together without parentheses.

EXAMPLE 4 Find $(-3) + (-2)$.

Solution Start at 0. Move 3 units left and then 2 more units left. The result is 5 units left of 0; that is,

$(-3) + (-2) = -5$

as shown in the following figure.

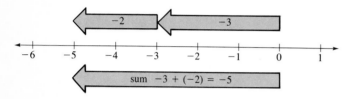

As you can see from Example 4, if we add numbers with the *same* sign (both $+$ or both $-$), the result is a number with this sign. Thus

$2 + 4 = 6$

and

$(-3) + (-2) = -5$

If we add numbers with *different* signs, the answer carries the sign of the number with the larger absolute value. Hence, in Example 3

$5 + (-3) = 2$

but note that

$-5 + 3 = -2$

The following rule summarizes this discussion.

TO ADD INTEGERS

1. If both numbers have the *same* sign, *add* their absolute values and give the sum the common sign.

2. If the numbers have *different* signs, *subtract* their absolute values and give the difference the sign of the number with the *larger* absolute value.

Thus

$8 + 5 = 13$

and

$-8 + (-5) = -13$

Problem 4 Find $(-1) + (-2)$.

To add $-8 + 5$, we first notice that the numbers have *different* signs. Thus, we subtract their absolute values and give the difference the sign of the number with the larger absolute value. Thus,

Use the sign of the number with the larger absolute value.

$$-8 + 5 = -(8 - 5) = -3$$

Subtract the smaller number from the larger one.

Similarly,

$$8 + (-5) = +(8 - 5) = 3$$

Here we have used the sign of the number with the larger absolute value, 8, which is understood to be $+$.

EXAMPLE 5 Find.

a. $(-14) + 6$ **b.** $14 + (-6)$

Solution

a. $(-14) + 6 = -8$
b. $14 + (-6) = 8$ ▲

Note the addition of integers is *commutative* and *associative*. Thus, $(-8) + 5 = 5 + (-8) = -3$. Also

$$-2 + (5 + 1) = (-2 + 5) + 1$$

since

$$-2 + 6 = 3 + 1$$
$$4 = 4$$

Problem 5 Find.
a. $(-12) + 5$ **b.** $12 + (-5)$

D.   SUBTRACTION OF INTEGERS

We are now ready to subtract integers. Suppose you use *positive* integers to indicate money *earned* and *negative* integers to indicate money *spent*. If you earn \$10 and then spend \$12, you owe \$2. Thus,

$$10 - 12 = 10 + (-12) = -2$$

Also,

$$-5 - 10 = -5 + (-10) = -15 \quad \text{To take away (subtract) earned money is the same as adding an expenditure.}$$

because if you spend \$5 and then spend \$10 more, you now owe \$15. What about $-10 - (-3)$? We claim that

$$-10 - (-3) = -7$$

because if you spend \$10 and then subtract (take away) a \$3 expenditure (represented by -3), you save \$3; that is,

$$-10 - (-3) = -10 + 3 = -7 \quad \text{When you take away (subtract) a \$3 expenditure, you save (add) \$3.}$$

In general, we have the following rule:

RULE FOR SUBTRACTING INTEGERS

$$a - b = a + (-b)$$

To subtract an integer, add its inverse.

Thus,

$$5 - 8 = 5 + (-8) = -3$$
$$7 - 3 = 7 + (-3) = 4$$
$$-4 - 2 = -4 + (-2) = -6$$
$$-6 - (-4) = -6 + 4 = -2$$

EXAMPLE 6 Find.

a. $17 - 6$ **b.** $-21 - 4$ **c.** $-11 - (-5)$ **d.** $-4 - (-6)$

Solution

a. $17 - 6 = 17 + (-6) = 11$
b. $-21 - 4 = -21 + (-4) = -25$
c. $-11 - (-5) = -11 + 5 = -6$
d. $-4 - (-6) = -4 + 6 = 2$

Problem 6 Find.
a. $11 - 5$ **b.** $-12 - 3$
c. $-9 - (-3)$ **d.** $-2 - (-5)$

E. ADDITION AND SUBTRACTION OF INTEGERS

Suppose you wish to find $18 - (-10) + 12 - 10 - 17$. Using the fact that $a - b = a + (-b)$, we write

$$18 - (-10) + 12 - 10 + (-17) = 18 + 10 + 12 + (-10) + (-17)$$
$$10 + (-10) = 0$$
$$= 18 + 12 + (-17)$$
$$= 30 + (-17)$$
$$= 13$$

EXAMPLE 7 Find $12 - (-13) + 10 - 25 - 13$.

Solution First, rewrite as an addition.

$$12 - (-13) + 10 - 25 - 13 = 12 + 13 + 10 + (-25) + (-13)$$
$$13 + (-13) = 0$$
$$= 12 + 10 + (-25)$$
$$= 22 + (-25)$$
$$= -3$$

Problem 7 Find
$17 - (-5) + 4 - 5 - 24$.

F. APPLICATIONS

EXAMPLE 8 The greatest temperature variation in a 24-hr period occurred in Browning, Montana, January 23–24, 1916. The temperature fell from $44°F$ to $-56°F$. How many degrees did the temperature fall?

Solution We have to find the difference between 44 and -56; that is, we have to find $44 - (-56)$:

$$44 - (-56) = 44 + 56$$
$$= 100$$

Thus, the temperature fell $100°F$.

Problem 8 The temperature in Verkhoyansk, Siberia, ranges from $98°F$ to $-94°F$. What is the temperature range?

ANSWERS (to problems on pages 94–95)
4. -3 **5. a.** -7 **b.** 7

NAME

CLASS

SECTION

A. Find the additive inverse (opposite) of the given number.

1. 4

2. 11

3. -49

4. -56

B. Find.

5. $|-2|$

6. $|-6|$

7. $|45|$

8. $|78|$

9. $|-(-3)|$

10. $|-(-17)|$

C. Find (verify your answer using the number line).

11. $3 + 3$

12. $2 + 1$

13. $(-5) + 1$

14. $(-4) + 3$

15. $6 + (-5)$

16. $5 + (-1)$

17. $(-2) + (-5)$

18. $(-3) + (-3)$

19. $3 + (-3)$

20. $(-4) + 4$

21. $(-18) + 21$

22. $(-3) + 5$

23. $19 + (-6)$

24. $8 + (-1)$

25. $(-9) + 11$

26. $(-8) + 13$

27. $-18 + 9$

28. $-17 + 4$

ANSWERS

1. _____

2. _____

3. _____

4. _____

5. _____

6. _____

7. _____

8. _____

9. _____

10. _____

11. _____

12. _____

13. _____

14. _____

15. _____

16. _____

17. _____

18. _____

19. _____

20. _____

21. _____

22. _____

23. _____

24. _____

25. _____

26. _____

27. _____

28. _____

29. $(-17) + (+5)$　　　　**30.** $(-4) + (+8)$

29. _____

30. _____

D. Find.

31. $-5 - 11$　　　　**32.** $-4 - 7$

31. _____

32. _____

33. $-4 - 16$　　　　**34.** $-9 - 11$

33. _____

34. _____

35. $7 - 13$　　　　**36.** $8 - 12$

35. _____

36. _____

37. $9 - (-7)$　　　　**38.** $8 - (-4)$

37. _____

38. _____

39. $0 - 4$　　　　**40.** $0 - (-4)$

39. _____

40. _____

E. Find.

41. $8 - (-10) + 5 - 20 - 10$　　　　**42.** $15 - (-9) + 8 - 2 - 9$

41. _____

42. _____

43. $-15 + 12 - 8 - (-15) + 5$　　　　**44.** $-12 + 14 - 7 - (-12) + 3$

43. _____

44. _____

45. $-10 + 9 - 14 - 3 - (-14)$　　　　**46.** $-7 + 2 - 6 - 8 - (-6)$

45. _____

46. _____

Fill in the blank.

47. $23 + \underline{\quad} = 0$　　　　**48.** $-13 + \underline{\quad} = 0$

47. _____

48. _____

49. $-28 + \underline{\quad} + 0$　　　　**50.** $-0 + \underline{\quad} = 0$

49. _____

50. _____

F. Applications

51. The temperature in the center core of the earth reaches $+5000°C$. In the thermosphere (a region in the upper atmosphere), the temperature is $+1500°C$. Find the difference in temperature between the center of the earth and the thermosphere.

51. _____

52. The record high temperature in Calgary, Alberta, is $+99°F$. The record low temperature is $-46°F$. Find the difference between these extremes.

52. _____

53. The price of a certain stock at the beginning of the week was $47. Here are the changes in price during the week: $+1, +2, -1, -2, -1$. What is the price of the stock at the end of the week?

53. _____

54. The price of a stock was $37. On Monday, the price went up $2; on Tuesday, it went down $3; and on Wednesday, it went down another $1. What was the price of the stock then?

54. _____

55. Here are the temperature changes (in degrees Celsius) by the hour in a certain city:

55. _____

1 P.M.	$+2$
2 P.M.	$+1$
3 P.M.	-1
4 P.M.	-3

If the temperature was initially 15°C, what was it at 4 P.M.?

✓ **SKILL CHECKER**

Find.

56. $5\frac{1}{4} + 3\frac{1}{8}$

57. $6\frac{1}{6} - 5\frac{7}{10}$

58. $6\frac{1}{4} \cdot \frac{20}{21}$

59. $\frac{18}{11} \cdot 5\frac{1}{2}$

60. $2\frac{1}{4} \div 1\frac{3}{8}$

56. _____

57. _____

58. _____

59. _____

60. _____

2.1 USING YOUR KNOWLEDGE

A Little History The accompanying chart contains some important historical dates.

IMPORTANT HISTORICAL DATES	
323 B.C.	Alexander the Great died
216 B.C.	Hannibal defeated the Romans
A.D. 476	Fall of the Roman Empire
A.D. 1492	Columbus discovered America
A.D. 1776	The Declaration of Independence signed
A.D. 1939	World War II started
A.D. 1988	Reagan–Gorbachev summit

We can use negative integers to represent years B.C. For example, the year Alexander the Great died can be written as -323, whereas the fall of the Roman Empire occurred in $+476$ (or simply 476). To find the number of years that elapsed between the fall of the Roman Empire and their defeat by Hannibal, we write

$$476 - (-216) = 476 + 216 = 692$$

Fall of the   Hannibal defeats   Years elapsed
Roman Empire   the Romans
(476 A.D.)   (216 B.C.)

Use these ideas to find the number of years elapsed between the following:

1. The fall of the Roman Empire and the death of Alexander the Great

2. Columbus's discovery of America and Hannibal's defeat of the Romans

3. The discovery of America and the signing of the Declaration of Independence

4. The year of the Reagan–Gorbachev summit and the signing of the Declaration of Independence

5. The start of World War II and the death of Alexander the Great

1. _____

2. _____

3. _____

4. _____

5. _____

CALCULATOR CORNER

Some calculators have a key that finds the additive inverse (opposite) of a given number. For example, to find the opposite of 5, as in Example 2, press $\boxed{5}$ $\boxed{+/-}$ or $\boxed{5}$ $\boxed{\text{CHS}}$ and the correct answer, -5, will be displayed. Here are the rest of the examples in this section done with a caculator.

EXAMPLE 3 $5 + (-3)$ Press $\boxed{5}$ $\boxed{+}$ $\boxed{-}$ $\boxed{3}$ $\boxed{=}$*   or

$\boxed{5}$ $\boxed{+}$ $\boxed{3}$ $\boxed{+/-}$ $\boxed{=}$

EXAMPLE 4 $(-3) + (-2)$ Press $\boxed{-}$ $\boxed{3}$ $\boxed{+}$ $\boxed{-}$ $\boxed{2}$ $\boxed{=}$ or

$\boxed{-}$ $\boxed{3}$ $\boxed{+}$ $\boxed{2}$ $\boxed{+/-}$ $\boxed{=}$

EXAMPLE 5 $(-14) + 6$ Press $\boxed{-}$ $\boxed{1}$ $\boxed{4}$ $\boxed{+}$ $\boxed{6}$ $\boxed{=}$

EXAMPLE 6

a. $17 - 6$ Press $\boxed{1}$ $\boxed{7}$ $\boxed{-}$ $\boxed{6}$ $\boxed{=}$

b. $-21 - 4$ Press $\boxed{-}$ $\boxed{2}$ $\boxed{1}$ $\boxed{-}$ $\boxed{4}$ $\boxed{=}$

c. $-11 - (-5)$ presents a different problem. Here you must know that $-(-5)$ must be entered as $+5$, or as -5 $\boxed{+/-}$. You then press

$\boxed{-}$ $\boxed{1}$ $\boxed{1}$ $\boxed{+}$ $\boxed{5}$ $\boxed{=}$ or $\boxed{-}$ $\boxed{1}$ $\boxed{1}$ $\boxed{-}$ $\boxed{5}$ $\boxed{+/-}$ $\boxed{=}$

d. $-4 - (-6)$ Press $\boxed{-}$ $\boxed{4}$ $\boxed{+}$ $\boxed{6}$ $\boxed{=}$

Note: The calculator is no substitute for knowledge. Even the best calculator will not get the correct answer for Example 6c unless you know the arithmetic involved.

---

*Some calculators will indicate an error if two signs of operation are pressed consecutively (for example, $\boxed{+}$ $\boxed{-}$).

2.2 MULTIPLICATION AND DIVISION OF INTEGERS

MOST ACTIVE STOCKS						
	OPEN	HIGH	LOW	CLOSE	CHG.	VOLUME
Citicorp	$26\frac{3}{4}$	$26\frac{3}{4}$	$24\frac{1}{2}$	$24\frac{1}{2}$	$-1\frac{1}{2}$	285,000
Am Medical	$3\frac{1}{8}$	$3\frac{1}{8}$	$2\frac{3}{4}$	3	$-\frac{1}{8}$	245,900
Southern Co	$10\frac{3}{8}$	$10\frac{3}{8}$	$10\frac{1}{8}$	$10\frac{1}{8}$	$-\frac{1}{4}$	211,700
FedNat Mtg	$12\frac{3}{4}$	$12\frac{7}{8}$	12	12	$-\frac{5}{8}$	191,900
FMC	$13\frac{1}{4}$	$13\frac{1}{4}$	$12\frac{1}{2}$	$12\frac{1}{2}$	$-\frac{5}{8}$	161,200
Kauf Broad	$2\frac{1}{2}$	$2\frac{5}{8}$	$2\frac{1}{4}$	$2\frac{1}{2}$	. . .	123,900
Dow Chem	$56\frac{3}{8}$	$56\frac{3}{8}$	$52\frac{3}{4}$	$52\frac{3}{4}$	$-3\frac{1}{8}$	122,900
Evans Pd	$3\frac{1}{4}$	$3\frac{1}{4}$	$2\frac{7}{8}$	3	. . .	119,400
MorganJP	$46\frac{3}{8}$	$46\frac{1}{2}$	44	44	$-2\frac{1}{8}$	113,200
East Kodak	$69\frac{7}{8}$	$69\frac{7}{8}$	$65\frac{7}{8}$	66	-3	110,500
Average closing price of most active stocks: 23.03.						

OBJECTIVES

REVIEW

Before starting this section you should know how to multiply and divide whole numbers.

OBJECTIVES

You should be able to:

A. Multiply two integers.
B. Evaluate expressions involving exponents.
C. Divide one integer by another (nonzero) integer.

Suppose you own 4 shares of Eastman Kodak (East Kodak). The closing price was *down* $3 (written as -3). Your loss that day would be

$$4 \cdot (-3)$$

Recall that 4 and -3 are called *factors*. Now, *multiplication* can be thought of as *repeated addition*. Thus,

$$4 \cdot (-3) = \underbrace{(-3) + (-3) + (-3) + (-3)}_{4 \text{ negative threes}} = -12$$

that is,

$$4 \cdot (-3) = -12$$

Also,

$$(-3) \cdot 4 = -12 \quad \text{(Assuming multiplication is commutative)}$$

From this, it looks as if the product of a *negative* and a *positive* integer is negative. What about the product of two negative integers, say $-4 \cdot (-3)$? Look for the pattern.

This number decreases by 1.   This number increases by 3.

$$4 \cdot (-3) = -12$$
$$3 \cdot (-3) = -9$$
$$2 \cdot (-3) = -6$$
$$1 \cdot (-3) = -3$$
$$0 \cdot (-3) = 0$$
$$-1 \cdot (-3) = 3$$
$$-2 \cdot (-3) = 6$$
$$-3 \cdot (-3) = 9$$
$$-4 \cdot (-3) = 12$$

Thus, $-4 \cdot (-3) = 12$. You can think of $-4 \cdot (-3)$ as subtracting -3 4 times; that is,

$$-(-3) - (-3) - (-3) - (-3) = 3 + 3 + 3 + 3 = 12$$

As you can see, when we multiply two integers with *different* (*unlike*) signs, the product is *negative*. If we multiply two integers with the *same* (*like*) signs, the product is *positive*. This is summarized next.

A. MULTIPLICATION OF INTEGERS

WHEN MULTIPLYING TWO NUMBERS WITH	THE PRODUCT IS
Like (same) signs	Positive $(+)$
Unlike (opposite) signs	Negative $(-)$

Here are some examples.

$$\left. \begin{array}{l} 9 \cdot 4 = 36 \\ (-9) \cdot (-4) = 36 \end{array} \right\}$$ 9 and 4 have the same sign $(+)$; -9 and -4 have the same sign $(-)$; thus, the answer is positive.

$$\left. \begin{array}{l} -9 \cdot 6 = -54 \\ 6 \cdot (-9) = -54 \end{array} \right\}$$ -9 and 6 have opposite signs; thus, the answer is negative.

EXAMPLE 1   Find.

a. $7 \cdot 8$ **b.** $-8 \cdot 6$ **c.** $4 \cdot (-3)$ **d.** $-7 \cdot (-9)$

Solution

a. $7 \cdot 8 = 56$

b. $-8 \cdot 6 = -48$

Opposite signs   Negative answer

c. $4 \cdot (-3) = -12$

Opposite signs   Negative answer

d. $-7 \cdot (-9) = 63$

Same signs   Positive answer

Note that the multiplication of integers is both commutative and associative. Thus, $-3 \cdot 4 = 4 \cdot (-3)$ (the result is -12) and $-3 \cdot (4 \cdot 5) = (-3 \cdot 4) \cdot 5$ (the result is -60).

B. USING EXPONENTS TO INDICATE MULTIPLICATION

Sometimes a number is used several times as a factor. Thus we may wish to find the products

$$3 \cdot 3 \quad \text{or} \quad 4 \cdot 4 \cdot 4 \quad \text{or} \quad 5 \cdot 5 \cdot 5 \cdot 5$$

Problem 1   Find.
a. $6 \cdot 9$ **b.** $-3 \cdot 11$
c. $9 \cdot (-10)$ **d.** $-5 \cdot (-11)$

In the expression $3 \cdot 3$, the 3 is used as a factor twice. In such cases it is easier to use **exponents** to indicate the multiplication. We then write

3^2 (read "3 squared") instead of $3 \cdot 3$

4^3 (read "4 cubed") instead of $4 \cdot 4 \cdot 4$

5^4 (read "5 to the fourth") instead of $5 \cdot 5 \cdot 5 \cdot 5$

The expression 3^2 uses the number **2** to indicate how many times the **base** 3 is used as a **factor.** Similarly, in the expression 5^4, the 5 is the **base** and the **4** is the **exponent.** Now,

$3^2 = 3 \cdot 3 = 9$

$4^3 = 4 \cdot 4 \cdot 4 = 64$

and

$5^4 = 5 \cdot 5 \cdot 5 \cdot 5 = 625$

What about $(-2)^2$? Using the definition of exponents, we have

$(-2)^2 = (-2) \cdot (-2) = 4$ -2 and -2 have the same sign; thus, their product is positive.

Moreover, -2^2 means $-(2 \cdot 2)$. To emphasize that the multiplication is to be done *first,* we put parentheses around the $2 \cdot 2$. Thus, the placing of the parentheses in the expression $(-2)^2$ is very important. Clearly, since $(-2)^2 = 4$ and $-2^2 = -(2 \cdot 2) = -4$,

$(-2)^2 \neq -2^2$ $(-2)^2$ is not equal to -2^2.

EXAMPLE 2 Find.

a. $(-4)^2$ **b.** -4^2

Solution

a. $(-4)^2 = (-4)(-4) = 16$
b. $-4^2 = -(4 \cdot 4) = -16$

EXAMPLE 3 Find.

a. $(-2)^3$ **b.** -2^3

Solution

a. $(-2)^3 = \underbrace{(-2) \cdot (-2)} \cdot (-2)$

$\quad\quad = \quad\quad 4 \quad\quad \cdot (-2)$

$\quad\quad = -8$

b. $-2^3 = -\underbrace{(2 \cdot 2 \cdot 2)}$

$\quad\quad = \quad -(8)$

$\quad\quad = -8$ ▲

Note that

$(-4)^2 = 16$ (Negative number raised to an even power; positive answer)

but

$(-2)^3 = -8$ (Negative number raised to an odd power; negative answer)

Problem 2 Find.
a. $(-9)^2$ **b.** -9^2

Problem 3 Find.
a. $(-7)^3$ **b.** -7^3

C. DIVISION OF INTEGERS

What about the rules for division? As you recall, a division problem can always be checked by multiplication. Thus the division

$$3\overline{)\,18\,}$$

$$\begin{array}{r} 6 \\ 3\overline{)\,18\,} \\ -18 \\ \hline 0 \end{array}$$

that is,

$$\frac{18}{3} = 6$$

is correct because $18 = 3 \cdot 6$. In general, if b is not 0,

$$\frac{a}{b} = c \quad \text{means} \quad a = b \cdot c$$

That is, the operation of division is defined using multiplication. Because of this, the same rules of sign that apply to the multiplication of integers also apply to the division of integers, that is, the division (quotient) of two integers with the same sign is positive, and the division (quotient) of two integers with opposite signs is negative. Here are some examples:

$$\frac{24}{6} = 4 \quad \Bigg\} \quad 24 \text{ and } 6 \text{ have the same sign; the quotient is positive.}$$

$$\frac{-18}{-9} = 2 \quad \Bigg\} \quad -18 \text{ and } -9 \text{ have the same sign; the quotient is positive.}$$

$$\frac{-32}{4} = -8 \quad \Bigg\} \quad -32 \text{ and } 4 \text{ have opposite signs; the quotient is negative.}$$

$$\frac{35}{-7} = -5 \quad \Bigg\} \quad 35 \text{ and } -7 \text{ have opposite signs; the quotient is negative.}$$

EXAMPLE 4 Find.

a. $48 \div 6$ **b.** $\dfrac{54}{-9}$ **c.** $\dfrac{-63}{-7}$ **d.** $-28 \div 4$ **e.** $5 \div 0$

Solution

a. $48 \div 6 = 8$ (48 and 6 have the same sign; the quotient is positive.)

b. $\dfrac{54}{-9} = -6$ (54 and -9 have opposite signs; the quotient is negative.)

c. $\dfrac{-63}{-7} = 9$ (-63 and -7 have the same sign; the quotient is positive.)

d. $-28 \div 4 = -7$ (-28 and 4 have opposite signs; the quotient is negative.)

e. $5 \div 0$ is not defined. Note that if you make $5 \div 0$ equal any number, such as a, we have

$$\frac{5}{0} = a \quad \text{which means} \quad 5 = a \cdot 0 = 0$$

This is *impossible*. Thus, $\frac{5}{0}$ is *not defined*.

Problem 4 Find.

a. $60 \div 10$ **b.** $\dfrac{48}{-3}$ **c.** $\dfrac{-18}{-2}$

d. $-14 \div 2$ **e.** $-4 \div 0$

NAME

CLASS

SECTION

A. Find.

1. $4 \cdot 9$

2. $16 \cdot 2$

3. $-10 \cdot 4$

4. $-7 \cdot 8$

5. $-9 \cdot 9$

6. $-2 \cdot 5$

7. $-6 \cdot (-3)$

8. $-4 \cdot (-5)$

9. $-9 \cdot (-2)$

10. $-7 \cdot (-10)$

B. Find.

11. -4^2

12. $(-4)^2$

13. $(-5)^2$

14. -5^2

15. -5^3

16. $(-5)^3$

17. $(-6)^4$

18. -6^4

19. -2^5

20. $(-2)^5$

C. Find.

21. $\dfrac{14}{2}$

22. $10 \div 2$

23. $-50 \div 10$

24. $\dfrac{-20}{5}$

25. $\dfrac{-30}{10}$

26. $-40 \div 8$

27. $\dfrac{-0}{3}$

28. $-0 \div 8$

ANSWERS

1. _____
2. _____
3. _____
4. _____
5. _____
6. _____
7. _____
8. _____
9. _____
10. _____

11. _____
12. _____
13. _____
14. _____
15. _____
16. _____
17. _____
18. _____
19. _____
20. _____

21. _____
22. _____
23. _____
24. _____
25. _____
26. _____
27. _____
28. _____

29. $-5 \div 0$

30. $\dfrac{-8}{0}$

31. $\dfrac{0}{7}$

32. $0 \div (-7)$

33. $-15 \div (-3)$

34. $-20 \div (-4)$

35. $\dfrac{-25}{-5}$

36. $\dfrac{-16}{-2}$

37. $\dfrac{18}{-9}$

38. $\dfrac{35}{-7}$

39. $30 \div (-5)$

40. $80 \div (-10)$

29. _____

30. _____

31. _____

32. _____

33. _____

34. _____

35. _____

36. _____

37. _____

38. _____

39. _____

40. _____

D. Applications

The tables below will be used in Problems 41 through 45.

1 all-beef frank	+45 calories
1 slice of bread	+65 calories

Running (1 min)	−15 calories
Swimming (1 min)	− 7 calories

41. If a person eats 2 beef franks and runs for 5 min, what is the caloric gain or loss?

42. If a person eats 2 beef franks and runs for 30 min, what is the caloric gain or loss?

43. If a person eats 2 beef franks with 2 slices of bread and then runs for 15 min, what is the caloric gain or loss?

44. If a person eats 2 beef franks with 2 slices of bread and then runs for 15 min and swims for 30 min, what is the caloric gain or loss?

45. If a person eats 2 beef franks, how many minutes does the person have to run to "burn off" the calories? *Hint:* You must *spend* the calories contained in the 2 beef franks.

41. _____

42. _____

43. _____

44. _____

45. _____

✓ **SKILL CHECKER**

46. Find the additive inverse of -5.

47. Find the additive inverse of -8.

46. _____

47. _____

ANSWER (to problem on page 104)
4. **a.** 6 **b.** -16 **c.** 9 **d.** -7
e. Not defined

48. Find $|-7|$.

49. Find $|-12|$.

50. Find the additive inverse of 6.

2.2 USING YOUR KNOWLEDGE

Splitting the Atom   The valence (or oxidation number) of a compound is found by using the *sum of the valences of each individual atom present* in the compound. For example, the valence of hydrogen (H) is $+1$, the valence of sulphur (S) is $+6$, and that of oxygen (O) is -2. Thus the valence of sulphuric acid is 0, since

$$H_2SO_4$$
$$2(\text{valence of H}) + (\text{valence of S}) + 4(\text{valence of 0})$$
$$= 2(+1) + (+6) + 4(-2)$$
$$= \quad 2 + \quad 6 + (-8) = 0$$

Use this idea to solve these problems.

1. Find the valence of phosphate, PO_4, if the valence of phosphorus (P) is $+5$ and that of oxygen (O) is -2.

2. Find the valence of nitrate, NO_3, if the valence of nitrogen (N) is $+5$ and that of oxygen (O) is -2.

3. Find the valence of sodium bromate, $NaBrO_3$, if the valence of sodium (Na) is $+1$, the valence of bromine (Br) is $+5$, and the valence of oxygen (O) is -2.

4. Find the valence of sodium dichromate, $Na_2Cr_2O_7$, if the valence of sodium (Na) is $+1$, the valence of chromium (Cr) is $+6$, and that of oxygen (O) is -2.

5. Find the valence of water, H_2O, if the valence of hydrogen (H) is $+1$ and that of oxygen (O) is -2.

1. _____

2. _____

3. _____

4. _____

5. _____

CALCULATOR CORNER

The $\boxed{+/-}$ or $\boxed{CHS}$ key is essential when multiplying or dividing integers. You will see why if we do Examples 1 and 4 using a calculator.

EXAMPLE 1

a. $7 \cdot 8$ Press $\boxed{7}\boxed{\times}\boxed{8}\boxed{=}$

b. $-8 \cdot 6$ Press $\boxed{-}\boxed{8}\boxed{\times}\boxed{6}\boxed{=}$

c. $4 \cdot (-3)$ presents a problem. If you press $\boxed{4}\boxed{\times}\boxed{-}\boxed{3}\boxed{=}$, the calculator gives 1 (or an error message) for an answer. To obtain the correct answer you must key in $\boxed{4}\boxed{\times}\boxed{3}\boxed{+/-}\boxed{=}$. This time the calculator multiplies 4 by the *opposite* of 3, or -3, to obtain the correct answer, -12.

d. $-7 \cdot (-9)$ Press $\boxed{-}\boxed{7}\boxed{\times}\boxed{9}\boxed{+/-}\boxed{=}$

EXAMPLE 4

a. $48 \div 6$ Press $\boxed{4}\boxed{8}\boxed{\div}\boxed{6}\boxed{=}$

b. $54 \div (-9)$ presents the same problem as Example 1c. If you key in $\boxed{5}\boxed{4}\boxed{\div}\boxed{-}\boxed{9}\boxed{=}$, you do not get the correct answer. As before, the proper procedure is to key in $\boxed{5}\boxed{4}\boxed{\div}\boxed{9}\boxed{+/-}\boxed{=}$, which gives the correct answer, -6.

c. $-63 \div (-7)$ Has to be done similarly to part b. The correct entries are $\boxed{-}\boxed{6}\boxed{3}\boxed{\div}\boxed{7}\boxed{+/-}\boxed{=}$.

d. $-28 \div 4$ is done by keying in $\boxed{-}\boxed{2}\boxed{8}\boxed{\div}\boxed{4}\boxed{=}$.

e. If you key in $\boxed{5}\boxed{\div}\boxed{0}$, an error message appears in the display.

2.3 THE RATIONAL NUMBERS

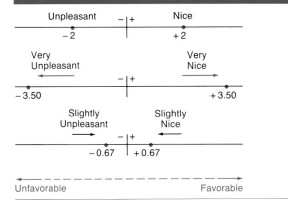

The picture shows the use of rational numbers to quantify adjectives (see 2.3 Using Your Knowledge). The **rational numbers** can be shown on the number line. They can be described as quotients of two integers as follows:

> The rational numbers consist of all numbers that can be written in the form $\frac{a}{b}$, where a and b are integers and b is not 0.

Since any number of the form $\frac{a}{b}$ can be written as a decimal (by dividing a by b), the rational numbers include all the corresponding *decimals* as well. Moreover, *integers* are rational numbers, since any integer a can be written as $\frac{a}{1}$. Fortunately, everything we have said about the integers works for the rational numbers. We now discuss some properties that apply to these rational numbers.

A. INVERSES

As with the integers, every rational number has an *additive inverse* (*opposite*). Here are some rational numbers and their additive inverses.

RATIONAL NUMBER	ADDITIVE INVERSE (OPPOSITE)
$\frac{9}{8}$	$-\frac{9}{8}$
$-\frac{3}{4}$	$\frac{3}{4}$
5.9	−5.9
−6.8	6.8

REVIEW

Before starting this section you should know:

1. How to find the additive inverse of an integer.
2. How to find the absolute value of an integer.
3. How to add, subtract, multiply, and divide integers

OBJECTIVES

You should be able to:

A. Find the additive inverse of a rational number.
B. Find the absolute value of a rational number.
C. Add two rational numbers written as decimals or as fractions.
D. Subtract two rational numbers.
E. Multiply two rational numbers.
F. Find the reciprocal of a rational number.
G. Divide one rational number by another.

EXAMPLE 1 Find the additive inverse (opposite).

a. $\dfrac{5}{2}$ **b.** -4.8 **c.** $-3\dfrac{1}{3}$ **d.** 7.2

Solution

a. $-\dfrac{5}{2}$

b. 4.8

c. $3\dfrac{1}{3}$

d. -7.2

Problem 1 Find the additive inverse.

a. $\dfrac{3}{4}$ **b.** -5.1 **c.** $-9\dfrac{1}{4}$ **d.** 3.9

B. ABSOLUTE VALUE

Every rational number has an absolute value, which is its distance from 0. Thus, $\left|-\frac{1}{2}\right| = \frac{1}{2}$, $|3.8| = 3.8$, and $\left|-1\frac{1}{7}\right| = 1\frac{1}{7}$, as shown on the following number line.

EXAMPLE 2 Find.

a. $\left|-\dfrac{3}{7}\right|$ **b.** $|2.1|$ **c.** $\left|-2\dfrac{1}{2}\right|$ **d.** $|-4.1|$

Solution

a. $\left|-\dfrac{3}{7}\right| = \dfrac{3}{7}$

b. $|2.1| = 2.1$

c. $\left|-2\dfrac{1}{2}\right| = 2\dfrac{1}{2}$

d. $|-4.1| = 4.1$

Problem 2 Find.

a. $\left|-\dfrac{1}{8}\right|$ **b.** $|3.4|$

c. $\left|-3\dfrac{1}{4}\right|$ **d.** $|-8.2|$

C. ADDITION

Addition of rational numbers uses the same rules for signs as the addition of integers; this is illustrated in Examples 3 and 4.

EXAMPLE 3 Find.

a. $-8.6 + 3.4$ **b.** $6.7 + (-9.8)$ **c.** $-2.3 + (-4.1)$

Solution

a. $-8.6 + 3.4 = -5.2$
b. $6.7 + (-9.8) = -3.1$
c. $-2.3 + (-4.1) = -6.4$

Problem 3 Find.
a. $-7.5 + 2.1$
b. $8.3 + (-9.7)$
c. $-1.4 + (-6.1)$

EXAMPLE 4 Find.

a. $-\dfrac{3}{7} + \dfrac{5}{7}$ **b.** $\dfrac{2}{5} + \left(-\dfrac{5}{8}\right)$

Problem 4 Find.

a. $-\dfrac{5}{9} + \dfrac{7}{9}$ **b.** $\dfrac{3}{4} + \left(-\dfrac{5}{3}\right)$

Solution

a. $-\dfrac{3}{7} + \dfrac{5}{7} = \dfrac{2}{7}$

b. As usual, we first find the LCM of 5 and 8, which is 40. We then write

$$\dfrac{2}{5} = \dfrac{16}{40} \quad \text{and} \quad -\dfrac{5}{8} = -\dfrac{25}{40}$$

Thus,

$$\dfrac{2}{5} + \left(-\dfrac{5}{8}\right) = \dfrac{16}{40} + \left(-\dfrac{25}{40}\right) = -\dfrac{9}{40}$$ ▲

As with the integers, the addition of rational numbers is commutative and associative. Thus,

$$-\dfrac{1}{7} + \dfrac{3}{7} = \dfrac{3}{7} + \left(-\dfrac{1}{7}\right) \quad \left(\text{The result is } \dfrac{2}{7}.\right)$$

and

$$-1.2 + (3.2 + 4) = (-1.2 + 3.2) + 4$$

since

$$-1.2 + (3.2 + 4) = -1.2 + 7.2 = 6$$

and

$$(-1.2 + 3.2) + 4 = 2.0 + 4 = 6$$

D. SUBTRACTION

As with the integers, we define subtraction as follows:

$$a - b = a + (-b) \qquad \text{for any rational numbers } a \text{ and } b$$

EXAMPLE 5 Find.

a. $-4.2 - (-3.1)$ **b.** $-2.5 - (-7.8)$

c. $\dfrac{2}{9} - \left(-\dfrac{4}{9}\right)$ **d.** $-\dfrac{5}{6} - \dfrac{7}{4}$

Problem 5 Find.
a. $-3.8 - (-2.5)$ **b.** $-4.7 - (-6.9)$
c. $\dfrac{3}{8} - \left(-\dfrac{1}{8}\right)$ **d.** $-\dfrac{7}{8} - \dfrac{5}{6}$

Solution   ——————— Note that $-(-3.1) = 3.1$.

a. $-4.2 - (-3.1) = -4.2 + 3.1 = -1.1$

b. $-2.5 - (-7.8) = -2.5 + 7.8 = 5.3$ Note that $-(-7.8) = 7.8$.

c. $\dfrac{2}{9} - \left(-\dfrac{4}{9}\right) = \dfrac{2}{9} + \dfrac{4}{9} = \dfrac{6}{9} = \dfrac{2}{3}$ Note that $-\left(-\dfrac{4}{9}\right) = \dfrac{4}{9}$.

d. The LCD is 12. Now,

$$-\frac{5}{6} = -\frac{10}{12} \quad \text{and} \quad \frac{7}{4} = \frac{21}{12}$$

Thus,

$$-\frac{5}{6} - \frac{7}{4} = -\frac{5}{6} + \left(-\frac{7}{4}\right) = -\frac{10}{12} + \left(-\frac{21}{12}\right) = -\frac{31}{12}$$

E. MULTIPLICATION

The multiplication of rational numbers uses the same rules for signs as the multiplication of integers.

WHEN MULTIPLYING TWO RATIONAL NUMBERS WITH	THE PRODUCT IS
Like (same) signs	Positive (+)
Unlike (opposite) signs	Negative (−)

EXAMPLE 6 Find.

a. $-3.1(4.2)$ **b.** $-1.2(-3.4)$ **c.** $-\frac{3}{4}\left(-\frac{5}{2}\right)$ **d.** $\frac{5}{6}\left(-\frac{4}{7}\right)$

Solution

a. -3.1 and 4.2 have *unlike* signs. The result is *negative*. Thus,

$$-3.1(4.2) = -13.02$$

b. -1.2 and -3.4 have *like* signs. The result is *positive*. Thus,

$$-1.2(-3.4) = 4.08$$

c. $-\frac{3}{4}$ and $-\frac{5}{2}$ have *like* signs. The result is *positive*. Thus,

$$-\frac{3}{4}\left(-\frac{5}{2}\right) = \frac{15}{8}$$

d. $\frac{5}{6}$ and $-\frac{4}{7}$ have *unlike* signs. The result is *negative*. Thus,

$$\frac{5}{6}\left(-\frac{4}{7}\right) = -\frac{20}{42} = -\frac{10}{21} \qquad \blacktriangle$$

Note that the multiplication of rational numbers is *commutative* and *associative*. Thus,

$$\frac{3}{4} \cdot \left(-\frac{1}{8}\right) = -\frac{1}{8} \cdot \frac{3}{4} \quad \left(\text{The result is } -\frac{3}{24}.\right)$$

and

$$-\frac{1}{2} \cdot \left(\frac{3}{4} \cdot \frac{2}{7}\right) = \left(-\frac{1}{2} \cdot \frac{3}{4}\right) \cdot \frac{2}{7} \quad \left(\text{The result is } \frac{-3}{28}.\right)$$

F. RECIPROCALS

The division of rational numbers is related to the *reciprocal* of a number. As you recall, two numbers are *reciprocals* (or *multiplicative*

Problem 6 Find.
a. $-2.2(3.2)$ **b.** $-1.3(-4.1)$

c. $-\frac{3}{7}\left(-\frac{4}{5}\right)$ **d.** $\frac{6}{7}\left(-\frac{2}{3}\right)$

inverses) if their product is 1. Here are some numbers and their reciprocals.

NUMBER	MULTIPLICATIVE INVERSE (RECIPROCAL)	CHECK
$\dfrac{3}{4}$	$\dfrac{4}{3}$	$\dfrac{3}{4} \cdot \dfrac{4}{3} = 1$
$-\dfrac{5}{2}$	$-\dfrac{2}{5}$	$-\dfrac{5}{2} \cdot \left(-\dfrac{2}{5}\right) = 1$
1	1	$1 \cdot 1 = 1$
-1	-1	$-1 \cdot (-1) = 1$

Note that 0 has *no* reciprocal, since there is *no* number n such that $0 \cdot n = 1$. In general, the reciprocal of any nonzero $\dfrac{a}{b}$ is $\dfrac{b}{a}$ and the reciprocal of $-\dfrac{a}{b}$ is $-\dfrac{b}{a}$. Thus, a number and its reciprocal have the same sign.

EXAMPLE 7 Find the reciprocal.

a. $\dfrac{3}{7}$ **b.** $-\dfrac{8}{3}$

Solution

a. The reciprocal of $\dfrac{3}{7}$ is $\dfrac{7}{3}$ because $\dfrac{3}{7} \cdot \dfrac{7}{3} = 1$

b. The reciprocal of $-\dfrac{8}{3}$ is $-\dfrac{3}{8}$ because $-\dfrac{8}{3} \cdot \left(-\dfrac{3}{8}\right) = 1$

Problem 7 Find the reciprocal.

a. $\dfrac{4}{5}$ **b.** $\dfrac{-7}{9}$

G. DIVISION

The idea of a reciprocal can be used to do division. Thus,

$$18 \div 3 = 6 \quad \text{and} \quad 18 \cdot \dfrac{1}{3} = 6$$

$$24 \div 8 = 3 \quad \text{and} \quad 24 \cdot \dfrac{1}{8} = 3$$

In each case, dividing by a number gives the same answer as multiplying by its reciprocal. Here is the general rule.

To divide $\dfrac{a}{b}$ by $\dfrac{c}{d}$, multiply $\dfrac{a}{b}$ by the reciprocal of $\dfrac{c}{d}$; that is,

$$\dfrac{a}{b} \div \dfrac{c}{d} = \dfrac{a}{b} \cdot \dfrac{d}{c} = \dfrac{ad}{bc}.$$

EXAMPLE 8 Find.

a. $\dfrac{2}{5} \div \left(-\dfrac{3}{4}\right)$ **b.** $-\dfrac{5}{6} \div \left(-\dfrac{7}{2}\right)$ **c.** $-\dfrac{3}{7} \div \dfrac{6}{7}$

Solution

a. $\dfrac{2}{5} \div \left(-\dfrac{3}{4}\right) = \dfrac{2}{5} \cdot \left(-\dfrac{4}{3}\right) = -\dfrac{8}{15}$

b. $-\dfrac{5}{6} \div \left(-\dfrac{7}{2}\right) = -\dfrac{5}{6} \cdot \left(-\dfrac{2}{7}\right) = \dfrac{10}{42} = \dfrac{5}{21}$

c. $-\dfrac{3}{7} \div \dfrac{6}{7} = -\dfrac{3}{7} \cdot \dfrac{7}{6} = -\dfrac{21}{42} = -\dfrac{1}{2}$ ▲

If the division involves rational numbers written as decimals, the rules of signs are, of course, unchanged. This is illustrated in the next example.

EXAMPLE 9 Find.

a. $\dfrac{-4.2}{2.1}$ **b.** $\dfrac{-8.1}{-16.2}$ **c.** $\dfrac{9.6}{-3.2}$

Solution

a. $\dfrac{-4.2}{2.1} = -2$

b. $\dfrac{-8.1}{-16.2} = \dfrac{1}{2}$

c. $\dfrac{9.6}{-3.2} = -3$ ▲

As you can see, a *negative* number divided by a *positive* number gives a *negative* answer and a *positive* number divided by a *negative* number also gives a *negative* answer. This idea can be generalized. Thus, we have the following.

For any numbers a and b, $b \neq 0$,

$$-\dfrac{a}{b} = \dfrac{-a}{b} = \dfrac{a}{-b} \quad \text{and} \quad -\dfrac{-a}{-b} = -\dfrac{a}{b}$$

We will use this concept in Section 2.7.

Problem 8 Find.

a. $\dfrac{3}{5} \div \left(-\dfrac{4}{7}\right)$ **b.** $-\dfrac{6}{7} \div \left(-\dfrac{3}{5}\right)$

c. $-\dfrac{4}{5} \div \dfrac{8}{5}$

Problem 9 Find.

a. $\dfrac{-3.6}{1.2}$ **b.** $\dfrac{-3.1}{-12.4}$ **c.** $\dfrac{6.5}{-1.3}$

A. Find the additive inverse (opposite).

1. $\dfrac{7}{3}$

2. $-\dfrac{8}{9}$

3. -6.4

4. -2.3

5. $3\dfrac{1}{7}$

6. $4\dfrac{1}{8}$

1. _____
2. _____
3. _____
4. _____
5. _____
6. _____

B. Find.

7. $\left|-\dfrac{4}{5}\right|$

8. $\left|-\dfrac{9}{2}\right|$

9. $|-3.4|$

10. $|-2.1|$

11. $\left|1\dfrac{1}{2}\right|$

12. $\left|3\dfrac{1}{4}\right|$

7. _____
8. _____
9. _____
10. _____
11. _____
12. _____

C. Find.

13. $-7.8 + (3.1)$

14. $-6.7 + (2.5)$

15. $3.2 + (-8.6)$

16. $4.1 + (-7.9)$

17. $-3.4 + (-5.2)$

18. $-7.1 + (-2.6)$

19. $-\dfrac{2}{7} + \dfrac{5}{7}$

20. $-\dfrac{5}{11} + \dfrac{7}{11}$

21. $-\dfrac{3}{4} + \dfrac{1}{4}$

22. $-\dfrac{5}{6} + \dfrac{1}{6}$

23. $\dfrac{3}{4} + \left(-\dfrac{5}{6}\right)$

24. $\dfrac{5}{6} + \left(-\dfrac{7}{8}\right)$

25. $-\dfrac{1}{6} + \dfrac{3}{4}$

26. $-\dfrac{1}{8} + \dfrac{7}{6}$

13. _____
14. _____
15. _____
16. _____
17. _____
18. _____
19. _____
20. _____
21. _____
22. _____
23. _____
24. _____
25. _____
26. _____

27. $-\dfrac{1}{3}+\left(-\dfrac{2}{7}\right)$

28. $-\dfrac{4}{7}+\left(-\dfrac{3}{8}\right)$

29. $-\dfrac{5}{6}+\left(-\dfrac{8}{9}\right)$

30. $-\dfrac{4}{5}+\left(-\dfrac{7}{8}\right)$

D. Find.

31. $-3.8-(-1.2)$

32. $-6.7-(-4.3)$

33. $-3.5-(-8.7)$

34. $-6.5-(-9.9)$

35. $4.5-8.2$

36. $3.7-7.9$

37. $\dfrac{3}{7}-\left(-\dfrac{1}{7}\right)$

38. $\dfrac{5}{6}-\left(-\dfrac{1}{6}\right)$

39. $-\dfrac{5}{4}-\dfrac{7}{6}$

40. $-\dfrac{2}{3}-\dfrac{3}{4}$

E. Find.

41. $-2.2(3.3)$

42. $-1.4(3.1)$

43. $-1.3(-2.2)$

44. $-1.5(-1.1)$

45. $\dfrac{5}{6}\left(-\dfrac{5}{7}\right)$

46. $\dfrac{3}{8}\left(-\dfrac{5}{7}\right)$

47. $-\dfrac{3}{5}\left(-\dfrac{5}{12}\right)$

48. $-\dfrac{4}{7}\left(-\dfrac{21}{8}\right)$

49. $-\dfrac{6}{7}\left(\dfrac{35}{8}\right)$

50. $-\dfrac{7}{5}\left(\dfrac{15}{28}\right)$

F. Find the reciprocal.

51. $\dfrac{3}{4}$

52. $\dfrac{5}{6}$

53. -6

54. -9

27. _____
28. _____
29. _____
30. _____
31. _____
32. _____
33. _____
34. _____
35. _____
36. _____
37. _____
38. _____
39. _____
40. _____
41. _____
42. _____
43. _____
44. _____
45. _____
46. _____
47. _____
48. _____
49. _____
50. _____
51. _____
52. _____
53. _____
54. _____

ANSWERS (to problems on page 114)

8. **a.** $\dfrac{-21}{20}$ **b.** $\dfrac{10}{7}$ **c.** $-\dfrac{1}{2}$

9. **a.** -3 **b.** $\dfrac{1}{4}$ **c.** -5

55. $-\dfrac{2}{7}$

56. $-\dfrac{3}{5}$

57. -1

58. 1

59. -7

60. -5

G. Find.

61. $\dfrac{3}{5} \div \left(-\dfrac{4}{7}\right)$

62. $\dfrac{4}{9} \div \left(-\dfrac{1}{7}\right)$

63. $-\dfrac{2}{3} \div \left(-\dfrac{7}{6}\right)$

64. $-\dfrac{5}{6} \div \left(-\dfrac{25}{18}\right)$

65. $-\dfrac{5}{8} \div \dfrac{7}{8}$

66. $-\dfrac{4}{5} \div \dfrac{8}{15}$

67. $\dfrac{-3.1}{6.2}$

68. $\dfrac{1.2}{-4.8}$

69. $\dfrac{-1.6}{-9.6}$

70. $\dfrac{-9.8}{-1.4}$

55. _____

56. _____

57. _____

58. _____

59. _____

60. _____

61. _____

62. _____

63. _____

64. _____

65. _____

66. _____

67. _____

68. _____

69. _____

70. _____

✓ **SKILL CHECKER**

Use the distributive law to remove parentheses.

71. $3(x + 6)$

72. $8(6 + y)$

Simplify.

73. $(8 + 7a) + 4$

74. $(2a + 5) + 5a$

71. _____

72. _____

73. _____

74. _____

2.3 USING YOUR KNOWLEDGE

Have you met anybody *nice* today or did you have an *unpleasant* experience? Perhaps the person you met was *very nice* or your experience *very unpleasant*. Psychologists and linguists have a numerical way to indicate the difference between *nice* and *very nice* or between *unpleasant* and *very unpleasant*. Suppose you assign a positive number ($+2$, for example) to the adjective *nice* and a negative number (say, -2) to *unpleasant* and a positive number greater than 1 (say $+1.75$) to *very*. Then, very nice means

Very nice

$(1.75) \cdot (2) = 3.50$ (1.75) and (2) are multiplied.

and very unpleasant means

Very unpleasant

$$(1.75) \cdot (-2) = -3.50$$

Here are some adverbs and adjectives and their average numerical values, as rated by a panel of college students. (Values differ from one panel to another.)

ADVERBS		ADJECTIVES	
Slightly	0.54	Wicked	−2.5
Rather	0.84	Disgusting	−2.1
Decidedly	0.16	Average	−0.8
Very	1.25	Good	3.1
Extremely	1.45	Lovable	2.4

Find the value of each.

1. Slightly wicked
2. Decidedly average
3. Extremely disgusting
4. Rather lovable
5. Very good

By the way, if you got all the answers correct, you are 4.495!

1. _____
2. _____
3. _____
4. _____
5. _____

2.4 USING THE DISTRIBUTIVE LAW

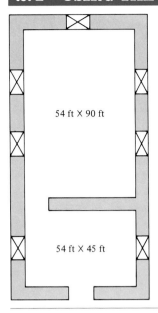

54 ft × 90 ft

54 ft × 45 ft

The drawing shows a plan for an office building. The total area of the building is the sum of the areas of the two rooms. Thus, the area is

$$54 \cdot 90 + 54 \cdot 45$$

This area is also

$$54 \cdot (90 + 45)$$

Thus,

$$54 \cdot (90 + 45) = 54 \cdot 90 + 54 \cdot 45$$

In either case, the area is 7290 ft^2.
This is an example of the distributive law, which we studied earlier. We restate this law here for your convenience.

> For any real numbers a, b, and c
>
> $a(b + c) = ab + ac$ distributive law

But there is another distributive law, the *distributive law of multiplication over subtraction*. This law is stated next.

> **DISTRIBUTIVE LAW OF MULTIPLICATION OVER SUBTRACTION**
>
> $a(b - c) = ab - ac$ for any real numbers a, b, and c

Thus, $3(5 - 4) = 3 \cdot 5 - 3 \cdot 4$, or 3. Similarly, $5(4 - 6) = 5 \cdot 4 - 5 \cdot 6$, or -10. As before, we write $ab - ac$ *without* parentheses with the understanding that *multiplications* are to be done *first*. Note that using the commutative law of multiplication, these two laws can also be written as follows.

REVIEW

Before starting this section you should know:

1. How to use the distributive law (Section 1.5).
2. The identity law for multiplication (Section 1.5).
3. How to multiply integers.

OBJECTIVES

You should be able to:

A. Remove parentheses using the distributive laws of multiplication.
B. Remove parentheses preceded by a minus (−) sign.
C. Remove parentheses in expressions with more than two terms inside the parentheses.
D. Remove parentheses when indicated in problems stated in words.

$$(a + b)c = ac + bc$$
$$(a - b)c = ac - bc$$

distributive laws

A. REMOVING PARENTHESES USING THE DISTRIBUTIVE LAWS

EXAMPLE 1   Remove parentheses (simplify).

a. $3(x - 2)$ **b.** $0.5(a - b)$

Solution

a. $3(x - 2) = 3x - 3 \cdot 2 = 3x - 6$
b. $0.5(a - b) = 0.5a - 0.5b$

B. REMOVING PARENTHESES PRECEDED BY A MINUS (−) SIGN

In expressions of the form

$$-(a + b)$$

or

$$-(a - b)$$

a and b are called **terms** (more about terms in Section 2.5). How do we remove parentheses in such expressions? We first recall that

$a = 1 \cdot a$ The number 1 is called the multiplicative identity.

So,

$$-a = -1 \cdot a$$

Hence,

$$-(a + b) = -1(a + b)$$
$$= -1 \cdot a + (-1 \cdot b)$$
$$= -a - b$$

Thus,

$$-(a + b) = -a - b$$

Similarly,

$$-(a - b) = -a + b$$

Problem 1   Remove parentheses (simplify).
a. $6(x - 3)$ **b.** $0.7(a - b)$

Note that

$$-(a - b) = -[a + (-b)]$$
$$= -1[a + (-b)]$$
$$= -1 \cdot a + (-1)(-b)$$
$$= -a + b$$

These rules tell us that to remove the parentheses in an expression preceded by a negative $(-)$ sign, we simply *change the sign of every term inside the parentheses,* or equivalently, multiply each term inside the parentheses by -1. For example,

Change sign

$$-(3 + x) = -3 - x$$

Change sign

Change sign

$$-(x - 2) = -x + 2$$

Change sign

EXAMPLE 2 Remove parentheses (simplify).

a. $-(3 + 7y)$ **b.** $-(x - 8y)$
c. $-(3x - 9y)$ **d.** $-(-8x + 7)$
e. $-(-2 - 3x)$

Solution To remove parentheses in an expression preceded by a negative $(-)$ sign we simply change the sign of each term inside the parentheses (or multiply each term by -1). Thus, we obtain

a. $-(3 + 7y) = -3 - 7y$
b. $-(x - 8y) = -x + 8y$
c. $-(3x - 9y) = -3x + 9y$
d. $-(-8x + 7) = 8x - 7$
e. $-(-2 - 3x) = 2 + 3x$

Problem 2 Remove parentheses (simplify).
a. $-(5 + 3a)$ **b.** $-(a - 4b)$
c. $-(5a - 8b)$ **d.** $-(-5a + 3)$
e. $-(-2 - 5a)$

C. REMOVING PARENTHESES IN EXPRESSIONS WITH MORE THAN TWO TERMS

At this point you might ask how you can remember all the rules involving the distributive law. Just remember these two facts:

1. If the factor in front of the parentheses has no written sign, simply multiply each term inside the parentheses by this factor, like this:

   $$a(b - c + d - e) = ab - ac + ad - ae$$

2. If the factor in front of the parentheses is preceded by a minus $(-)$ sign, multiply this factor by each of the terms inside the parentheses and *change* the sign of each of these terms; that is,

   $$-a(b - c + d - e) = -ab + ac - ad + ae$$

EXAMPLE 3   Remove parentheses (simplify).

a. $7(x + 2y - 5)$
b. $-3(x - 2y + z - 4)$
c. $-4(-2x + y - z - 3)$

Solution

a. $7(x + 2y - 5) = 7x + 7 \cdot 2y - 7 \cdot 5$
$$= 7x + 14y - 35$$
b. $-3(x - 2y + z - 4) = -3x + 3 \cdot 2y - 3z + 12$
$$= -3x + 6y - 3z + 12$$
c. $-4(-2x + y - z - 3) = 8x - 4y + 4z + 12$

D. APPLICATIONS

Sometimes, the distributive law is used to state relationships between variables representing real numbers. We illustrate this in the next example.

EXAMPLE 4   In electronics, the resistance R_t (read "R sub t") equals the resistance R_0 multiplied by the sum of 1 and αt.

a. Write this relationship using parentheses.
b. Remove parentheses in the expression obtained in a.

Solution

a. We first translate the given information.

The resistance R_t equals R_0 multiplied by the sum of 1 and αt

$$R_t = R_0(1 + \alpha t)$$

b. Using the distributive law, we obtain

$$R_t = R_0 \cdot 1 + R_0 \cdot \alpha t = R_0 + R_0 \alpha t$$

Problem 3   Remove parentheses (simplify).
a. $6(2a + b - 2)$
b. $-5(2a - b + c - 3)$
c. $-7(-3a + b - c - 2)$

Problem 4   In physics, the tension T of a cable equals the product of m and the difference of g and a.
a. Write this relationship using parentheses.
b. Remove parentheses in the expression obtained in a.

ANSWERS

A. Remove parentheses (simplify).

1. $4(x - y)$ **2.** $3(a - b)$

1. _____

2. _____

3. $9(a - b)$ **4.** $6(x - y)$

3. _____

4. _____

5. $3(4x - 2)$ **6.** $2(3a - 9)$

5. _____

6. _____

B. Remove parentheses (simplify).

7. $-\left(\dfrac{3a}{2} - \dfrac{6}{7}\right)$ **8.** $-\left(\dfrac{2x}{3} - \dfrac{1}{5}\right)$

7. _____

8. _____

9. $-(2x - 6y)$ **10.** $-(3a - 6b)$

9. _____

10. _____

11. $-(2.1 + 3y)$ **12.** $-(5.4 + 4b)$

11. _____

12. _____

13. $-4(a + 5)$ **14.** $-6(x + 8)$

13. _____

14. _____

15. $-x(6 + y)$ **16.** $-y(2x + 3)$

15. _____

16. _____

17. $-8(x - y)$ **18.** $-9(a - b)$

17. _____

18. _____

C. Remove parentheses (simplify).

19. $-3(2a - 7b)$ **20.** $-4(3x - 9y)$

19. _____

20. _____

21. $0.5(x + y - 2)$ **22.** $0.8(a + b - 6)$

21. _____

22. _____

23. $\dfrac{6}{5}(a - b + 5)$ **24.** $\dfrac{2}{3}(x - y + 4)$

23. _____

24. _____

25. $-2(x - y + 4)$ **26.** $-4(a - b + 8)$

25. _____

26. _____

27. $-0.3(x + y - 6)$ **28.** $-0.2(a + b - 3)$

27. _____

28. _____

EXERCISE 2.4

29. $-\dfrac{5}{2}(a - 2b + c - 1)$ **30.** $-\dfrac{4}{7}(2a - b + 3c - 5)$

29. _____

30. _____

D. Applications

In Problems 31 and 32, write the formula using parentheses.

31. The formula for converting degrees Fahrenheit (°F) to degrees Celsius (°C) is obtained by letting C equal $\frac{5}{9}$ times the difference of F and 32.

32. The perimeter P of a rectangle equals twice the sum of its length L and its width W.

31. _____

32. _____

✓ SKILL CHECKER

Find.

33. 3^4 **34.** 2^3

33. _____

34. _____

35. $(-3)^4$ **36.** $(-2)^4$

35. _____

36. _____

37. $(-3)^3$ **38.** $(-2)^3$

37. _____

38. _____

39. $(-2)^6$ **40.** $(-3)^6$

39. _____

40. _____

2.4 USING YOUR KNOWLEDGE

Many problems associated with the distributive law are written in words. Use your knowledge of the distributive law to solve the following problems.

1. If an automobile is accelerated so that it speeds up at a constant rate, its average velocity v_a is one-half the sum of its initial velocity v_1 and its final velocity v_2. Write the expression for the average velocity v_a.
   a. Using parentheses
   b. Without using parentheses

1. **a.** _____

   b. _____

2. The voltage drop V across a resistor is the current I times the resistance R. If two resistors of resistances R_1 and R_2 are connected in series, the voltage drop across both is the sum of the voltage drops across each. Write the expression for the voltage V.
   a. Using parentheses
   b. Without using parentheses

2. **a.** _____

   b. _____

ANSWERS (to problems on page 122)
3. a. $12a + 6b - 12$
   b. $-10a + 5b - 5c + 15$
   c. $21a - 7b + 7c + 14$
4. a. $T = m(g - a)$
   b. $T = mg - ma$

3. The momentum M of a moving object is the product of its mass m and its velocity v. If two billiard balls of equal mass m and moving in the same straight line collide, the total momentum is the sum of the two momentums. Suppose one ball has velocity v_1 and the other has velocity v_2. Write the expression for the total momentum M.
   a. Using parentheses
   b. Without using parentheses

3. a. _____
   b. _____

4. The kinetic energy (K.E.) of a moving object is given by the expression $\frac{1}{2}mv^2$, where m is the mass and v is the velocity of the object. Refer to Problem 3 and write the expression for the total K.E. of the two billiard balls.
   a. Using parentheses
   b. Without using parentheses

4. a. _____
   b. _____

5. The approximate length of the belt needed for the pulleys shown in the margin is given by the formula

   $$L = \pi(r_1 + r_2) + 2l$$

   Write a formula for L without using parentheses.

5. _____

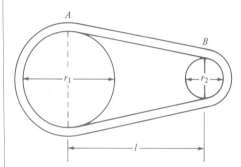

CALCULATOR CORNER

Did you know that the order of operations used in the distributive law is programmed into most calculators? As you recall $5(4 + 6) = 5 \cdot 4 + 5 \cdot 6$, with the provision that the multiplications on the right side ($5 \cdot 4$ and $5 \cdot 6$) must be performed *first*. This is because without a specific set of rules $5 \cdot 4 + 5 \cdot 6$ could have several meanings. If you have a set of parentheses keys, $\boxed{(}$ and $\boxed{)}$, find the answer for each of several possible meanings of $5 \cdot 4 + 5 \cdot 6$:

   a. $5 \times (4 + 5) \times 6 =$ _____
   b. $(5 \times 4) + (5 \times 6) =$ _____
   c. $(5 \times 4 + 5) \times 6 =$ _____
   d. $5 \times [4 + (5 \times 6)] =$ _____ (Here, we need *two* sets of parentheses. If you key in $5 \times (4 + 5 \times 6)$, some calculators will not get the correct answer.)

   Now, try $5 \times 4 + 5 \times 6$. Which of the interpretations did your calculator choose? You can tell by looking at the answer you obtain. If it was answer b, the distributive law is programmed into your calculator. If the answer was different from that obtained in b, when using your calculator to find expressions such as $p \times q + r \times s$ you must enter the sequence

   $\boxed{(}\ \boxed{p}\ \boxed{\times}\ \boxed{q}\ \boxed{)}\ \boxed{+}\ \boxed{(}\ \boxed{r}\ \boxed{\times}\ \boxed{s}\ \boxed{)}\ \boxed{=}$

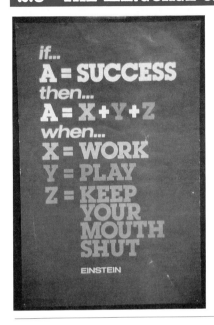

REVIEW

Before starting this section you should know:

1. How to evaluate expressions involving exponents (Section 2.2).
2. The words used to indicate addition and subtraction.

OBJECTIVES

You should be able to:

A. Write a given sum or difference in symbols.
B. Write an indicated product without using the multiplication sign.
C. Write an indicated product using exponents.
D. Write an indicated product using parentheses.
E. Write an indicated quotient in symbols.
F. Find the number of terms in an expression.

The poster uses the language of algebra to tell you how to be successful! The letters X, Y, and Z are used as *placeholders*.

In algebra, it is customary to use the last few letters of the alphabet as *placeholders, unknowns,* or *variables.* Thus the letters t, u, v, w, x, y, and z are frequently used for unknowns and the letter x is used most often. (x, y, and z are used in algebra because they are seldom used in ordinary words.) Of course, we still use the same symbols as in arithmetic to denote the operations of addition ($+$) and subtraction ($-$).

A. SUMS AND DIFFERENCES

There are many words that indicate an addition or a subtraction. We list some of these here for your convenience.

ADD ($+$)	SUBTRACT ($-$)
Plus, sum, increase, more than	Minus, difference, decrease, less than
$a + b$ (read "a plus b") means: 1. The sum of a and b 2. a increased by b 3. b more than a	$a - b$ (read "a minus b") means: 1. The difference of a and b 2. a decreased by b 3. b less than a

With this in mind, try the next example.

EXAMPLE 1 Write in symbols.

a. The sum of x and y
b. x minus y
c. $7x$ plus $2a$ minus 3

Solution

a. $x + y$
b. $x - y$
c. $7x + 2a - 3$ ▲

Here is another example illustrating the use of these words.

EXAMPLE 2

If You Want IRS To Figure Your Tax, See Page 29 of the Instructions.		
Caution: If you are under age 14 and have more than $1,000 of investment income, check here . . . . ▶ ☐ Also see page 30 to see if you have to use Form 8615 to figure your tax.		
20 Find the tax on the amount on line 19. Check if from: ☐ Tax Table (pages 37–42) or ☐ Form 8615	20	T
21 Credit for child and dependent care expenses. Complete and attach Schedule 1, Part I.	21	C
22 Subtract line 21 from line 20. Enter the result. (If line 21 is more than line 20, enter -0-.) This is your **total tax**. ▶ 22		
23a Total Federal income tax withheld—from Box 9 of your W-2 form(s). (If any is from Form(s) 1099, check here ▶ ☐ .) 23a	F	
b Earned income credit, from the worksheet on page 35 of the instructions. Also see page 34. 23b	E	
24 Add lines 23a and 23b. Enter the total. These are your **total payments**. ▶ 24		

The preceding material is part of Form 1040A of the federal income tax return.

a. What symbol is used to represent the tax on the amount on line 19?
b. Which symbol is used to represent credit for child- and dependent-care expenses?
c. In line 22 we must enter the difference between lines 20 and 21. What would that be in symbols?

Solution

a. T
b. C
c. $T - C$ ▲

Sometimes it is necessary to translate English sentences containing the words we have studied into symbols. For example, the sentence, the price P of an item is obtained by adding the cost C and the tax T, can be translated as

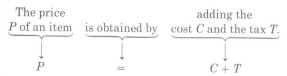

Let us practice how to do this in the next example.

Problem 1 Write in symbols.
a. The sum of p and q
b. q minus p
c. $3q$ plus $5y$ minus 2

Problem 2 In the form of Example 2, line 24 is the sum of lines 23a and 23b. What entry must be made on line 24?

EXAMPLE 3 Translate into symbols.

a. The selling price S of an item is obtained by adding the cost C to the markup M.

b. The amount of money A you have left at the end of each month equals the difference between your salary S and your expenses E.

Solution

a.

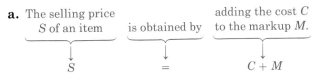

b.

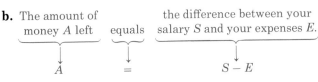

B. PRODUCTS

How do we write multiplication problems in algebra? We use the raised dot ($\cdot$) or parentheses, ( ). Here are some ways of writing the product of a and b.

A raised dot:	$a \cdot b$
Parentheses:	$a(b)$, $(a)b$, or $(a)(b)$
Juxtaposition (writing a and b next to each other):	ab

In each of these cases, a and b (the things to be multiplied) are called **factors.** Of course, the last notation must not be used when multiplying specific numbers because then "5 times 8" would be written as "58," which looks like fifty-eight. Here are some words that indicate a multiplication.

MULTIPLY ($\times$, or $\cdot$)
times, of, product

We will use them in Example 4.

EXAMPLE 4 Write the indicated products using juxtaposition.

a. 8 times x
b. $-x$ times y times z
c. 4 times x times x
d. $\frac{1}{5}$ of x
e. The product of 3 and x

Solution

a. 8 times x is written as $8x$.
b. $-x$ times y times z is written as $-xyz$.

Problem 3 Translate into symbols.
a. The gross national product G equals the net national product N plus the depreciation D.
b. The sale price S of an item equals the regular price R less the discount D.

Problem 4 Write using juxtaposition.
a. -3 times x
b. a times b times c
c. 5 times a times a
d. $\frac{1}{2}$ of a
e. the product of 5 and a

c. 4 times x times x is written as $4xx$.

d. $\frac{1}{5}x$

e. $3x$

C. EXPONENTS

In Example 4c, we wrote 4 times x times x as $4xx$. In algebra (as in arithmetic) we use **exponential notation** to indicate the number of times a quantity is to be used as a factor. As you recall from Section 2.2

$$\underbrace{2 \times 2 \times 2 \times 2}_{\text{4 factors of 2}} = 2^4 \leftarrow \text{Exponent}$$

Base

2^4 is read as "two to the fourth power."

Here the exponent **4** tells us that the base **2** is used as a factor **4** times. Similarly,

$2 \times 2 \times 2 = 2^3$ The product of 3 factors of 2

$2 \times 2 = 2^2$ The product of 2 factors of 2

$2 = 2^1$ The product of 1 factor of 2

2^3 is read as "two to the third power," or "two cubed."

2^2 is read as "two to the second power," or "two squared."

The same technique works for unknowns. Thus,

$x \cdot x \cdot x \cdot x \cdot x = x^5$

$x \cdot x \cdot x \cdot x = x^4$

$x \cdot x \cdot x = x^3$

$x \cdot x = x^2$

$x = x^1$

EXAMPLE 5   Write using exponents.

a. $8 \cdot 8 \cdot 8 \cdot 8$

b. $x \cdot x \cdot y \cdot y \cdot y$

c. $3 \cdot 3 \cdot c \cdot a \cdot c \cdot a \cdot c$

d. The square of $2x$

e. The cube of $3y$

Problem 5   Write using exponents.

a. $3 \cdot 3 \cdot 3 \cdot 3 \cdot 3$

b. $a \cdot a \cdot a \cdot b \cdot b$

c. $5 \cdot 5 \cdot a \cdot a \cdot b \cdot a \cdot b$

d. The square of $-3y$

e. The cube of $5x$

Solution

a. $8 \cdot 8 \cdot 8 \cdot 8 = 8^4$

b. $x \cdot x \cdot y \cdot y \cdot y = x^2 y^3$

c. $3 \cdot 3 \cdot c \cdot a \cdot c \cdot a \cdot c = 3^2 c^3 a^2 = 3^2 a^2 c^3$, where the result is written in alphabetical order.

d. The square of $2x$ is written as $(2x)^2$. Note that we *do not* write $2x^2$ because this means two times x squared.

e. The cube of $3y$ is written as $(3y)^3$. Note that $3y^3$ is *not* the same as $(3y)^3$, since $(3y)^3 = (3y)(3y)(3y) = 27y^3$.

D. GROUPING SYMBOLS IN MULTIPLICATION

As you can see from Example 5d and e, more complicated products such as "the square of $2x$" require parentheses, ( ), as a grouping symbol. Brackets, [ ], are also used as grouping symbols in the same manner. Thus, 9 times the sum of a and b can be written as

$9(a + b)$

$9[a + b]$

EXAMPLE 6 Write.

a. The product of $(a + b)$ and c
b. The product of $(a - b)$ and $(a + b)$
c. The product of $(a - c)$ and $(a - c)$

Solution

a. $(a + b)c$
b. $(a - b)(a + b)$
c. $(a - c)(a - c) = (a - c)^2$

Problem 6 Write.
a. The product of $(a - b)$ and c
b. The product of $(x + y)$ and $(x - y)$
c. The product of $(x + y)$ and $(x + y)$

E. DIVISION

What about division? In arithmetic we use the division ($\div$) sign to indicate division. In algebra we usually use fractions to indicate division. Thus in arithmetic we write $15 \div 3$ (or $3\overline{)15}$) to indicate the quotient of 15 and 3. However, in algebra usually we write

$$\frac{15}{3}$$

Similarly, "the quotient of x and y" is written as

$$\frac{x}{y}$$

(We avoid writing $\frac{x}{y}$ as x/y because more complicated expressions such

as $\dfrac{x}{y + z}$ need to be written as $x/(y + z)$.)

Here are some words that indicate a division.

DIVIDE ($\div$, or the bar —)
Divided by, quotient

We will use them in Example 7.

EXAMPLE 7 Write in symbols.

a. The quotient of x and 7
b. The quotient of 7 and x
c. The quotient of $(x + y)$ and z
d. The sum of a and b, divided by the difference of a and b

Solution

a. $\dfrac{x}{7}$

b. $\dfrac{7}{x}$

Problem 7 Write in symbols.
a. The quotient of a and b
b. The quotient of b and a
c. The quotient of $(x - y)$ and z
d. The difference of x and y, divided by the sum of x and y

c. $\dfrac{x+y}{z}$

d. $\dfrac{a+b}{a-b}$

F. EXPRESSIONS

How do we express mathematical ideas in the language of algebra? By using expressions, of course! An **expression** is a collection of numbers and letters connected by operation signs. Here are some expressions.

1. xy^2
2. $x + y$
3. $3x^2 - 2y + z$

In these expressions, the parts that are to be added or subtracted are called **terms.** Thus the expression in (1) has one term, xy^2; the expression in (2) has two terms, x and y; and the expression in (3) has three terms, $3x^2$, $-2y$, and z. In most applications these expressions are given in words and have to be translated into mathematical symbols. Review the different words used to indicate a multiplication or a division in this section. (We gave the words indicating an addition or a subtraction in the preceding section.)

EXAMPLE 8 Write in symbols and indicate the number of terms to the right of the equals sign.

a. The area A of a circle is obtained by multiplying π by the square of the radius r.

b. The perimeter P of a rectangle is obtained by adding twice the length L to twice the width W.

c. The current I in a circuit is given by the quotient of the voltage V and the resistance R.

Solution

a. The area A of a circle $\quad$ is obtained by $\quad$ multiplying π by the square of the radius r.

$$A \quad = \quad \pi r^2$$

There is one term, πr^2, to the right of the equals sign.

b. The perimeter P of a rectangle $\quad$ is obtained by $\quad$ adding twice the length L to twice the width W.

$$P \quad = \quad 2L + 2W$$

There are two terms, $2L$ and $2W$, to the right of the equals sign.

c. The current I in a circuit $\quad$ is given by $\quad$ the quotient of the voltage V and the resistance R.

$$I \quad = \quad \dfrac{V}{R}$$

There is one term to the right of the equals sign.

Problem 8 Write in symbols and indicate the number of terms to the right of the equals sign.
a. The area A of a rectangle is obtained by multiplying the length L by the width W.
b. The amount A of money in an account at the end of a year is the sum of the principal P and the interest I.
c. The radius r of a circle is the quotient of the circumference C and 2π.

ANSWERS (to problems on pages 130–131)
5. a. 3^5 **b.** $a^3 b^2$ **c.** $5^2 a^3 b^2$ **d.** $(-3y)^2$ **e.** $(5x)^3$
6. a. $(a-b)c$ **b.** $(x+y)(x-y)$ **c.** $(x+y)^2$
7. a. $\dfrac{a}{b}$ **b.** $\dfrac{b}{a}$ **c.** $\dfrac{x-y}{z}$ **d.** $\dfrac{x-y}{x+y}$

A. Write in symbols.

ANSWERS

1. The sum of a and c

1. _____

2. The sum of u and v

2. _____

3. The sum of $3x$ and y

3. _____

4. The sum of 8 and x

4. _____

5. $9x$ plus $17y$

5. _____

6. $5a$ plus $2b$

6. _____

7. The difference of $3a$ and $2b$

7. _____

8. The difference of $6x$ and $3y$

8. _____

9. $-2x$ less 5

9. _____

10. $-7y$ less $3x$

10. _____

B. Write the products using juxtaposition.

11. 7 times a

11. _____

12. -9 times y

12. _____

13. $\frac{1}{7}$ of a

13. _____

14. $\frac{1}{9}$ of y

14. _____

15. The product of b and d

15. _____

16. The product of 4 and c

16. _____

17. xy multiplied by z

17. _____

18. $-a$ multiplied by bc

18. _____

19. $-b$ times the sum of c and d

19. _____

20. The sum of p and q, multiplied by r

20. _____

21. $(a - b)$ times x

21. _____

22. $(a + d)$ times $(x - y)$

22. _____

23. The product of $(x - 3y)$ and $(x + 7y)$

23. _____

24. The product of $(a - 2b)$ and $(2a - 3b)$

24. _____

25. $(c - 4d)$ times $(x + y)$

25. _____

C. Write the products using exponents.

26. The product of x, y, and x

26. _____

27. The product of $-a$, a, and c

27. _____

28. x times y times x times y

28. _____

29. The product of $(a + c)$ and $(a + c)$

29. _____

30. The product of $(3x - y)$ and $(3x - y)$

30. _____

31. The product of $2x$ times itself

31. _____

32. The square of a number x

32. _____

33. The cube of a number n

33. _____

D. Write the products using parentheses.

34. The square of the expression $(a + b)$

34. _____

35. The square of the expression $(2x - y)$

35. _____

36. The third power of yz

36. _____

37. The fourth power of yz

37. _____

38. The fifth power of xyz

38. _____

39. The fourth power of the sum of x and y

39. _____

40. The third power of the difference of a and b

40. _____

E. Write in symbols.

41. The quotient of a and the sum of x and y

41. _____

42. The quotient of $(a + b)$ and c

42. _____

43. The quotient of the difference of a and b, and c

43. _____

44. The sum of a and b divided by the difference of x and y

44. _____

45. The quotient when x is divided into y

45. _____

46. The quotient when y is divided into x

46. _____

47. The quotient when the sum of p and q is divided into the difference of p and q

47. _____

48. The quotient when the difference of $3x$ and y is divided into the sum of x and $3y$

48. _____

49. The quotient obtained when the sum of x and $2y$ is divided by the difference of x and $2y$

49. _____

50. The quotient obtained when the difference of x and $3y$ is divided by the sum of x and $3y$

50. _____

F. Write in symbols and indicate the number of terms to the right of the equals sign.

51. The number of hours H a growing child should sleep is the difference between 17 and one-half the child's age A.

51. _____

ANSWER (to problem on page 132)
8. **a.** $A = LW$. There is one term to the right of the equals sign.
   b. $A = P + I$. There are two terms to the right of the equals sign.
   c. $r = \dfrac{C}{2\pi}$. There is one term to the right of the equals sign.

52. The surface area S of a cone is the sum of πr^2 and πrs.

53. The reciprocal of f is the sum of the reciprocal of u and the reciprocal of v.

54. The height h attained by a rocket is the sum of its height a at burnout and $r^2/20$, where r is the speed at burnout.

55. The Fahrenheit temperature F can be obtained by adding 40 to the quotient of n and 4, where n is the number of cricket chirps in 1 min.

52. _____

53. _____

54. _____

55. _____

✓ **SKILL CHECKER**

Find.

56. $-3 + (-2)$

57. $-4 + (-7)$

58. $-5 + 3$

59. $-6 + 2$

60. $3 + (-8)$

61. $2 + (-7)$

62. $3 - 10$

63. $4 - 17$

64. $-3 - (-18)$

65. $-2 - (-19)$

66. $-18 - (-3)$

67. $-15 - (-6)$

68. $-12 - 23$

69. $-10 - 18$

70. $-15 - 18$

56. _____

57. _____

58. _____

59. _____

60. _____

61. _____

62. _____

63. _____

64. _____

65. _____

66. _____

67. _____

68. _____

69. _____

70. _____

2.5 USING YOUR KNOWLEDGE

The words we have studied are used in many different fields. Perhaps you have seen some of the following material in one of your classes! Use the knowledge gained in the last two sections to write it in symbols. The word in italics indicates the field from which the material is taken.

1. *Electricity*   The voltage V across any part of a circuit is the product of the current I and the resistance R.

2. *Economics*   The total profit TP equals the total revenue TR minus the total cost TC.

3. *Chemistry*   The total pressure P in a container filled with gases A, B, and C is equal to the sum of the partial pressures P_A, P_B, and P_C.

1. _____

2. _____

3. _____

4. *Psychology*   The intelligence quotient (IQ) for a child is obtained by multiplying his/her mental age M by 100 and dividing the result by his/her chronological age C.

4. _____

5. *Physics*   The distance D traveled by an object moving at a constant rate R is the product of R and the time T.

5. _____

6. *Astronomy*   The square of the period P of a planet equals the product of a constant C and the cube of the planet's distance R from the sun.

6. _____

7. *Relativity*   The energy E of an object equals the product of its mass m and the square of the speed of light c.

7. _____

8. *Engineering*   The depth h of a gear tooth is found by dividing the difference between the major diameter D and the minor diameter d by 2.

8. _____

9. *Geometry*   The square of the hypotenuse h of a right triangle equals the sum of the squares of the sides a and b.

9. _____

10. *Auto Mechanics*   The horsepower hp of an engine is obtained by multiplying 0.4 by the square of the diameter D of each cylinder and by the number N of cylinders.

10. _____

REVIEW

Before starting this section you should know:

1. How to add and subtract integers (Section 2.1).
2. How to remove parentheses in expressions of the form $+(a + b)$ or $-(a + b)$ (Section 2.4).
3. How to use the distributive law (Section 2.4).

OBJECTIVES

You should be able to:

A. Simplify expressions by adding like terms.
B. Simplify expressions by subtracting terms.
C. Remove parentheses preceded by a plus ($+$) sign and then combine like terms.
D. Remove parentheses preceded by a minus ($-$) sign and then combine like terms.
E. Remove parentheses and other grouping symbols and then combine like terms.

What would be the price of the 9 tacos? If 3 tacos cost \$1.39 and 6 tacos cost \$2.39, then 3 tacos + 6 tacos will cost \$1.39 + \$2.39; that is, 9 tacos will cost \$3.78. Note that

$$3 \ tacos + 6 \ tacos = 9 \ tacos$$

and

$$\$1.39 + \$2.39 = \$3.78$$

The term **3** *tacos* uses the number **3** to tell "how many." The number 3 is called the **numerical coefficient** (or simply the *coefficient*) of the term. Similarly, the terms $5x$, $7y$, and $-8xy$ have numerical coefficients of **5, 7,** and $-$**8,** respectively. When two or more terms are *exactly alike* (except possibly for their coefficients or the order in which the factors are multiplied), they are called **like terms.** Thus, like terms contain the *same variables* with the *same exponents,* but their coefficients may be different. Thus, 3 tacos and 6 tacos are like terms, $3x$ and $-5x$ are like terms, and $-xy^2$ and $7xy^2$ are like terms. On the other hand, $2x$ and $2x^2$ or $2xy^2$ and $2x^2y$ are *not* like terms.

A.   ADDING LIKE TERMS

When an algebraic expression contains like terms, these terms can be combined into a single term, just as we combined 3 tacos and 6 tacos into the single term 9 tacos. Combining like terms is really simple; you just have to be sure that the variable parts of the terms to be combined are identical and then add (or subtract) the numerical coefficients. This can be done since, by the distributive law,

$$ac + bc = (a + b)c$$

Thus,

$$\underbrace{3x}_{(x + x + x)} + \underbrace{5x}_{(x + x + x + x + x)} = \underbrace{(3 + 5)x = 8x.}_{(x + x + x + x + x + x + x + x)}$$

$$\underbrace{2x^2}_{(x^2 + x^2)} + \underbrace{4x^2}_{(x^2 + x^2 + x^2 + x^2)} = \underbrace{(2 + 4)x^2 = 6x^2}_{(x^2 + x^2 + x^2 + x^2 + x^2 + x^2)}$$

$$\underbrace{3ab}_{(ab + ab + ab)} + \underbrace{2ab}_{(ab + ab)} = \underbrace{(3 + 2)ab = 5ab}_{(ab + ab + ab + ab + ab)}$$

But what about another one, like combining $-3x$ and $-2x$? We first write

$$\overbrace{(-3x)}^{} \quad + \quad \overbrace{(-2x)}^{}$$
$$\underbrace{(-x) + (-x) + (-x) + (-x) + (-x)}$$

The answer is $-5x$. Thus

$$(-3x) + (-2x) = [-3 + (-2)]x = -5x$$

Note that we write the addition of $-3x$ and $-2x$ as $(-3x) + (-2x)$, using parentheses around the $-3x$ and $-2x$. This is done to avoid the confusion of writing

$$-3x + -2x$$

Never use two signs of operation together without parentheses.

As you can see, if both quantities to be added are preceded by minus $(-)$ signs, the result is preceded by a minus $(-)$ sign. If they are both preceded by plus $(+)$ signs, the result is preceded by a plus $(+)$ sign. But what about the next expression

$$(-5x) + (3x)$$

Here we write

$$\underbrace{(-x) + (-x) + (-x) + (-x) + (-x)}_{(-5x)} + \underbrace{x + x + x}_{3x}$$

Since we have 2 more negative x's than positive x's, the result is $-2x$. Thus

$$(-5x) + (3x) = (-5 + 3)x = -2x \quad \text{(Using the distributive law)}$$

On the other hand, $5x + (-3x)$ can be written as

$$\underbrace{x + x + x + x + x}_{5x} + \underbrace{(-x) + (-x) + (-x)}_{(-3x)}$$

and the answer is $2x$; that is,

$$5x + (-3x) = [5 + (-3)]x = 2x \quad \text{(Using the distributive law)}$$

Now, here is an easy one: What is $x + x$? Since $1 \cdot x = x$, the coefficient of x is assumed to be 1. Thus,

$$\begin{aligned} x + x &= 1 \cdot x + 1 \cdot x \\ &= (1 + 1)x \\ &= 2x \end{aligned}$$

In all the examples that follow, make sure you use the fact

$$ac + bc = (a + b)c$$

to combine the numerical coefficients.

EXAMPLE 1 Combine like terms.
a. $-7x + 2x$ **b.** $-4x + 6x$ **c.** $(-2x) + (-5x)$ **d.** $x + (-5x)$

Solution

a. $-7x + 2x = (-7 + 2)x = -5x$

b. $-4x + 6x = (-4 + 6)x = 2x$

Problem 1 Combine like terms.
a. $-4x + 3x$ **b.** $-5x + 8x$
c. $(-8x) + (-2x)$ **d.** $x + (-3x)$

c. $(-2x) + (-5x) = [-2 + (-5)]x = -7x$

d. First, recall that $x = 1 \cdot x$. Thus,

$$x + (-5x) = 1x + (-5x)$$
$$= [1 + (-5)]x = -4x$$

B. SUBTRACTING LIKE TERMS

Subtraction of like terms is defined in terms of addition. This is because the definition of subtraction of rational numbers states the following:

DEFINITION OF SUBTRACTION

To subtract a number b from another number a, add the *additive inverse* (*opposite*) of b to a; that is, $a - b = a + (-b)$.

As before, to subtract like terms we use the fact that $ac + bc = (a + b)c$. Thus,

$$3x - 5x = 3x + (-5x) = [3 + (-5)]x = -2x$$
$$-3x - 5x = -3x + (-5x) = [(-3) + (-5)]x = -8x$$
$$3x - (-5x) = 3x + (5x) = (3 + 5)x = 8x$$
$$-3x - (-5x) = -3x + (5x) = (-3 + 5)x = 2x$$

Note that

$$-3x - (-5x) = -3x + 5x$$

In general,

$a - (-b) = a + b$ $-(-b)$ is replaced by $+b$.

We can now combine like terms involving subtraction.

EXAMPLE 2   Combine like terms.

a. $7ab - 9ab$ **b.** $8x^2 - 3x^2$
c. $-5ab^2 - (-8ab^2)$ **d.** $-6a^2b - (-2a^2b)$

Solution

a. $7ab - 9ab = 7ab + (-9ab) = [7 + (-9)]ab = -2ab$
b. $8x^2 - 3x^2 = 8x^2 + (-3x^2) = [8 + (-3)]x^2 = 5x^2$
c. $-5ab^2 - (-8ab^2) = -5ab^2 + 8ab^2 = (-5 + 8)ab^2 = 3ab^2$
d. $-6a^2b - (-2a^2b) = -6a^2b + 2a^2b = (-6 + 2)a^2b = -4a^2b$

C. REMOVING PARENTHESES

Sometimes it is necessary to remove parentheses before combining like terms. For example, to combine like terms in

$$(3x + 5) + (2x - 2)$$

Problem 2   Combine like terms.
a. $8a^2b - 12a^2b$
b. $13a^2 - 6a^2$
c. $-6ab^3 - (-3ab^3)$
d. $-2a^3b - (-7a^3b)$

we have to remove the parentheses *first*. If there is a plus $(+)$ sign (or no sign) in front of the parentheses, we can simply remove the parentheses; that is,

$$+(a + b) = a + b \quad \text{and} \quad +(a - b) = a - b$$

With this in mind, we have

$$(3x + 5) + (2x - 2) = 3x + 5 + 2x - 2$$
$$= 3x + 2x + 5 - 2 \qquad \text{(By the commutative law)}$$
$$= (3x + 2x) + (5 - 2) \quad \text{(By the associative law)}$$
$$= 5x + 3$$

Note that we used the properties we studied to rearrange like terms together.

Once you understand the use of these laws, you can then see that the simplification of $(3x + 5) + (2x - 2)$ consists of just adding $3x$ to $2x$ and 5 to -2. The work done can then be shown like this:

Like terms

$$(3x + 5) + (2x - 2) = 3x + 5 + 2x - 2$$

Like terms

$$= 5x + 3$$

Use this idea in the next example.

EXAMPLE 3 Remove parentheses and combine like terms.

a. $(4x - 5) + (7x - 3)$ **b.** $(3a + 5b) + (4a - 9b)$

Solution We first remove parentheses; then we add like terms.

a. $(4x - 5) + (7x - 3) = 4x - 5 + 7x - 3$
$$= 11x - 8$$

b. $(3a + 5b) + (4a - 9b) = 3a + 5b + 4a - 9b$
$$= 7a - 4b$$

D. REMOVING PARENTHESES PRECEDED BY A MINUS $(-)$ SIGN

To remove parentheses in expressions preceded by a minus $(-)$ sign, we recall that

$$-(a + b) = -1 \cdot (a + b)$$
$$= -1 \cdot a + (-1)(b)$$
$$= -a + (-b)$$
$$= -a - b$$

Problem 3 Remove parentheses and combine like terms.
a. $(8y - 7) + (3y - 2)$
b. $(4t + u) + (5t - 7u)$

Also,

$$-(a - b) = -1 \cdot (a - b)$$
$$= -1 \cdot a + (-1)(-b)$$
$$= -a + b$$

Thus,

$$-(a + b) = -a - b$$

and

$$-(a - b) = -a + b$$

To simplify

$$4x + 3(x - 2) - (x + 5)$$

we proceed as follows:

1. Use the distributive law in the expression $3(x - 2) = 3x - 6$.

$$4x + 3(x - 2) - (x + 5)$$
$$= 4x + 3x - 6 - (x + 5)$$

2. Remove parentheses in the expression $-(x + 5) = -x - 5$.

$$= 4x + 3x - 6 - x - 5$$

3. Associate like terms.

$$= (4x + 3x - x) + (-6 - 5)$$

4. Simplify.

$$= 6x - 11$$

EXAMPLE 4 Remove parentheses and combine like terms:

$$8x - 2(x - 1) - (x + 3)$$

Solution

1. Remove parentheses: $-2(x - 1) = -2x + 2$.

$$8x - 2(x - 1) - (x + 3)$$
$$= 8x - 2x + 2 - (x + 3)$$

2. Remove parentheses: $-(x + 3) = -x - 3$.

$$= 8x - 2x + 2 - x - 3$$

3. Associate like terms.

$$= (8x - 2x - x) + (2 - 3)$$

4. Simplify.

$$= 5x + (-1) \quad \text{or} \quad 5x - 1$$

Problem 4 Remove parentheses and combine like terms:
$$9x - 3(x - 2) - (2x + 7)$$

E. REMOVING OTHER GROUPING SYMBOLS

Sometimes parentheses occur within other parentheses. To avoid confusion, we use different grouping symbols. Thus, we usually do not write $((x + 5) + 3)$. Instead, we use a different grouping symbol and write $[(x + 5) + 3]$. To simplify (combine like terms) in such expressions, the *innermost grouping symbols are removed first*. The procedure is illustrated in the next example.

EXAMPLE 5 Remove grouping symbols and simplify:

$[(x^2 - 1) + (2x + 5)] + [(x - 2) - (3x^2 + 3)]$

Solution We first remove the innermost parentheses and then combine like terms. Thus,

$[(x^2 - 1) + (2x + 5)] + [(x - 2) - (3x^2 + 3)]$

$= [x^2 - 1 + 2x + 5] + [x - 2 - 3x^2 - 3]$ Remove parentheses. Note that $-(3x^2 + 3) = -3x^2 - 3$.

$= [x^2 + 2x + 4] + [-3x^2 + x - 5]$ Combine like terms inside the brackets.

$= x^2 + 2x + 4 - 3x^2 + x - 5$ Remove brackets.

$= -2x^2 + 3x - 1$ Combine like terms.

Problem 5 Remove grouping symbols and simplify:

$[(2x^2 - 3) + (3x + 1)] + [(x - 1) - (x^2 + 2)]$

NAME

CLASS

SECTION

A. Combine like terms (simplify).

1. $19a + (-8a)$

2. $-2b + 5b$

3. $-8c + 3c$

4. $-5d + (-7d)$

5. $4n^2 + 8n^2$

6. $3x^2 + (-9x^2)$

7. $-3ab^2 + (-4ab^2)$

8. $9ab^2 + (-3ab^2)$

9. $-4abc + 7abc$

10. $6xyz + (-9xyz)$

11. $7ab + (-3ab) + 9ab$

12. $-5x^2y + 8x^2y + 3x^2y$

13. $-3xy^2 + 2x^2y + (-6xy^2)$

14. $2x^2y + (-3xy^2) + 4xy^2$

15. $8abc^2 + 3ab^2c + (-8abc^2)$

16. $3xy + 5ab + 2xy + (-3ab)$

17. $-8ab + 9xy + 2ab + (-2xy)$

18. $7 + \frac{1}{2}x + 3 + \frac{1}{2}x$

19. $\frac{1}{5}a + \frac{3}{7}a^2b + \frac{2}{5}a + \frac{1}{7}a^2b$

20. $\frac{4}{9}ab + \frac{4}{5}ab^2 + \left(-\frac{1}{9}ab\right) + \left(-\frac{1}{5}a^2b\right)$

B. Combine like terms (simplify).

21. $13x - 2x$

22. $8x - (-2x)$

23. $6ab - (-2ab)$

24. $4.2xy - (-3.7xy)$

25. $-4a^2b - (3a^2b)$

26. $-8ab^2 - 4a^2b$

ANSWERS

1. _____

2. _____

3. _____

4. _____

5. _____

6. _____

7. _____

8. _____

9. _____

10. _____

11. _____

12. _____

13. _____

14. _____

15. _____

16. _____

17. _____

18. _____

19. _____

20. _____

21. _____

22. _____

23. _____

24. _____

25. _____

26. _____

27. $3.1t^2 - 3.1t^2$

28. $-4.2ab - 3.8ab$

29. $0.3x^2 - 0.3x^2$

30. $0 - (-0.8xy^2)$

C. Remove parentheses and combine like terms.

31. $(3xy + 5) + (7xy - 9)$

32. $(8ab - 9) + (7 - 2ab)$

33. $(7R - 2) + (8 - 9R)$

34. $(5xy - 3ab) + (9ab - 8xy)$

35. $(5L - 3W) + (W - 6L)$

36. $(2ab - 2ac) + (ab - 4ac)$

D. Remove parentheses and combine like terms.

37. $5x - (8x + 1)$

38. $3x - (7x + 2)$

39. $\dfrac{2x}{9} - \left(\dfrac{x}{9} - 2\right)$

40. $\dfrac{5x}{7} - \left(\dfrac{2x}{7} - 3\right)$

41. $4a - (a + b) + 3(b + a)$

42. $8x - 3(x + y) - (x - y)$

43. $7x - 3(x + y) - (x + y)$

44. $4(b - a) + 3(b + a) - 2(a + b)$

45. $-(x + y - 2) + 3(x - y + 6) - (x + y - 16)$

E. Remove grouping symbols and simplify.

46. $[(a^2 - 4) + (2a^3 - 5)] + [(4a^3 + a) + (a^2 + 9)]$

47. $(x^2 + 7 - x) + [-2x^3 + (8x^2 - 2x) + 5]$

48. $[(0.4x - 7) + 0.2x^2] + [(0.3x^2 - 2) - 0.8x]$

49. $\left[\left(\dfrac{1}{4}x^2 + \dfrac{1}{5}x\right) - \dfrac{1}{8}\right] + \left[\left(\dfrac{3}{4}x^2 - \dfrac{3}{5}x\right) + \dfrac{5}{8}\right]$

50. $[3(x + 2) - 10] + [5 + 2(5 + x)]$

51. $[3(2a - 4) + 5] - [2(a - 1) + 6]$

52. $[6(a - b) + 2a] - [3b - 4(a - b)]$

53. $[4a - (3 + 2b)] - [6(a - 2b) + 5a]$

54. $-[-(x + y) + 3(x - y)] - [4(x + y) - (3x - 5y)]$

55. $-[-(0.2x + y) + 3(x - y)] - [2(x + 0.3y) - 5]$

27. _____

28. _____

29. _____

30. _____

31. _____

32. _____

33. _____

34. _____

35. _____

36. _____

37. _____

38. _____

39. _____

40. _____

41. _____

42. _____

43. _____

44. _____

45. _____

46. _____

47. _____

48. _____

49. _____

50. _____

51. _____

52. _____

53. _____

54. _____

55. _____

ANSWER (to problem on page 142)

5. $x^2 + 4x - 5$

✓ SKILL CHECKER

Find.

56. $-3 \cdot 8$

57. $-4 \cdot 7$

58. $-2(-4)$

59. $-3 \cdot (-8)$

60. $\dfrac{-6}{2}$

61. $\dfrac{-18}{9}$

62. $\dfrac{-18}{-3}$

63. $\dfrac{-28}{-7}$

56. _____

57. _____

58. _____

59. _____

60. _____

61. _____

62. _____

63. _____

2.6 USING YOUR KNOWLEDGE

The ideas in this section are used to simplify formulas. For example, the **perimeter** (distance around) of the rectangle given in the margin is found by following the color line. The perimeter is

$$P = W + L + W + L$$
$$= 2W + 2L$$

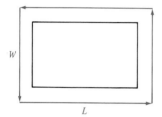

Use this idea to find the perimeter of the given figure.

1. The square of side S (see margin)

2. The parallelogram of base b and side s (see margin)

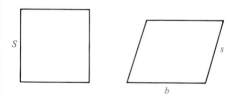

When obtaining the actual measurement of certain perimeters, we have to add like terms if the measurements are given in feet and inches. For example, the perimeter of the rectangle in the margin is

$$P = (2 \text{ ft} + 7 \text{ in.}) + (4 \text{ ft} + 1 \text{ in.}) + (2 \text{ ft} + 7 \text{ in.}) + (4 \text{ ft} + 1 \text{ in.})$$
$$= 12 \text{ ft} + 16 \text{ in.}$$

Since 16 in. = 1 ft + 4 in.,

$$P = 12 \text{ ft} + (1 \text{ ft} + 4 \text{ in.})$$
$$= 13 \text{ ft} + 4 \text{ in.}$$

Use these ideas to obtain the perimeter of the given rectangle.

1. _____

2. _____

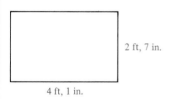

2 ft, 7 in.

4 ft, 1 in.

3.

3 ft, 1 in.

6 ft, 2 in.

3. _____

4.

4 ft, 5 in.

8 ft, 2 in.

4. _____

5. The U.S. Postal Service has a regulation stating that "the sum of the length and girth of a package may be no more than 100 inches." What is the sum of the length and girth of the rectangular package in the margin? *Hint:* The girth of the package is obtained by measuring the length of the colored line.

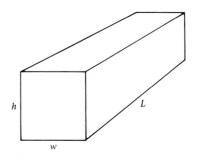

h   *L*

w

5. _____

6. Write in simplified form the height of the step block whose picture is shown in the margin.

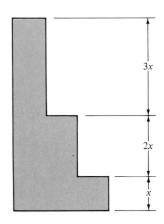

$3x$

$2x$

x

6. _____

7. Write a simplified expression for the length of the metal plate shown.

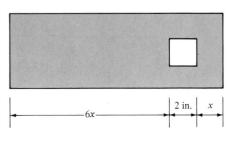

$6x$ 2 in. x

7. _____

2.7 MULTIPLYING AND DIVIDING EXPRESSIONS

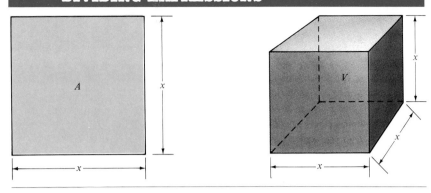

OBJECTIVES

REVIEW

Before starting this section you should know:

1. How to multiply and divide integers (Section 2.2).
2. The meaning of the commutative and associative laws of multiplication.

OBJECTIVES

You should be able to:

A. Multiply expressions using the proper law of exponents.
B. Divide expressions using the proper law of exponents.

As we mentioned in Section 2.5, exponential notation is used to indicate how many times a quantity is to be used as a *factor*. For example, the area A of the *square* in the figure can be written as

$A = x \cdot x = x^2$ Read "*x squared*."

The exponent 2 indicates that the x is used as a factor twice. Similarly, the volume V of the *cube* is

$V = x \cdot x \cdot x = x^3$ Read "*x cubed*."

This time, the exponent 3 indicates that the x is used as a factor three times.

If a symbol x (called the **base**) is to be used n times as a factor, we use the following definition:

$$\underbrace{x \cdot x \cdot x \cdot \cdots \cdot x}_{n \text{ factors}} = x^{\overset{\curvearrowleft \text{exponent}}{n}}_{\underset{\uparrow}{\text{base}}}$$

When n is a natural number, some of the powers of x are

$x \cdot x \cdot x \cdot x \cdot x = x^5$ x to the fifth power
$x \cdot x \cdot x \cdot x = x^4$ x to the fourth power
$x \cdot x \cdot x = x^3$ x to the third power (also read "x cubed")
$x \cdot x = x^2$ x to the second power (also read "x squared")
$x = x^1$ x to the first power

Note that if the base carries no exponent, the exponent is assumed to be 1, that is,

$a = a^1, \quad b = b^1, \quad \text{and} \quad c = c^1$

A. MULTIPLYING EXPRESSIONS

We are now ready to multiply expressions involving exponents. For example, to multiply x^2 by x^3, we first write

$$\underbrace{x^2}_{} \cdot \underbrace{x^3}_{}$$
$$\underbrace{x \cdot x \cdot x \cdot x \cdot x}_{}$$
or x^5

Clearly, we have just added the exponents of x^2 and x^3 to find the exponent of the result. Similarly, $2^2 \cdot 2^3 = 2^{2+3} = 2^5$ and $3^2 \cdot 3^3 = 3^{2+3} = 3^5$. Thus,

$$a^3 \cdot a^4 = a^{3+4} = a^7$$

and

$$b^2 \cdot b^4 = b^{2+4} = b^6$$

From these and similar examples, we have the following law of exponents in multiplication:

$x^m \cdot x^n = x^{m+n}$ **First Law of Exponents**

This law means that to multiply expressions with the *same base*, we keep the base and *add* the exponents.

Of course, some expressions may have numerical coefficients other than 1. For example, the expression $3x^2$ has the numerical coefficient 3. Similarly, the numerical coefficient of $5x^3$ is 5. If we now decide to multiply $3x^2$ by $5x^3$, we just multiply numbers by numbers (coefficients) and letters by letters. This procedure is possible because of the *commutative and associative laws of multiplication* we have studied. Using these two laws, we then write

$(3x^2)(5x^3) = (3 \cdot 5)(x^2 \cdot x^3)$ We use parentheses to indicate
$\qquad\qquad = 15x^{2+3}$ multiplication.
$\qquad\qquad = 15x^5$

and

$(8x^2y)(4xy^2)(2x^5y^3)$
$\quad = (8 \cdot 4 \cdot 2) \cdot (x^2 \cdot x^1 \cdot x^5)(y^1 \cdot y^2 \cdot y^3)$ Note that $x = x^1$ and
$\quad = 64x^8y^6$ $y = y^1$.

EXAMPLE 1 Find.

a. $(5x^4)(3x^7)$ **b.** $(3ab^2c^3)(4a^2b)(2bc)$

Solution

a. We use the commutative and associative laws to write the coefficients and the letters together:

$(5x^4)(3x^7) = (5 \cdot 3)(x^4 \cdot x^7)$
$\qquad\qquad = 15x^{4+7}$
$\qquad\qquad = 15x^{11}$

b. Using the commutative and associative laws, and the fact that $a = a^1$, $b = b^1$, and $c = c^1$, we write

$(3ab^2c^3)(4a^2b)(2bc) = (3 \cdot 4 \cdot 2)(a^1 \cdot a^2)(b^2 \cdot b^1 \cdot b^1)(c^3 \cdot c^1)$
$\qquad\qquad\qquad = 24a^{1+2}b^{2+1+1}c^{3+1}$
$\qquad\qquad\qquad = 24a^3b^4c^4$ ▲

Problem 1 Find.
a. $(5a^3)(6a^8)$
b. $(3x^2yz)(2xy^2)(8xz^3)$

To multiply expressions involving signed coefficients in general, we recall the rule of signs for multiplication:

WHEN MULTIPLYING TWO NUMBERS WITH	THE PRODUCT IS
Like (same) signs	Positive $(+)$
Unlike (opposite) signs	Negative $(-)$

Thus to multiply $(-3x^5)$ by $(8x^2)$, we first note that the expressions have *unlike* signs; the product should have a negative coefficient; that is,

$$(-3x^5)(8x^2) = (-3 \cdot 8)(x^5 \cdot x^2)$$
$$= -24x^{5+2}$$
$$= -24x^7$$

Of course, if the expressions have like signs, the product should have a positive coefficient. Thus,

$$(-2x^3)(-7x^5) = (-2)(-7)(x^3 \cdot x^5)$$

Note that the product of -2 and -7 was written as $(-2)(-7)$ and *not* as $(-2 \cdot -7)$ to avoid confusion.

$$= +14x^{3+5}$$
$$= 14x^8$$

Recall that $+14 = 14$.

EXAMPLE 2 Find.

a. $(7a^2bc)(-3ac^4)$ **b.** $(-2xyz^3)(-4x^3yz^2)$

Solution

a. Since the coefficients have unlike signs, the result must have a negative coefficient. Hence

$$(7a^2bc)(-3ac^4) = (7)(-3)(a^2 \cdot a^1)(b^1)(c^1 \cdot c^4)$$
$$= -21a^{2+1}b^1c^{1+4}$$
$$= -21a^3bc^5$$

b. The expressions have like signs, so the result must have a positive coefficient. That is,

$$(-2xyz^3)(-4x^3yz^2) = [(-2)(-4)](x^1 \cdot x^3)(y^1 \cdot y^1)(z^3 \cdot z^2)$$
$$= +8x^{1+3}y^{1+1}z^{3+2}$$
$$= 8x^4y^2z^5$$

Problem 2 Find.
a. $(-3x^3yz)(4xz^4)$
b. $(-3ab^2c^4)(-5a^4bc^3)$

B. DIVIDING EXPRESSIONS

We are now ready to discuss the division of one expression by another. As you recall, the same rule of signs that applies to the multiplication of integers applies to the division of integers. We write this rule for easy reference.

WHEN DIVIDING TWO NUMBERS WITH	THE QUOTIENT IS
Like (same) signs	Positive $(+)$
Unlike (opposite) signs	Negative $(-)$

Now we know what to do with the numerical coefficients when we divide one expression by another. But what about the exponents? In order to divide expressions involving exponents, we need to develop a rule to handle these exponents. For example, to divide x^5 by x^3, we first use the definition of exponent and write

$$\frac{x^5}{x^3} = \frac{x \cdot x \cdot x \cdot x \cdot x}{x \cdot x \cdot x} \quad (x \neq 0)$$

Since $(x \cdot x \cdot x)$ is common to the numerator and denominator, we have

$$\frac{x^5}{x^3} = \frac{(x \cdot x \cdot x) \cdot x \cdot x}{(x \cdot x \cdot x)} = x \cdot x = x^2$$

Here the color x's mean that we divided the numerator and denominator by the common factor $(x \cdot x \cdot x)$. Of course, you can immediately see that the exponent 2 in the answer is simply the difference of the original two exponents; that is,

$$\frac{x^5}{x^3} = x^{5-3} = x^2$$

Similarly,

$$\frac{x^7}{x^4} = x^{7-4} = x^3 \quad (x \neq 0)$$

and

$$\frac{y^4}{y^1} = y^{4-1} = y^3 \quad (y \neq 0)$$

We can now state a law for dividing expressions involving exponents:

$$\frac{x^m}{x^n} = x^{m-n} \quad (x \neq 0)$$

where m is greater than n \qquad Second Law of Exponents

But what if m is not greater than n? For example, how do we simplify

$$\frac{x^3}{x^5}?$$

We first write

$$\frac{x^3}{x^5} = \frac{(x \cdot x \cdot x) \cdot 1}{(x \cdot x \cdot x) \cdot x \cdot x} = \frac{1}{x \cdot x} = \frac{1}{x^2}$$

Since $x \cdot x \cdot x$ was common to the numerator and denominator, we have

$$\frac{x^3}{x^5} = \frac{1}{x^{5-3}} = \frac{1}{x^2}$$ Since 5 is greater than 3, the answer is $\frac{1}{x^2}$, not $\frac{x^2}{1}$.

Here is the law:

$$\frac{x^m}{x^n} = \frac{1}{x^{n-m}} \quad (x \neq 0)$$

where m is less than n

Of course, if $m = n$, then we have

$$\frac{x^m}{x^n} = \frac{x^m}{x^m} = 1, \quad x \neq 0$$

This law says that any number (x^m) divided by itself (x^m) equals 1. But note that if you subtract exponents to simplify $\frac{x^m}{x^m}$, you obtain

$$\frac{x^m}{x^m} = x^{m-m} = x^0$$

Since $\frac{x^m}{x^m} = 1$, it follows that

$$x^0 = 1, \qquad x \neq 0$$

We now combine the last two laws and solve more complicated problems. For example, to simplify

$$\frac{-6x^4}{12x^2}$$

we first note that the numerator and denominator have *unlike signs,* so the result must have a *negative* sign. Moreover, $6 = 2 \cdot 3$ and $12 = 2 \cdot 2 \cdot 3$. We then write

$$\frac{-6x^4}{12x^2} = \frac{-2 \cdot 3 \cdot x \cdot x \cdot x \cdot x}{2 \cdot 2 \cdot 3 \cdot x \cdot x}$$

$$= \frac{-(2 \cdot 3) \cdot (x \cdot x) \cdot x \cdot x}{2 \cdot (2 \cdot 3) \cdot (x \cdot x)} = \frac{-x^2}{2} = -\frac{x^2}{2}$$

Of course, we usually save space and time by simply writing

$$\frac{-6x^4}{12x^2} = \frac{-6}{12} \cdot \frac{x^4}{x^2} \qquad$$ Note that we are dividing coefficients by coefficients (-6 by 12) and letters by letters (x^4 by x^2).

$$= \frac{-1}{2} \cdot x^{4-2}$$

$$= \frac{-x^2}{2}$$

$$= -\frac{x^2}{2}$$

EXAMPLE 3   Find.

a. $\dfrac{24x^2y^6}{-6xy^4}$ **b.** $\dfrac{-3x^3y^4z^6}{9x^5y^7z^8}$ **c.** $\dfrac{-18a^3b^2c}{-12ab^7c}$

Solution

a. $\dfrac{24x^2y^6}{-6xy^4} = \dfrac{2 \cdot 2 \cdot (2 \cdot 3 \cdot x) \cdot x \cdot (y \cdot y \cdot y \cdot y) \cdot y \cdot y}{-(2 \cdot 3 \cdot x) \cdot (y \cdot y \cdot y \cdot y)}$

$$= -2 \cdot 2 \cdot x \cdot y \cdot y$$

$$= -4xy^2$$

Problem 3   Find.

a. $\dfrac{15x^4y^7}{-3x^2y}$ **b.** $\dfrac{-5x^2y^3z^2}{15x^5y^6z^6}$

c. $\dfrac{-3a^2b^3c^2}{-12ab^9c^2}$

To save time, we could divide 24 by -6, x^2 by x, and y^6 by y^4, like this:

$$\frac{24x^2y^6}{-6xy^4} = \frac{24}{-6} \cdot \frac{x^2}{x} \cdot \frac{y^6}{y^4}$$

$$= -4 \cdot x^{2-1} \cdot y^{6-4}$$

$$= -4xy^2$$

b. $\dfrac{-3x^3y^4z^6}{9x^5y^7z^8} = \dfrac{-3}{9} \cdot \dfrac{x^3}{x^5} \cdot \dfrac{y^4}{y^7} \cdot \dfrac{z^6}{z^8}$

$$= \frac{-1}{3 \cdot x^{5-3} \cdot y^{7-4} \cdot z^{8-6}}$$

$$= \frac{-1}{3x^2y^3z^2}$$

c. $\dfrac{-18a^3b^2c}{-12ab^7c} = \dfrac{-18}{-12} \cdot \dfrac{a^3}{a} \cdot \dfrac{b^2}{b^7} \cdot \dfrac{c}{c}$

$$= \frac{3 \cdot a^{3-1} \cdot 1}{2 \cdot b^{7-2}}$$

$$= \frac{3a^2}{2b^5}$$

ANSWERS

A. Find.

1. $(4x)(6x^2)$

2. $(2a^2)(3a^3)$

3. $(5ab^2)(6a^3b)$

4. $(-2xy)(x^2y)$

5. $(-xy^2)(-3x^2y)$

6. $(x^2y)(-5xy)$

7. $b^3\left(\dfrac{-b^2c}{5}\right)$

8. $-a^3b\left(\dfrac{ab^4c}{3}\right)$

9. $\left(\dfrac{-3xy^2z}{2}\right)\left(\dfrac{-5x^2yz^5}{5}\right)$

10. $(-2x^2yz^3)(4xyz)$

11. $(-2xyz)(3x^2yz^3)(5x^2yz^4)$

12. $(-a^2b)(-0.4b^2c^3)(1.5abc)$

13. $(a^2c^3)(-3b^2c)(-5a^2b)$

14. $(ab^2c)(-ac^2)(-0.3bc^3)(-2.5a^3c^2)$

15. $(-2abc)(-3a^2b^2c^2)(-4c)(-b^2c)$

16. $(xy)(yz)(xz)(yz)$

B. Find.

17. $\dfrac{x^7}{x^3}$

18. $\dfrac{8a^3}{4a^2}$

19. $\dfrac{-8a^4}{16a^2}$

20. $\dfrac{9y^5}{6y^2}$

21. $\dfrac{12x^5y^3}{6x^2y}$

22. $\dfrac{18x^6y^2}{9xy}$

23. $\dfrac{-6x^2y}{12x^3y}$

24. $\dfrac{8x^3y^4}{4x^5y^2}$

25. $\dfrac{-14a^3y^6}{-21a^5y^2}$

26. $\dfrac{-2a^2y^3}{-6a^2y^3}$

27. $\dfrac{-27a^2b^3c}{-36ab^5c^2}$

28. $\dfrac{-5x^6y^2z^5}{10x^2y^5z^2}$

1. _____
2. _____
3. _____
4. _____
5. _____
6. _____
7. _____
8. _____
9. _____
10. _____
11. _____
12. _____
13. _____
14. _____
15. _____
16. _____

17. _____
18. _____
19. _____
20. _____
21. _____
22. _____
23. _____
24. _____
25. _____
26. _____
27. _____
28. _____

29. $\dfrac{3a^3 \cdot a^5}{2a^4}$

30. $\dfrac{y^2 \cdot y^8}{y \cdot y^3}$

31. $\dfrac{(2x^2y^3)(-3x^5y)}{6xy^3}$

32. $\dfrac{(-3x^3y^2z)(4xy^3z)}{6xy^2z}$

33. $\dfrac{(-x^2y)(x^3y^2)}{x^3y}$

34. $\dfrac{(-8x^2y)(-7x^5y^3)}{-2x^2y^3}$

29. _____

30. _____

31. _____

32. _____

33. _____

34. _____

C. Applications

35. The first manufactured object to leave the solar system was the spacecraft *Pioneer 10*. On its way to Jupiter this spacecraft reached a velocity of 5.1×10^4 kilometers per hour (km/hr) when leaving the earth. After 10 mo it traveled $(5.1 \times 10^4) \times (7.2 \times 10^3) = (5.1 \times 7.2)(10^4 \times 10^3)$ km. Write $(5.1 \times 7.2) \times (10^4 \times 10^3)$ km as a whole number.

35. _____

36. The top speed of *Pioneer 10* was 1.31×10^5 km/hr. In 10 yr $(8.7 \times 10^4$ hr), *Pioneer 10* would travel $(1.31 \times 10^5) \times (8.7 \times 10^4)$ km. Write $(1.31 \times 10^5) \times (8.7 \times 10^4)$ as a whole number.

36. _____

37. The estimated reserves of petroleum and natural gas in the United States amount to about 2.8×10^{17} kilocalories (kcal) of energy. Unfortunately, we are consuming these fuels at the rate of about 1.4×10^{16} kcal per year, which means that our reserves will last

$$\dfrac{2.8 \times 10^{17}}{1.4 \times 10^{16}} \text{ yr}$$

How many years is that? (Write the answer as a whole number.)

37. _____

38. The world reserves of petroleum and natural gas amount to about 420×10^{16} kcal. Each year, 5.4×10^{16} kcal are used. At this rate of consumption, the world reserves would last

$$\dfrac{420 \times 10^{16}}{5.4 \times 10^{16}} \text{ yr}$$

How many years is that? (Answer to the nearest year.)

38. _____

39. Nuclear fusion might be the ideal energy source. One of these reactions (using deuterium and tritium) would produce

$$\dfrac{4 \times 10^8}{5} \text{ kcal of energy per gram of fuel}$$

How many kilocalories is that? (Write the answer as a whole number.)

39. _____

40. The mass of the earth is about 1.2×10^{25} lb. Jupiter is about 3×10^2 times more massive than the earth. What is the mass of Jupiter? Write the answer $(1.2 \times 10^{25}) \times (3 \times 10^2)$ using exponents.

40. _____

41. Our solar system is about 3×10^4 light-years from the center of the Milky Way Galaxy. Since a light year is about 6×10^{12} miles, our distance from the center of the galaxy is $(3 \times 10^4)(6 \times 10^{12})$ miles. How many miles is that? (Write the answer using exponents.)

42. The orbital velocity of the planet Mars is 5.4×10^4 mi/hr. How far would Mars go in 8.7×10^3 hr (about 1 yr)? (Write the answer using exponents.)

43. The orbital velocity of Mercury is 1.1×10^5 mi/hr. How far would Mercury go in 8.7×10^3 hr (about 1 yr)? (Write the answer using exponents.)

44. The U.S. population is about 2.5×10^8. If the average American eats about 600 lb of dairy products a year, what is the total amount of dairy products consumed each year in the United States? (Write the answer using exponents.)

45. The U.S. population is about 2.5×10^8. If the average American eats 200 lb of fruits and vegetables a year, what is the total amount of fruits and vegetables consumed each year in the United States? (Write the answer using exponents.)

41. _____

42. _____

43. _____

44. _____

45. _____

✓ **SKILL CHECKER**

Find.

46. 6^2

47. 4^3

Simplify.

48. $[x - (x + 6)] + [3 + 4(x - 1)]$

49. $[x^2 - (x^2 + 1)] + [5 + 4(x^2 - 1)]$

50. $[3x^2 - (x^2 - 1)] - [6 + 3(x^2 - 2)]$

46. _____

47. _____

48. _____

49. _____

50. _____

2.7 USING YOUR KNOWLEDGE

There are many interesting patterns involving exponents. Use your knowledge to find the answers to the following questions.

1. $1^2 = 1$

  $(11)^2 = 121$

  $(111)^2 = 12{,}321$

  $(1111)^2 = 1{,}234{,}321$

  a. Find $(11{,}111)^2$.
  b. Find $(111{,}111)^2$.

2. $1^2 = 1$
  $2^2 = 1 + 2 + 1$
  $3^2 = 1 + 2 + 3 + 2 + 1$
  $4^2 = 1 + 2 + 3 + 4 + 3 + 2 + 1$

  a. Use this pattern to write 5^2.
  b. Use this pattern to write 6^2.

1. a. _____
   b. _____

2. a. _____
   b. _____

3.
$$1 + 3 = 2^2$$
$$1 + 3 + 5 = 3^2$$
$$1 + 3 + 5 + 7 = 4^2$$

a. Find $1 + 3 + 5 + 7 + 9$.
b. Find $1 + 3 + 5 + 7 + 9 + 11 + 13$.

4. In this problem, discover your own pattern. What is the largest number you can construct by using the number 9 three times? (It is not 999!)

3. a. _____
   b. _____

4. _____

2.8 INTEGERS AS EXPONENTS

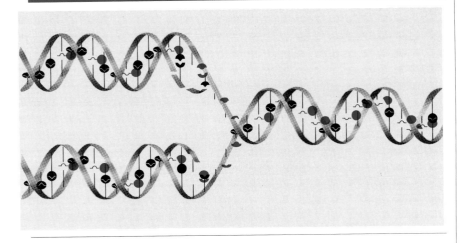

OBJECTIVES

REVIEW

Before starting this section you should know:

1. How to raise numbers to a power.
2. The laws of exponents given on pages 148, 150–151.

OBJECTIVES

You should be able to:

A. Write a number involving a negative exponent as a fraction and vice versa.
B. Write a fraction involving exponents as a number with a negative power.
C. Multiply expressions involving negative exponents.
D. Divide expressions involving negative exponents.
E. Use negative exponents when raising a power to a power.
F. Solve applications using the concepts studied.

A. NEGATIVE EXPONENTS

In science and technology, negative numbers are used as exponents. For example, the diameter of a DNA molecule is 10^{-8} meter, and the time it takes for an electron to go from source to screen in a TV tube is 10^{-6} second. What do 10^{-8} and 10^{-6} mean? Look at the pattern obtained by *dividing by 10* in each step.

$10^3 = 1000$ Note that the exponents decrease by 1 at each step.

$10^2 = 100$

$10^1 = 10$

$10^0 = 1$

$$10^{-1} = \frac{1}{10} = \frac{1}{10}$$

$$10^{-2} = \frac{1}{100} = \frac{1}{10^2}$$

$$10^{-3} = \frac{1}{1000} = \frac{1}{10^3}$$

As you can see, this procedure yields $10^0 = 1$. In general, we make the following definition.

For $x \neq 0$, $x^0 = 1$

Thus, $5^0 = 1$, $8^0 = 1$, and $6^0 = 1$.

Now look at the numbers in color. Note that we obtained

$$10^{-1} = \frac{1}{10}, \qquad 10^{-2} = \frac{1}{10^2}, \quad \text{and} \quad 10^{-3} = \frac{1}{10^3}$$

Thus, we make the following definition:

If n is a positive integer

$$x^{-n} = \frac{1}{x^n}, \qquad x \neq 0$$

This definition says that x^{-n} and x^n are reciprocals, since $x^{-n} \cdot x^n = \frac{1}{x^n} \cdot x^n = 1$. By definition,

$$5^{-2} = \frac{1}{5^2} = \frac{1}{5 \cdot 5} = \frac{1}{25}$$

and

$$2^{-3} = \frac{1}{2^3} = \frac{1}{2 \cdot 2 \cdot 2} = \frac{1}{8}$$

Similarly,

$$\frac{1}{4^2} = 4^{-2}$$

and

$$\frac{1}{3^4} = 3^{-4}$$

EXAMPLE 1   Write as a fraction.

a. 6^{-2} **b.** 4^{-3}

Solution

a. $6^{-2} = \frac{1}{6^2} = \frac{1}{6 \cdot 6} = \frac{1}{36}$

b. $4^{-3} = \frac{1}{4^3} = \frac{1}{4 \cdot 4 \cdot 4} = \frac{1}{64}$

Problem 1   Write as a fraction.
a. 5^{-2} **b.** 3^{-3}

B.   FROM FRACTIONS TO NEGATIVE EXPONENTS

EXAMPLE 2   Write using negative exponents.

a. $\frac{1}{5^4}$ **b.** $\frac{1}{7^5}$

Solution   Using the definition, we have the following.

a. $\frac{1}{5^4} = 5^{-4}$

b. $\frac{1}{7^5} = 7^{-5}$

Problem 2   Write using negative exponents.
a. $\frac{1}{7^4}$ **b.** $\frac{1}{6^5}$

C. MULTIPLYING EXPRESSIONS WITH NEGATIVE EXPONENTS

In the preceding section, we multiplied expressions containing positive exponents. For example,

$$x^5 \cdot x^8 = x^{5+8} = x^{13}$$

and

$$y^2 \cdot y^3 = y^{2+3} = y^5$$

Can we multiply expressions involving negative exponents using the same idea? Let us see.

$$x^5 \cdot x^{-2} = x^5 \cdot \frac{1}{x^2} = \frac{x^5}{x^2} = x^3$$

Adding exponents,

$$x^5 \cdot x^{-2} = x^{5+(-2)} = x^3 \quad \text{Same answer!}$$

Similarly,

$$x^{-3} \cdot x^{-2} = \frac{1}{x^3} \cdot \frac{1}{x^2} = \frac{1}{x^{3+2}} = \frac{1}{x^5} = x^{-5}$$

Adding exponents,

$$x^{-3} \cdot x^{-2} = x^{-3+(-2)} = x^{-5} \quad \text{Same answer again!}$$

We state the resulting law here for your convenience.

FIRST LAW OF EXPONENTS

If m and n are integers,

$$x^m \cdot x^n = x^{m+n}$$

This law says that when *multiplying* expressions with the same bases, we keep the base and *add* exponents. Note that the law does *not* apply to $x^7 \cdot y^6$ because the bases are different.

EXAMPLE 3 Multiply and simplify.

a. $2^6 \cdot 2^{-4}$ **b.** $4^3 \cdot 4^{-5}$ **c.** $y^{-2} \cdot y^{-3}$ **d.** $a^{-5} \cdot a^5$

Solution

a. $2^6 \cdot 2^{-4} = 2^{6+(-4)} = 2^2 = 4$

b. $4^3 \cdot 4^{-5} = 4^{3+(-5)} = 4^{-2} = \frac{1}{4^2} = \frac{1}{16}$

Note that we wrote the answer *without* using negative exponents. In algebra, it is customary to write the answer without negative exponents.

c. $y^{-2} \cdot y^{-3} = y^{-2+(-3)} = y^{-5} = \frac{1}{y^5}$

Again, we wrote the answer without negative exponents.

d. $a^{-5} \cdot a^5 = a^{-5+5} = a^0 = 1$

Problem 3 Multiply.
a. $2^7 \cdot 2^{-3}$ **b.** $3^4 \cdot 3^{-7}$
c. $x^{-4} \cdot x^{-2}$ **d.** $y^3 \cdot y^{-3}$

D. DIVIDING EXPRESSIONS WITH NEGATIVE EXPONENTS

In the preceding section, we divided expressions with the same base. Thus,

$$\frac{7^5}{7^2} = 7^{5-2} = 7^3 \quad \text{and} \quad \frac{8^3}{8} = 8^{3-1} = 8^2$$

The law used then can be extended to any integers.

SECOND LAW OF EXPONENTS

If m and n are integers,

$$\frac{x^m}{x^n} = x^{m-n}$$

This law says that when *dividing* expressions with the *same* bases, we keep the base and *subtract* exponents. Note that

$$\frac{x^m}{x^n} = x^m \cdot \frac{1}{x^n} = x^m \cdot x^{-n} = x^{m-n}$$

This second law of exponents can be used *instead* of the three laws given on pages 150–151.

EXAMPLE 4 Find.

a. $\dfrac{6^5}{6^{-2}}$ **b.** $\dfrac{x}{x^5}$ **c.** $\dfrac{y^{-2}}{y^{-2}}$ **d.** $\dfrac{z^{-3}}{z^{-4}}$

Solution

a. $\dfrac{6^5}{6^{-2}} = 6^{5-(-2)} = 6^{5+2} = 6^7$

b. $\dfrac{x}{x^5} = x^{1-5} = x^{-4} = \dfrac{1}{x^4}$

c. $\dfrac{y^{-2}}{y^{-2}} = y^{-2-(-2)} = y^{-2+2} = y^0 = 1$

d. $\dfrac{z^{-3}}{z^{-4}} = z^{-3-(-4)} = z^{-3+4} = z^1 = z$

E. RAISING A POWER TO A POWER

Suppose we wish to find $(5^3)^2$. By definition,

$$(5^3)^2 = 5^3 \cdot 5^3 = 5^{3+3}, \quad \text{or} \quad 5^6$$

We could get this answer by multiplying exponents in $(5^3)^2$. Similarly,

$$(4^{-2})^3 = \frac{1}{4^2} \cdot \frac{1}{4^2} \cdot \frac{1}{4^2} = \frac{1}{4^6} = 4^{-6}$$

Again, we could have multiplied exponents in $(4^{-2})^3$ to get 4^{-6}. We use these ideas to state the following law.

Problem 4 Find.

a. $\dfrac{7^5}{7^{-3}}$ **b.** $\dfrac{y}{y^6}$ **c.** $\dfrac{x^{-3}}{x^{-3}}$ **d.** $\dfrac{z^{-4}}{z^{-5}}$

If m and n are integers,

$$(x^m)^n = x^{mn}$$

This law says that when raising a *power* to a *power,* we keep the base and *multiply* exponents.

EXAMPLE 5 Simplify.

a. $(2^3)^4$ **b.** $(x^{-2})^3$ **c.** $(y^4)^{-5}$ **d.** $(z^{-2})^{-3}$

Solution

a. $(2^3)^4 = 2^{3 \cdot 4} = 2^{12}$

b. $(x^{-2})^3 = x^{-2 \cdot 3} = x^{-6} = \dfrac{1}{x^6}$

c. $(y^4)^{-5} = y^{4(-5)} = y^{-20} = \dfrac{1}{y^{20}}$

d. $(z^{-2})^{-3} = z^{-2(-3)} = z^6$ ▲

Sometimes we need to raise several factors inside parentheses to a power, such as in $(x^2y^3)^3$. We use the definition of cubing and write:

$$
\begin{aligned}
(x^2y^3)^3 &= x^2y^3 \cdot x^2y^3 \cdot x^2y^3 \\
&= (x^2 \cdot x^2 \cdot x^2)(y^3 \cdot y^3 \cdot y^3) \\
&= (x^2)^3(y^3)^3 \\
&= x^6y^9
\end{aligned}
$$

We could get the same answer by multiplying each of the exponents in x^2y^3 by 3, obtaining $x^{2 \cdot 3}y^{3 \cdot 3} = x^6y^9$. Thus, to raise several factors inside parentheses to a power, we raise each factor to the given power, as shown next.

If m, n, and k are integers,

$$(x^my^n)^k = (x^m)^k(y^n)^k = x^{mk}y^{nk}$$

EXAMPLE 6 Simplify.

a. $(3x^2y^{-2})^3$ **b.** $(2x^{-2}y^3)^3$ **c.** $(3x^{-2}y^3)^{-2}$ **d.** $(-2x^{-3}y^2)^{-3}$

Solution

a. $(3x^2y^{-2})^3 = 3^3(x^2)^3(y^{-2})^3$ Note that since 3 is a *factor* in $3x^2y^{-2}$,
$$= 27x^6y^{-6} \qquad \text{it is also raised to the third power.}$$
$$= \dfrac{27x^6}{y^6}$$

b. $(2x^{-2}y^3)^3 = 2^3(x^{-2})^3(y^3)^3$
$$= 8x^{-6}y^9$$
$$= \dfrac{8y^9}{x^6}$$

Problem 5 Simplify.
a. $(5^3)^4$ **b.** $(x^{-3})^4$
c. $(y^3)^{-6}$ **d.** $(z^{-3})^{-5}$

Problem 6 Simplify.
a. $(2x^2y^{-2})^3$ **b.** $(3x^{-3}y^2)^3$
c. $(2x^3y^2)^{-2}$ **d.** $(-3x^{-4}y^3)^{-3}$

c. $(3x^{-2}y^3)^{-2} = 3^{-2}(x^{-2})^{-2}(y^3)^{-2}$

$$= \frac{1}{3^2} x^4 y^{-6}$$

$$= \frac{1x^4}{9y^6}$$

$$= \frac{x^4}{9y^6}$$

d. $(-2x^{-3}y^2)^{-3} = (-2)^{-3}(x^{-3})^{-3}(y^2)^{-3}$

$$= \frac{1}{(-2)^3} x^9 y^{-6}$$

$$= \frac{x^9}{(-2)^3 y^6}$$

$$= \frac{x^9}{-8y^6}$$

$$= -\frac{x^9}{8y^6}$$

F. APPLICATIONS: COMPOUND INTEREST

Suppose you invest P dollars at 10% compounded annually. At the end of 1 yr you will have your original principal P plus the interest (10%)P, that is, $P + 0.10P = (1 + 0.10)P = 1.10P$. At the end of 2 yr, you will have $1.10P$ plus the interest earned on the $1.10P$, that is, $1.10P + 0.10(1.10P) = 1.10P(1 + 0.10) = 1.10P(1.10)$, or $(1.10)^2 P$. If you follow this pattern, at the end of 3 yr you will have $(1.10)^3 P$, and so on. Here is the general formula:

> If the principal P is invested at rate r, compounded annually, in n years the compound amount A will be
>
> $A = P(1 + r)^n$

EXAMPLE 7 If $1000 is invested at 10% compounded annually, how much will be in the account at the end of 3 yr?

Solution Here $P = \$1000$, $r = 10\% = 0.10$, and $n = 3$. We have

$$A = 1000(1 + 0.10)^3$$
$$= 1000(1.10)^3 = 1000(1.331) = \$1331$$

Problem 7 If $500 is invested at 10% compounded annually, how much will be in the account at the end of 2 yr?

NAME

CLASS

SECTION

ANSWERS

A. Write as a fraction.

1. 4^{-2} 2. 2^{-3}

3. 5^{-3} 4. 7^{-2}

5. 3^{-4} 6. 6^{-3}

1. _____

2. _____

3. _____

4. _____

5. _____

6. _____

B. Write using negative exponents.

7. $\dfrac{1}{2^3}$ 8. $\dfrac{1}{3^4}$

9. $\dfrac{1}{4^5}$ 10. $\dfrac{1}{5^6}$

11. $\dfrac{1}{3^5}$ 12. $\dfrac{1}{7^4}$

7. _____

8. _____

9. _____

10. _____

11. _____

12. _____

C. Multiply and simplify. (Write answers without negative exponents.)

13. $3^5 \cdot 3^{-4}$ 14. $4^{-6} \cdot 4^8$

15. $2^{-5} \cdot 2^7$ 16. $3^8 \cdot 3^{-5}$

17. $4^{-6} \cdot 4^4$ 18. $5^{-4} \cdot 5^2$

19. $6^{-1} \cdot 6^{-2}$ 20. $3^{-2} \cdot 3^{-1}$

21. $2^{-4} \cdot 2^{-2}$ 22. $4^{-1} \cdot 4^{-2}$

23. $x^6 \cdot x^{-4}$ 24. $y^7 \cdot y^{-2}$

25. $y^{-3} \cdot y^5$ 26. $x^{-7} \cdot x^8$

13. _____

14. _____

15. _____

16. _____

17. _____

18. _____

19. _____

20. _____

21. _____

22. _____

23. _____

24. _____

25. _____

26. _____

27. $a^3 \cdot a^{-8}$

28. $b^4 \cdot b^{-7}$

29. $x^{-5} \cdot x^3$

30. $y^{-6} \cdot y^2$

31. $x \cdot x^{-3}$

32. $y \cdot y^{-5}$

33. $a^{-2} \cdot a^{-3}$

34. $b^{-5} \cdot b^{-2}$

35. $b^{-3} \cdot b^3$

36. $a^6 \cdot a^{-6}$

27. _____

28. _____

29. _____

30. _____

31. _____

32. _____

33. _____

34. _____

35. _____

36. _____

D. Divide and simplify.

37. $\dfrac{3^4}{3^{-1}}$

38. $\dfrac{2^2}{2^{-2}}$

39. $\dfrac{4^{-1}}{4^2}$

40. $\dfrac{3^{-2}}{3^3}$

41. $\dfrac{y}{y^3}$

42. $\dfrac{x}{x^4}$

43. $\dfrac{x}{x^{-2}}$

44. $\dfrac{y}{y^{-3}}$

45. $\dfrac{x^{-3}}{x^{-1}}$

46. $\dfrac{x^{-4}}{x^{-2}}$

47. $\dfrac{x^{-3}}{x^4}$

48. $\dfrac{y^{-4}}{y^5}$

49. $\dfrac{x^{-2}}{x^{-5}}$

50. $\dfrac{y^{-3}}{y^{-6}}$

37. _____

38. _____

39. _____

40. _____

41. _____

42. _____

43. _____

44. _____

45. _____

46. _____

47. _____

48. _____

49. _____

50. _____

E. Simplify.

51. $(3^2)^2$

52. $(2^3)^2$

53. $(3^{-1})^2$

54. $(2^{-2})^2$

55. $(2^{-2})^{-3}$

56. $(3^{-1})^{-2}$

51. _____

52. _____

53. _____

54. _____

55. _____

56. _____

ANSWER   (to problem on page 162)
7. $605

57. $(3^2)^{-1}$ **58.** $(2^3)^{-2}$

59. $(x^3)^{-3}$ **60.** $(y^2)^{-4}$

61. $(y^{-3})^2$ **62.** $(x^{-4})^3$

63. $(a^{-2})^{-3}$ **64.** $(b^{-3})^{-5}$

65. $(2x^3y^{-2})^3$ **66.** $(3x^2y^{-3})^2$

67. $(2x^{-2}y^3)^2$ **68.** $(3x^{-4}y^4)^3$

69. $(-3x^3y^2)^{-3}$ **70.** $(-2x^5y^4)^{-4}$

71. $(x^{-6}y^{-3})^2$ **72.** $(y^{-4}z^{-3})^5$

73. $(x^{-4}y^{-4})^{-3}$ **74.** $(y^{-5}z^{-3})^{-4}$

57. _____
58. _____
59. _____
60. _____
61. _____
62. _____
63. _____
64. _____
65. _____
66. _____
67. _____
68. _____
69. _____
70. _____
71. _____
72. _____
73. _____
74. _____

F. Applications

75. If $1000 is invested at 8% compounded annually, how much will be in the account at the end of 3 yr?

76. If $500 is invested at 10% compounded annually, how much will be in the account at the end of 3 yr?

77. Suppose $1000 is invested at 10% compounded annually. How much will be in the account at the end of 4 yr?

75. _____

76. _____

77. _____

✓ **SKILL CHECKER**

Find.

78. 7.31×10^1

79. 7.31×10^2

80. 7.31×10^3

78. _____
79. _____
80. _____

2.8 USING YOUR KNOWLEDGE

The idea of exponents can be used to measure population growth. Thus, if we assume that the world population is increasing about 2% each year (experts say the rate is between 2% and 4%), we can predict the population of the world next year by *multiplying* the present world population by 1.02 ($100\% + 2\% = 102\% = 1.02$). If we let the world population be P, we have:

Population in 1 year $= 1.02P$

Population in 2 years $= 1.02(1.02P) = (1.02)^2 P$

Population in 3 years $= 1.02(1.02)^2 P = (1.02)^3 P$

1. If the population P in 1988 is 5.128 billion people, what will it be in 2 yr, in 1990? (Give the answer to three decimal places.)

1. _____

2. What will the population be in 5 yr? (Give the answer to three decimal places.)

2. _____

To find the population 1 yr hence, we multiply by 1.02. What shall we do to find the population one year *ago*? We divide by 1.02. Thus, if the population today is P:

Population 1 year ago $= \dfrac{P}{1.02} = P \cdot 1.02^{-1}$

Population 2 years ago $= \dfrac{P \cdot 1.02^{-1}}{1.02} = P \cdot 1.02^{-2}$

Population 3 years ago $= \dfrac{P \cdot 1.02^{-2}}{1.02} = P \cdot 1.02^{-3}$

3. If the population P in 1988 is 5.128 billion people, what was it 2 yr ago? (Give the answer to three decimal places.)

3. _____

4. What was the population 5 yr ago? (Give the answer to three decimal places.)

4. _____

2.9 SCIENTIFIC NOTATION

REVIEW

Before starting this section you should know the laws of exponents.

OBJECTIVES

You should be able to:

A. Convert between ordinary decimal notation and scientific notation.
B. Multiply and divide numbers in scientific notation.
C. Solve applications using the concepts studied.

How many facts do you know about the sun? Here is some information taken from an encyclopedia article.

Mass: 2.19×10^{27} tons

Temperature: 1.8×10^6 degrees Fahrenheit

Energy per minute: 2.4×10^4 hp

All numbers involved are written as products of a number between 1 and 10 and an appropriate power of 10. This form is called *scientific notation*.

A. SCIENTIFIC NOTATION

A number in **scientific notation** is written as

$M \times 10^n$ M = a number between 1 and 10
               n = an integer

How do we change a whole number to scientific notation? First, recall that when we *multiply* a number by a power of 10 ($10^1 = 10$, $10^2 = 100$, and so on) we simply move the decimal point as many places to the *right* as indicated by the exponent of 10. Thus,

$7.31 \times 10^1 = 7.31 = 73.1$ Exponent 1; move 1 place right

$72.813 \times 10^2 = 72.813 = 7281.3$ Exponent 2; move 2 places right

$160.7234 \times 10^3 = 160.7234 = 160,723.4$ Exponent 3; move 3 places right

On the other hand, if we divide a number by a power of 10, we move the decimal point as many places to the left as indicated by the exponent of 10. Thus,

$$\frac{7}{10} = 0.7 = 7 \times 10^{-1}$$

$$\frac{8}{100} = 0.08 = 8 \times 10^{-2}$$

and

$$\frac{4.7}{100,000} = 0.000047 = 4.7 \times 10^{-5}$$

Remembering these rules makes it easy to write a number in scientific notation.

TO WRITE A WHOLE NUMBER IN SCIENTIFIC NOTATION

($M \times 10^n$)

1. Move the decimal point in the given number so that there is only one digit to its left. The resulting number is M.

2. Count how many places you have to move the decimal point in Step 1. If the decimal point must be moved to the *left*, n is *positive;* if it must be moved to the *right*, n is *negative*.

3. Write $M \times 10^n$.

For example,

$5.3 = 5.3 \times 10^0$	The decimal point in 5.3 must be moved 0 places.
$87 = 8.7 \times 10^1 = 8.7 \times 10$	The decimal point in 87 must be moved 1 place *left* to get 8.7.
$68,000 = 6.8 \times 10^4$	The decimal point in 68,000 must be moved 4 places *left* to get 6.8.
$0.49 = 4.9 \times 10^{-1}$	The decimal point in 0.49 must be moved 1 place *right* to get 4.9.
$0.072 = 7.2 \times 10^{-2}$	The decimal point in 0.072 must be moved 2 places *right* to get 7.2.

EXAMPLE 1 The approximate distance to the sun is 93,000,000 mi and the wavelength of its ultraviolet light is 0.000035 cm. Write 93,000,000 and 0.000035 in scientific notation.

Solution

$$93,000,000 = 9.3 \times 10^7$$

$$0.000035 = 3.5 \times 10^{-5}$$

EXAMPLE 2 A jumbo jet weighs 7.75×10^5 lb, whereas a house spider weighs 2.2×10^{-4} lb. Write 7.75×10^5 and 2.2×10^{-4} in standard notation.

Solution

$$7.75 \times 10^5 = 775,000$$

$$2.2 \times 10^{-4} = 0.00022$$

Problem 1 The distance to the moon is about 239,000 mi and its mass is 0.123456 times that of the earth. Write 239,000 and 0.123456 in scientific notation.

Problem 2 The Concorde weighs 4.08×10^5 lb and a cricket weighs 3.125×10^{-4} lb. Write 4.08×10^5 and 3.125×10^{-4} in standard notation.

B. MULTIPLYING AND DIVIDING USING SCIENTIFIC NOTATION

Consider the product $300 \cdot 2000 = 600{,}000$. In scientific notation, we would write

$$(3 \times 10^2) \cdot (2 \times 10^3) = 6 \times 10^5$$

To find the answer, we can multiply 3 by 2 to obtain 6 and 10^2 by 10^3, obtaining 10^5. To multiply numbers in scientific notation, we proceed in a similar manner:

1. Multiply the decimal parts first and write the result in scientific notation.
2. Multiply the powers of 10.
3. The answer is the product obtained in Steps 1 and 2 after simplification.

EXAMPLE 3 Multiply.

a. $(5 \times 10^3) \times (8.1 \times 10^4)$ **b.** $(3.2 \times 10^2) \times (4 \times 10^{-5})$

Solution

a. We multiply the decimal part first.

$5 \times 8.1 = 40.5 = 4.05 \times 10$

Then multiply the powers of 10.

$10^3 \times 10^4 = 10^7$ (Adding exponents)

The answer is $(4.05 \times 10) \times 10^7$, or 4.05×10^8.

b. Multiply the decimals.

$3.2 \times 4 = 12.8 = 1.28 \times 10$

Multiply the powers of 10.

$10^2 \times 10^{-5} = 10^{2-5} = 10^{-3}$

The answer is $(1.28 \times 10) \times 10^{-3}$, or $1.28 \times 10^{1+(-3)} = 1.28 \times 10^{-2}$.

▲

Division is done in the same manner. For example, $\dfrac{3.2 \times 10^5}{1.6 \times 10^2}$ is found by dividing 3.2 by 1.6 (yielding 2) and 10^5 by 10^2, which is 10^3. The answer is 2×10^3.

EXAMPLE 4 Find $(1.24 \times 10^{-2}) \div (3.1 \times 10^{-3})$.

Solution First divide 1.24 by 3.1, obtaining $0.4 = 4 \times 10^{-1}$. Now divide powers of 10:

$$10^{-2} \div 10^{-3} = 10^{-2-(-3)}$$
$$= 10^{-2+3}$$
$$= 10^1$$

The answer is $(4 \times 10^{-1}) \times 10^1 = 4 \times 10^0 = 4$.

Problem 3 Multiply.
a. $(6 \times 10^4) \times (2.2 \times 10^3)$
b. $(4.1 \times 10^2) \times (3 \times 10^{-5})$

Problem 4 Find
$(2.52 \times 10^{-2}) \div (4.2 \times 10^{-3})$.

C. APPLICATIONS

EXAMPLE 5   The total energy received from the sun each minute is 1.02×10^{19} calories. Since the area of the earth is 5.1×10^{18} cm^2, the amount of energy received per square centimeter of earth surface every minute (the solar constant) is:

$$\frac{1.02 \times 10^{19}}{5.1 \times 10^{18}}$$

Simplify this expression.

Solution   Dividing 1.02 by 5.1, we obtain $0.2 = 2 \times 10^{-1}$. Now, $10^{19} \div 10^{18} = 10^{19-18} = 10^1$. Thus, the final answer is

$$(2 \times 10^{-1}) \times 10^1 = 2 \times 10^0 = 2$$

This means that the earth receives about 2 calories of heat per square centimeter each minute.

Problem 5   The width of the asteroid belt is 2.8×10^8 km. The speed of *Pioneer 10* in passing through this belt was 1.4×10^5 km/hr. Thus, *Pioneer 10* took $\dfrac{2.8 \times 10^8}{1.4 \times 10^5}$ hr to go through the belt. How many hours is that?

EXERCISE 2.9

ANSWERS

A. Write in scientific notation.

1. 54,000,000 (working women in the United States)

2. 68,000,000 (working men in the United States)

3. 248,000,000 (U.S. population now)

4. 268,000,000 (estimated U.S. population in the year 2000)

5. 1,900,000,000 (dollars spent on water beds and accessories in 1 yr)

6. 0.035 (ounces in a gram)

7. 0.00024 (probability of four-of-a-kind in poker)

8. 0.000005 (the gram-weight of an amoeba)

9. 0.000000002 (the gram-weight of one liver cell)

10. 0.00000009 (wavelength of an X ray in centimeters)

1. _____

2. _____

3. _____

4. _____

5. _____

6. _____

7. _____

8. _____

9. _____

10. _____

Write in standard notation.

11. 1.53×10^2 (pounds of meat consumed per person per year in the United States)

12. 5.96×10^2 (pounds of dairy products consumed per person per year in the United States)

13. 8×10^6 (bagels eaten per day in the United States)

14. 2.01×10^6 (estimated number of jobs created in service industries between now and the year 2000)

15. 6.85×10^9 (estimated worth, in dollars, of the five wealthiest women)

16. 1.962×10^{10} (estimated worth, in dollars, of the five wealthiest men)

17. 2.3×10^{-1} (kilowatts per hour used by your TV)

18. 4×10^{-2} (inches in 1 mm)

19. 2.5×10^{-4} (thermal conductivity of glass)

20. 4×10^{-11} (energy, in joules, released by splitting one uranium atom)

11. _____

12. _____

13. _____

14. _____

15. _____

16. _____

17. _____

18. _____

19. _____

20. _____

B. Perform the indicated operations (give answer in scientific notation).

21. $(3 \times 10^4) \times (5 \times 10^5)$ **22.** $(5 \times 10^2) \times (3.5 \times 10^3)$

21. _____

22. _____

23. $(6 \times 10^{-3}) \times (5.1 \times 10^6)$ **24.** $(3 \times 10^{-2}) \times (8.2 \times 10^5)$

25. $(4 \times 10^{-2}) \times (3.1 \times 10^{-3})$ **26.** $(3.1 \times 10^{-3}) \times (4.2 \times 10^{-2})$

27. $\dfrac{4.2 \times 10^5}{2.1 \times 10^2}$ **28.** $\dfrac{5 \times 10^6}{2 \times 10^3}$

29. $\dfrac{2.2 \times 10^4}{8.8 \times 10^6}$ **30.** $\dfrac{2.1 \times 10^3}{8.4 \times 10^5}$

23. _____

24. _____

25. _____

26. _____

27. _____

28. _____

29. _____

30. _____

C. Applications

31. The average American eats 80 lb of vegetables each year. Since there are about 250 million Americans, the number of pounds of vegetables consumed each year should be $(8 \times 10^1) \times (2.5 \times 10^8)$.
   a. Write this number in scientific notation.
   b. Write this number in standard notation.

31. a. _____
   b. _____

32. The average American drinks 44.8 gal of soft drinks each year. Since there are about 250 million Americans, the number of gallons of soft drinks consumed each year should be $(4.48 \times 10^2) \times (2.5 \times 10^8)$.
   a. Write this number in scientific notation.
   b. Write this number in standard notation.

32. a. _____
   b. _____

33. America produces 148.5 million tons of garbage each year. Since a ton is 2000 lb, and there are about 360 days in a year and 250 million Americans, the number of pounds of garbage produced each day of the year for every man, woman, and child in America is

$$\frac{(1.485 \times 10^8) \times (2 \times 10^3)}{(2.5 \times 10^8) \times (3.6 \times 10^2)}$$

Write this number in standard notation.

33. _____

34. The velocity of light can be measured by dividing the distance from the sun to the earth $(1.47 \times 10^{11}$ m) by the time it takes for sunlight to reach the earth $(4.9 \times 10^2$ sec). Thus, the velocity of light is

$$\frac{1.47 \times 10^{11}}{4.9 \times 10^2}$$

How many meters per second is that?

34. _____

35. Nuclear fission is used as an energy source. Do you know how much energy a gram of uranium 235 gives? The answer is

$$\frac{4.7 \times 10^9}{235} \text{ kilocalories}$$

Write this number in scientific notation.

35. _____

ANSWER (to problem on page 170)
5. $2 \times 10^3 = 2000$

EXERCISE 2.9

✓ SKILL CHECKER

Find.

36. $\dfrac{1}{4} + \dfrac{1}{2}$ **37.** $\dfrac{1}{2} + \dfrac{3}{5}$

38. $\dfrac{0.9}{-0.3}$ **39.** $\dfrac{0.16}{-0.04}$

40. $\dfrac{-0.8}{0.4}$

36. _____

37. _____

38. _____

39. _____

40. _____

2.9 USING YOUR KNOWLEDGE

Scientific notation is especially useful when very large quantities are involved. For example, in astronomy we find that the speed of light is 299,792,458 m/sec.

1. Write 299,792,458 in scientific notation.

1. _____

Astronomical distances are so large that they are measured in astronomical units (A.U.). An astronomical unit is defined as the average separation (distance) of the earth and the sun, that is, 150,000,000 km.

2. Write 150,000,000 in scientific notation.

2. _____

3. Distances in astronomy are also measured in **parsecs:** 1 parsec = 2.06×10^5 A.U. Thus, 1 parsec = $(2.06 \times 10^5) \times (1.5 \times 10^8)$ km. Written in scientific notation, how many kilometers is that?

3. _____

4. Astronomers also measure distances in **light years,** the distance light travels in 1 yr: 1 light year = 9.46×10^{12} km. The closest star, Proxima Centauri, is 4.22 light years away. In scientific notation using two decimal places, how many kilometers is that?

4. _____

5. Since 1 parsec = 3.09×10^{13} km (see Exercise 3) and 1 light year = 9.46×10^{12} km, the number of light years in a parsec is

$$\frac{3.09 \times 10^{13}}{9.46 \times 10^{12}}$$

Write this number in standard notation using two decimal places.

5. _____

CALCULATOR CORNER

If you have a scientific calculator and you multiply 9,800,000 by 4,500,000, the display will show

    4.41   13

This means that the answer is 4.41×10^{13}.

1. The display on a calculator shows

    3.34   5

Write this number in scientific notation.

1. _____

2. The display on a calculator shows

$$-9.97 \quad -6$$

Write this number in scientific notation.

2. _____

3. To enter large or small numbers in a calculator with scientific notation, you must write the number using this notation first. Thus, to enter the number 8,700,000,000 in the calculator, you must know that 8,700,000,000 is 8.7×10^9; then you can key in

$\boxed{8}\ \boxed{.}\ \boxed{7}\ \boxed{EE\downarrow}\ \boxed{9}$

The calculator displays

$$8.7 \quad 09$$

a. What would the display read when you enter the number 73,000,000,000?

b. What would the display read when you enter the number 0.000000123?

3. a. _____

b. _____

SUMMARY

SECTION	ITEM	MEANING	EXAMPLE
2.1A	Additive inverse	The additive inverse of any integer a is $-a$.	The additive inverse of 5 is -5 and the additive inverse of -8 is 8.
2.1B	Absolute value of n, denoted by $|n|$	The absolute value of a number n is the distance from n to 0.	$|-7| = 7$ and $|13| = 13$
2.1C	Adding integers	If both integers have the *same* sign, add their absolute values and give the sum the common sign. If the numbers have *different* signs, subtract their absolute values and give the difference the sign of the number with the larger absolute value.	$-3 + (-5) = -8$ $-3 + 5 = +2$ $-3 + 1 = -2$
2.1D	Subtraction of integers	$a - b = a + (-b)$	$3 - 5 = 3 + (-5) = -2$ and $4 - (-2) = 4 + 2 = 6$
2.2A	Multiplication and division of integers	When multiplying or dividing two number with *like* signs, the answer is *positive*. With *unlike* signs, the answer is negative.	$(-3)(-5) = 15$ $\frac{-15}{-3} = 5$ $(-3)(5) = -15$ $\frac{-15}{3} = -5$
2.2B	Exponents	In the expression 3^2, 2 is the exponent.	3^2 means $3 \cdot 3$
2.2B	Base	In the expression 3^2, 3 is the base.	
2.3A	Rational number	A number that can be written in the form $\frac{a}{b}$, where a and b are integers and b is not 0.	$\frac{3}{4}, \frac{-7}{2}, \frac{9}{-3}$, and $\frac{0}{8}$ are rational numbers.
2.3E	Reciprocals	The reciprocal of $\frac{a}{b}$ is $\frac{b}{a}$.	The reciprocal of $\frac{3}{4}$ is $\frac{4}{3}$. The reciprocal of $\frac{-5}{3}$ is $\frac{3}{-5}$.

(Continued)

SECTION	ITEM	MEANING	EXAMPLE
2.4	Distributive Law of multiplication over addition	For any real numbers a, b, and c, $a(b + c) = ab + ac$.	$3(4 + 5) = 3 \cdot 4 + 3 \cdot 5$
2.4	Distributive Law of multiplication over subtraction	For any real numbers a, b, and c, $a(b - c) = ab - ac$.	$3(5 - 4) = 3 \cdot 5 - 3 \cdot 4$
2.5A	Sums and differences	The sum of a and b is $a + b$. The difference of a and b is $a - b$.	The sum of 3 and x is $3 + x$. The difference of 6 and x is $6 - x$.
2.5B	Product	The product of a and b is $a \cdot b$, $(a)(b)$, or ab.	
2.5E	Quotient	The quotient of a and b is $\frac{a}{b}$.	
2.5F	Expressions	A collection of numbers and letters connected by operation signs	xy^2, $x + y$, $x - y$, and $xy^2 - y$ are expressions.
2.6	Like terms	Terms with the same variable with the same exponents	$3x$ and $-5x$ are like terms. $-xy^2$ and $3xy^2$ are like terms.
2.7A	Multiplying expressions	$x^m \cdot x^n = x^{m+n}$	$x^3 \cdot x^5 = x^{3+5} = x^8$
2.7B	Dividing expressions	$\dfrac{x^m}{x^n} = x^{m-n}$	$\dfrac{x^8}{x^3} = x^{8-3} = x^5$
2.8A	x^0	For x not 0, $x^0 = 1$.	$10^0 = 1$, $3^0 = 1$, $\left(\dfrac{1}{2}\right)^0 = 1$
2.8A	x^{-n}, n a positive integer	$x^{-n} = \dfrac{1}{x^n}$	
2.8E	$(x^m)^n$	$(x^m)^n = x^{m \cdot n}$	$(x^2)^3 = x^{2 \cdot 3} = x^6$
2.9	Scientific notation	A number is in scientific notation when written in the form $M \times 10^n$, where M is a number between 1 and 10 and n is an integer.	3×10^{-3} and 2.7×10^5 are in scientific notation.

NAME

CLASS

SECTION

(If you need help with these exercises, look in the section indicated in brackets.)

ANSWERS

1. [2.1A] Find the additive inverse (opposite).
   a. -5
   b. -3
   c. -8

   1. a. _____
      b. _____
      c. _____

2. [2.1B] Find.
   a. $|-7|$
   b. $|-4|$
   c. $|-9|$

   2. a. _____
      b. _____
      c. _____

3. [2.1C] Find.
   a. (i) $6 + (-2)$
      (ii) $8 + (-3)$
      (iii) $7 + (-5)$
   b. (i) $(-2) + (-3)$
      (ii) $(-4) + (-5)$
      (iii) $(-6) + (-2)$
   c. (i) $-8 + 2$
      (ii) $-8 + 3$
      (iii) $-9 + 6$
   d. (i) $9 + (-6)$
      (ii) $11 + (-4)$
      (iii) $7 + (-3)$

   3. a. (i) _____
         (ii) _____
         (iii) _____
      b. (i) _____
         (ii) _____
         (iii) _____
      c. (i) _____
         (ii) _____
         (iii) _____
      d. (i) _____
         (ii) _____
         (iii) _____

4. [2.1D] Find.
   a. (i) $-16 - 4$
      (ii) $-10 - 5$
      (iii) $-8 - 2$
   b. (i) $-16 - (-3)$
      (ii) $-8 - (-2)$
      (iii) $-6 - (-3)$
   c. (i) $-8 - (-9)$
      (ii) $-6 - (-8)$
      (iii) $-11 - (-12)$

   4. a. (i) _____
         (ii) _____
         (iii) _____
      b. (i) _____
         (ii) _____
         (iii) _____
      c. (i) _____
         (ii) _____
         (iii) _____

5. [2.2A] Find.
   a. (i) $3 \cdot 4$
      (ii) $5 \cdot 6$
      (iii) $3 \cdot 9$
   b. (i) $-5 \cdot 7$
      (ii) $-3 \cdot 9$
      (iii) $-8 \cdot 6$
   c. (i) $7 \cdot (-3)$
      (ii) $9 \cdot (-4)$
      (iii) $6 \cdot (-9)$

   5. a. (i) _____
         (ii) _____
         (iii) _____
      b. (i) _____
         (ii) _____
         (iii) _____
      c. (i) _____
         (ii) _____
         (iii) _____

6. [2.2B] Find.
  a. **(i)** $(-3)^2$
      (ii) $(-4)^2$
      (iii) $(-5)^2$
  b. **(i)** -3^2
      (ii) -4^2
      (iii) -5^2

7. [2.2C] Find.

  a. **(i)** $\dfrac{-50}{10}$

      (ii) $\dfrac{-48}{6}$

      (iii) $\dfrac{-36}{9}$

  b. **(i)** $\dfrac{-8}{-2}$

      (ii) $\dfrac{-10}{-5}$

      (iii) $\dfrac{-42}{-6}$

  c. **(i)** $-8 \div (-4)$
      (ii) $-10 \div (-5)$
      (iii) $-20 \div (-4)$

8. [2.3A] Find the additive inverse (opposite).
  a. -3.4
  b. -5.6
  c. -7.2

9. [2.23B] Find.

  a. $\left| -1\dfrac{1}{2} \right|$

  b. $\left| -5\dfrac{1}{4} \right|$

  c. $\left| -9\dfrac{3}{7} \right|$

10. [2.3C] Find.
  a. **(i)** $-6.3 + 2.1$
      (ii) $-7.4 + 3.2$
      (iii) $-8.5 + 2.1$

  b. **(i)** $-\dfrac{3}{4} + \left(-\dfrac{1}{2} \right)$

      (ii) $-\dfrac{3}{5} + \left(-\dfrac{1}{8} \right)$

      (iii) $-\dfrac{3}{8} + \left(-\dfrac{1}{2} \right)$

6. a. (i) _____
    (ii) _____
    (iii) _____
  b. (i) _____
    (ii) _____
    (iii) _____

7. a. (i) _____
    (ii) _____
    (iii) _____
  b. (i) _____
    (ii) _____
    (iii) _____
  c. (i) _____
    (ii) _____
    (iii) _____

8. a. _____
  b. _____
  c. _____

9. a. _____
  b. _____
  c. _____

10. a. (i) _____
    (ii) _____
    (iii) _____
  b. (i) _____
    (ii) _____
    (iii) _____

11. [2.3D] Find.

    a. **(i)** $-7.6 - (-5.2)$

        (ii) $-4.3 - (-1.1)$

        (iii) $-9.4 - (-3.2)$

    b. **(i)** $\dfrac{5}{6} - \dfrac{9}{4}$

        (ii) $\dfrac{1}{6} - \dfrac{9}{10}$

        (iii) $\dfrac{3}{8} - \dfrac{7}{6}$

11. a. (i) _____
    (ii) _____
    (iii) _____
  b. (i) _____
    (ii) _____
    (iii) _____

12. [2.3E] Find.

    a. **(i)** $-6.1(3.2)$

        (ii) $-7.1(3.2)$

        (iii) $-8.1(3.2)$

    b. **(i)** $-\dfrac{2}{7}\left(-\dfrac{3}{5}\right)$

        (ii) $-\dfrac{3}{7}\left(-\dfrac{4}{5}\right)$

        (iii) $-\dfrac{4}{7}\left(-\dfrac{2}{5}\right)$

12. a. (i) _____
    (ii) _____
    (iii) _____
  b. (i) _____
    (ii) _____
    (iii) _____

13. [2.3G] Find.

    a. **(i)** $\dfrac{3}{5} \div \left(-\dfrac{9}{20}\right)$

        (ii) $\dfrac{2}{7} \div \left(-\dfrac{4}{21}\right)$

        (iii) $\dfrac{3}{8} \div \left(-\dfrac{9}{16}\right)$

    b. **(i)** $-\dfrac{5}{6} \div \left(-\dfrac{9}{2}\right)$

        (ii) $-\dfrac{5}{6} \div \left(\dfrac{-5}{2}\right)$

        (iii) $-\dfrac{3}{4} \div \left(-\dfrac{8}{9}\right)$

13. a. (i) _____
    (ii) _____
    (iii) _____
  b. (i) _____
    (ii) _____
    (iii) _____

14. [2.4A, B, C] Simplify.

    a. **(i)** $2(x - 1)$

        (ii) $3(x - 2)$

        (iii) $4(x - 3)$

    b. **(i)** $-(2x - 3y)$

        (ii) $-(3x - 4y)$

        (iii) $-(4x - 5y)$

    c. **(i)** $-2(x + 2y - 1)$

        (ii) $-3(x + 2y - 2)$

        (iii) $-4(x + 2y - 3)$

14. a. (i) _____
    (ii) _____
    (iii) _____
  b. (i) _____
    (ii) _____
    (iii) _____
  c. (i) _____
    (ii) _____
    (iii) _____

15. [2.5A] Write in symbols.
    a. The quotient of $(x - y)$ and z
    b. The quotient of $(a - b)$ and c
    c. The quotient of $(p - q)$ and r

15. a. _____
    b. _____
    c. _____

16. [2.5B] Write the indicated products using juxtaposition.
    a. **(i)** $-a$ times b times c
        (ii) $-p$ times q times r
        (iii) $-b$ times c times d
    b. **(i)** $3 \cdot 3 \cdot x \cdot x \cdot y \cdot x$
        (ii) $4 \cdot 4 \cdot a \cdot a \cdot b \cdot a \cdot b$
        (iii) $2 \cdot 2 \cdot a \cdot b \cdot b \cdot a \cdot b \cdot b$

16. a. (i) _____
    (ii) _____
    (iii) _____
    b. (i) _____
    (ii) _____
    (iii) _____

17. [2.6A] Combine like terms.
    a. **(i)** $(-3x) + (-4x)$
        (ii) $(-2x) + (-5x)$
        (iii) $(-3x) + (-7x)$
    b. **(i)** $-4ab^2 - (-2ab^2)$
        (ii) $-5xy^2 - (-3xy^2)$
        (iii) $-4x^2y - (-2x^2y)$

17. a. (i) _____
    (ii) _____
    (iii) _____
    b. (i) _____
    (ii) _____
    (iii) _____

18. [2.6B, C, D, E] Simplify.
    a. **(i)** $3x - 2(x - 1) - (x + 1)$
        (ii) $5x - 2(x - 1) - (x + 2)$
        (iii) $6x - 2(x - 1) - (x + 1)$
    b. **(i)** $[x - (x + 5)] + [2 + 2(x - 1)]$
        (ii) $[x - (x + 4)] + [3 + 3(x - 1)]$
        (iii) $[x - (x + 6)] + [4 + 4(x - 1)]$

18. a. (i) _____
    (ii) _____
    (iii) _____
    b. (i) _____
    (ii) _____
    (iii) _____

19. [2.7A] Find.
    a. **(i)** $(3a^2b)(-5ab^3)$
        (ii) $(4a^2b)(-6ab^4)$
        (iii) $(5a^2b)(-7ab^3)$
    b. **(i)** $(-2xy^2z)(-3x^2yz^4)$
        (ii) $(-3x^2yz^2)(-4xy^3z)$
        (iii) $(-4xyz)(-5xy^2z^3)$

19. a. (i) _____
    (ii) _____
    (iii) _____
    b. (i) _____
    (ii) _____
    (iii) _____

20. [2.7B] Find.

    a. **(i)** $\dfrac{16x^6y^8}{-8xy^4}$

        (ii) $\dfrac{24x^7y^6}{-4xy^3}$

        (iii) $\dfrac{-18x^8y^7}{9xy^4}$

    b. **(i)** $\dfrac{-8xy^7}{-16x^4y}$

        (ii) $\dfrac{-5xy^8}{-10x^6y}$

        (iii) $\dfrac{-3xy^7}{-9x^8y}$

20. a. (i) _____
    (ii) _____
    (iii) _____
    b. (i) _____
    (ii) _____
    (iii) _____

21. [2.8C] Multiply and simplify.
    a. **(i)** $2^8 \cdot 2^{-5}$
        (ii) $2^7 \cdot 2^{-4}$
        (iii) $2^6 \cdot 2^{-3}$
    b. **(i)** $y^{-3} \cdot y^{-5}$
        (ii) $y^{-2} \cdot y^{-3}$
        (iii) $y^{-4} \cdot y^{-2}$

21. a. (i) _____
      (ii) _____
      (iii) _____
  b. (i) _____
      (ii) _____
      (iii) _____

22. [2.8D] Find.

    a. **(i)** $\dfrac{x}{x^5}$

        (ii) $\dfrac{x}{x^7}$

        (iii) $\dfrac{x}{x^9}$

    b. **(i)** $\dfrac{a^{-2}}{a^{-2}}$

        (ii) $\dfrac{a^{-4}}{a^{-4}}$

        (iii) $\dfrac{a^{-10}}{a^{-10}}$

    c. **(i)** $\dfrac{x^{-2}}{x^{-3}}$

        (ii) $\dfrac{x^{-5}}{x^{-8}}$

        (iii) $\dfrac{x^{-7}}{x^{-9}}$

22. a. (i) _____
      (ii) _____
      (iii) _____
  b. (i) _____
      (ii) _____
      (iii) _____
  c. (i) _____
      (ii) _____
      (iii) _____

23. [2.8E] Simplify.
    a. **(i)** $(a^{-2})^{-3}$
        (ii) $(a^{-3})^{-4}$
        (iii) $(a^{-4})^{-5}$
    b. **(i)** $(2x^2y^{-3})^2$
        (ii) $(3x^2y^{-2})^3$
        (iii) $(2x^2y^{-3})^4$
    c. **(i)** $(2x^{-2}y^4)^{-2}$
        (ii) $(3x^{-2}y^4)^{-2}$
        (iii) $(4x^{-2}y^3)^{-3}$

23. a. (i) _____
      (ii) _____
      (iii) _____
  b. (i) _____
      (ii) _____
      (iii) _____
  c. (i) _____
      (ii) _____
      (iii) _____

24. [2.9A] Write in scientific notation.
    a. **(i)** 44,000,000
        (ii) 4,500,000
        (iii) 460,000
    b. **(i)** 0.0014
        (ii) 0.00015
        (iii) 0.000016

24. a. (i) _____
      (ii) _____
      (iii) _____
  b. (i) _____
      (ii) _____
      (iii) _____

25. [2.9B] Perform the indicated operations and write the answer in scientific notation.

    a. **(i)** $(2 \times 10^2) \times (1.1 \times 10^3)$

       (ii) $(3 \times 10^2) \times (3.1 \times 10^4)$

      (iii) $(4 \times 10^2) \times (3.1 \times 10^5)$

    b. **(i)** $\dfrac{1.15 \times 10^{-3}}{2.3 \times 10^{-4}}$

      (ii) $\dfrac{1.38 \times 10^{-3}}{2.3 \times 10^{-4}}$

     (iii) $\dfrac{1.61 \times 10^{-3}}{2.3 \times 10^{-4}}$

25. a. (i) _____

    (ii) _____

   (iii) _____

  b. (i) _____

    (ii) _____

   (iii) _____

NAME

CLASS

SECTION

(Answers on pages 186–187)

ANSWERS

1. Find the additive inverse (opposite) of -6.

   1. _____

2. Find $|-8|$.

   2. _____

3. Find.
   a. $9 + (-4)$
   b. $(-2) + (-6)$
   c. $(-16) + 9$
   d. $10 + (-8)$

   3. a. _____
      b. _____
      c. _____
      d. _____

4. Find.
   a. $-26 - 2$
   b. $-19 - (-8)$
   c. $-6 - (-8)$

   4. a. _____
      b. _____
      c. _____

5. Find.
   a. $2 \cdot 6$
   b. $-5 \cdot 6$
   c. $8 \cdot (-6)$
   d. $-3 \cdot (-6)$

   5. a. _____
      b. _____
      c. _____
      d. _____

6. Find.
   a. $(-6)^2$
   b. -6^2

   6. a. _____
      b. _____

7. Find.

   a. $\dfrac{-48}{8}$

   b. $\dfrac{-36}{-4}$

   c. $-60 \div -10$

   d. $\dfrac{6}{0}$

   7. a. _____
      b. _____
      c. _____
      d. _____

8. Find the additive inverse (opposite) of -6.8.

   8. _____

9. Find $\left| -9\dfrac{1}{6} \right|$.

   9. _____

10. Find.
    a. $-8.9 + 6.2$

    b. $-\dfrac{3}{5} + \left(-\dfrac{3}{8} \right)$

    10. a. _____
        b. _____

11. Find.
    a. $-8.9 - (-6.7)$

    b. $\dfrac{5}{6} - \dfrac{7}{4}$

12. Find.
    a. $-5.1(3.2)$

    b. $-\dfrac{3}{7}\left(-\dfrac{4}{5}\right)$

13. Find.
    a. $\dfrac{2}{5} \div \dfrac{-4}{15}$

    b. $\dfrac{-5}{6} \div \left(-\dfrac{7}{2}\right)$

14. Simplify.
    a. $6(x - 2)$
    b. $-(6x - 8y)$
    c. $-3(x + 2y - 4)$

15. Write in symbols.
    a. The quotient of $(x + y)$ and z.
    b. The markup M of an item is obtained by subtracting the cost C from the selling price S.

16. Write the indicated products using juxtaposition.
    a. $-x$ times y times z
    b. $6 \cdot 6 \cdot b \cdot b \cdot a \cdot b \cdot a$

17. Combine like terms.
    a. $(-8x) + (-2x)$
    b. $-8ab^2 - (-3ab^2)$

18. Simplify.
    a. $9x - 3(x - 3) - (x + 2)$
    b. $[x - (x + 6)] + [3 + 3(x - 1)]$

19. Find.
    a. $(2a^3b)(-6ab^3)$
    b. $(-2x^2yz)(-6xy^3z^4)$

20. Find.
    a. $\dfrac{18x^5y^7}{-9xy^3}$

    b. $\dfrac{-8xy^6}{-24x^5y}$

21. Multiply and simplify.
    a. $2^9 \cdot 2^{-6}$
    b. $y^{-2} \cdot y^{-6}$

11. a. _____	b. _____
12. a. _____	b. _____
13. a. _____	b. _____
14. a. _____	b. _____
	c. _____
15. a. _____	b. _____
16. a. _____	b. _____
17. a. _____	b. _____
18. a. _____	b. _____
19. a. _____	b. _____
20. a. _____	b. _____
21. a. _____	b. _____

22. Find.

    a. $\dfrac{x}{x^4}$

    b. $\dfrac{x^{-6}}{x^{-6}}$

    c. $\dfrac{x^{-6}}{x^{-7}}$

23. Simplify.
    a. $(x^2)^{-6}$
    b. $(3x^2y^{-4})^3$
    c. $(2x^{-3}y^4)^{-2}$

24. Write in scientific notation.
    a. 48,000,000
    b. 0.00000037

25. Perform the indicated operations.
    a. $(3 \times 10^4) \times (7.1 \times 10^6)$

    b. $\dfrac{2.84 \times 10^{-2}}{7.1 \times 10^{-3}}$

22. a. _____
    b. _____
    c. _____

23. a. _____
    b. _____
    c. _____

24. a. _____
    b. _____

25. a. _____
    b. _____

IF YOU MISSED QUESTION	SECTION	EXAMPLES	PAGE	ANSWERS
1	2.1	1	92	1. 6
2	2.1	2	93	2. 8
3a	2.1	3	93	3. a. 5
3b	2.1	4	94	b. -8
3c	2.1	5a	95	c. -7
3d	2.1	5b	95	d. 2
4a	2.1	6b	96	4. a. -28
4b	2.1	6c	96	b. -11
4c	2.1	6d	96	c. 2
5a	2.2	1a	102	5. a. 12
5b	2.2	1b	102	b. -30
5c	2.2	1c	102	c. -48
5d	2.2	1d	102	d. 18
6a	2.2	2a	103	6. a. 36
6b	2.2	2b	103	b. -36
7a	2.2	4b	104	7. a. -6
7b	2.2	4c	104	b. 9
7c	2.2	4d	104	c. 6
7d	2.2	4e	104	d. Not defined
8	2.3	1	110	8. 6.8
9	2.3	2	110	9. $9\frac{1}{6}$
10a	2.3	3	110	10. a. -2.7
10b	2.3	4	111	b. $-\dfrac{39}{40}$
11a	2.3	5a, b	111	11. a. -2.2
11b	2.3	5c, d	111–112	b. $-\dfrac{11}{12}$
12a	2.3	6a	112	12. a. -16.32
12b	2.3	6b	112	b. $\dfrac{12}{35}$
13a	2.3	8a	114	13. a. $-\dfrac{3}{2}$
13b	2.3	8b	114	b. $\dfrac{5}{21}$

IF YOU MISSED QUESTION	SECTION	EXAMPLES	PAGE	ANSWERS
14a	2.4	1	120	**14. a.** $6x - 12$
14b	2.4	2	121	**b.** $-6x + 8y$
14c	2.4	3	122	**c.** $-3x - 6y + 12$
15a	2.5	1, 7	128, 131	**15. a.** $\dfrac{x + y}{z}$
15b	2.5	3	129	**b.** $M = S - C$
16a	2.5	4	129	**16. a.** $-xyz$
16b	2.5	5	130	**b.** $6^2 a^2 b^3 = 36a^2 b^3$
17a	2.6	1	138	**17. a.** $-10x$
17b	2.6	2	139	**b.** $-5ab^2$
18a	2.6	4	141	**18. a.** $5x + 7$
18b	2.6	5	142	**b.** $3x - 6$
19a	2.7	1, 2	148, 149	**19. a.** $-12a^4 b^4$
19b	2.7	2	149	**b.** $12x^3 y^4 z^5$
20a	2.7	3a, b	151	**20. a.** $-2x^4 y^4$
20b	2.7	3c	151	**b.** $\dfrac{y^5}{3x^4}$
21a	2.8	3a, b	159	**21. a.** 8
21b	2.8	3c, d	159	**b.** $\dfrac{1}{y^8}$
22a	2.8	4b	160	**22. a.** $\dfrac{1}{x^3}$
22b	2.8	4c	160	**b.** 1
22c	2.8	4d	160	**c.** x
23a	2.8	5	161	**23. a.** $\dfrac{1}{x^{12}}$
23b	2.8	6b	161	**b.** $\dfrac{27x^6}{y^{12}}$
23c	2.8	6c	161	**c.** $\dfrac{x^6}{4y^8}$
24a	2.9	1, 2	168	**24. a.** 4.8×10^7
24b	2.9	1, 2	168	**b.** 3.7×10^{-7}
25a	2.9	3	169	**25. a.** 2.13×10^{11}
25b	2.9	4	169	**b.** 4

CHAPTER 3

SOLVING EQUATIONS AND INEQUALITIES

In this chapter we use the properties of the number systems we have studied to solve equations and inequalities. We also introduce the addition, subtraction, multiplication, and division properties for equations and inequalities. We then apply these techniques to the solution of word problems.

NAME _____

CLASS _____

SECTION _____

(Answers on page 190*)*

ANSWERS

1. Does the number 8 satisfy the equation $1 = 9 - x$?

    1. _____

2. Solve $x - \dfrac{3}{7} = \dfrac{2}{7}$.

    2. _____

3. Solve $-2x + \dfrac{5}{8} + 5x - \dfrac{1}{8} = \dfrac{7}{8}$.   4. Solve $3 = 3(x - 1) + 6 - 2x$.

    3. _____

    4. _____

5. Solve $\dfrac{2}{3}x = -6$.   6. Solve $-\dfrac{2}{3}x = -4$.

    5. _____

    6. _____

7. Solve $\dfrac{x}{2} + \dfrac{2x}{5} = 9$.   8. Solve $\dfrac{x}{3} - \dfrac{x}{7} = 4$.

    7. _____

    8. _____

9. Solve $\dfrac{x - 2}{6} - \dfrac{x + 1}{9} = 0$.   10. Solve $6 - \dfrac{x}{3} = \dfrac{2(x + 4)}{5}$.

    9. _____

    10. _____

11. Solve for h in the equation $V = \dfrac{4}{3}\pi r^2 h$.

    11. _____

12. Graph $5x - 1 < 3(x + 1)$.

    12. _____

13. Graph $2(x - 2) \geq 5x + 2$.   14. Graph $-\dfrac{x}{3} + \dfrac{x}{5} \leq \dfrac{x - 5}{5}$.

    13. _____

    14. _____

15. Solve and graph the compound statement $x + 3 \leq 6$ and $-2x < 6$.

    15. _____

16. The sum of two numbers is 95. If one of the numbers is 35 more than the other, what are the numbers?

    16. _____

17. A man has invested a certain amount of money in stocks and bonds. His annual return from these investments is $780. If the stocks produce $240 more in returns than the bonds, how much money does the man receive annually from each investment?

    17. _____

18. A freight train leaves a station traveling at 30 mi/hr. One hour later, a passenger train leaves the same station in the same direction at 40 mi/hr. How long does it take for the passenger train to catch the freight train?

    18. _____

19. How many pounds of coffee selling for $1.40 per pound should be mixed with 20 lb of coffee selling for $1.80 per pound to obtain a mixture selling for $1.60 per pound?

    19. _____

20. An investor bought some municipal bonds yielding 6% annually and some certificates of deposit yielding 7% annually. If her total investment amounts to $20,000 and her annual income from the investments is $1280, how much money is invested in bonds and how much in certificates of deposit?

    20. _____

IF YOU MISSED QUESTION	SECTION	EXAMPLES	PAGE	ANSWERS
1	3.1	1	191	1. Yes
2	3.1	2	192	2. $x = \dfrac{5}{7}$
3	3.1	3	194	3. $x = \dfrac{1}{8}$
4	3.1	4	196	4. $x = 0$
5	3.2	1, 2, 3	204, 205, 208	5. $x = -9$
6	3.2	1, 2, 3	204, 205, 208	6. $x = 6$
7	3.2	4	210	7. $x = 10$
8	3.2	4	210	8. $x = 21$
9	3.2	5	211	9. $x = 8$
10	3.3	1, 2	220, 223	10. $x = 6$
11	3.3	3	224	11. $h = \dfrac{3V}{4\pi r^2}$
12	3.4	1, 2, 3	232, 233, 235	12.
13	3.4	4, 5	237, 239	13.
14	3.4	6	239	14.
15	3.4	7	241	15.
16	3.5	1, 2	249	16. 30 and 65
17	3.5	3	250	17. $270 and $510
18	3.6	1, 2, 3	258, 259, 260	18. 3 hr
19	3.6	4	261	19. 20 lb
20	3.6	5	264	20. $12,000 in bonds; $8000 in certificates

3.1 THE ADDITION AND SUBTRACTION PROPERTY OF EQUALITY

REVIEW

Before starting this section you should know how to add and subtract rational numbers.

OBJECTIVES

You should be able to:

A. Determine if a number is a solution of (satisfies) an equation.

B. Solve an equation such as $x - a = b$ by adding a on both sides.

C. Solve an equation such as $x + a = b$ by subtracting a on both sides.

D. Solve an equation using the three-step procedure given in the text.

Courtesy of Princeton University Library.

Do you know the man in the photo? It is Albert Einstein. He discovered the equation $E = mc^2$. In the equation $E = mc^2$, E is the energy, m is the mass, and c is the velocity of light. An **equation** is a sentence with equals ($=$) for its verb. Some equations are *true* ($1 + 1 = 2$, $5 - 2 = 3$), some are *false* ($1 + 1 = 11$, $5 - 2 = -3$), and some are neither true nor false. The equation $x + 1 = 5$ is neither true nor false. In this equation, the variable x can be replaced by many numbers, but only one number will make the resulting statement *true*. This number is called the *solution* of the equation.

A. SOLUTIONS

> The **solutions** of an equation are the replacements of the variable that make the equation a *true* statement. When we find the solution of an equation, we say that we have **solved** the equation.

How do we know if a given number is a solution of (or *satisfies*) an equation? Write the number in place of the variable in the given equation and see if the result is true. For example, to tell if 4 is a solution of the equation

$x + 1 = 5$

replace x by 4. This gives the true statement:

$4 + 1 = 5$

so 4 is a solution of the equation $x + 1 = 5$.

EXAMPLE 1 Determine if the given number is a solution of the equation.

a. $9;\ x - 4 = 5$ **b.** $8;\ 5 = 3 - y$ **c.** $10;\ \frac{1}{2}z - 5 = 0$

Problem 1 Determine if the given number is a solution of the equation.
a. $3;\ x - 1 = 2$ **b.** $9;\ 6 = 3 - y$
c. $18;\ \frac{1}{3}z - 6 = 0$

Solution

a. If x is 9, $x - 4 = 5$ becomes $9 - 4 = 5$, a *true* statement. Thus, 9 is a solution of the equation.

b. If y is 8, $5 = 3 - y$ becomes $5 = 3 - 8$, a *false* statement. Hence, 8 is not a solution of the equation.

c. If z is 10, $\frac{1}{2}z - 5 = 0$ becomes $\frac{1}{2}(10) - 5 = 0$, a *true* statement. Thus, 10 is a solution of the equation. ▲

We have learned how to determine if a number satisfies an equation. To find this number, we must find an *equivalent* equation whose solution is obvious.

> Two equations are **equivalent** if their solutions are the same.

How do we find these equivalent equations? We use properties of equations.

B. THE ADDITION PROPERTY OF EQUALITY

Look at the ad in the margin. It says that you can cut $5 off the price of a gallon of paint and pay $6.69, the sale price, for it. What was the old price p of the paint? Since the old price p was cut by $5, the new price is $p - 5$. The ad says that the new price is $6.69. Thus,

$$p - 5 = 6.69$$

To find the old price p, we simply add back the $5 we cut. That is,

$$p - 5 + 5 = 6.69 + 5 \quad \text{Here we added 5 to both sides of the equation to}$$
$$p = 11.69 \qquad \text{obtain an equivalent equation.}$$

Thus the old price was $11.69, as can be easily verified, since $11.69 - 5 = 6.69$. Note that by adding 5 to both sides of the equation $p - 5 = 6.69$, we produce an equivalent equation, $p = 11.69$, whose solution is obvious. This example illustrates the fact that we can *add* the same number on both sides of an equation and produce an *equivalent* equation—that is, an equation whose solution is identical to the solution of the original one. Here is the idea.

THE ADDITION PROPERTY OF EQUALITY

The equation $a = b$ is equivalent to

$a + c = b + c$ for any number c

We use this property in the next example.

EXAMPLE 2 Solve.

a. $x - 3 = 9$ b. $x - \dfrac{1}{7} = \dfrac{5}{7}$

Problem 2 Solve.

a. $y - 5 = 7$ b. $y - \dfrac{2}{5} = \dfrac{1}{5}$

Solution

a. This problem is similar to the example. To solve the equation, we want to get x by itself on one side of the equation. We can do this by adding 3 (the additive inverse of -3) on both sides of the equation. The procedure looks like this:

$$x - 3 = 9$$
$$x - 3 + 3 = 9 + 3$$
$$x = 12$$

Thus, 12 is the solution of $x - 3 = 9$.

Check: Substituting **12** for x in the original equation, we have $12 - 3 = 9$, a true statement.

b.
$$x - \frac{1}{7} = \frac{5}{7}$$
$$x - \frac{1}{7} + \frac{1}{7} = \frac{5}{7} + \frac{1}{7}$$
$$x = \frac{6}{7}$$

Thus, $\frac{6}{7}$ is the solution of $x - \frac{1}{7} = \frac{5}{7}$.

Check: $\frac{6}{7} - \frac{1}{7} = \frac{5}{7}$ is a true statement. Thus, $\frac{6}{7}$ is the solution. ▲

Sometimes it is necessary to simplify an equation before we add the same number on both sides. For example, to solve the equation

$$3x + 5 - 2x - 9 = 6x + 5 - 6x$$

we first simplify both sides of the equation by collecting like terms as shown:

$$3x + 5 - 2x - 9 = 6x + 5 - 6x$$

$(3x - 2x) + (5 - 9) = (6x - 6x) + 5$	Grouping like terms
$x + (-4) = 0 + 5$	Combining like terms
$x - 4 = 5$	Rewriting $x + (-4)$ as $x - 4$
$x - 4 + 4 = 5 + 4$	Adding 4 to both sides
$x = 9$	Solving for x

Thus, 9 is the solution of the equation.

Check: We substitute **9** for x in the original equation. To save time we use the following diagram, substituting 9 for x.

$$3x + 5 - 2x - 9 \overset{?}{=} 6x + 5 - 6x$$

$3(9) + 5 - 2(9) - 9$	$6(9) + 5 - 6(9)$
$27 + 5 - 18 - 9$	$54 + 5 - 54$
$32 - 18 - 9$	5
$32 - 27$	
5	

Since both sides yield 5, our result is correct.

C.   THE SUBTRACTION PROPERTY OF EQUALITY

Now, suppose that the price of an article is increased by \$3 and the article is selling for \$8 now. What was its old price, p? The equation here is

$$p + 3 = 8$$

Old price   Went up \$3   Cost \$8 now

To solve this equation, we have to bring the price down, that is, subtract 3 on both sides of the equation, obtaining

$$p + 3 - 3 = 8 - 3$$
$$p = 5$$

Thus the old price was \$5. We have *subtracted* 3 on both sides of the equation. Here is the idea.

THE SUBTRACTION PROPERTY OF EQUALITY

The equation $a = b$ is equivalent to

$a - c = b - c$ for any number c

This property tells us that we can *subtract* the same number on both sides of an equation to produce an *equivalent* equation. Note that since $a - c = a + (-c)$, you can think of *subtracting* c as *adding* $(-c)$.

EXAMPLE 3   Solve.

a. $2x + 4 - x + 2 = 10$ **b.** $-3x + \dfrac{5}{7} + 4x - \dfrac{3}{7} = \dfrac{6}{7}$

Solution

a. We need to solve the equation by getting x by itself on the left—that is, by obtaining $x = \boxed{}$, where $\boxed{}$ is a number. We proceed as follows:

$$2x + 4 - x + 2 = 10$$
$$x + 6 = 10 \qquad \text{Simplifying}$$
$$x + 6 - 6 = 10 - 6 \quad \text{Subtracting } 6 \text{ from both sides}$$
$$x = 4$$

Thus, 4 is the solution of the equation.

Check: $2x + 4 - x + 2 \overset{?}{=} 10$

$2(4) + 4 - 4 + 2$	10
$8 + 2$	
10	

Problem 3   Solve.

a. $5y + 2 - 4y + 5 = 9$

b. $-4y + \dfrac{2}{5} + 5y - \dfrac{1}{5} = \dfrac{3}{5}$

ANSWER   (to problem on page 192)

2. a. $y = 12$ **b.** $y = \dfrac{3}{5}$

b. $-3x + \dfrac{5}{7} + 4x - \dfrac{3}{7} = \dfrac{6}{7}$

$\qquad\qquad x + \dfrac{2}{7} = \dfrac{6}{7}$ Simplifying

$\qquad x + \dfrac{2}{7} - \dfrac{2}{7} = \dfrac{6}{7} - \dfrac{2}{7}$ Subtracting $\dfrac{2}{7}$ from both sides

$\qquad\qquad\qquad x = \dfrac{4}{7}$

Thus, $\frac{4}{7}$ is the solution of the equation.

Check: $\qquad\qquad -3x + \dfrac{5}{7} + 4x - \dfrac{3}{7} \overset{?}{=} \dfrac{6}{7}$

$$\begin{array}{c|c} -3\left(\dfrac{4}{7}\right) + \dfrac{5}{7} + 4\left(\dfrac{4}{7}\right) - \dfrac{3}{7} & \dfrac{6}{7} \\[2ex] -\dfrac{12}{7} + \dfrac{5}{7} + \dfrac{16}{7} - \dfrac{3}{7} & \\[2ex] -\dfrac{7}{7} + \dfrac{13}{7} & \\[2ex] \dfrac{6}{7} & \end{array}$$

D. USING BOTH PROPERTIES TOGETHER

Can we solve the equation $2x - 7 = x + 2$? We can try by adding 7 on both sides, obtaining

$\qquad 2x - 7 + 7 = x + 2 + 7$

$\qquad\qquad\quad 2x = x + 9$

In order to solve this equation, we must get x by itself on the left—that is, $x = \boxed{}$, where $\boxed{}$ is a number. How do we do this? We want variables on one side of the equation (and we have them: $2x$) but only specific numbers on the other (there we are in trouble, we have an x on the right). To "get rid of" this x, we subtract x on both sides to obtain

$\qquad 2x - x = x - x + 9$ Remember $x = 1x$

$\qquad\qquad x = 9$

Thus, 9 is the solution of the equation.

Check: $\qquad 2x - 7 \overset{?}{=} x + 2$

$$\begin{array}{c|c} 2(9) - 7 & 9 + 2 \\ 11 & 11 \end{array}$$

Now, remember, to solve equations by adding or subtracting, we can proceed as follows:

THREE-STEP RULE TO SOLVE EQUATIONS BY ADDING OR SUBTRACTING

1. Simplify both sides if necessary.

2. Add or subtract the same numbers on both sides of the equation so that one side contains only variables.

3. Add or subtract the same expressions on both sides of the equation so that the other side contains only numbers.

Use these three steps in solving the next example.

EXAMPLE 4 Solve.

a. $3 = 8 + x$ **b.** $4y - 3 = 3y + 8$
c. $0 = 3(z - 2) + 4 - 2z$ **d.** $2(x + 1) = 3x + 5$

Solution

a. Given: $3 = 8 + x$

   1. Both sides of the equation are $3 = 8 + x$
      already simplified.

   2. Subtract 8 on both sides. $3 - 8 = 8 - 8 + x$

$$-5 = x$$
$$x = -5$$

The solution is -5.

Check: $3 \overset{?}{=} 8 + x$

$$\begin{array}{c|c} 3 & 8 + (-5) \\ \hline & 3 \end{array}$$

b. Given: $4y - 3 = 3y + 8$

   1. Both sides of the equation are $4y - 3 = 3y + 8$
      already simplified.

   2. Add 3 on both sides. $4y - 3 + 3 = 3y + 8 + 3$
$$4y = 3y + 11$$

   3. Subtract $3y$ on both sides. $4y - 3y = 3y - 3y + 11$
$$y = 11$$

The solution is 11.

Check: $4y - 3 \overset{?}{=} 3y + 8$

$$\begin{array}{cc|c} 4(11) - 3 & & 3(11) + 8 \\ 44 - 3 & & 33 + 8 \\ 41 & & 41 \end{array}$$

c. Given: $0 = 3(z - 2) + 4 - 2z$

   1. Simplify by using the distributive $0 = 3z - 6 + 4 - 2z$
      law and combining like terms $0 = z - 2$

   2. Add 2 to both sides. $0 + 2 = z - 2 + 2$
$$2 = z$$
$$z = 2$$

The solution is 2.

Problem 4 Solve.
a. $5 = 7 + y$
b. $3x - 2 = 2x + 5$
c. $0 = 2(x - 3) + 2 - x$
d. $3(z + 2) = 4z + 8$

Check: $0 \overset{?}{=} 3(z - 2) + 4 - 2z$

0	$3(2 - 2) + 4 - 2(2)$
0	$3(0) + 4 - 4$
0	$0 + 4 - 4$
	0

d. Given: $\qquad\qquad\qquad\qquad\qquad 2(x + 1) = 3x + 5$

1. Simplify. $\qquad\qquad\qquad\qquad 2x + 2 = 3x + 5$

2. Subtract 2 from both sides. $\qquad 2x + 2 - 2 = 3x + 5 - 2$

$\qquad\qquad\qquad\qquad\qquad\qquad\qquad 2x = 3x + 3$

3. Subtract $3x$ from both sides so all $\qquad 2x - 3x = 3x - 3x + 3$
the variables are on the left. $\qquad\qquad\qquad -x = 3$

If $-x = 3$, then $x = -3$ because the opposite of a number is the number with its sign changed; that is, if the opposite of x is 3, then x itself must be -3. Thus the solution is -3.

Check: $2(x + 1) \overset{?}{=} 3x + 5$

$2(-3 + 1)$	$3(-3) + 5$
$2(-2)$	$3(-3) + 5$
-4	$-9 + 5$
	-4

EXAMPLE 5   Solve $8x + 7 = 7x + 3$.

Solution

Given: $\qquad\qquad\qquad\qquad\qquad 8x + 7 = 7x + 3$

1. The equation is already simplified.

2. Subtract 7 from both sides. $\qquad 8x + 7 - 7 = 7x + 3 - 7$

$\qquad\qquad\qquad\qquad\qquad\qquad\qquad 8x = 7x - 4$

3. Subtract $7x$ from both sides. $\qquad 8x - 7x = 7x - 7x - 4$

$\qquad\qquad\qquad\qquad\qquad\qquad\qquad x = -4$

The solution is -4.

Check: $8x + 7 \overset{?}{=} 7x + 3$

$8(-4) + 7$	$7(-4) + 3$
$-32 + 7$	$-28 + 3$
-25	-25

Problem 5   Solve $4x + 6 = 3x + 2$.

A. Determine if the given number is a solution of the equation. (Do not solve.)

1. $x = 3$; $x - 1 = 2$

2. $x = 4$; $6 = x - 10$

3. $y = -2$; $3y + 6 = 0$

4. $z = -3$; $-3z + 9 = 0$

5. $n = 2$; $12 - 3n = 6$

6. $m = 3\frac{1}{2}$; $3\frac{1}{2} + m = 7$

7. $d = 10$; $\frac{2}{5}d + 1 = 3$

8. $c = 2.3$; $3.4 = 2c - 1.4$

9. $a = 2.1$; $4.6 = 11.9 - 3a$

10. $x = \frac{1}{10}$; $0.2 = \frac{7}{10} - 5x$

B. Solve.

11. $x - 5 = 9$

12. $y - 3 = 6$

13. $11 = m - 8$

14. $6 = n - 2$

15. $y - 2 = 3$

16. $R - 4 = 9$

17. $k - 16 = 5$

18. $n - 2 = 7$

19. $\frac{1}{4} = z - \frac{2}{3}$

20. $\frac{7}{2} = v - \frac{1}{5}$

21. $0 = x - \frac{7}{2}$

22. $0 = y - \frac{3}{4}$

C. Solve.

23. $8 = c + 8$

24. $0 = b + 9$

25. $3 + x = 0$

26. $y + 4 = 0$

27. $\frac{1}{4} + y = \frac{1}{4}$

28. $\frac{2}{3} + y = \frac{1}{3}$

1. _____
2. _____
3. _____
4. _____
5. _____
6. _____
7. _____
8. _____
9. _____
10. _____
11. _____
12. _____
13. _____
14. _____
15. _____
16. _____
17. _____
18. _____
19. _____
20. _____
21. _____
22. _____
23. _____
24. _____
25. _____
26. _____
27. _____
28. _____

29. $3.4 = C + 0.4$

30. $1.7 = C + 0.7$

D. Solve.

31. $6p + 9 = 5p$

32. $7q + 4 = 6q$

33. $3x + 3 + 2x = 4x$

34. $2y + 4 + 6y = 7y$

35. $4(m - 2) + 2 - 3m = 0$

36. $3(n + 4) + 2 = 2n$

37. $5(y - 2) = 4y + 8$

38. $3(z - 1) = 4z + 1$

39. $2(a - 4) = 3a - 1$

40. $5b - 3 = 4(b + 1)$

41. $6c - 2 = 5(c - 2)$

42. $5 - 3R + 8 = -4R + 6$

43. $3x + 5 - 2x + 1 = 6x + 4 - 6x$

44. $6f - 2 - 4f = -2f + 5 + 3f$

45. $-2g + 4 - 5g = 6g + 1 - 14g$

46. $-2x + 3 + 9x = 6x - 1$

47. $6(x + 4) + 4 - 3x = 4x$

48. $6(y - 1) - 2 + y = 8y + 4$

49. $10(z - 2) - 6 - z = 8z + 10$

50. $8(a - 4) - 3 - a = 6a + 3$

51. $3b + 6 - 2b = 2(b - 2) + 4$

52. $3b + 2 - b = 3(b - 2) + 5$

53. $2p + \dfrac{2}{3} - 5p = -4p + 7\dfrac{1}{3}$

54. $4q + \dfrac{2}{7} - 6q = -3q + 2\dfrac{2}{7}$

55. $5r + \dfrac{3}{8} - 9r = -5r + 1\dfrac{1}{2}$

E. Applications

56. The price of an item is increased by \$7. The item is selling for \$23 now. What was the old price of the item?

57. In a certain year, the average hourly earnings were \$9.81, an increase of 40¢ over the previous year. What were the average hourly earnings the previous year?

29. _____

30. _____

31. _____

32. _____

33. _____

34. _____

35. _____

36. _____

37. _____

38. _____

39. _____

40. _____

41. _____

42. _____

43. _____

44. _____

45. _____

46. _____

47. _____

48. _____

49. _____

50. _____

51. _____

52. _____

53. _____

54. _____

55. _____

56. _____

57. _____

58. The consumer price index for housing in a recent year was 115.3, a 4.7 point increase over the previous year. What was the consumer price index for housing the previous year?

59. The cost of medical care increased 142.2 points in a 6-yr period. If the cost of medical care reached the 326.9 mark, what was it 6 yr ago?

60. In the last 10 yr mathematics scores in the Scholastic Aptitude Test (SAT) have declined 16 points, to 476. What was the mathematics score 10 yr ago?

✓ **SKILL CHECKER**

Find.

61. $4(-5)$

62. $6(-3)$

63. $-\dfrac{2}{3}\left(\dfrac{3}{4}\right)$

64. $-\dfrac{5}{7}\left(\dfrac{7}{10}\right)$

Find the reciprocal of each number.

65. $\dfrac{3}{2}$

66. $\dfrac{2}{5}$

Find the LCM of each pair of numbers.

67. 6 and 16

68. 9 and 12

69. 10 and 8

70. 30 and 18

58. _____

59. _____

60. _____

61. _____

62. _____

63. _____

64. _____

65. _____

66. _____

67. _____

68. _____

69. _____

70. _____

3.1 USING YOUR KNOWLEDGE

Some Detective Work In this section, we learned how to determine if a given number *satisfies* an equation. This idea can be used to do some detective work! Suppose the police department finds a femur bone from a female. The relationship between the length f of the femur and the height H of a female (in centimeters, cm) is given by

$$H = 1.95f + 72.85$$

1. If the length of the femur is 40 cm and a missing female is known to be 120 cm tall, can the bone belong to the missing female?

2. If the length of the femur is 40 cm and a missing female is known to be 150.85 cm tall, can the bone belong to the missing female?

1. _____

2. _____

3.   Have you seen a police officer measuring the length of a skid mark after an accident? There is a formula relating the velocity V_a at the time of an accident, the length L_a of the skid mark at the time of the accident, and the velocity and length of the skid mark obtained by performing a test consisting of driving the car at a predetermined speed, V_t, skidding to a stop, and then measuring the length of the skid L_t. The formula is

$$V_a^2 = \frac{L_a V_t^2}{L_t}$$

If $L_t = 36$, $L_a = 144$, $V_t = 30$, and the driver claims that at the time of the accident his velocity V_a was 50 mi/hr, can you believe him?

4.   Can you believe him if he says he was going 90 mi/hr?

3. _____

4. _____

CALCULATOR CORNER

Calculators are ideal to check the problems in this section. Thus, to check if $x = \frac{1}{10}$ is a solution of the equation $0.2 = \frac{7}{10} - 5x$ (Problem 10), we key in

$$\boxed{7} \; \boxed{\div} \; \boxed{10} \; \boxed{-} \; \boxed{(} \; \boxed{5} \; \boxed{\times} \; \boxed{1} \; \boxed{\div} \; \boxed{10} \; \boxed{)}$$

and obtain 0.2. Note that we used parentheses when subtracting the $5x$ because some calculators will not obtain the correct answer unless the parentheses are used. (These calculators will divide 7 by 10; then when you key in -5, they will subtract 5 before you have a chance to multiply by $\frac{1}{10}$). If you can key in

$$\boxed{7} \; \boxed{\div} \; \boxed{10} \; \boxed{-} \; \boxed{5} \; \boxed{\times} \; \boxed{1} \; \boxed{\div} \; \boxed{10} \; \boxed{=}$$

and obtain 0.2, you have a calculator with algebraic hierarchy and things may be a little bit easier.

    If your instructor permits it, use your calculator to check the problems in this section and later in Sections 3.2 and 3.3.

3.2 THE MULTIPLICATION AND DIVISION PROPERTY OF EQUALITY

Aramid is a revolutionary lightweight belting material that gives exceptional strength to the tire for excellent impact resistance

28^{00} plus $1.84 F.E.T.

OBJECTIVES

REVIEW

Before starting this section you should know:

1. How to multiply and divide signed numbers.
2. How to find the reciprocal of a number.
3. How to find the LCM of two or more numbers.
4. How to write a fraction as a percent, and vice versa.

OBJECTIVES

You should be able to:

A. Solve an equation such as $\dfrac{x}{a} = b$ by multiplying both sides by a.

B. Solve an equation such as $ax = b$ by dividing both sides by a.

C. Solve an equation such as $\dfrac{a}{b}x = c$ by multiplying both sides by $\dfrac{b}{a}$, the reciprocal of $\dfrac{a}{b}$.

D. Solve equations involving addition or subtraction of fractions by first multiplying by the LCM.

E. Solve applications involving percent problems.

The tire in the ad is on sale at half-price. It now costs $28. What was its old price, p? Since you are paying half-price for the tire, the new price is $\frac{1}{2}$ of p, that is, $\frac{1}{2}p$, or $\dfrac{p}{2}$. Since this price is $28, we have

$$\frac{p}{2} = 28$$

What was the old price? Twice as much, of course. Thus, to obtain the old price p we multiply both sides of the equation by **2,** the reciprocal of $\frac{1}{2}$, obtaining

$$2 \cdot \frac{p}{2} = 2 \cdot 28$$
$$p = 56$$

Hence the old price was $56, as can be easily checked, since $\frac{56}{2} = 28$.

A. THE MULTIPLICATION PROPERTY OF EQUALITY

This example illustrates the fact that you can *multiply* both sides of an equation by a nonzero number and obtain an *equivalent* equation—that is, an equation whose solution is the same as the original one. Here is the principle involved:

THE MULTIPLICATION PROPERTY OF EQUALITY

The equation $a = b$ is equivalent to $ac = bc$ for any nonzero number c.

We use this property next.

EXAMPLE 1 Solve.

a. $\dfrac{x}{3} = 2$ **b.** $\dfrac{y}{5} = -3$

Problem 1 Solve.

a. $\dfrac{y}{4} = 3$ **b.** $\dfrac{x}{2} = -7$

Solution

a. We multiply both sides of the equation by **3**, the reciprocal of $\frac{1}{3}$:

$$3 \cdot \frac{x}{3} = 3 \cdot 2$$

$$\overset{1}{\cancel{3}} \cdot \frac{x}{\cancel{3}} = 6 \quad \text{Note that } 3 \cdot \frac{1}{3} = 1, \text{ since 3 and } \frac{1}{3} \text{ are reciprocals.}$$

$$x = 6$$

Thus, the solution is 6.

Check: $\dfrac{x}{3} \overset{?}{=} 2$

$$\begin{array}{c|c} \dfrac{6}{3} & 2 \end{array}$$

b. We multiply both sides by **5**, the reciprocal of $\frac{1}{5}$:

$$5 \cdot \frac{y}{5} = 5(-3)$$

$$\overset{1}{\cancel{5}} \cdot \frac{y}{\cancel{5}} = -15 \quad \text{Recall that } 5 \cdot (-3) = -15.$$

$$y = -15$$

Thus, the solution is -15.

Check: $\dfrac{y}{5} \overset{?}{=} -3$

$$\begin{array}{c|c} \dfrac{-15}{5} & -3 \end{array}$$

B. THE DIVISION PROPERTY OF EQUALITY

Suppose the price of an article is doubled and it now sells for $50. What was its original price, p? Half as much, right? Here is the equation:

$$2p = 50$$

We solve it by dividing both sides by **2** (to find half as much):

$$\frac{2p}{2} = \frac{50}{2}$$

$$\frac{\overset{1}{\cancel{2}}p}{\cancel{2}} = 25$$

Thus the original price p is $25, as you can check:

$$2 \cdot 25 = 50$$

Note that dividing both sides by 2 is the same as *multiplying* by $\frac{1}{2}$. Thus, you can also solve

$$2p = 50$$

by *multiplying* by $\frac{1}{2}$, the reciprocal of 2, obtaining

$$2p = 50$$
$$\frac{1}{2} \cdot 2p = \frac{1}{2} \cdot 50$$
$$p = 25$$

This example suggests that we can *divide* both sides of an equation by a (nonzero) number and obtain an *equivalent* equation.

We now state this principle and use it in the next example.

THE DIVISION PROPERTY OF EQUALITY

The equation $a = b$ is equivalent to $\dfrac{a}{c} = \dfrac{b}{c}$ for any nonzero number c.

Note that we can also *multiply* both sides of $a = b$ by the *reciprocal* of c, $\dfrac{1}{c}$, obtaining

$$\frac{1}{c} \cdot a = \frac{1}{c} \cdot b$$

or

$$\frac{a}{c} = \frac{b}{c} \quad \text{(Same result!)}$$

EXAMPLE 2 Solve.

a. $8x = 24$ **b.** $5x = -20$ **c.** $-3x = 7$

Solution

a. We need to get x by itself on the left. That is, we need $x = \boxed{}$, where $\boxed{}$ is a number. Thus we divide both sides of the equation by 8 (the coefficient of x), obtaining

$$\frac{8x}{8} = \frac{24}{8}$$

$$\frac{\overset{1}{\cancel{8}}x}{\cancel{8}} = 3$$

$$x = 3$$

The solution is 3.

Check: $8x \overset{?}{=} 24$

$$8 \cdot 3 \mid 24$$

Problem 2 Solve.
a. $7y = 21$ **b.** $3y = -18$
c. $-7y = 16$

You can also solve this problem by *multiplying* both sides by $\frac{1}{8}$, the reciprocal of 8.

$$\frac{1}{8} \cdot 8x = \frac{1}{8} \cdot 24$$

$$1x = \frac{1}{\cancel{8}} \cdot \frac{\overset{3}{\cancel{24}}}{1}$$

$$x = 3$$

b. Here, we divide by 5 (the coefficient of x) so that we have x by itself on the left. Thus,

$$\frac{5x}{5} = \frac{-20}{5}$$

$$x = -4$$

The solution is -4.

Check: $\qquad 5x \overset{?}{=} 20$

$$\overline{\qquad 5 \cdot (-4) \mid -20 \qquad}$$

You can also solve this problem by *multiplying* both sides by $\frac{1}{5}$, the reciprocal of 5.

$$\frac{1}{5} \cdot 5x = \frac{1}{5} \cdot (-20)$$

$$1x = \frac{1}{\cancel{5}} \cdot \frac{\overset{-4}{\cancel{-20}}}{1}$$

$$x = -4$$

c. In this case we divide by -3 (the coefficient of x). We then have

$$\frac{\overset{1}{\cancel{-3}}x}{\cancel{-3}} = \frac{7}{-3}$$

$$x = -\frac{7}{3} \qquad \text{Recall that the quotient of two numbers with unlike signs is negative.}$$

The solution is $-\frac{7}{3}$.

Check: $\qquad -3x \overset{?}{=} 7$

$$\overline{\qquad -3\left(-\dfrac{7}{3}\right) \mid -7 \qquad}$$

$$-7 \mid$$

You can also solve this problem by multiplying both sides by $-\frac{1}{3}$, the reciprocal of -3.

$$-\frac{1}{3} \cdot (-3x) = -\frac{1}{3} \cdot 7$$

$$1x = -\frac{1}{3} \cdot \frac{7}{1}$$

$$x = -\frac{7}{3}$$

C. MULTIPLYING BY RECIPROCALS

Now suppose a person buys one of the items advertised at $\frac{1}{3}$ off and pays \$10 for it. What was the original price of the item? Since the item is $\frac{1}{3}$ off, the person is paying $\frac{2}{3}p$, or \$10, for the item. Thus,

$$\frac{2}{3}p = 10$$

To solve this equation, we need to multiply (or divide) the left member by a number that will leave the answer in the form $p = \boxed{}$; that is, we need a number that multiplied by $\frac{2}{3}$ will yield 1. This number is the **reciprocal** (or **multiplicative inverse**) of $\frac{2}{3}$, and it is simply obtained by interchanging the numerator and denominator of $\frac{2}{3}$.

In general, the reciprocal (or multiplicative inverse) of

$$\frac{a}{b} \text{ is } \frac{b}{a} \quad \text{and} \quad \frac{a}{b} \cdot \frac{b}{a} = 1$$

Thus the reciprocal of $\frac{2}{3}$ is $\frac{3}{2}$. Multiplying both sides by $\frac{3}{2}$, we obtain

$$\frac{3}{2}\left(\frac{2}{3}p\right) = \frac{3}{2}(10) \quad \text{Note that since } \frac{3}{2} \text{ and } \frac{2}{3} \text{ are reciprocals, } \frac{3}{2} \cdot \frac{2}{3} = 1.$$

$$1 \cdot p = \frac{3}{2} \cdot \frac{10}{1}$$

$$p = \frac{30}{2} = 15$$

Hence the original price p was \$15, which can easily be checked, since

$$\frac{2}{3}(15) = \frac{2}{3} \cdot \frac{15}{1} = \frac{30}{3} = 10$$

What does this example tell us? It suggests that if we have an equation of the form

$$\frac{a}{b}x = c$$

we simply multiply by the reciprocal (multiplicative inverse) of $\frac{a}{b}$, that is, by $\frac{b}{a}$, to obtain

$$\frac{b}{a}\left(\frac{a}{b}x\right) = \frac{b}{a}(c)$$

$$1 \cdot x = \frac{b}{a} \cdot \frac{c}{1}$$

$$x = \frac{bc}{a}.$$

Note that you can use the idea of multiplying by a reciprocal when solving equations of the form $\frac{x}{a} = b$ $\left(\text{multiply by } a, \text{ the reciprocal of } \frac{1}{a}\right)$ or equations of the form $ax = b$ $\left(\text{multiply by } \frac{1}{a}, \text{ the reciprocal of } a\right)$.

EXAMPLE 3 Solve.

a. $\dfrac{3}{4}x = 18$ **b.** $-\dfrac{2}{5}x = 8$ **c.** $-\dfrac{3}{8}x = -15$

Problem 3   Solve.

a. $\dfrac{4}{5}y = 8$ **b.** $-\dfrac{3}{4}y = 6$

c. $-\dfrac{2}{7}y = -4$

Solution

a. Multiply both sides by the reciprocal of $\frac{3}{4}$ (which is $\frac{4}{3}$). We then get

$$\frac{4}{3}\left(\frac{3}{4}x\right) = \frac{4}{3}(18)$$

$$1 \cdot x = \frac{4}{\cancel{3}} \cdot \frac{\overset{6}{\cancel{18}}}{1} = \frac{24}{1}$$

$$x = 24$$

Hence the solution is 24.

Check: $\dfrac{3}{4}x \overset{?}{=} 18$

$$
\begin{array}{c|c}
\dfrac{3}{4}(24) & 18 \\[2ex]
\dfrac{3}{\cancel{4}} \cdot \dfrac{\overset{6}{\cancel{24}}}{1} & \\[1ex]
18 &
\end{array}
$$

b. Multiplying both sides by the reciprocal of $-\frac{2}{5}$, that is, by $-\frac{5}{2}$, we have

$$-\frac{5}{2}\left(-\frac{2}{5}x\right) = -\frac{5}{2}(8)$$

$$1 \cdot x = \frac{-5}{\cancel{2}} \cdot \frac{\overset{4}{\cancel{8}}}{1} = -\frac{20}{1}$$

$$x = -20$$

The solution is -20.

Check: $-\dfrac{2}{5}x \overset{?}{=} 8$

$$
\begin{array}{c|c}
-\dfrac{2}{5}(-20) & 8 \\[2ex]
-\dfrac{2}{\cancel{5}}\left(\dfrac{\overset{-4}{\cancel{-20}}}{1}\right) & \\[1ex]
8 &
\end{array}
$$

c. Here we multiply by $-\frac{8}{3}$, the reciprocal of $-\frac{3}{8}$, obtaining

$$-\frac{8}{3}\left(\frac{-3}{8}x\right) = -\frac{8}{3}(-15)$$

$$x = \frac{-8(-15)}{3} = \frac{-8(\overset{-5}{-15})}{\underset{1}{3}} = 40$$

The solution is 40.

Check: $\qquad -\frac{3}{8}x \overset{?}{=} -15$

$$-\frac{3}{8}(40) \quad\bigg|\quad -15$$

$$-\frac{3}{\underset{1}{8}}\overset{5}{(40)}$$

$$-15 \quad\bigg|$$

D. MULTIPLYING BY THE LCM

Finally, if the equation to be solved contains sums or differences of fractions, we first eliminate these fractions by multiplying each term in the equation by the smallest number that is a multiple of each of the denominators. This number is called the *lowest common multiple* (or LCM for short) of the denominators. Do you remember how to find this number? If you don't, there is a quick way of doing it. Suppose you wish to solve the equation

$$\frac{x}{6} + \frac{x}{16} = 22$$

To find the LCM of 6 and 16, write the denominators in a horizontal row (as shown in Step 1 in the margin) and divide each of them by the largest number that will divide *both* of them. In this case the number is **2.** The quotients are **3** and **8,** as shown in Step 3. Since there are no other numbers that will divide both 3 and 8, the LCM is the product of **2** and the final quotients **3** and **8,** as indicated in Step 4. Now that we have the LCM, 48, we multiply each side of the equation by this LCM, obtaining

STEP 1. $\quad \big|\, 6 \quad 16$

STEP 2. $\;2\,\big|\, 6 \quad 16$

STEP 3. $\;2\,\big|\, 6 \quad 16$
$\qquad\qquad\;\; 3 \quad\; 8$

STEP 4. $\;2\,\big|\, 6 \quad 16$
$\qquad\qquad \underset{}{\llcorner}\, 3 \!-\! 8 \to 2 \cdot 3 \cdot 8 = 48$
$\qquad\qquad\qquad\quad$ is the LCM.

$$48\left(\frac{x}{6} + \frac{x}{16}\right) = 48 \cdot 22$$

$$\overset{8}{48} \cdot \frac{x}{\underset{1}{6}} + \overset{3}{48} \cdot \frac{x}{\underset{1}{16}} = 48 \cdot 22 \qquad \textit{Note:} \quad \text{Do not multiply } 48 \cdot 22 \text{ yet,}$$
$$\text{we will simplify this later.}$$

$$8x + 3x = 48 \cdot 22$$

$$11x = 48 \cdot 22$$

$$x = \frac{48 \cdot \overset{2}{22}}{\underset{1}{11}}$$

$$x = 96$$

The solution is 96.

Check: $\dfrac{x}{6} + \dfrac{x}{16} \overset{?}{=} 22$

$$\dfrac{96}{6} + \dfrac{96}{16} \quad\bigg|\quad 22$$

$$16 + 6$$

$$22 \quad\bigg|$$

▲

Note that if we wish to eliminate fractions in

$$\dfrac{a}{b} + \dfrac{c}{d} = \dfrac{e}{f}$$

by multiplying both sides by the LCM (which we call L), we get

$$L\left(\dfrac{a}{b} + \dfrac{c}{d}\right) = L\dfrac{e}{f}$$

or

$$\dfrac{La}{b} + \dfrac{Lc}{d} = \dfrac{Le}{f}$$

Thus, we multiplied each *term* by L.

EXAMPLE 4 Solve.

a. $\dfrac{x}{10} + \dfrac{x}{8} = 9$ **b.** $\dfrac{x}{3} - \dfrac{x}{8} = 10$

Solution

a. The LCM of 10 and 8 is 40, since the first three multiples of 10 are 20, 30, 40 and 8 divides 40. You can also find the LCM by writing

$$\begin{array}{c|cc} 2 & 10 & 8 \\ \hline & 5 & 4 \end{array} \rightarrow 2 \cdot 5 \cdot 4 = 40$$

Multiplying each term by **40,** we have

$$40 \cdot \dfrac{x}{10} + 40 \cdot \dfrac{x}{8} = 40 \cdot 9$$

$$4x + 5x = 40 \cdot 9 \quad \text{Simplifying}$$

$$9x = 40 \cdot 9 \quad \text{Combining like terms}$$

$$x = 40 \qquad \text{Dividing by 9}$$

The solution is 40.

Check: $\dfrac{x}{10} + \dfrac{x}{8} \overset{?}{=} 9$

$$\dfrac{40}{10} + \dfrac{40}{8} \quad\bigg|\quad 9$$

$$4 + 5$$

$$9 \quad\bigg|$$

b. The LCM of 3 and 8 is 24, since the largest number that divides 3 and 8 is 1.

$$\begin{array}{c|cc} 1 & 3 & 8 \\ \hline & 3 & 8 \end{array} \rightarrow 1 \cdot 3 \cdot 8 = 24$$

Problem 4 Solve.

a. $\dfrac{y}{6} + \dfrac{y}{10} = 8$ **b.** $\dfrac{y}{4} - \dfrac{y}{7} = 3$

Multiplying each term by 24 yields

$$24 \cdot \frac{x}{3} - 24 \cdot \frac{x}{8} = 24 \cdot 10$$

$$8x - 3x = 24 \cdot 10 \quad \text{Simplifying}$$

$$5x = 24 \cdot 10 \quad \text{Combining like terms}$$

$$x = \frac{24 \cdot \overset{2}{\cancel{10}}}{\underset{1}{\cancel{5}}} \quad \text{Dividing by 5}$$

$$x = 48$$

The solution is 48.

Check: $\quad \dfrac{x}{3} - \dfrac{x}{8} \overset{?}{=} 10$

$$
\begin{array}{c|c}
\dfrac{48}{3} - \dfrac{48}{8} & 10 \\[2mm]
16 - 6 & \\[1mm]
10 &
\end{array}
$$

▲

    In some cases, the numerators of the fractions involved contain more than one term. However, the procedure to solve equations is still the same. We illustrate such an occurrence in Example 5.

EXAMPLE 5 Solve.

a. $\dfrac{x+1}{3} + \dfrac{x-1}{10} = 5$ **b.** $\dfrac{x+1}{3} - \dfrac{x-1}{8} = 4$

Problem 5 Solve.

a. $\dfrac{x+2}{4} + \dfrac{x-1}{5} = 3$

b. $\dfrac{x+3}{2} - \dfrac{x-2}{3} = 5$

Solution

a. The LCM of 3 and 10 is $3 \cdot 10 = 30$, since 3 and 10 do not have any common factors. Multiplying each term by 30, we have

$$\overset{10}{\cancel{30}}\left(\frac{x+1}{3}\right) + \overset{3}{\cancel{30}}\left(\frac{x-1}{10}\right) = 30 \cdot 5$$

$$10(x+1) + 3(x-1) = 150$$

$$10x + 10 + 3x - 3 = 150 \quad \text{Using the distributive law}$$

$$13x + 7 = 150 \quad \text{Combining like terms}$$

$$13x = 143 \quad \text{Subtracting 7}$$

$$x = 11 \quad \text{Dividing by 13}$$

The check is left to the student.

b. Here the LCM is $3 \cdot 8 = 24$. Multiplying each term by 24, we obtain

$$\overset{8}{\cancel{24}}\left(\frac{x+1}{\cancel{3}}\right) - \overset{8}{\cancel{24}}\left(\frac{x-1}{\cancel{8}}\right) = 24 \cdot 4$$

$$8(x+1) - 3(x-1) = 96$$

$$8x + 8 - 3x + 3 = 96 \quad \text{Using the distributive law}$$

$$5x + 11 = 96 \quad \text{Combining like terms}$$

$$5x = 85 \quad \text{Subtracting 11}$$

$$x = 17 \quad \text{Dividing by 5}$$

Be *sure* you *check* this answer in the original equation.

E. APPLICATIONS: PERCENT PROBLEMS

Among the most common types of problems in mathematics and other fields are percent problems. Basically, there are three types of percent problems.

Type 1 asks you to find a percent of a number.

EXAMPLE   20% (read "20 percent") of 80 is what number?

Type 2 asks you what percent of a number is another given number.

EXAMPLE   What percent of 20 is 5?

Type 3 asks you to find a number when it is known that a given percent of the number equals another given number.

EXAMPLE   10 is 40% of what number?

To do these problems, you need only recall how to translate words into the language of algebra and how to write percents as fractions (Sections 1.4 and 2.5).

Now, do you remember what 20% means? Recall the symbol % is read as "percent," which means "for every hundred." Thus % means "for every hundred." As you recall,

$$20\% = \frac{20}{100} = \frac{\overset{1}{\cancel{20}}}{\underset{5}{\cancel{100}}} = \frac{1}{5}$$

Similarly

$$60\% = \frac{60}{100} = \frac{\overset{3}{\cancel{60}}}{\underset{5}{\cancel{100}}} = \frac{3}{5}$$

$$17\% = \frac{17}{100}$$

We are now ready to do the next examples.

EXAMPLE 6   20% of 80 is what number?

Solution

$$\frac{20}{100} \cdot 80 = n \qquad \text{Translating}$$

$$\frac{1}{5} \cdot 80 = n \qquad \text{Since } \frac{20}{100} = \frac{1}{5}$$

$$\frac{80}{5} = n \qquad \text{Multiplying } \frac{1}{5} \text{ by } \frac{80}{1}$$

$$n = 16 \qquad \text{Reducing } \frac{80}{5}$$

Thus 20% of 80 is 16.

20% means "20 for every hundred," that is, $\frac{20}{100}$.

Problem 6   40% of 60 is what number?

ANSWERS   (to problems on pages 210–211)
4. a. $y = 30$ **b.** $y = 28$
5. a. $x = 6$ **b.** $x = 17$

EXAMPLE 7   What percent of 20 is 5?

Solution
$$x \cdot 20 = 5 \quad \text{Translating}$$

$$\frac{x \cdot 20}{20} = \frac{5}{20} \quad \text{Dividing by 20}$$

$$x = \frac{1}{4} \quad \text{Reducing } \frac{5}{20}$$

Thus $x = \frac{1}{4} = \frac{25}{100}$ or 25%.

EXAMPLE 8   10 is 40% of what number?

Solution
$$10 = \frac{40}{100} \cdot n \quad \text{Translating}$$

$$\frac{40}{100} \cdot n = 10 \quad \text{Rearranging}$$

$$\frac{2}{5} \cdot n = 10 \quad \text{Reducing } \frac{40}{100}$$

$$\frac{\overset{1}{\cancel{5}}}{\overset{}{\cancel{2}}} \cdot \frac{\cancel{2}}{\cancel{5}} n = \frac{5}{\cancel{2}} \cdot \overset{5}{\cancel{10}} \quad \text{Multiplying by } \frac{5}{2}$$

$$n = 25$$

Problem 7   What percent of 40 is 8?

Problem 8   20 is 40% of what number?

ANSWERS (to problems on pages 212–213)

6. 24   **7.** 20%   **8.** 50

NAME

CLASS

SECTION

A. Solve.

ANSWERS

1. $\dfrac{x}{7} = 5$

2. $\dfrac{y}{2} = 9$

3. $-4 = \dfrac{x}{2}$

4. $\dfrac{a}{5} = -6$

5. $\dfrac{b}{-3} = 5$

6. $7 = \dfrac{c}{-4}$

7. $-3 = \dfrac{f}{-2}$

8. $\dfrac{g}{-4} = -6$

9. $\dfrac{v}{4} = \dfrac{1}{3}$

10. $\dfrac{w}{3} = \dfrac{2}{7}$

11. $\dfrac{x}{5} = \dfrac{-3}{4}$

12. $\dfrac{-8}{9} = \dfrac{y}{2}$

B. Solve.

13. $3z = 33$

14. $4y = 32$

15. $-42 = 6x$

16. $7b = -49$

17. $-8c = 56$

18. $-5d = 45$

19. $-5x = -35$

20. $-12 = -3x$

21. $-3y = 11$

22. $-5z = 17$

23. $-2a = 1.2$

24. $-3b = 1.5$

25. $3t = 4\dfrac{1}{2}$

26. $4r = 6\dfrac{2}{3}$

C. Solve.

27. $\dfrac{1}{3}x = -0.75$

28. $\dfrac{1}{4}y = 0.25$

ANSWERS

1. _____
2. _____
3. _____
4. _____
5. _____
6. _____
7. _____
8. _____
9. _____
10. _____
11. _____
12. _____
13. _____
14. _____
15. _____
16. _____
17. _____
18. _____
19. _____
20. _____
21. _____
22. _____
23. _____
24. _____
25. _____
26. _____
27. _____
28. _____

29. $-6 = \dfrac{3}{4}C$

30. $-2 = \dfrac{2}{9}F$

31. $\dfrac{5}{6}a = 10$

32. $24 = \dfrac{2}{7}z$

33. $-\dfrac{4}{5}y = 0.4$

34. $0.5x = \dfrac{-1}{4}$

35. $\dfrac{-2}{11}p = 0$

36. $\dfrac{-4}{9}q = 0$

37. $-18 = \dfrac{3}{5}t$

38. $-8 = \dfrac{2}{7}R$

39. $\dfrac{7x}{0.02} = -7$

40. $-6 = \dfrac{6y}{0.03}$

D. Solve.

41. $\dfrac{y}{2} + \dfrac{y}{3} = 10$

42. $\dfrac{a}{4} + \dfrac{a}{3} = 14$

43. $\dfrac{x}{7} + \dfrac{x}{3} = 10$

44. $\dfrac{z}{6} + \dfrac{z}{4} = 20$

45. $\dfrac{x}{5} + \dfrac{x}{10} = 6$

46. $\dfrac{r}{2} + \dfrac{r}{6} = 8$

47. $\dfrac{t}{6} + \dfrac{t}{8} = 7$

48. $\dfrac{f}{9} + \dfrac{f}{12} = 14$

49. $\dfrac{x}{2} + \dfrac{x}{5} = \dfrac{7}{10}$

50. $\dfrac{a}{3} + \dfrac{a}{7} = \dfrac{20}{21}$

51. $\dfrac{c}{3} - \dfrac{c}{5} = 2$

52. $\dfrac{F}{4} - \dfrac{F}{7} = 3$

53. $\dfrac{W}{6} - \dfrac{W}{8} = \dfrac{5}{12}$

54. $\dfrac{m}{6} - \dfrac{m}{10} = \dfrac{4}{3}$

55. $\dfrac{x}{5} - \dfrac{3}{10} = \dfrac{1}{2}$

56. $\dfrac{3y}{7} - \dfrac{1}{14} = \dfrac{1}{14}$

57. $\dfrac{x+4}{4} - \dfrac{x+2}{3} = -\dfrac{1}{2}$

58. $\dfrac{w-1}{2} + \dfrac{w}{8} = \dfrac{7w+1}{16}$

59. $\dfrac{x}{6} + \dfrac{3}{4} = x - \dfrac{7}{4}$

60. $\dfrac{x}{6} + \dfrac{4}{3} = x - \dfrac{1}{3}$

E. Applications

61. 30% of 40 is what number?

29. _____
30. _____
31. _____
32. _____
33. _____
34. _____
35. _____
36. _____
37. _____
38. _____
39. _____
40. _____
41. _____
42. _____
43. _____
44. _____
45. _____
46. _____
47. _____
48. _____
49. _____
50. _____
51. _____
52. _____
53. _____
54. _____
55. _____
56. _____
57. _____
58. _____
59. _____
60. _____
61. _____

62. 17% of 80 is what number?

63. 40% of 70 is what number?

64. What percent of 40 is 8?

65. What percent of 30 is 15?

66. What percent of 40 is 4?

67. 30 is 20% of what number?

68. 20 is 40% of what number?

69. 12 is 60% of what number?

70. 24 is 50% of what number?

71. An item is selling for half of its original price. It now sells for $12. How much was the original price?

72. The price of an item has tripled and it now sells for $36. What was its original price?

73. An item is on sale at $\frac{1}{3}$ off. It now costs $8. What was its original price?

74. An item can be bought for $\frac{1}{3}$ of its original price. If it is now selling for $17, what was its original price?

75. A popular light beer advertises that it has $\frac{1}{3}$ less calories than their regular beer. If this light beer has 100 calories, how many calories does the regular beer have?

62. _____

63. _____

64. _____

65. _____

66. _____

67. _____

68. _____

69. _____

70. _____

71. _____

72. _____

73. _____

74. _____

75. _____

✓ **SKILL CHECKER**

Use the distributive law to multiply.

76. $3(6 - x)$

77. $5(8 - y)$

78. $6(8 - 2y)$

79. $9(6 - 3y)$

80. $-3(4x - 2)$

81. $-5(3x - 4)$

Find.

82. $20 \cdot \dfrac{3}{4}$

83. $24 \cdot \dfrac{1}{6}$

84. $-5 \cdot \left(-\dfrac{4}{5}\right)$

85. $-7 \cdot \left(-\dfrac{3}{7}\right)$

76. _____

77. _____

78. _____

79. _____

80. _____

81. _____

82. _____

83. _____

84. _____

85. _____

3.2 USING YOUR KNOWLEDGE

Have you ever been on strike? If so, after negotiating a salary increase, did you come out ahead? How can we find out? Let us take an example. If an employee earns $10 an hour and works 250 8-h days a year, her salary would be $10 \cdot 8 \cdot 250 = \$20,000$. What percent increase in salary is needed to make up for the lost time during the strike? First, we must know how many days were lost (the average strike in the United States lasts about 100 days). Let us assume that 100 days were lost. The wages for 100 days would be $10 \cdot 8 \cdot 100 = \$8000$. The employee needs \$20,000 plus \$8000 or \$28,000 to make up the lost wages. Let x be the multiple of the old wages needed to make up the loss within 1 yr. We then have:

$$20,000 \cdot x = 28,000$$

$$\frac{20,000 \cdot x}{20,000} = \frac{28,000}{20,000} \quad \text{Dividing by 20,000}$$

$$x = 1.40 \quad \text{(or 140\%)}$$

Thus, the employee needs a 40% increase to recuperate the lost salary within 1 yr.

1. Assume that an employee works 8 hr a day for 250 days. If the employee makes \$20 an hour, what is the yearly salary?

2. If the employee goes on strike for 100 days, find the wages lost.

3. What percent increase will be necessary to make up for the lost wages within 1 yr?

1. _____

2. _____

3. _____

TYPE	MILEAGE RATE			FLAT RATE	
	DAY	WEEK	MILE	DAY	WEEK
Economy	$9.00	$49.00	.10	$15.00	$79.00
Subcompact	10.00	55.00	.10	17.00	89.00
Compact	11.00	60.00	.11	17.00	89.00
Intermediate sedan	12.00	65.00	.12	19.00	99.00
Sedans	13.00	70.00	.13	17.95	99.00
Premium sedans	15.00	85.00	.15	20.00	109.00
Station wagon	18.00	90.00	.18	25.00	135.00

You pay only for gas you use!
Minimum rental 24 hours.

OBJECTIVES

REVIEW

Before starting this section you should know:

1. How to find the LCM of three numbers.
2. How to add, subtract, multiply, and divide rational numbers.
3. How to use the distributive law to remove parentheses.

OBJECTIVES

You should be able to:

A. Solve linear equations using the six-step procedure given in the text.
B. Solve a literal equation for one of the unknowns using the six-step procedure in the book.

Suppose you want to rent a subcompact. Which is the better deal, to take the mileage rate (10¢ per mile, $10 a day) or the $17 flat rate? Well, it depends on how many miles you travel! First, let us see how many miles you have to travel before the costs are identical. The per-day cost C based on traveling m miles is

$$C = \overbrace{0.10m}^{\text{Mileage}} + \overbrace{10}^{\text{Fixed}}$$

Note that the cost C consists of two parts: the $10 fixed charge and the $0.10m$ mileage charge, which is the additional charge when you travel m miles. This cost is the same as the $17 flat rate when $C = 17$. Thus we write

$$\underset{17}{\underbrace{\text{The flat rate}}} \quad \underset{=}{\underbrace{\text{is the same as}}} \quad \underset{0.10m + 10}{\underbrace{\text{the per-mile rate}}}$$

If we solve for m we will know how many miles we have to go before the mileage rate and the flat rate are the same. Here is the equation again:

$$0.10m + 10 = 17$$
$$0.10m + 10 - 10 = 17 - 10 \quad \text{Subtracting } \mathbf{10}$$
$$0.10m = 7$$
$$\frac{0.10m}{0.10} = \frac{7}{0.10} \quad \text{Dividing by } \mathbf{0.10:}$$
$$m = 70$$

$$\begin{array}{r} 70. \\ 0.10\overline{)7.00} \\ \underline{7\ 0} \\ 00 \end{array}$$

Thus when you travel **70** mi, the cost is $C = 0.10 \cdot \mathbf{70} + 10 = 17$ (which is the same as the flat rate). Of course, if you travel more than 70 mi per day, you pay more than $17. So, here is the decision: If you plan to travel *less* than 70 mi per day, use the mileage rate. If you plan to travel *more* than 70 mi, use the flat rate.

Now, let us concentrate on the idea we used in solving $0.10m + 10 = 17$. The main objective is to have a solution of the form $m = \boxed{}$, where $\boxed{}$ is a number. Because of this, we first subtracted 10 and then divided by 0.10. This technique works for any equation of the form

$$ax + b = c$$

a, *b*, and *c* are numbers.

A. SOLVING LINEAR EQUATIONS

Equations that can be written in the form

$$ax + b = c$$

are called **linear equations.** How do we solve them? The same way we solved for m before. We use the subtraction and division properties of equality that we studied in Section 3.2.

1. Given: $\qquad\qquad ax + b = c$

2. Subtract b: $\qquad ax + b - b = c - b$

$$ax = c - b$$

3. Dividing by a: $\qquad \dfrac{ax}{a} = \dfrac{c - b}{a}$

$$x = \dfrac{c - b}{a}$$

Linear equations may have *one* solution (as in the next examples), *no* solution ($x + 1 = x + 2$ has *no* solution), or *many* solutions ($2x + 2 = 2(x + 1)$ has many solutions).

EXAMPLE 1 Solve.

a. $3x + 7 = 13$ $\qquad$ **b.** $-5x - 3 = 1$

Problem 1 Solve.
a. $3x + 8 = 11$ $\qquad$ **b.** $-4x - 5 = 2$

Solution

a. 1. Given: $\qquad\qquad\qquad\qquad\qquad 3x + 7 = 13$

$\qquad$ **2.** Subtract 7: $\qquad\qquad\qquad 3x + 7 - 7 = 13 - 7$

$$3x = 6$$

$\qquad$ **3.** Divide by 3
$\qquad$ (or multiply by the reciprocal of 3): $\qquad \dfrac{3x}{3} = \dfrac{6}{3}$

$$x = 2$$

The solution is 2.

Check: $\qquad 3x + 7 \overset{?}{=} 13$

$$\begin{array}{c|c} 3(2) + 7 & 13 \\ 6 + 7 & \\ 13 & \end{array}$$

b. 1. Given: $-5x - 3 = 1$

2. Add 3: $-5x - 3 + 3 = 1 + 3$

$$-5x = 4$$

3. Divide by -5
(or multiply by the reciprocal of -5):

$$\frac{-5x}{-5} = \frac{4}{-5}$$

$$x = -\frac{4}{5}$$

The solution is $-\dfrac{4}{5}$.

Check: $\quad -5x - 3 = 1$

$$
\begin{array}{c|c}
-5\left(-\dfrac{4}{5}\right) - 3 & 1 \\
4 - 3 & \\
1 &
\end{array}
$$

▲

Note that the main idea is to place the variables on one side of the equation until you can write the solution in the form $x = \boxed{}$ (or $\boxed{} = x$), where $\boxed{}$ is a number (a constant). Can we solve the equation $5(x + 2) = 3(x + 1) + 9$? This equation is *not* of the form $ax + b = c$, but we can write it in this form if we use the distributive law to remove parentheses. Here is the way we do it.

$$5(x + 2) = 3(x + 1) + 9 \qquad \text{Given}$$

$$5x + 10 = 3x + 3 + 9 \qquad \text{Using the distributive law to}$$
$$5x + 10 = 3x + 12 \qquad\qquad \text{remove parentheses}$$

$$5x + 10 - 10 = 3x + 12 - 10 \qquad \text{Subtracting \textbf{10} on both sides}$$

$$5x = 3x + 2$$

$$5x - 3x = 3x - 3x + 2 \qquad \text{Subtracting \textbf{3x} on both sides}$$

$$2x = 2$$

$$\frac{2x}{2} = \frac{2}{2} \qquad\qquad \text{Dividing both sides by \textbf{2}}$$
$$\qquad\qquad\qquad\qquad \text{(or multiplying by the reciprocal of 2)}$$

$$x = 1$$

The solution is 1.

Check: $5(x + 2) \overset{?}{=} 3(x + 1) + 9$

$$
\begin{array}{c|c}
5(1 + 2) & 3(1 + 1) + 9 \\
5(3) & 3(2) + 9 \\
15 & 6 + 9 \\
 & 15
\end{array}
$$

What if we have fractions in the equation? No problem. We can clear them by multiplying by the LCM, as we did before. For example, to solve

$$\frac{3}{4} + \frac{x}{10} = 1$$

we first multiply each term by **20,** the LCM of 4 and 10. You can obtain the LCM by noting that the first multiple of 10 is 20 and 20 is divisible by 4, or by writing

$$\frac{2\;\lfloor\; 4 \qquad 10}{\qquad 2 \text{——} 5} \to 2 \cdot 2 \cdot 5 = 20 \quad \text{is the LCM}$$

In either case:

Multiply by the LCM:
$$\overset{5}{\cancel{20}} \cdot \frac{3}{\cancel{4}} + \overset{2}{\cancel{20}} \cdot \frac{x}{\cancel{10}} = 20 \cdot 1$$

Simplify: $\qquad\qquad\qquad\qquad 15 + 2x = 20$

Subtract 15: $\qquad\qquad\qquad\qquad 2x = 5$

Divide by 2
(or multiply by the reciprocal of 2): $\qquad \dfrac{2x}{2} = \dfrac{5}{2}$

$$x = \frac{5}{2}$$

The solution is $\frac{5}{2}$.

Check: $\qquad \dfrac{3}{4} + \dfrac{x}{10} \overset{?}{=} 1$

$$
\begin{array}{c|c}
\dfrac{3}{4} + \dfrac{\frac{5}{2}}{10} & 1 \\[2ex]
\dfrac{3}{4} + \dfrac{5}{2} \cdot \dfrac{1}{10} & \\[2ex]
\dfrac{3}{4} + \dfrac{5}{20} & \\[2ex]
\dfrac{3}{4} + \dfrac{1}{4} & \\[2ex]
1 &
\end{array}
$$

The procedure we have used in solving the preceding examples can be generalized so that it can be used to solve any linear equation. As before, what we need to do is have the variables on one side of the equation and the numbers on the other so that we can write the solution in the form $x = \boxed{}$ or $\boxed{} = x$. Here is the way to accomplish this, with a step-by-step example shown in the margin. (If you have forgotten how to find the LCM for three numbers, see Section 1.2.)

PROCEDURE FOR SOLVING LINEAR EQUATIONS

1. If there are fractions, multiply each term on both sides of the equation by the LCM of the fractions.

2. Remove parentheses and collect like terms (simplify) if necessary.

3. Add or subtract the same number on both sides of the equation so that one side has only variables.

4. Add or subtract the same expression on both sides of the equation so that the other side has only numbers.

5. If the coefficient of the variable is not 1, divide both sides of the equation by this coefficient (or, equivalently, multiply by the reciprocal of the coefficient of the variable).

6. Be sure to check your answer in the original equation.

Given: $\quad \dfrac{x}{4} - \dfrac{1}{6} = \dfrac{7}{12}(x - 2)$

This is one term.

1. $12 \cdot \dfrac{x}{4} - 12 \cdot \dfrac{1}{6} = 12 \cdot \overbrace{\left[\dfrac{7}{12}(x - 2)\right]}$

2. $3x - 2 = 7(x - 2)$
$\qquad\qquad = 7x - 14$

3. $3x - 2 + 2 = 7x - 14 + 2$
$\qquad\quad 3x = 7x - 12$

4. $3x - 7x = 7x - 7x - 12$
$\qquad -4x = -12$

5. $\dfrac{-4x}{-4} = \dfrac{-12}{-4}$

$\qquad x = 3$

The solution is 3.

6. Check: $\quad \dfrac{x}{4} - \dfrac{1}{6} \overset{?}{=} \dfrac{7}{12}(x - 2)$

$$
\begin{array}{c|c}
\dfrac{3}{4} - \dfrac{1}{6} & \dfrac{7}{12}(3 - 2) \\[2ex]
\dfrac{9}{12} - \dfrac{2}{12} & \dfrac{7}{12} \cdot 1 \\[2ex]
\dfrac{7}{12} & \dfrac{7}{12}
\end{array}
$$

EXAMPLE 2 Solve.

a. $\dfrac{7}{24} = \dfrac{x}{8} + \dfrac{1}{6}$ **b.** $\dfrac{1}{5} - \dfrac{x}{4} = \dfrac{7(x+3)}{10}$

Problem 2 Solve.

a. $\dfrac{7}{12} = \dfrac{x}{4} + \dfrac{1}{3}$ **b.** $\dfrac{1}{3} - \dfrac{x}{5} = \dfrac{8(x+2)}{15}$

Solution

a. We use the six steps just given to solve $\dfrac{7}{24} = \dfrac{x}{8} + \dfrac{1}{6}$.

1. Multiply by 24, the LCM:

$$\overset{}{24} \cdot \frac{7}{24} = \overset{3}{24} \cdot \frac{x}{8} + \overset{4}{24} \cdot \frac{1}{6}$$

2. Simplify: $\quad 7 = 3x + 4$

3. Subtract 4: $\quad 7 - 4 = 3x + 4 - 4$

4. The left side has numbers only: $\quad 3 = 3x$

5. Divide by 3
(or multiply by the reciprocal of 3): $\quad \dfrac{3}{3} = \dfrac{3x}{3}$

$$1 = x$$
$$x = 1$$

The solution is 1.

6. Check: $\dfrac{7}{24} \overset{?}{=} \dfrac{x}{8} + \dfrac{1}{6}$

$$
\begin{array}{c|c}
\dfrac{7}{24} & \dfrac{1}{8} + \dfrac{1}{6} \\[2ex]
 & \dfrac{3}{24} + \dfrac{4}{24} \\[2ex]
 & \dfrac{7}{24}
\end{array}
$$

b. Given: $\quad \dfrac{1}{5} - \dfrac{x}{4} = \dfrac{7(x+3)}{10}$

1. Multiply by 20, the LCM: $\quad \overset{4}{20} \cdot \dfrac{1}{5} - \overset{5}{20} \cdot \dfrac{x}{4} = \overset{2}{20} \cdot \dfrac{7(x+3)}{10}$

2. Simplify and use the distributive law:
$$4 - 5x = 14(x + 3)$$
$$= 14x + 42$$

3. Subtract 4:
$$4 - 4 - 5x = 14x + 42 - 4$$
$$-5x = 14x + 38$$

4. Subtract $14x$:
$$-5x - 14x = 14x - 14x + 38$$
$$-19x = 38$$

5. Divide by -19
(or multiply by the reciprocal of -19):
$$\frac{-19x}{-19} = \frac{38}{-19}$$
$$x = -2$$

The solution is -2.

ANSWER (to problem on page 220)

1. a. $x = 1$ **b.** $x = -\dfrac{7}{4}$

6. Check:

$$\dfrac{1}{5} - \dfrac{x}{4} \overset{?}{=} \dfrac{7(x+3)}{10}$$

$$\begin{array}{c|c}
\dfrac{1}{5} - \dfrac{(-2)}{4} & \dfrac{7(-2+3)}{10} \\[2mm]
\dfrac{1}{5} + \dfrac{1}{2} & \dfrac{7(1)}{10} \\[2mm]
\dfrac{2}{10} + \dfrac{5}{10} & \dfrac{7}{10} \\[2mm]
\dfrac{7}{10} &
\end{array}$$

B. SOLVING LITERAL EQUATIONS

The procedures we have described to solve linear equations can also be used to solve some *literal* equations. A **literal equation** is an equation that contains letters other than the variable for which we wish to solve. In business, science, and engineering, these literal equations are usually given as formulas such as the area of a circle of radius r $(A = \pi r^2)$, the interest earned on a principal P at a given rate r for a given time t $(I = Prt)$, and so on. Unfortunately, these formulas are sometimes not in the form needed to solve the problem at hand. This is where the first five steps of our procedure come in! As it turns out, to solve for a particular variable in one of these formulas, we can use the same methods we have learned. For example, if we wish to solve for P in the formula $I = Prt$, we write

$$I = Prt$$

We need to solve for P, so we first circle P in color.

$$I = \textcircled{P}rt$$

Then divide by rt:
$$\dfrac{I}{rt} = \dfrac{\textcircled{P}rt}{rt}$$

Simplify:
$$\dfrac{I}{rt} = \textcircled{P}$$

Rewrite:
$$\textcircled{P} = \dfrac{I}{rt}$$

Now, go on to the next example.

EXAMPLE 3 A *trapezoid* is a 4-sided figure in which two of the sides are parallel. The area of the trapezoid shown is

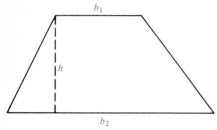

$$A = \dfrac{h}{2}(b_1 + b_2) \quad \text{where } h \text{ is the altitude, } b_1 \text{ and } b_2 \text{ the bases}$$

Solve for b_2.

Problem 3 Solve for b_1 in the equation

$$A = \dfrac{h}{2}(b_1 + b_2)$$

Solution We circle b_2 so you remember we are solving for $\boxed{b_2}$.

Multiply by the LCM 2: $\qquad\qquad 2A = \cancel{2} \cdot \dfrac{h}{\cancel{2}}(b_1 + \boxed{b_2})$

Simplify: $\qquad\qquad\qquad\qquad 2A = h(b_1 + \boxed{b_2})$

Use the distributive property $\qquad 2A = hb_1 + h\boxed{b_2}$
to multiply h by $(b_1 + b_2)$:

Subtract hb_1: $\qquad\qquad 2A - hb_1 = hb_1 - hb_1 + h\boxed{b_2}$

$\qquad\qquad\qquad\qquad\qquad 2A - hb_1 = h\boxed{b_2}$

Divide by h (or multiply by $\qquad \dfrac{2A - hb_1}{h} = \dfrac{h\boxed{b_2}}{h}$
the reciprocal of h):

$\qquad\qquad\qquad\qquad\qquad \dfrac{2A - hb_1}{h} = \boxed{b_2}$

$$\boxed{b_2} = \frac{2A - hb_1}{h} \qquad \blacktriangle$$

Good luck with the exercises.

NAME

CLASS

SECTION

A. Solve.

ANSWERS

1. $3x - 12 = 0$

2. $5a + 10 = 0$

1. _____

2. _____

3. $2y + 6 = 8$

4. $4b - 5 = 3$

3. _____

4. _____

5. $-3z - 4 = -10$

6. $-4r - 2 = 6$

5. _____

6. _____

7. $-5y + 1 = -13$

8. $-3x + 1 = -9$

7. _____

8. _____

9. $3x + 4 = x + 10$

10. $4x + 4 = x + 7$

9. _____

10. _____

11. $5x - 12 = 6x - 8$

12. $5x + 7 = 7x + 19$

11. _____

12. _____

13. $4v - 7 = 6v + 9$

14. $8t + 4 = 15t - 10$

13. _____

14. _____

15. $6m - 3m + 12 = 0$

16. $10k + 15 - 5k = 25$

15. _____

16. _____

17. $10 - 3z = 8 - 6z$

18. $8 - 4y = 10 + 6y$

17. _____

18. _____

19. $5(x + 2) = 3(x + 3) + 1$

20. $y - (4 - 2y) = 7(y - 1)$

19. _____

20. _____

21. $5(4 - 3a) = 7(3 - 4a)$

22. $\frac{3}{4}y - 4.5 = \frac{1}{4}y + 1.3$

21. _____

22. _____

23. $-\frac{7}{8}c + 5.6 = -\frac{5}{8}c - 3.3$

24. $x + \frac{2}{3}x = 10$

23. _____

24. _____

25. $-2x + \frac{1}{4} = 2x + \frac{4}{5}$

26. $6x + \frac{1}{7} = 2x - \frac{2}{7}$

25. _____

26. _____

27. $\frac{x - 1}{2} + \frac{x - 2}{2} = 3$

28. $\frac{3x + 5}{3} + \frac{x + 3}{3} = 12$

27. _____

28. _____

29. $\frac{x}{5} - \frac{x}{4} = 1$

30. $\frac{x}{3} - \frac{x}{2} = 1$

29. _____

30. _____

31. $\dfrac{x+1}{4} - \dfrac{2x-2}{3} = 3$

32. $\dfrac{z+4}{3} = \dfrac{z+6}{4}$

33. $\dfrac{2h-1}{3} = \dfrac{h-4}{12}$

34. $\dfrac{5-6y}{7} - \dfrac{-7-4y}{3} = 2$

35. $\dfrac{2w+3}{2} - \dfrac{3w+1}{4} = 1$

36. $\dfrac{7r+2}{6} + \dfrac{1}{2} = \dfrac{r}{4}$

37. $\dfrac{8x-23}{6} + \dfrac{1}{3} = \dfrac{5}{2}x$

38. $\dfrac{x+1}{2} + \dfrac{x+2}{3} + \dfrac{x+4}{4} = -8$

39. $\dfrac{x-5}{2} - \dfrac{x-4}{3} = \dfrac{x-3}{2} - (x-2)$

40. $\dfrac{x+1}{2} + \dfrac{x+2}{3} + \dfrac{x+3}{4} = 16$

41. $-4x + \dfrac{1}{2} = 6\left(x - \dfrac{1}{8}\right)$

42. $-6x + \dfrac{2}{3} = 4\left(x - \dfrac{1}{5}\right)$

43. $\dfrac{1}{2}(8x+4) - 5 = \dfrac{1}{4}(4x+8) + 1$

44. $\dfrac{1}{3}(3x+9) + 2 = \dfrac{1}{9}(9x+18) + 3$

45. $x + \dfrac{x}{2} - \dfrac{3x}{5} = 9$

46. $\dfrac{5x}{3} - \dfrac{3x}{4} + \dfrac{11}{6} = 0$

47. $\dfrac{4x}{9} - \dfrac{3}{2} = \dfrac{5x}{6} - \dfrac{3x}{2}$

48. $\dfrac{7x}{2} - \dfrac{4x}{3} + \dfrac{2x}{5} = -\dfrac{11}{6}$

49. $\dfrac{3x+4}{2} - \dfrac{1}{8}(19x-3) = 1 - \dfrac{7x+18}{12}$

50. $\dfrac{11x-2}{3} - \dfrac{1}{2}(3x-1) = \dfrac{17x+7}{6} - \dfrac{2}{9}(7x-2)$

B. Solve each equation for the indicated letter.

51. $C = 2\pi r$; solve for r.

52. $A = bh$; solve for h.

53. $A = \pi r^2$; solve for r^2.

54. $V = \pi r^2 h$; solve for h.

55. $A = \pi(r^2 + rs)$; solve for s.

56. $T = 2\pi(r^2 + rh)$; solve for h.

57. $\dfrac{V_2}{V_1} = \dfrac{P_1}{P_2}$; solve for V_2.

58. $\dfrac{a}{b} = \dfrac{c}{d}$; solve for a.

31. _____

32. _____

33. _____

34. _____

35. _____

36. _____

37. _____

38. _____

39. _____

40. _____

41. _____

42. _____

43. _____

44. _____

45. _____

46. _____

47. _____

48. _____

49. _____

50. _____

51. _____

52. _____

53. _____

54. _____

55. _____

56. _____

57. _____

58. _____

59. $S = \dfrac{f}{H - h}$; solve for H.   **60.** $I = \dfrac{E}{R + nr}$; solve for R.

59. _____

60. _____

✓ **SKILL CHECKER**

Write in symbols.

61. The quotient of $(a + b)$ and c

62. The quotient of a and $(b - c)$

63. The product of a and the sum of b and c

64. The product of the difference of a and b times c

65. The difference of a and the product of b and c.

61. _____

62. _____

63. _____

64. _____

65. _____

3.3 USING YOUR KNOWLEDGE

At the beginning of this section, we discussed a method that would tell us which was a better deal: to rent a car using the mileage rate or to rent it using the flat rate. In the next few problems we are going to ask similar questions based on the rates appearing at the beginning of the section. Follow the procedures used in the text to solve these problems.

1. Suppose you wish to rent a subcompact for a week. Write a formula for the cost C based on traveling m miles.

2. How many miles do you have to travel so that the per-mile cost C is the same as the flat rate, that is, 89?

3. If you were planning to travel 300 mi during the week, would you use the mileage rate or the flat rate?

4. In our American economy, merchants sell goods to make a profit. These profits are obtained by selling goods at a **markup** in price. This markup M can be based on the cost C or on the selling price S. The formula relating the selling price S, the cost C, and the markup M is

$S = C + M$

A merchant plans to have a 20% markup on cost.
a. What would the selling price S be?
b. Use the formula obtained in (a) to find the selling price of an article that cost the merchant $8.

5. a. If the markup on an article is 25% of the selling price, use the formula $S = C + M$ to find the cost C of the article.
b. Use the formula obtained in part (a) to find the cost of an article that sells for $80.

1. _____

2. _____

3. _____

4. a. _____
   b. _____

5. a. _____

   b. _____

3.4 SOLVING INEQUALITIES

Women's Macrame and Canvas Sandals

Cut

$1 to $3

OBJECTIVES

REVIEW

Before starting this section you should know:

1. How to add, subtract, multiply, and divide integers.
2. The procedure used to solve linear equations (page 222).

OBJECTIVES

You should be able to:

A. Determine which of two given numbers is the greater one and write the result by using the $>$ or the $<$ symbol.
B. Solve and graph a linear inequality.
C. Write and graph a compound inequality.

The ad says that you can cut $1 to $3 when buying sandals. If x is the amount of money you can cut, what can x be? Well, it can be at least $1 and as much as $3, that is, $x = 1$ or $x = 3$ or x can be between 1 and 3. In the language of algebra we write this fact as

$1 \leq x \leq 3$ Read "1 is less than or equal to x and x is less than or equal to 3" or "x is between 1 and 3, inclusive."

The statement $1 \leq x \leq 3$ is made up of two parts:

$1 \leq x$ which means that 1 is *less than or equal to x* (or that x is *greater than or equal to* 1), and

$x \leq 3$ which means that x is *less than or equal to* 3 (or that 3 is *greater than or equal to x*).

The statements $1 \leq x$, $x \leq 3$, and $1 \leq x \leq 3$ are examples of **inequalities**. In algebra, an inequality is a statement with $>$, $<$, $\geq$, or $\leq$ as its verb. Inequalities can be represented on a number line. Here is the way to do it. As you recall, a number line is constructed by drawing a line, selecting a point on this line, and calling it 0, the origin (see the figure).

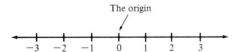

We then locate equally spaced points to the right of the origin on the line and label them with the *positive* integers 1, 2, 3, and so on. The corresponding points to the left of 0 are labeled -1, -2, -3, and so on (the *negative* integers). This construction allows the *association of numbers with points on the line*. The number associated with a point is called the **coordinate** of that point. For example, there are points associated with the numbers -2.5, $-1\frac{1}{2}$, $\frac{3}{4}$, and 2.5 (see the figure).

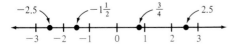

All these numbers are real numbers. As we have mentioned, the real numbers include whole numbers, integers, fractions, and decimals as well as the irrational numbers (which are discussed in more detail later). Thus, the real numbers are represented by the numbers included on the number line.

A. ORDER

As you can see, the numbers are placed in order on the number line. *Greater* numbers are always to the *right* of *smaller* ones. (The farther to the *right,* the *greater* the number.) Thus any number to the *right* of a second number is said to be **greater than,** $>$, the second number. We also say that the second number is **less than,** $<$, the first number. For example, since 3 is to the right of 1, we write

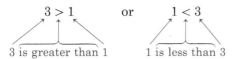

$$3 > 1 \qquad \text{or} \qquad 1 < 3$$

3 is greater than 1 $\qquad$ 1 is less than 3

Similarly,

$$-1 > -3 \quad \text{or} \quad -3 < -1$$
$$0 > -2 \quad \text{or} \quad -2 < 0$$
$$3 > -1 \quad \text{or} \quad -1 < 3$$

Note that the inequality signs $>$ and $<$ always point to the smaller number.

EXAMPLE 1 Fill in the blank with $>$ or $<$ so that the resulting statement is true.

a. 3 _____ 4 $\quad$ **b.** -4 _____ -3 $\quad$ **c.** -2 _____ -3

Solution We first construct a number line containing these numbers. (Of course, we could just think about the number line without actually drawing one.)

a. Since 3 is to the left of 4, $3 < 4$.
b. Since -4 is to the left of -3, $-4 < -3$.
c. Since -2 is to the right of -3, $-2 > -3$.

Problem 1 Fill in the blank with $>$ or $<$ so that the resulting statement is true.
a. 5 _____ 7
b. -2 _____ -1
c. -4 _____ -5

B. SOLVING INEQUALITIES

Just as we solved equations, we can also solve inequalities. We say that we have *solved* a given inequality when we obtain an inequality equivalent to the one given and in the form $x < \boxed{}$ or $x > \boxed{}$. For example, consider the inequality

$$x < 3$$

There are many real numbers that will make this inequality a true statement. A few of these are shown in the table.

$x < 3$
$2 < 3$
$1\frac{1}{2} < 3$
$1 < 3$
$0 < 3$
$-\frac{1}{2} < 3$

2 is a solution of $x < 3$ because $2 < 3$. Similarly, $-\frac{1}{2}$ is a solution of $x < 3$ because $-\frac{1}{2} < 3$.

As you can see from this table, 2, $1\frac{1}{2}$, 1, 0, and $-\frac{1}{2}$ are *solutions* of the inequality $x < 3$. Of course, we cannot list all the real numbers satisfying the inequality $x < 3$ because there are infinitely many numbers that do, but we can certainly show all the solutions of $x < 3$ *graphically* by using the number line (see the figure).

This type of representation is called the **graph** of the solutions of $x < 3$ and is indicated by the heavy line in the figure. Note that there is an open circle at $x = 3$ to indicate that 3 is *not* part of the graph of $x < 3$ (since 3 is not less than 3). Also, the colored arrowhead points to the left (just as the $<$ in $x < 3$ points to the left) to indicate that the heavy line continues to the left without end. On the other hand, the graph of $x \geq 2$ should continue to the right without end. (See the figure.)

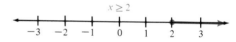

Moreover, since $x = 2$ is included in the graph, a solid dot appears at the point $x = 2$.

EXAMPLE 2   Graph the inequality on a number line.

a. $x \geq -1$ **b.** $x < -2$

Solution

a. The numbers that satisfy the inequality $x \geq -1$ are the numbers that are *greater than or equal to* -1, that is, the number -1 and all the numbers to the right of -1 (note that $\geq$ points to the right). The graph is shown in the following figure.

b. The numbers that satisfy the inequality $x < -2$ are the numbers that are *less than* -2, that is, the numbers to the *left of but not including* -2 (note that $<$ points to the left). The graph of these points is shown in the following figure.

Problem 2   Graph the inequality on a number line.

a. $x \geq -2$

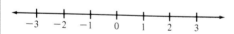

b. $x < 1$

We solve more complicated inequalities the same way we solved an equality, by finding an equivalent inequality whose solution is obvious. Remember, we have *solved* a given inequality when we obtain an inequality equivalent to the one given and in the form $x < \square$ or $x > \square$. Thus to solve the inequality $2x - 1 < x + 3$ we try to find an equivalent inequality of the form $x < \square$ or $x > \square$. As before, we need to have some rules to go by. The first of these rules are the addition and subtraction properties. We illustrate these rules next.

If $3 < 4$, then

$3 + 5 < 4 + 5$ Adding 5

    $8 < 9$ True

Similarly, if $3 > -2$, then

$3 + 7 > -2 + 7$ Adding 7

  $10 > 5$ True

Also, if $3 < 4$, then

$3 - 1 < 4 - 1$ Subtracting 1

    $2 < 3$ True

Similarly, if $3 > -2$, then

$3 - 5 > -2 - 5$ Subtracting 5

  $-2 > -7$ True because -2 is to the right of -7.

In general, we have the following rules.

ADDITION AND SUBTRACTION PROPERTIES OF INEQUALITIES

You can *add* or *subtract* the same number on both sides of an inequality and obtain an equivalent inequality. In symbols:

If	$x < y$		If	$x > y$
then	$x + a < y + a$		then	$x + a > y + a$
or	$x - b < y - b$		or	$x - b > y - b$

Since $x - b = x + (-b)$, subtracting b from both sides is the same as adding the inverse of b, $(-b)$.

Now let us return to the inequality $2x - 1 < x + 3$. To solve this inequality, we need to have all the variables by themselves on one side. Thus we proceed as follows:

Given:	$2x - 1 < x + 3$
Add 1:	$2x - 1 + 1 < x + 3 + 1$
Simplify:	$2x < x + 4$
Subtract x:	$2x - x < x - x + 4$
Simplify:	$x < 4$

Any number less than 4 is a solution. The graph of this inequality is shown in the figure.

ANSWERS (to problems on pages 232–233)

1. a. $5 < 7$ **b.** $-2 < -1$ **c.** $-4 > -5$

2. a.

$x \geq -2$

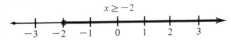

b.

$x < 1$

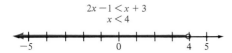
$$2x - 1 < x + 3$$
$$x < 4$$

You can check that this solution is correct, by selecting any number from the graph (say **0**) and replacing x with the number in the original inequality. For $x = 0$, we have $2(0) - 1 < 0 + 3$, or $-1 < 3$, a true statement. Of course, this is only a "partial" check, since we did not try *all* the numbers in the graph. You can check a little further by selecting a number *not* on the graph, making sure the result is false. For example, when $x = 5$,

$$2x - 1 < x + 3$$

becomes

$$2(5) - 1 < 5 + 3$$
$$10 - 1 < 8$$
$$9 < 8 \quad \text{False}$$

EXAMPLE 3 Solve and graph the inequality on a number line.

a. $3x - 2 < 2(x - 2)$ **b.** $4(x + 1) \geq 3x + 7$

Solution

a. Given: $\qquad\qquad 3x - 2 < 2(x - 2)$

Simplify: $\qquad\qquad 3x - 2 < 2x - 4$

Add **2**: $\qquad 3x - 2 + 2 < 2x - 4 + 2$

Simplify: $\qquad\qquad\quad 3x < 2x - 2$

Subtract **2x** $\qquad 3x - 2x < 2x - 2x - 2$
(or add $-2x$):

Simplify: $\qquad\qquad\qquad\quad x < -2$

Any number less than -2 is a solution. The graph of this inequality appears in the following figure.

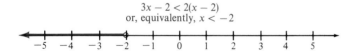
$$3x - 2 < 2(x - 2)$$
or, equivalently, $x < -2$

b. Given: $\qquad\qquad 4(x + 1) \geq 3x + 7$

Simplify: $\qquad\qquad 4x + 4 \geq 3x + 7$

Subtract **4**: $\qquad 4x + 4 - 4 \geq 3x + 7 - 4$

Simplify: $\qquad\qquad\qquad 4x \geq 3x + 3$

Subtract **3x** $\qquad 4x - 3x \geq 3x - 3x + 3$
(or add $-3x$):

Simplify: $\qquad\qquad\qquad\quad x \geq 3$

Any number greater than or equal to 3 is a solution. The graph of this inequality appears in the following figure.

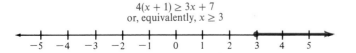
$$4(x + 1) \geq 3x + 7$$
or, equivalently, $x \geq 3$

▲

Problem 3 Solve and graph on a number line.
a. $4x - 7 < 3(x - 2)$
b. $3(x + 1) \geq 2x + 5$

How do we solve an inequality such as $\dfrac{x}{2} < 3$? If half a number is less than 3, the number must be less than 6. This suggests that you can *multiply* (or *divide*) both sides of an inequality by a *positive* number and obtain an equivalent inequality.

Given:	$\dfrac{x}{2} < 3$
Multiply by 2:	$2 \cdot \dfrac{x}{2} < 2 \cdot 3$
Simplify:	$x < 6$
To solve	$2x < 8$
Divide by 2 (or multiply by the reciprocal of 2):	$\dfrac{2x}{2} < \dfrac{8}{2}$
Simplify:	$x < 4$

Any number less than 4 is a solution. Let us try some more examples.
If $3 < 4$, then

$5 \cdot 3 < 5 \cdot 4$
$15 < 20$ True

If $-2 > -10$, then

$5 \cdot (-2) > 5 \cdot (-10)$
$-10 > -50$ True

Also, if $6 < 8$, then

$\dfrac{6}{2} < \dfrac{8}{2}$

$3 < 4$ True

Similarly, if $-6 > -10$, then

$-\dfrac{6}{2} > -\dfrac{10}{2}$

$-3 > -5$ True

Shown next is the general rule we have used.

MULTIPLICATION AND DIVISION PROPERTIES OF INEQUALITIES

You can *multiply* or *divide* both sides of an inequality by any *positive* number and obtain an equivalent inequality. In symbols:

If	$x < y$, and a is positive,		If	$x > y$, and a is positive,
then	$ax < ay$		then	$ax > ay$
or	$\dfrac{x}{a} < \dfrac{y}{a}$		or	$\dfrac{x}{a} > \dfrac{y}{a}$

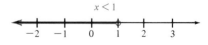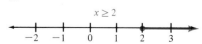

EXAMPLE 4 Solve and graph the inequality on a number line.

a. $5x + 3 \leq 2x + 9$ **b.** $4(x - 1) > 2x + 6$

Problem 4 Solve and graph on a number line.
a. $4x + 5 \leq x + 11$
b. $3(x - 1) > x + 3$

Solution

a. Given: $\qquad\qquad\qquad\qquad\qquad 5x + 3 \leq 2x + 9$

Subtract 3 (or add -3): $\qquad\qquad 5x + 3 - 3 \leq 2x + 9 - 3$

Simplify: $\qquad\qquad\qquad\qquad\qquad 5x \leq 2x + 6$

Subtract $2x$ (or add $-2x$): $\qquad 5x - 2x \leq 2x - 2x + 6$

Simplify: $\qquad\qquad\qquad\qquad\qquad 3x \leq 6$

Divide by 3
(or multiply by the reciprocal of 3): $\qquad \dfrac{3x}{3} \leq \dfrac{6}{3}$

Simplify: $\qquad\qquad\qquad\qquad\qquad x \leq 2$

Any number less than or equal to 2 is a solution. The graph is shown in the following figure.

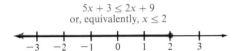

$5x + 3 \leq 2x + 9$
or, equivalently, $x \leq 2$

b. Given: $\qquad\qquad\qquad\qquad\qquad 4(x - 1) > 2x + 6$

Simplify: $\qquad\qquad\qquad\qquad\qquad 4x - 4 > 2x + 6$

Add 4: $\qquad\qquad\qquad\qquad 4x - 4 + 4 > 2x + 6 + 4$

Simplify: $\qquad\qquad\qquad\qquad\qquad 4x > 2x + 10$

Subtract $2x$ (or add $-2x$): $\qquad 4x - 2x > 2x - 2x + 10$

Simplify: $\qquad\qquad\qquad\qquad\qquad 2x > 10$

Divide by 2
(or multiply by the reciprocal of 2): $\qquad \dfrac{2x}{2} > \dfrac{10}{2}$

Simplify: $\qquad\qquad\qquad\qquad\qquad x > 5$

Any number greater than 5 is a solution. The graph appears in the following figure.

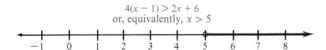

$4(x - 1) > 2x + 6$
or, equivalently, $x > 5$

You may have noticed that the multiplication (or division) principle allowed us to *multiply* or *divide* only by a *positive* number. However, to solve the inequality $-2x < 4$, we need to divide by -2, a number that is *not* positive. Let us first see what happens when we divide both sides of an inequality by a *negative* number.

Consider: $\qquad\qquad\qquad\qquad\qquad 2 < 4$

If we divide both sides of this
inequality by -2, we get: $\qquad \dfrac{2}{-2} < \dfrac{4}{-2}$

or: $\qquad\qquad\qquad\qquad\qquad -1 < -2$

which is not true.

In order to obtain a true statement, we must *reverse the inequality sign* and write:

$$-1 > -2$$

Similarly, consider

$$-6 > -8$$
$$\frac{-6}{-2} > \frac{-8}{-2}$$
$$3 > 4$$

which is *not* true. Thus, we must reverse the inequality sign and write $3 < 4$. Thus if we *divide* both sides of an inequality by a *negative* number, we must reverse the inequality sign to obtain an equivalent inequality. Similarly, if we *multiply* both sides of an inequality by a *negative* number we reverse the inequality sign to obtain an equivalent inequality. Thus:

Given: $\qquad\qquad\qquad\qquad 2 < 4$

If we multiply by $-3,$ reverse $\qquad -3 \cdot 2 > -3 \cdot 4$
the inequality sign: $\qquad\qquad\qquad\quad -6 > -12$

Here are some more examples.

$3 < 12$	$8 > 4$
$-2 \cdot 3 > -2 \cdot 12$	$\dfrac{8}{-2} < \dfrac{4}{-2}$
Reverse	Reverse
$-6 > -24$	$-4 < -2$

The rule summarizing this discussion is next.

MULTIPLICATION AND DIVISION PROPERTIES OF INEQUALITIES

You can *multiply* or *divide* both sides of an inequality by a *negative* number and obtain an equivalent inequality provided you *reverse* the inequality sign. In symbols:

If $\qquad x < y$ and $\qquad\qquad a$ is negative	If $\qquad x > y$ and $\qquad\qquad a$ is negative
then $\quad$ (1) $ax > ay$	then $\quad$ (1) $ax < ay$
Reverse	or $\quad$ (2) $\dfrac{x}{a} < \dfrac{y}{a}$
or $\quad$ (2) $\dfrac{x}{a} > \dfrac{y}{a}$	
Reverse	

We shall use this rule in the next example.

ANSWER (to problem on page 237)
4. a.

$x \le 2$

b.

$x > 3$

EXAMPLE 5 Solve.

a. $-3x < 15$ **b.** $\dfrac{-x}{4} > 2$ **c.** $3(x - 2) \le 5x + 2$

Solution

a. To solve this inequality we need to have the x by itself on the left, that is, we have to divide both sides by -3. Of course, when we do this, we must reverse the inequality sign.

Given: $\qquad\qquad\qquad\qquad -3x < 15$

Divide by -3 and reverse the sign: $\qquad \dfrac{-3x}{-3} > \dfrac{15}{-3}$

Simplify: $\qquad\qquad\qquad\qquad x > -5$

Any number greater than -5 is a solution.

b. Here we multiply both sides by -4 and reverse the inequality sign.

Given: $\qquad\qquad\qquad\qquad \dfrac{-x}{4} > 2$

Multiply by -4 and reverse
the inequality sign: $\qquad\qquad -4\left(\dfrac{-x}{4}\right) < -4 \cdot 2$

Simplify: $\qquad\qquad\qquad\qquad x < -8$

Thus, any number less than -8 is a solution.

c. Given: $\qquad\qquad\qquad\quad 3(x - 2) \le 5x + 2$

Simplify: $\qquad\qquad\qquad\quad 3x - 6 \le 5x + 2$

Add **6**: $\qquad\qquad\qquad 3x - 6 + 6 \le 5x + 2 + 6$

Simplify: $\qquad\qquad\qquad\qquad 3x \le 5x + 8$

Subtract $5x$ (or add $-5x$): $\qquad 3x - 5x \le 5x - 5x + 8$

Simplify: $\qquad\qquad\qquad\qquad -2x \le 8$

Divide by -2 (or multiply
by the reciprocal of -2) and $\qquad \dfrac{-2x}{-2} \ge \dfrac{8}{-2}$
reverse the inequality sign:

Simplify: $\qquad\qquad\qquad\qquad x \ge -4$

Thus, any number greater than or equal to -4 is a solution.   ▲

Of course, to solve more complicated inequalities (for example, those involving fractions), we simply follow the procedure given to solve linear equations shown on page 222. This is done in the next example.

EXAMPLE 6 Solve.

$$\dfrac{-x}{4} + \dfrac{x}{6} < \dfrac{x - 3}{6}$$

Solution We follow the 6-step procedure given on page 222.

Given: $\qquad\qquad\qquad\qquad \dfrac{-x}{4} + \dfrac{x}{6} < \dfrac{x - 3}{6}$

Problem 5 Solve.

a. $-2x < 10$ **b.** $-\dfrac{x}{3} > 1$

c. $4(x - 1) \le 6x + 8$

Problem 6 Solve.

$$\dfrac{-x}{3} + \dfrac{x}{4} < \dfrac{x - 4}{4}$$

1. Multiply by **12**, the LCM of 4 and 6:

$$12 \cdot \left(\frac{-x}{4}\right) + 12 \cdot \frac{x}{6} < 12\left(\frac{x-3}{6}\right)$$

2. Simplify:

$$-3x + 2x < 2(x-3)$$
$$-x < 2x - 6$$

3. There are no numbers on the left, only the variable $-x$.

4. Subtract $2x$ (or add $-2x$):

$$-x - 2x < 2x - 2x - 6$$
$$-3x < -6$$

5. Divide by the coefficient of x, -3 (or multiply by the reciprocal of -3), and reverse the inequality sign:

$$\frac{-3x}{-3} > \frac{-6}{-3}$$

$$x > 2$$

Thus, any number greater than 2 is a solution.

6. Check: Try $x = 12$ (it is a good idea to try the LCM).

$$\frac{-x}{4} + \frac{x}{6} \overset{?}{<} \frac{x-3}{6}$$

$\dfrac{-12}{4} + \dfrac{12}{6}$	$\dfrac{12-3}{6}$
$-3 + 2$	$\dfrac{9}{6}$
-1	$\dfrac{3}{2}$

Since $-1 < \frac{3}{2}$, the inequality is true. Of course, this only partially *verifies* the answer, since we are unable to try *every* solution.

C. COMPOUND INEQUALITIES

What about inequalities like the one at the beginning of this section? Inequalities such as

$$1 \le x \le 3$$

are called **compound inequalities** because they are equivalent to two other inequalities. For example, $1 \le x \le 3$ means $1 \le x$ and $x \le 3$. Thus if we are asked to solve the compound inequality

$$2 \le x \quad \text{and} \quad x \le 4$$

we write

$$2 \le x \le 4 \quad 2 \le x \le 4 \text{ is a compound inequality.}$$

The graph of this inequality consists of all the points between 2 and 4, inclusive, and is shown in the figure.

The *key* to solving these compound inequalities is to write the inequality in the form $a \le x \le b$ (or $a < x < b$), where a and b are real numbers. This form is regarded as the *required* solution. Of course, if we try to write $x < 3$ and $x > 7$ in this form, we get $7 < x < 3$, which is not true.

EXAMPLE 7 Solve and graph on a number line.

a. $1 < x$ and $x < 3$ **b.** $5 \geq -x$ and $x \leq -3$
c. $x + 1 \leq 5$ and $-2x < 6$

Solution

a. Given: $1 < x$ and $x < 3$. $1 < x$ and $x < 3$ is written as $1 < x < 3$. Thus, the solution consists of the numbers *between* 1 and 3, as shown.

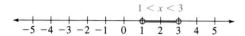

b. Since we wish to write $5 \geq -x$ and $x \leq 3$ in the form $a \leq x \leq b$, we multiply both sides of $5 \geq -x$ by -1, obtaining

$$-1 \cdot 5 \leq -1 \cdot (-x)$$
$$-5 \leq x$$

We now have

$$-5 \leq x \quad \text{and} \quad x \leq -3$$

that is,

$$-5 \leq x \leq -3$$

Thus, the solution consists of all the numbers between -5 and -3 *inclusive,* as shown.

c. We solve $x + 1 \leq 5$ by subtracting 1 from both sides, obtaining

$$x + 1 - 1 \leq 5 - 1$$
$$x \leq 4$$

We now have

$$x \leq 4 \quad \text{and} \quad -2x < 6$$

We then divide both sides of $-2x < 6$ by -2; obtaining $x > -3$. We now have

$$x \leq 4 \quad \text{and} \quad x > -3$$

Rearranging these inequalities, we write

$$-3 < x \quad \text{and} \quad x \leq 4$$

that is,

$$-3 < x \leq 4$$

Here, the solution consists of all numbers between -3 and 4 and the number 4 itself, as shown in the figure.

Problem 7 Solve and graph on a number line.
a. $2 < x$ and $x < 4$
b. $3 \geq -x$ and $x \leq -1$
c. $x + 2 \leq 6$ and $-3x < 9$

NAME

CLASS

SECTION

A. Fill in the blank with $>$ or $<$ so that the resulting statement is true.

ANSWERS

1. $8 \underline{\hspace{1cm}} 9$

2. $-8 \underline{\hspace{1cm}} -9$

3. $-4 \underline{\hspace{1cm}} -9$

4. $7 \underline{\hspace{1cm}} 3$

5. $\dfrac{1}{4} \underline{\hspace{1cm}} \dfrac{1}{3}$

6. $\dfrac{1}{5} \underline{\hspace{1cm}} \dfrac{1}{2}$

7. $-\dfrac{2}{3} \underline{\hspace{1cm}} -1$

8. $-\dfrac{1}{5} \underline{\hspace{1cm}} -1$

9. $-3\dfrac{1}{4} \underline{\hspace{1cm}} -3$

10. $-4\dfrac{1}{5} \underline{\hspace{1cm}} -4$

1. _____
2. _____
3. _____
4. _____
5. _____
6. _____
7. _____
8. _____
9. _____
10. _____

B. Solve and graph on a number line.

11. $2x + 6 \leq 8$

12. $4y - 5 \leq 3$

13. $-3y - 4 \geq -10$

14. $-4z - 2 \geq 6$

15. $-5x + 1 \leq -14$

16. $-3x + 1 \leq -8$

17. $3a + 4 \leq a + 10$

18. $4b + 4 \leq b + 7$

19. $5z - 12 \geq 6z - 8$

20. $5z + 7 \geq 7z + 19$

21. $10 - 3x \leq 7 - 6x$

22. $8 - 4y \leq -12 + 6y$

23. $5(x + 2) < 3(x + 3) + 1$

24. $5(4 - 3x) < 7(3 - 4x) + 12$

25. $-2x + \dfrac{1}{4} \geq 2x + \dfrac{4}{5}$

26. $6x + \dfrac{1}{7} \geq 2x - \dfrac{2}{7}$

27. $\dfrac{x}{5} - \dfrac{x}{4} \leq 1$

28. $\dfrac{x}{3} - \dfrac{x}{2} \leq 1$

11. _____
12. _____
13. _____
14. _____
15. _____
16. _____
17. _____
18. _____
19. _____
20. _____
21. _____
22. _____
23. _____
24. _____
25. _____
26. _____
27. _____
28. _____

29. $\dfrac{7x+2}{6}+\dfrac{1}{2}\geq\dfrac{3}{4}x$

30. $\dfrac{8x-23}{6}+\dfrac{1}{3}\geq\dfrac{5}{2}x$

29. _____

30. _____

C. Solve and graph on a number line.

31. $x<3$ and $-x<-2$

32. $-x<5$ and $x<2$

31. _____

32. _____

33. $x+1<4$ and $-x<-1$

34. $x-2<1$ and $-x<2$

33. _____

34. _____

35. $x-2<3$ and $2>-x$

36. $x-3<1$ and $1>-x$

35. _____

36. _____

37. $x+2<3$ and $-4<x+1$

38. $x+4<5$ and $-1<x+2$

37. _____

38. _____

39. $x-1>2$ and $x+7<12$

40. $x-2>1$ and $-x>-5$

39. _____

40. _____

D. Applications

Write the given information as an inequality.

41. The temperature t in your refrigerator is between 20°F and 40°F.

41. _____

42. The height h (in feet) of any mountain is always less than or equal to that of Mt. Everest, 29,028 ft.

42. _____

43. My salary s for this year will be between $12,000 and $13,000.

43. _____

44. The gas mileage m (in miles) per gallon of gas is between 18 and 22, depending on your driving.

44. _____

45. The number of possible eclipses e in a year varies from 2 to 7, inclusive.

45. _____

46. The assets a of the Dupont family are in excess of $150,000 million.

46. _____

47. The cost c of ordinary hardware (tools, mowers, etc.) is between $3.50 and $4.00 per pound.

47. _____

48. The range r (in miles) of a rocket is always less than 19,000 mi.

48. _____

49. The altitude a (in feet) attained by the first liquid fueled rocket was less than 41 ft.

49. _____

50. The number of days d a person has remained in the weightlessness of space has not exceeded 85.

50. _____

✓ **SKILL CHECKER**

Solve.

51. $(w+3)+w=35$

52. $(x+5)+x=16$

51. _____

52. _____

53. $3n+6=126$

54. $5n+6=106$

53. _____

54. _____

55. $5n - 12 = n + 4$ **56.** $3n - 10 = n + 6$

55. _____

56. _____

3.4 USING YOUR KNOWLEDGE

Crock by Rechin, Parker and Wilder, Field Enterprises, Inc., 1976. Courtesy of Field
Newspaper Syndicate.

In this section we shall solve the problem in the cartoon. Let

    J be Joe's height
    B be Bill's height
    F be Frank's height
    S be Sam's height

Translate each statement into an equation or an inequality:

1. Joe is 5 ft (60 in.) tall.

2. Bill is taller than Frank.

3. Frank is 3 in. shorter than Sam.

4. Frank is taller than Joe.

5. Sam is 6 ft 5 in. (77 in.)

6. According to the statement in Problem 2, Bill is taller than Frank and according to the statement in 4, Frank is taller than Joe. Write these two statements as an inequality of the form $a > b > c$.

7. Based on the answer to Problem 6 and the fact that you can obtain Frank's height by using the results of Problems 3 and 5, what can you really say about Bill's height?

1. _____

2. _____

3. _____

4. _____

5. _____

6. _____

7. _____

The first ten problems in this section can be done by using the following definition:

$$a > b \quad \text{means} \quad a - b > 0$$

Thus, to do problem 6, assume that

$$\frac{1}{5} > \frac{1}{2}$$

By definition

$$\frac{1}{5} - \frac{1}{2} > 0 \quad \text{(If your assumption is right)}$$

Key in

$$\boxed{1} \boxed{\div} \boxed{5} \boxed{-} \boxed{(} \boxed{1} \boxed{\div} \boxed{2} \boxed{)} \boxed{=}$$

which gives -0.3, a *negative* number. Thus, our assumption was wrong and $\frac{1}{5} \not> \frac{1}{2}$. Thus,

$$\frac{1}{5} \boxed{<} \frac{1}{2}$$

To do problem 8, we can assume that

$$-\frac{1}{5} > -1$$

By definition, $-\frac{1}{5} - (-1)$ must be positive. Now,

$$\boxed{-} \boxed{1} \boxed{\div} \boxed{5} \boxed{-} \boxed{1} \boxed{+/-} \boxed{=}$$

gives 0.8, which is *positive*. Thus

$$-\frac{1}{5} \boxed{>} -1$$

You can also use your calculator to check the solution of an inequality by substituting several values taken from the graph and several not in the graph. For example, the solution to $3x - 2 < 2(x - 2)$ (Example 3a), is $x < -2$. Let us try $x = -3$ to see if it satisfies the original inequality. Key in

$$\boxed{3} \boxed{\times} \boxed{3} \boxed{+/-} \boxed{-} \boxed{2} \boxed{=} \quad \text{yielding } -11$$

and

$$\boxed{2} \boxed{\times} \boxed{(} \boxed{3} \boxed{+/-} \boxed{-} \boxed{2} \boxed{)} \boxed{=} \quad \text{or } -10$$

Since $-11 < -10$, the inequality is satisfied. If your instructor permits it, check the solution to the inequality problems in this section.

3.5 WORD PROBLEMS

From the *Guinness Book of World Records.*

REVIEW

Before starting this section, you should know:

1. How to translate sentences into the language of algebra (Section 2.6).
2. How to solve linear equations (Section 3.3).

OBJECTIVES

You should be able to:

A. Solve word problems involving integers using the five-step procedure given in the text.
B. Solve other problems dealing with whole numbers or integers using the five-step procedure given in the text.

Now that you have learned how to solve equations and inequalities, you need to know how to apply this knowledge to solve problems in algebra. These problems are usually stated in words and consequently are called **word,** or **story, problems.** This is an area in which many students encounter difficulties, but don't panic; we will give you a sure-fire method of tackling word problems. To start, let us look at a problem that may be familiar to you. Look at the picture at the beginning of this section. It shows the Stimson twins, Mary and Margaret. At birth, Margaret was 3 oz heavier than Mary, and together they weighed 35 oz. Can you find their weights? There you have it, a word problem! Here is our way of solving this problem. It is as easy as 1-2-3-4-5.

RSTUV PROCEDURE FOR SOLVING WORD PROBLEMS

1. **Read** the problem carefully and decide what is asked for (the unknown).

2. **Select** a variable to represent this unknown.

3. **Translate** the problem into the language of algebra.

4. **Use** the rules of algebra to solve for the unknown.

5. **Verify** the answer.

If you really want to learn how to do word problems this is the procedure you have to master. Study it carefully, and then use it. *It works!* How do we remember all of these steps? Easy. Look at the first letter in each sentence. We call this method the **RSTUV** method.

Read the problem.
    Select the unknown.
        Translate the problem.
            Use algebra to solve the equation.
                Verify your answer.

Here we go.

1. *Read* the problem slowly—not once, but two or three times. (Reading algebra is *not* like reading a magazine; algebra problems may have to be read several times before you understand them.)
2. *Select* the variable w to be the weight for Mary (which makes Margaret $w + 3$ ounces, since she is 3 ounces heavier).
3. *Translate* the problem into the language of algebra: Together

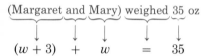

$$(w + 3) \quad + \quad w \quad = \quad 35$$

4. *Use* the rules of algebra to solve this equation.

Given:	$(w + 3) + w = 35$
Remove parentheses:	$w + 3 + w = 35$
Combine like terms:	$2w + 3 = 35$
Subtract 3:	$2w + 3 - 3 = 35 - 3$
Simplify:	$2w = 32$
Divide by 2:	$\dfrac{2w}{2} = \dfrac{32}{2}$
	$w = 16$ Mary's weight

Thus, Mary weighed 16 oz and Margaret $16 + 3 = 19$ oz.
5. *Verify* your answer.
Do they weigh 35 oz together? Yes, $16 + 19$ is 35.

A. INTEGER PROBLEMS

Sometimes we use words that may be new to you. For example, a popular problem in algebra is the *integer* problem. You remember the integers, they are the positive integers 1, 2, 3, and so on; the negative integers -1, -2, -3, and so on; and 0. Now, if you are given any integer, can you find the integer that comes right after it? Of course, you simply add 1 to the given integer and you have the answer. For example, the integer that comes after 4 is $4 + 1 = 5$ and the one that comes after -4 is $-4 + 1 = -3$. In general, if n is any integer, the integer that follows n is $n + 1$. We usually say that n and $n + 1$ are **consecutive** integers. We have illustrated this idea in the figure.

Now, suppose you are given an *even* integer (an integer divisible by 2) such as 8 and you are asked to find the next even integer (which is

10). This time you must add 2 to 8 to obtain the answer. What is the next even integer after 24? $24 + 2 = 26$. If n is even, the next even integer after n is $n + 2$. What about the next *odd* integer after 5? First, recall that the odd integers are $\ldots -3, -1, 1, 3 \ldots$. The next odd integer after 5 is $5 + 2 = 7$. Similarly, the next odd integer after 21 is $21 + 2 = 23$. In general, if n is odd, the next odd integer after n is $n + 2$. Thus, if you need two consecutive even or odd integers and you call the first one n, the next one will be $n + 2$. We shall use all these ideas as well as the RSTUV method in the next example.

EXAMPLE 1   The sum of three consecutive even integers is 126. Find the integers.

Solution   We use the RSTUV method.

1. Read the problem and note that we are asking for three consecutive *even integers*.
2. Select n to be the first of the integers. Since we want three consecutive even integers, we need to find the next two consecutive even integers. The next even integer after n is $n + 2$ and the one after $n + 2$ is $n + 4$. Thus the three consecutive even integers are $n, n + 2$, and $n + 4$.
3. Translate the problem into the language of algebra. If n is the first even integer, $n + 2$ is the second and $n + 4$ is the third. Now,

The sum of 3 consecutive even integers   is   126.

$$n + (n + 2) + (n + 4) = 126$$

4. Use algebra to solve this equation.

   1. Given: $\qquad\qquad\qquad n + (n + 2) + (n + 4) = 126$
   2. Remove parentheses: $\qquad n + n + 2 + n + 4 = 126$
   3. Combine like terms: $\qquad\qquad\quad 3n + 6 = 126$
   4. Subtract 6: $\qquad\qquad\quad 3n + 6 - 6 = 126 - 6$
   5. Simplify: $\qquad\qquad\qquad\qquad 3n = 120$
   6. Divide by 3: $\qquad\qquad\qquad \dfrac{3n}{3} = \dfrac{120}{3}$
      $$n = 40$$

   Thus the three consecutive even integers are 40, 42, and 44.
5. Verify that the sum of the three integers is 126. Since $40 + 42 + 44 = 126$, our result is correct. ▲

The method we have used to solve word problems works for numbers other than consecutive integers. We solve a different type of problem in the next example.

EXAMPLE 2   12 less than 5 times a number is the same as the number increased by 4. Find the number.

Solution   We use the five steps in the RSTUV method.

1. Read the problem. Note that we are asking for a certain number.
2. Select n to be the number.

Problem 1   The sum of three consecutive odd integers is 249. Find the integers.

Problem 2   3 less than 4 times a number is the same as the number increased by 9. Find the number.

3. Translate the problem into algebra.

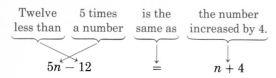

Twelve less than	5 times a number	is the same as	the number increased by 4.
$5n - 12$		$=$	$n + 4$

4. Use algebra to solve the equation.

    1. Given: $\qquad\qquad\qquad 5n - 12 = n + 4$

    2. Add 12: $\qquad\qquad 5n - 12 + 12 = n + 4 + 12$

    3. Simplify: $\qquad\qquad\qquad\; 5n = n + 16$

    4. Subtract n: $\qquad\quad\; 5n - n = n - n + 16$

    5. Simplify: $\qquad\qquad\qquad\; 4n = 16$

    6. Divide by 4: $\qquad\qquad\quad \dfrac{4n}{4} = \dfrac{16}{4}$

$$n = 4$$

Thus, the number is 4.

5. Verify that 12 less than 5 times 4 is 4 increased by 4; that is, is $5(4) - 12 = 4 + 4$?

$$5(4) - 12 = 4 + 4$$
$$20 - 12 = 8$$

Yes, this sentence is true.   ▲

There are many interesting problems that can be solved using the RSTUV method. Go on to the next example to see how it is done.

B. MORE WORD PROBLEMS

EXAMPLE 3 Have you eaten at a fast-food restaurant lately? If you eat a cheeseburger and fries, you would consume 1070 calories. As a matter of fact, the fries contain 30 more calories than the cheeseburger. How many calories in each?

Solution

1. Read the problem. We are asked for the number of calories in the cheeseburger and in the fries.
2. Select the variable c to represent the number of calories in the cheeseburger. This makes the number of calories in the fries $c + 30$, that is, 30 more calories. (We could also do the problem by letting f be the number of calories in the french fries; then $f - 30$ is the number of calories in the cheeseburger.)
3. Translate the problem. The total number of calories:

(calories in fries)	and	(calories in cheeseburger)	is	1070
$(c + 30)$	$+$	c	$=$	1070

Problem 3 If you eat a single slice of a 16-in. mushroom pizza and a 10-oz chocolate shake, you would consume 530 calories. If the shake has 70 more calories than the pizza, how many calories in each?

ANSWERS (to problems on page 249)
1. 81, 83, 85
2. $n = 4$

4. Use algebra to solve the equation.

    1. Given: $(c + 30) + c = 1070$

    2. Remove parentheses: $c + 30 + c = 1070$

    3. Combine like terms: $2c + 30 = 1070$

    4. Subtract **30**: $2c + 30 - 30 = 1070 - 30$

    5. Simplify: $2c = 1040$

    6. Divide by **2**: $\dfrac{2c}{2} = \dfrac{1040}{2}$

                                       $c = 520$

    Thus, the cheeseburger has 520 calories. Since the fries have 30 calories more than the cheeseburger, the fries have $520 + 30 = 550$ calories.

5. Verify that the total number of calories—that is, $520 + 550$—equals 1070. Since $520 + 550 = 1070$, our results are correct.

NAME

CLASS

SECTION

A. Solve the given problem.

ANSWERS

1. The sum of three consecutive even integers is 138. Find the integers.

    1. _____

2. The sum of three consecutive odd integers is 135. Find the integers.

    2. _____

3. The sum of three consecutive even integers is -24. Find the integers.

    3. _____

4. The sum of three consecutive odd integers is -27. Find the integers.

    4. _____

5. The sum of two consecutive integers is -25. Find the integers.

    5. _____

6. The sum of two consecutive integers is -9. Find the integers.

    6. _____

7. Find three consecutive integers $(n, n + 1,$ and $n + 2)$ such that the last added to twice the first is 23.

    7. _____

8. Find three consecutive integers $(n, n + 1,$ and $n + 2)$ such that the last added to twice the first is 47.

    8. _____

9. Find three consecutive integers $(n, n + 1,$ and $n + 2)$ such that the sum of the first and the last is the same as twice the middle one.

    9. _____

10. One number is 27 greater than another and their sum is 141. What are the numbers?

    10. _____

11. One number is 24 greater than another. Their sum is 64. What are the numbers?

    11. _____

12. The sum of two numbers is 179 and one of them is 5 more than the other. Find the numbers.

    12. _____

13. The sum of two numbers is 133 and one of them is 55 more than the other. Find the numbers.

    13. _____

14. Find two numbers such that one is two-fifths of the other, and their sum is 210.

    14. _____

15. Find two numbers such that one is 4 greater than three-eighths of the other, and their sum is 147.

    15. _____

16. The second of three numbers is 2 greater than the first, and the third is 2 greater than the second. What are the numbers if their sum is 261?

    16. _____

17. The sum of three numbers is 254. The second is 3 times the first, and the third is 5 less than the second. Find the numbers.

    17. _____

18. The sum of three consecutive integers is 17 less than 4 times the smallest of the three integers. Find the integers.

    18. _____

19. The larger of two numbers is 6 times the smaller. Their sum is 147. Find the numbers.

19. _____

20. Five times a certain fraction yields the same as 3 times 1 more than the fraction. Find the fraction.

20. _____

B. Solve the given problem.

21. The number of rooms in the Detroit Plaza Hotel exceeds the number of rooms in the Peachtree Plaza by 300. If the combined number of rooms in these hotels is 2500, how many rooms are there in each hotel?

21. _____

22. The Manhattan, a U.S.-built tanker, can carry 50 times as much oil as the first oceangoing tanker, the Gluckhauf. If the Manhattan can carry 115,000 tons of oil, how much oil can the Gluckhauf carry?

22. _____

23. Norway has three times as many merchant ships as Sweden, a total of 2811 ships. How many merchant ships does Sweden have?

23. _____

24. Russia and Japan have the most merchant ships in the world. Their combined total is 15,426. If Japan has 2276 more ships than Russia, how many does each have?

24. _____

25. The height of the Empire State building and its antenna is 1472 ft. The building is 1250 ft high. How high is the antenna?

25. _____

26. The cost C for renting a car is given by $C = 0.10m + 10$, where m is the number of miles traveled. If the total cost amounted to $17.50, how many miles were traveled?

26. _____

27. A toy rocket goes vertically upwards with an initial velocity of 96 feet per second (ft/sec). After t seconds, the velocity of the rocket is given by the formula $v = 96 - 32t$, where we are neglecting air resistance. In how many seconds will the rocket reach its highest point? (*Hint:* At the highest point, $v = 0$.)

27. _____

28. Refer to Problem 27 and find the number of seconds t that must elapse before the velocity decreases to 16 ft/sec.

28. _____

29. In a recent school election, 980 votes were cast. The winner received 372 votes more than the loser. How many votes did each of them receive?

29. _____

30. The combined cost of the U.S. and Soviet space programs has been estimated at $71 billion. The U.S. program is cheaper; it costs $19 billion less than the Soviet program. What is the cost of each program?

30. _____

✓ **SKILL CHECKER**

Solve.

31. $55T = 100$

32. $88R = 3240$

33. $15T = 120$

34. $81T = 3240$

31. _____

32. _____

33. _____

34. _____

35. $-75x = -600$

36. $-45x = -900$

37. $-0.02P = -70$

38. $-0.05R = -100$

39. $-0.04x = -40$

40. $-0.03x = 30$

35. _____

36. _____

37. _____

38. _____

39. _____

40. _____

3.5 USING YOUR KNOWLEDGE

1. If you are planning to become an algebraist (an expert in algebra), you might not be very famous. As a matter of fact, very little is known about one of the best algebraists of all time, the Greek, Diophantus. According to a legend, the following problem is in the inscription on Diophantus' tomb: One-sixth of his life God granted him youth. After a twelfth more, he grew a beard. After an additional seventh, he married, and 5 yr later, he had a son. Alas, the unfortunate son's life span was only one-half that of his father, who consoled his grief in the remaining 4 yr of his life. Can you use your knowledge to find how many years Diophantus lived? (*Hint:* Let x be the number of years Diophantus lived.)

1. _____

3.6 MOTION, MIXTURE, AND INVESTMENT PROBLEMS

B.C. by permission of Johnny Hart and Field Enterprises.

OBJECTIVES

REVIEW

Before starting this section, you should know:

1. How to translate sentences into the language of algebra (Section 2.6).
2. How to solve linear equations (Section 3.3).

OBJECTIVES

Using the RSTUV method and the charts we discuss in this section, you should be able to solve:

A. Motion problems.
B. Mixture problems.
C. Investment problems.

In the cartoon, the bird is trying an impossible task, unless he turns around! If he does, how does Curly know that it will take him less than 2 hr to fly 100 mi? Because there is a formula to figure it out! If an object moves at a constant rate R for a time T, the distance D traveled by the object is given by

$$D = RT \quad \text{In this formula, } D \text{ is the distance, } R \text{ the average rate,} \quad (1)$$
and T the time.

The object may be your car moving at a constant rate R of 55 mi/hr for $T = 2$ hr. In this case, you would have traveled a distance $D = 55 \times 2 = 110$ mi. (Note that the rate is in miles per *hour* and the time is in hours. Units have to be consistent!) Similarly, if you jog at a constant rate of 4 mi/hr for 2 hr, you would travel $4 \times 2 = 8$ mi. (Note that when the units are in miles per hour, the time must be in hours.) In working with Equation (1), and especially in more complicated problems, it may be helpful to write the formula in a chart. For example, to figure out how far you jogged, you would write:

	R	×	**T**	=	**D**
Jogger	4 mi/hr		2 hr		8 mi

A. MOTION PROBLEMS

Now let us go back to the bird problem, which is a **motion** problem. If the bird turns around and is flying at 5 mi/hr with a tail wind of 50 mi/hr, his rate R would be $50 + 5 = 55$ mi/hr. The wind is helping the bird, so the wind speed must be added to his rate. Curly wants to know how long it would take the bird to fly a distance of 100 mi; that is, he wants to find the time T. We write the information in a table, substituting 55 for R and 100 for D.

	R	×	**T**	=	**D**
Bird	55		T		100

Since

$$R \times T = D$$

we have

$$55T = 100$$

$$T = \frac{100}{55} = \frac{20}{11} = 1\frac{9}{11} \text{ hr} \quad \text{Dividing by } 55$$

Curly was right; it does take less than 2 hr!

EXAMPLE 1 The longest regularly scheduled bus route is Greyhound's "Supercruiser" Miami–to–San Francisco route, a distance of 3240 mi. If this distance is covered in about 82 hr, what is the average speed R of the bus?

Solution We use the RSTUV method.

1. Read the problem. We want to know the rate.
2. Select the variable R to represent this rate.
3. Translate the problem and write the information in a chart.

	R	×	**T**	=	**D**
Supercruiser	R		82		3240

The equation is

$$R \times 82 = 3240$$

or

$$82R = 3240$$

4. Solve the equation.

1. Given. $82R = 3240$

2. Divide by 82 $\dfrac{82R}{82} = \dfrac{3240}{82}$

3. By long division (to the nearest tenth). $R = 39.5$

```
         39.51
    82 | 3240.0
         246
         ---
         780
         738
         ---
          42 0
          41 0
          ----
           1 00
             82
           ----
             18
```

5. Verify that $R \times T = D$, that is,

39.5×82 has to be 3240

Problem 1 The distance from Miami to Tampa is about 250 mi. This distance can be covered in 5 hr by car. What is the average speed?

However, $39.5 \times 82 = 3239$, *not* 3240. This happens because we approximated the answer to the nearest tenth. ▲

Sometimes we have two objects in motion. In such cases we simply make two lines to record the information in the chart. We illustrate the procedure in the next example.

EXAMPLE 2 The Supercruiser bus leaves Miami traveling at an average rate of 40 mi/hr. Three hours later a car leaves Miami for San Francisco traveling on the same route at 55 mi/hr. How long does it take for the car to overtake the bus?

Solution

1. Read the problem carefully. We need to know the time it takes the car to overtake the bus.
2. Select the variable T to represent this time.
3. Translate the given information and write it in a chart. Note that if the car goes for T hr, the bus goes for $(T + 3)$ hr (since it left 3 hr earlier).

	R	$\times$	T	$=$	D
Car	55		T		$55T$
Bus	40		$T + 3$		$40(T + 3)$

When the car overtakes the bus, they will have traveled the same distance. According to the chart, the car has traveled $55T$ and the bus $40(T + 3)$ miles. Thus:

$$\underbrace{55T}_{\substack{\text{Distance} \\ \text{traveled by car}}} = \underbrace{40(T + 3)}_{\substack{\text{Distance traveled} \\ \text{by Supercruiser}}}$$

4. Solve the equation.

   1. Given. $\qquad\qquad\qquad 55T = 40(T + 3)$
   2. Simplify. $\qquad\qquad\quad 55T = 40T + 120$
   3. Subtract **40T**. $\quad 55T - 40T = 40T - 40T + 120$
   4. Simplify. $\qquad\qquad\quad 15T = 120$
   5. Divide by **15**. $\qquad\quad \dfrac{15T}{15} = \dfrac{120}{15}$
   $$T = 8$$

   Thus it takes the car 8 hr to overtake the bus.
5. Verify the answer.
   The car travels for 8 hr at 55 mi/hr; thus, it travels $55 \times 8 = 440$ mi, whereas the bus went at 40 mi/hr for 11 hr, a total of $40 \times 11 = 440$ mi. ▲

In the preceding problems, the motion of the two vehicles was in the *same* direction. A variation of this type of problem involves motion in *opposite* directions, as shown in the next example.

Problem 2 A bus leaves Los Angeles traveling at 50 mi/hr. An hour later a car leaves at 60 mi/hr to try to catch the bus. How long does it take before the car catches the bus?

EXAMPLE 3 The Supercruiser leaves Miami for San Francisco 3240 mi away, traveling at an average speed of 40 mi/hr. At the same time a slightly faster bus leaves San Francisco for Miami traveling at 41 mi/hr. After how many hours do the buses meet?

Solution

1. Read the problem. We need to know how many hours it takes for the buses to meet.
2. Select T to represent the hours each bus travels before they meet.
3. Translate the information and write it in a chart.

	R $\times$	T $=$	D
Supercruiser	40	T	$40T$
Bus	41	T	$41T$

The distance the Supercruiser travels is $40T$, whereas the bus travels $41T$, as shown in the diagram.

When they meet, the combined distance traveled by *both* buses is 3240 mi. This distance is also $40T + 41T$. Thus, we have

$$40T \quad + \quad 41T \quad = \quad 3240$$

Traveled by   Traveled by   Total
Supercruiser   other bus   distance

4. Solve the equation.

   1. Given. $40T + 41T = 3240$

   2. Combine like terms. $81T = 3240$

   3. Divide by 81. $\dfrac{81T}{81} = \dfrac{3240}{81}$

                             $T = 40$

Thus, each bus traveled 40 hr before they met.

5. Verify that 40 hr is correct. The Supercruiser travels $40 \times 40 = 1600$ mi in 40 hr, whereas the other bus travels $40 \times 41 = 1640$ mi. Clearly, the total distance traveled is $1600 + 1640 = 3240$ mi.

B. MIXTURE PROBLEMS

Another type of problem that can be solved by using a chart is the **mixture** problem. In this type of problem two or more things are put together to form a mixture. For example, dental-supply houses mix pure gold and platinum to make white gold for dental fillings. Suppose one of these houses wishes to make 10 troy oz of white gold to sell for $415 per ounce. If pure gold sells for $400 per ounce and platinum for $475 per ounce, how much of each should the supplier mix?

Problem 3 Two trains are 300 mi apart, traveling toward each other. One is traveling at 40 mi/hr and the other at 35 mi/hr. After how many hours do they pass?

This problem can be solved by the RSTUV method. After reading the problem, note that we are looking for the amount of each material we have to mix in order to get 10 oz of a mixture selling for $415 per ounce. If we let x be the total number of ounces of gold, then $10 - x$, the balance, must be the number of ounces of platinum. Note that if a quantity T is split into two parts, one part may be x and the other $T - x$. This can be checked by adding $T - x$ and x, obtaining $T - x + x = T$. We then write all the information in a chart. The top line of the chart should tell us that if we multiply the price of the item by the amount used we will get the total price. (Look at the chart that follows.) After this, the first entry (following the word *gold*) tells us that the price of gold ($400) times the amount being used (x oz) gives us the total price ($400x$). In the second line (following the word *platinum*), we write the price of platinum ($475) times the amount being used ($10 - x$ oz) to get a total price of $475(10 - x)$.

Finally, the third line tells us that the price of the mixture is $415, that we wish to use 10 oz of it, and that the total price will be $4150. The completed chart is shown.

	PRICE/OUNCE	×	AMOUNT	=	TOTAL PRICE
Gold	400		x		$400x$
Platinum	475		$10 - x$		$475(10 - x)$
Mixture	415		10		4150

Since the sum of the total prices of gold and platinum in the last column must be equal to the total price of the mixture, it follows that:

$$400x + 475(10 - x) = 4150$$

Simplifying.
$$400x + 4750 - 475x = 4150$$

$$4750 - 75x = 4150$$

Subtracting **4750.**
$$4750 - 4750 - 75x = 4150 - 4750$$

Simplifying.
$$-75x = -600$$

Dividing by **−75.**
$$\frac{-75x}{-75} = \frac{-600}{-75}$$

$$x = 8$$

Thus the supplier must use 8 oz of gold and $10 - 8 = 2$ oz of platinum. You can verify that this is correct! (8 oz of gold at $400 per ounce and 2 oz of platinum at $475 per ounce make 10 oz costing $4150.)

EXAMPLE 4 How many ounces of a 50% acetic acid solution should a photographer add to 32 oz of a 5% acetic acid solution to obtain a 10% acetic acid solution?

Solution We use the RSTUV method.

1. Read the problem and note that we are asking for the number of ounces of the 50% solution (a solution consisting of 50% acetic acid) that should be added.

Problem 4 How many gallons of a 10% salt solution should be added to 15 gal of a 20% salt solution to obtain a 16% solution?

2. Select x to stand for the number of ounces of 50% solution to be added.

3. To translate the problem, we first use a chart. In this case, the heading for the chart should contain the percent of acetic acid and the amount to be mixed. The product of these two numbers would give us the amount of pure acetic acid. This information does indeed appear as the heading of the chart below.

	%	×	AMOUNT	=	AMOUNT OF PURE ACID IN FINAL MIXTURE
50% solution	0.50		x		$0.50x$
5% solution	0.05		32		1.60
10% solution	0.10		$x + 32$		$0.10(x + 32)$

The percents have been converted to decimals.

Since we have x oz of one and 32 oz of the other, we have $(x + 32)$ of the mixture.

Since the sum of the total amounts of pure acetic acid should be the same as the amount of pure acetic acid in the final mixture, we have

$$0.50x + 1.60 = 0.10(x + 32)$$

4. We now solve the equation $0.50x + 1.60 = 0.10(x + 32)$.

Multiply by **10** to get rid of the decimals.

$$10 \cdot 0.50x + 10 \cdot 1.60 = 10 \cdot [0.10(x + 32)]$$
$$5x + 16 = 1(x + 32)$$
$$5x + 16 = x + 32$$

Subtract **16**.
$$5x + 16 - 16 = x + 32 - 16$$

Simplify.
$$5x = x + 16$$

Subtract x.
$$5x - x = x - x + 16$$

Simplify.
$$4x = 16$$

Divide by **4**.
$$\frac{4x}{4} = \frac{16}{4}$$
$$x = 4$$

Thus the photographer must add 4 oz of the 50% solution.

5. The verification of this fact is left to the student.

C. INVESTMENT PROBLEMS

Finally, there is another problem that is very similar to the mixture problem—the **investment** problem. Investment problems depend on the fact that the interest I you can earn (or pay) on principal P invested at rate r for 1 yr is given by the formula

$$I = Pr \qquad (2)$$

ANSWERS   (to problems on pages 260–261)

3. 4 hr   **4.** 10 gal

Now suppose you have a total of $10,000 invested. Part of the money is invested at 6% and the rest at 8%. If the bank tells you that you have earned $730 interest for 1 yr, how much do you have invested at each rate?

In this case, we need to know how much is invested at each rate. Thus, if we say that we have invested P dollars at 6%, the rest of the money, that is, $10,000 - P$, would be invested at 8%. Note that for this to work, the sum of the amount invested at 6%, P dollars, plus the rest, $10,000 - P$, must have a sum of $10,000. Since $P + 10,000 - P = 10,000$, we have done it correctly. All this information is entered in a chart, as shown next:

	P	$\times$	r	$=$	I
6% invest.	P		0.06		$0.06P$
8% invest.	$10,000 - P$		0.08		$0.08(10,000 - P)$

This column must add to 10,000.   The total interest is the sum of the two expressions.

From the chart we can see that the total interest is the sum of the expressions in the last column. Or, $0.06P + 0.08(10,000 - P)$. Of course, the total interest is also $730. Thus,

$$0.06P + 0.08(10,000 - P) = 730$$

You can solve this equation by first multiplying each term by **100** to get rid of the decimals, obtaining

$$100 \cdot 0.06P + 100 \cdot 0.08(10,000 - P) = 100 \cdot 730$$
$$6P + 8(10,000 - P) = 73,000$$
$$6P + 80,000 - 8P = 73,000$$
$$-2P = -7000$$
$$P = 3500$$

You can also solve

$$0.06P + 0.08(10,000 - P) = 730$$

as follows:

Simplify. $0.06P + 800 - 0.08P = 730$
                    $800 - 0.02P = 730$

Subtract **800** $800 - 800 - 0.02P = 730 - 800$

Simplify. $-0.02P = -70$

Divide by **−0.02** $\dfrac{-0.02P}{-0.02} = \dfrac{-70}{-0.02}$

$$P = 3500$$

$$\begin{array}{r} 35\ 00 \\ 0.02\overline{\smash{)}70.00} \\ 6 \\ \hline 10 \\ 10 \\ \hline 0 \end{array}$$

Thus $3500 is invested at 6% and the rest, $10,000 - 3500$, or $6500, is invested at 8%. You can verify that 6% of $3500 added to 8% of $6500 yields $730.

EXAMPLE 5 A woman has some stocks yielding 5% annually and some bonds that yield 10%. If her investment totals $6000 and her annual income from the investment is $500, how much does she have invested in stocks and how much in bonds?

Solution

1. Read the problem. We want to find how much she has invested in stocks and how much in bonds.
2. Let s be the amount she has invested in stocks. This makes the amount invested in bonds $(6000 - s)$.
3. We enter the information in the following chart:

	P	$\times$	r	$=$	I
Stocks	s		0.05		$0.05s$
Bonds	$6000 - s$		0.10		$0.10(6000 - s)$

The total interest is the sum of the entries in the last column, that is, $0.05s$ and $0.10(6000 - s)$. This amount must be $500:

$$0.05s + 0.10(6000 - s) = 500$$

4. Solve the equation.

1. Given.	$0.05s + 0.10(6000 - s) = 500$
2. Simplify.	$0.05s + 600 - 0.10s = 500$
	$600 - 0.05s = 500$
3. Subtract **600.**	$600 - 600 - 0.05s = 500 - 600$
4. Simplify.	$-0.05s = -100$
5. Divide by -0.05.	$\dfrac{-0.05s}{-0.05} = \dfrac{-100}{-0.05}$
	$s = 2000$

Thus the woman has $2000 in stocks and the rest, $4000, in bonds.
5. To verify the answer, note that 5% of 2000 is 100 and 10% of 4000 is 400. Thus the total interest is indeed $500.

Problem 5 A man has two investments totaling $8000. One investment yields 5% and the other 10%. If the total annual interest is $650, how much is invested at each rate?

NAME

CLASS

SECTION

A. Solve the following problems (use the chart).

ANSWERS

1. The distance from Los Angeles to Sacramento is about 400 mi. A bus covers this distance in about 8 hr. What is the average speed of the bus?

1. _____

	$R \times T = D$
Bus	

2. The distance from Boston to New Haven is 260 mi. A car leaves Boston at 7 in the morning and gets to New Haven just in time for lunch, at exactly 12 noon. What is the speed of the car?

2. _____

	$R \times T = D$
Car	

3. A small plane leaves San Antonio for San Francisco, a distance of 1632 mi. If it takes the pilot 8 hr to get there, how fast is the plane going?

3. _____

	$R \times T = D$
Plane	

4. A jet takes off from San Francisco to San Antonio, a distance of 1632 mi. If the jet flies at an average speed of 408 mi/hr, how long does the trip take?

4. _____

	$R \times T = D$
Jet	

5. The distance from Miami to Tampa is about 200 mi. If a jet flies at an average speed of 400 mi/hr, how long does it take to go from Tampa to Miami?

5. _____

	$R \times T = D$
Jet	

6. A freight train leaves the station traveling at 30 mi/hr. One hour later, a passenger train leaves the same station traveling at 60 mi/hr. How long does it take for the passenger train to overtake the freight train?

6. _____

	R × T = D
Pass.	
Freight	

7. A bus leaves the station traveling at 60 kilometers per hour (km/hr). Two hours later, the wife of one of the passengers shows up at the station with a briefcase belonging to an absent-minded professor riding the bus. If she immediately starts after the bus at 90 km/hr, how long would it be before she reunites the briefcase with the professor?

7. _____

	R × T = D
Car	
Bus	

8. An accountant and his boss have to travel to a nearby town. The accountant catches a train traveling at 50 mi/hr, whereas the boss leaves 1 hr later in a car traveling at 60 mi/hr. They have decided to meet at the train station and, strangely enough, they get there at exactly the same time! If the train and the car traveled in a straight line on parallel paths, how far is it from one town to the other?

8. _____

	R × T = D
Car	
Train	

9. The basketball coach at a local high school left for work on his bicycle traveling at 15 mi/hr. Half an hour later, his wife noticed that he had left his lunch. She got in her car and took his lunch to him, traveling at 60 mi/hr. Luckily, she got to school at exactly the same time as her husband. How far is it from the house to the school?

9. _____

	R × T = D
Wife	
Coach	

10. A jet traveling 480 mi/hr leaves San Antonio for San Francisco, a distance of 1632 mi. An hour later another plane, going at the same speed, leaves San Francisco for San Antonio. How long would it be before the planes pass each other?

10. _____

	R × T = D
Plane	
Jet	

ANSWER (to problem on page 264)
5. $3000 at 5%; $5000 at 10%

11. A car leaves town A going toward B at 50 mi/hr. At the same time, another car leaves B going toward A at 55 mi/hr. How long will it be before the two cars meet if the distance from A to B is 630 mi?

11. _____

	R × T = D
Car 1	
Car 2	

12. A contractor has two jobs that are 275 km apart. Her headquarters, by sheer luck, happen to be on a straight road between the two construction sites. Her first crew left for one job traveling at 70 km/hr. Two hours later, she left for the other job, traveling at 65 km/hr. If the contractor and her crew arrived at their job sites simultaneously, how far did the first crew have to drive?

12. _____

	R × T = D
Cont.	
Crew	

13. A plane has 7 hr to reach a target and come back to base. It flies out to the target at 480 mi/hr and returns at 640 mi/hr. How many miles from the base is the target?

13. _____

	R × T = D
Out	
Return	

14. A man left home driving at 40 mi/hr. When his car broke down, he walked home at a rate of 5 mi/hr; the whole trip took him $2\frac{1}{4}$ hr. How far from his house did his car break down?

14. _____

	R × T = D
Car	
Walk	

B. Solve the following problems.

15. How many liters (L) of a 40% glycerin solution must be mixed with 10 L of an 80% glycerin solution to obtain a 65% solution?

15. _____

	% × AMOUNT = TOTAL
40%	
80%	
65%	

16. How many parts of glacial acetic acid (99.5% acetic acid) must be added to 100 parts of a 10% solution of acetic acid to give a 28% solution?

16. _____

	% × AMOUNT = TOTAL
99.5%	
10%	
28%	

17. If the price of copper is 65¢ per pound and the price of zinc is 30¢ per pound, how many pounds of copper and zinc should be mixed to make 70 lb of brass selling for 45¢ per pound?

17. _____

	PRICE × AMOUNT = TOTAL
Copper	
Zinc	
Brass	

18. Oolong tea sells for $19 per pound. How many pounds of Oolong should be mixed with regular tea selling at $4 per pound to produce 50 lb of tea selling for $7 per pound?

18. _____

	PRICE × AMOUNT = TOTAL
Oolong	
Reg.	
Mixture	

19. Do you think the price of coffee is high? You have not seen anything yet! Blue Jamaican coffee sells for about $5 per pound! How many pounds of Blue Jamaican should be mixed with 80 lb of regular coffee selling at $2 per pound so that the result is a mixture selling for $2.60 per pound? (You can cleverly advertise this mixture as "Containing the incomparable Blue Jamaican coffee.")

19. _____

	PRICE × AMOUNT = TOTAL
Blue	
Reg.	
Mix	

NAME

20. How many ounces of vermouth containing 10% alcohol should be added to 20 oz of gin containing 60% alcohol so that the resulting pitcher of martinis will contain 30% alcohol?

20. _____

	% × AMOUNT = TOTAL
10%	
60%	
30%	

21. Do you know how to make manhattans? They are mixed by combining bourbon and sweet vermouth. How many ounces of manhattans containing 40% vermouth should a bartender mix with manhattans containing 20% vermouth so that she can obtain a half gallon (64 oz) of manhattans containing 30% vermouth?

21. _____

	% × AMOUNT = TOTAL
40%	
20%	
30%	

22. A car radiator contains 30 qt of 50% antifreeze solution. How many quarts of this solution should be drained and replaced with pure antifreeze so that the new solution is 70% antifreeze?

22. _____

	% × AMOUNT = TOTAL
50%	
100%	
30%	

23. A car radiator contains 30 qt of 50% antifreeze solution. How many quarts of this solution should be drained and replaced with water so that the new solution is 30% antifreeze?

23. _____

	% × AMOUNT = TOTAL
50%	
0%	
30%	

C. Solve the following problems.

24. Two sums of money totaling $15,000 earn, respectively, 5% and 7% annual interest. If the interest from both investments amounts to $870, how much is invested at each rate?

24. _____

	P × r = I
5%	
7%	

25. An investor invested $20,000, part at 6% and the rest at 8%. Find the amount invested at each rate if the annual income from the two investments is $1500.

25. _____

	P × r = I
6%	
8%	

26. A woman invested $25,000, part at 7.5% and the rest at 6%. If her annual interest from these two investments amounted to $1620, how much money did she have invested at each rate?

26. _____

	P × r = I
7.5%	
6%	

27. A man has a savings account that pays 5% annual interest and some certificates of deposit paying 7% annually. His total interest from the two investments is $1100, and the total amount of money in the two investments is $18,000. How much money does he have in the savings account?

27. _____

	P × r = I
5%	
7%	

28. A woman invested $20,000 at 8%. What additional amount must she invest at 6% so that her annual income is $2200?

28. _____

	P × r = I
8%	
6%	

29. A sum of $10,000 is split and the two parts are invested at 5% and 6%, respectively. If the interest in the 5% investment exceeds the interest in the 6% investment by $60, how much is invested at each rate?

29. _____

	P × r = I
5%	
6%	

30. An investor receives $600 annually from two investments. He has $500 more invested at 8% than at 6%. Find the amount invested at each rate.

30. _____

	P × r = I
8%	
6%	

✓ **SKILL CHECKER**

Find.

31. $-16(2)^2 + 118$

32. $-8(3)^2 + 80$

33. $-4 \cdot 8 \div 2 + 20$

34. $-5 \cdot 6 \div 2 + 25$

31. _____

32. _____

33. _____

34. _____

CALCULATOR CORNER

Some of the mixture problems given in this section can be solved using the *guess-and-correct* procedure developed by Dr. Harrison A. Gesselmann of Cornell University. The procedures depends on taking a guess at the answer and then using the calculator to *correct* this guess. For example, to solve the very first mixture problem presented in this section (pages 260–261), we have to mix gold and platinum to obtain 10 oz of a mixture selling for $415 per ounce. Our first guess is to use *equal amounts* (5 oz each) of gold and platinum. This gives a mixture with a price per pound equal to the average price of gold ($400) and platinum ($475), that is,

$$\frac{400 + 475}{2} = \$437.50 \text{ per pound}$$

As you can see from the figure, more gold must be used in order to bring the $437.50 average down to the desired $415:

5 oz	Desired	Average	5 oz
$400	$415	$437.50	$475

22.50

37.50

Thus, the correction for the additional amount of gold that must be used is

$$\frac{22.50}{37.50} \times 5 \text{ oz}$$

This expression can be obtained by the keystroke sequence

| 22.50 | ÷ | 37.50 | × | 5 | = |

which gives the correction 3. The correct amount is:

First guess	5 oz of gold
+ Correction	3 oz of gold
Total =	8 oz of gold

and the remaining 2 oz is platinum.

   If your instructor wishes, use this method to work Problems 17 and 21.

   The guess-and-correct method also works for investment problems in which we must find how much is invested at certain rates when the *final* amount of interest is known. For example, the very first investment problem (page 263) supposes that $10,000 has been invested, part at 6% and the rest at 8%. We know the final interest for the year, $730, and we must find how much was invested at each rate. Our first guess is that equal amounts ($5000 each) were invested at each rate. This would give a rate of return of

$$\frac{6+8}{2} = 7\%$$

The actual rate of return is

$$\frac{730}{10,000} = 7.3\%$$

so we must use some of the money invested at 8% to bring the average up to the desired amount (see the figure).

$5000	Average	Desired	$5000
6%	7%	7.3%	8%

Thus, the correction would be

$$\frac{0.3}{1} \times 5000$$

more dollars invested at 8%, that is,

| . | 3 | ÷ | 1 | × | 5000 | = |

or $1500 more dollars than the original $5000 guess must be invested at 8%. Thus, we must invest $5000 + $1500 or $6500 at 8% and the rest ($3500) at 6%.

   If your instructor approves, try Problems 25 and 27.

SUMMARY

SECTION	ITEM	MEANING	EXAMPLE
3.1A	Solutions	The solutions of an equation are the replacements of the variable that make the equation a true statement.	4 is a solution of $x + 1 = 5$.
3.1A	Equivalent equations	Two equations are equivalent if their solutions are the same.	$x + 1 = 4$ and $x = 3$ are equivalent.
3.1B	The addition property of equality	$a = b$ is equivalent to $a + c = b + c$.	$x - 1 = 2$ is equivalent to $x - 1 + 1 = 2 + 1$.
3.1C	The subtraction property of equality	$a = b$ is equivalent to $a - c = b - c$.	$x + 1 = 2$ is equivalent to $x + 1 - 1 = 2 - 1$.
3.2A	The multiplication property of equality	$a = b$ is equivalent to $ac = bc$ if c is not 0.	$\dfrac{x}{2} = 3$ is equivalent to $2 \cdot \dfrac{x}{2} = 2 \cdot 3$.
3.2B	The division property of equality	$a = b$ is equivalent to $\dfrac{a}{c} = \dfrac{b}{c}$ if c is not 0.	$2x = 6$ is equivalent to $\dfrac{2x}{2} = \dfrac{6}{2}$
3.2C	Reciprocal	The reciprocal of $\dfrac{a}{b}$ is $\dfrac{b}{a}$.	The reciprocal of $\dfrac{5}{2}$ is $\dfrac{2}{5}$.
3.2D	LCM (lowest common multiple)	The smallest number that is a multiple of each of the given numbers.	The LCM of 3, 8, and 9 is 72.
3.3A	Linear equations	Equations that can be written in the form $ax + b = c$	$5x + 5 = 2x + 6$ is a linear equation (it can be written as $3x + 5 = 6$).
3.4	Inequality	A statement with $>$, $<$, $\geq$, or $\leq$ for its verb	$2x + 1 > 5$ and $3x - 5 \leq 7 - x$ are inequalities.
3.4B	The addition property of inequality	$a < b$ is equivalent to $a + c < b + c$	$x - 1 < 2$ is equivalent to $x - 1 + 1 < 2 + 1$.
3.4B	The subtraction property of inequality	$a < b$ is equivalent to $a - c < b - c$.	$x + 1 < 2$ is equivalent to $x + 1 - 1 < 2 - 1$.
3.4B	The multiplication property of inequality	$a < b$ is equivalent to: (1) $ac < bc$ if $c > 0$ (2) $ac > bc$ if $c < 0$	$\dfrac{x}{2} < 3$ is equivalent to $2 \cdot \dfrac{x}{2} < 2 \cdot 3$.
3.4B	The division property of inequality	$a < b$ is equivalent to: (1) $\dfrac{a}{c} < \dfrac{b}{c}$, if $c > 0$ (2) $\dfrac{a}{c} > \dfrac{b}{c}$, if $c < 0$	$-2x < 6$ is equivalent to $\dfrac{-2x}{-2} > \dfrac{6}{-2}$.
3.5	RSTUV Procedure	**R**ead, **S**elect a variable, **T**ranslate, **U**se algebra, and **V**erify your answer when solving word problems.	

NAME

CLASS

SECTION

(If you need help with these exercises, look in the section indicated in brackets.)

ANSWERS

1. [3.1A] Determine if the given number satisfies the equation.
   a. $5; 7 = 14 - x$
   b. $4; 13 = 17 - x$
   c. $-2; 8 = 6 - x$

   1. a. _____
   b. _____
   c. _____

2. [3.1B] Solve the given equation.

   a. $x - \dfrac{1}{3} = \dfrac{1}{3}$

   b. $x - \dfrac{5}{7} = \dfrac{2}{7}$

   c. $x - \dfrac{5}{9} = \dfrac{1}{9}$

   2. a. _____
   b. _____
   c. _____

3. [3.1D] Solve the given equation.

   a. $-3x + \dfrac{5}{9} + 5x - \dfrac{2}{9} = \dfrac{5}{9}$

   b. $-2x + \dfrac{4}{7} + 6x - \dfrac{2}{7} = \dfrac{6}{7}$

   c. $-4x + \dfrac{5}{6} + 7x - \dfrac{1}{6} = \dfrac{5}{6}$

   3. a. _____
   b. _____
   c. _____

4. [3.1D] Solve the given equation.
   a. $3 = 4(x - 1) + 2 - 3x$
   b. $4 = 5(x - 1) + 9 - 4x$
   c. $5 = 6(x - 1) + 8 - 5x$

   4. a. _____
   b. _____
   c. _____

5. [3.2A] Solve the given equation.

   a. $\dfrac{3}{5}x = -9$

   b. $\dfrac{2}{7}x = -4$

   c. $\dfrac{5}{6}x = -10$

   5. a. _____
   b. _____
   c. _____

6. [3.2C] Solve the given equation.

   a. $-\dfrac{3}{4}x = -9$

   b. $-\dfrac{1}{5}x = -3$

   c. $-\dfrac{2}{3}x = -6$

   6. a. _____
   b. _____
   c. _____

7. [3.2D] Solve the given equation.

    a. $\dfrac{x}{3} + \dfrac{2x}{4} = 5$

    b. $\dfrac{x}{4} + \dfrac{3x}{2} = 6$

    c. $\dfrac{x}{5} + \dfrac{3x}{10} = 10$

7. a. _____
   b. _____
   c. _____

8. [3.2D] Solve the given equation.

    a. $\dfrac{x}{3} - \dfrac{x}{4} = 1$

    b. $\dfrac{x}{2} - \dfrac{x}{7} = 10$

    c. $\dfrac{x}{4} - \dfrac{x}{5} = 2$

8. a. _____
   b. _____
   c. _____

9. [3.2D] Solve the given equation.

    a. $\dfrac{x-1}{4} - \dfrac{x+1}{6} = 1$

    b. $\dfrac{x-1}{6} - \dfrac{x+1}{8} = 0$

    c. $\dfrac{x-1}{8} - \dfrac{x+1}{10} = 0$

9. a. _____
   b. _____
   c. _____

10. [3.3A] Solve the given equation.

    a. $3 - \dfrac{x}{2} = \dfrac{2(x+1)}{3}$

    b. $5 - \dfrac{x}{3} = \dfrac{3(x+1)}{7}$

    c. $8 - \dfrac{x}{4} = \dfrac{7(x+1)}{5}$

10. a. _____
    b. _____
    c. _____

11. [3.3B] Solve for the indicated letter in the given equation.

    a. $A = \dfrac{1}{2}bh$; solve for h.

    b. $C = 2\pi r$; solve for r.

    c. $V = \dfrac{bh}{3}$; solve for b.

11. a. _____
    b. _____
    c. _____

12. [3.4B] Graph the given inequality.
    a. $4x - 2 < 2(x + 2)$
    b. $5x - 4 < 2(x + 1)$
    c. $7x - 1 < 3(x + 1)$

12. a. _____
    b. _____
    c. _____

13. [3.4B] Graph the given inequality.
    a. $6(x - 1) \geq 4x + 2$
    b. $5(x - 1) \geq 2x + 1$
    c. $4(x - 2) \geq 2x + 2$

13. a. _____
   b. _____
   c. _____

14. [3.4B] Graph the given inequality.

    a. $-\dfrac{x}{3} + \dfrac{x}{6} \leq \dfrac{x - 1}{6}$

    b. $-\dfrac{x}{4} + \dfrac{x}{7} \leq \dfrac{x - 1}{7}$

    c. $-\dfrac{x}{5} + \dfrac{x}{3} \leq \dfrac{x - 1}{3}$

14. a. _____
   b. _____
   c. _____

15. [3.4C] Solve and graph the compound inequality.
    a. $x + 2 \leq 4$ and $-2x < 6$
    b. $x + 3 \leq 5$ and $-3x < 9$
    c. $x + 1 \leq 2$ and $-4x < 8$

15. a. _____
   b. _____
   c. _____

16. [3.5A] Solve.
    a. The sum of two numbers is 84 and one of the numbers is 20 more than the other. What are the numbers?
    b. The sum of two numbers is 47 and one of the numbers is 19 more than the other. What are the numbers?
    c. The sum of two numbers is 81 and one of the numbers is 23 more than the other. What are the numbers?

16. a. _____
   b. _____
   c. _____

17. [3.5B] The annual return from two investments is $980. The second investment produces $280 more than the first.
    a. How much money does each investment produce?
    b. Repeat the problem if the investments produce $920 and the second investment produces $200 more than the first.
    c. Repeat the problem if the investments produce $673 and the second investment produces $147 more than the first.

17. a. _____
   b. _____
   c. _____

18. [3.6A] Solve.
    a. A car leaves a town traveling at 40 mi/hr. An hour later, another car leaves the same town in the same direction traveling at 50 mi/hr. How long does it take the second car to overtake the first one?
    b. Repeat the problem if the first car was traveling at 30 mi/hr and the second one at 50 mi/hr.
    c. Repeat the problem if the first car was traveling at 40 mi/hr and the second one at 60 mi/hr.

18. a. _____
   b. _____
   c. _____

19. [3.6B] Solve.
    a. How many pounds of a product selling at $1.50 per pound should be mixed with 15 lb of another product selling at $3 per pound to obtain a mixture selling at $2.40 per pound?
    b. Repeat the problem if the products sell for $2, $3, and $2.50, respectively.
    c. Repeat the problem if the products sell for $6, $2, and $4.50, respectively.

19. a. _____
   b. _____
   c. _____

20. [3.6C] Solve.

   a. A woman invested $30,000, part at 5% and part at 6%. Her annual interest amounts to $1600. How much does she have invested at each rate?

   b. Repeat the problem if the rates are 7% and 9%, respectively, and her annual return amounts to $2300.

   c. Repeat the problem if the rates are 6% and 10%, respectively, and her annual return amounts to $2000.

20. a. _____

     b. _____

     c. _____

NAME

CLASS

SECTION

ANSWERS

(*Answers on page* 280)

1. Does the number 3 satisfy the equation $6 = 9 - x$?

1. _____

2. Solve $x - \dfrac{2}{7} = \dfrac{3}{7}$.

2. _____

3. Solve $-2x + \dfrac{3}{8} + 5x - \dfrac{1}{8} = \dfrac{5}{8}$. 　　4. Solve $2 = 3(x - 1) + 5 - 2x$.

3. _____

4. _____

5. Solve $\dfrac{2}{3}x = -4$. 　　6. Solve $-\dfrac{2}{3}x = -6$.

5. _____

6. _____

7. Solve $\dfrac{x}{4} + \dfrac{2x}{3} = 11$. 　　8. Solve $\dfrac{x}{3} - \dfrac{x}{5} = 2$.

7. _____

8. _____

9. Solve $\dfrac{x-2}{5} - \dfrac{x+1}{8} = 0$. 　　10. Solve $6 - \dfrac{x}{3} = \dfrac{2(x+4)}{5}$.

9. _____

10. _____

11. Solve for h in the equation $V = \dfrac{4}{3}\pi r^2 h$.

11. _____

12. Graph $5x - 1 < 2(x + 1)$.

12. _____

13. Graph $3(x - 2) \geq 5x + 2$. 　　14. Graph $-\dfrac{x}{3} + \dfrac{x}{5} \leq \dfrac{x-5}{5}$.

13. _____

14. _____

15. Solve and graph the compound statement.

$x + 1 \leq 3$ 　and　 $-2x < 6$

15. _____

16. The sum of two numbers is 75. If one of the numbers is 15 more than the other, what are the numbers?

16. _____

17. A man has invested a certain amount of money in stocks and bonds. His annual return from these investments is $780. If the stocks produce $230 more in returns than the bonds, how much money does the man receive annually from each investment?

17. _____

18. A freight train leaves a station traveling at 30 mi/hr. Two hours later, a passenger train leaves the same station in the same direction at 40 mi/hr. How long does it take for the passenger train to catch the freight train?

18. _____

19. How many pounds of coffee selling for $1.20 per pound should be mixed with 20 lb of coffee selling for $1.80 per pound to obtain a mixture selling for $1.60 per pound?

19. _____

20. An investor bought some municipal bonds yielding 5% annually and some certificates of deposit yielding 7% annually. If her total investment amounts to $20,000 and her annual income is $1160, how much money is invested in bonds and how much in certificates of deposit?

20. _____

IF YOU MISSED QUESTION	SECTION	EXAMPLES	PAGE	ANSWERS
1	3.1	1	191	1. Yes
2	3.1	2	192	2. $x = \dfrac{5}{7}$
3	3.1	3	194	3. $x = \dfrac{1}{8}$
4	3.1	4	196	4. $x = 0$
5	3.2	1, 2, 3	204, 205, 208	5. $x = -6$
6	3.2	1, 2, 3	204, 205, 208	6. $x = 9$
7	3.2	4	210	7. $x = 12$
8	3.2	4	210	8. $x = 15$
9	3.2	5	211	9. $x = 7$
10	3.3	1, 2	220, 223	10. $x = 6$
11	3.3	3	224	11. $h = \dfrac{3V}{4\pi r^2}$
12	3.4	1, 2, 3	232, 233, 235	12.
13	3.4	4, 5	237, 239	13.
14	3.4	6	239	14.
15	3.4	7	241	15. $-3 < x \le 2$
16	3.5	1, 2	249	16. 30 and 45
17	3.5	3	250	17. $275 and $505
18	3.6	1, 2, 3	258, 259, 260	18. 6 hr
19	3.6	4	261	19. 10 lb
20	3.6	5	264	20. $12,000 in bonds, $8000 in certificates

CHAPTER 4

POLYNOMIALS

In this chapter we study one of the most important concepts in introductory algebra: polynomials. Polynomials are formed by adding, subtracting, or multiplying numbers and variables. We will learn how to add, subtract, multiply, and divide polynomials.

NAME

CLASS

SECTION

ANSWERS

(*Answers on page* 284)

1. Classify as a monomial (M), binomial (B), or trinomial (T):
   a. $8x + 5$
   b. $6x^3$
   c. $9x^2 - 2 + 6x$

2. Find the degree of the polynomial $6x - 8x^2 + 9$.

3. Write the polynomial $-3x + 7 + 9x^2$ in descending order of exponents.

4. Find the value of $-16t^2 + 100$ when $t = 3$.

5. Add $-6x + 3x^2 - 8$ and $-6x^2 - 4 + 3x$.

6. Subtract $6x - 2 + 8x^2$ from $2x^2 - 3x$.

7. Remove parentheses (simplify): $-6x^2(x + 8y)$.

8. Find $(x + 6)(x + 7)$.

9. Find $(x + 6)(x - 3)$.

10. Find $(x + 4)(x - 8)$.

11. Find $(5x - 3y)(4x - 3y)$.

12. Expand $(2x + 5y)^2$.

13. Expand $(3x - 7y)^2$.

14. Find $(3x - 5y)(3x + 5y)$.

15. Find $(x + 2)(x^2 + 3x + 2)$.

16. Find $2x(x + 2)(x + 5)$.

17. Expand $(x + 6)^3$.

18. Expand $\left(2x^2 - \dfrac{1}{2}\right)^2$.

19. Find $(2x^2 + 7)(2x^2 - 7)$.

20. Divide $2x^3 - 19x + 5$ by $x - 3$.

1. a. _____
   b. _____
   c. _____

2. _____

3. _____

4. _____

5. _____

6. _____

7. _____

8. _____

9. _____

10. _____

11. _____

12. _____

13. _____

14. _____

15. _____

16. _____

17. _____

18. _____

19. _____

20. _____

IF YOU MISSED QUESTION	SECTION	EXAMPLES	PAGE	ANSWERS
1	4.1	1	286	**1. a.** B
				b. M
				T
2	4.1	2	286	**2.** 2
3	4.1	3	287	**3.** $9x^2 - 3x + 7$
4	4.1	4	287	**4.** -44
5	4.2	1, 2	293, 294	**5.** $-3x^2 - 3x - 12$
6	4.2	3	295	**6.** $-6x^2 - 9x + 2$
7	4.3	1, 2	301	**7.** $-6x^3 - 48x^2y$
8	4.3	3	303	**8.** $x^2 + 13x + 42$
9	4.3	3, 4	303	**9.** $x^2 + 3x - 18$
10	4.3	3, 4	303	**10.** $x^2 - 4x - 32$
11	4.3	4, 5	303, 304	**11.** $20x^2 - 27xy + 9y^2$
12	4.4	1	310	**12.** $4x^2 + 20xy + 25y^2$
13	4.4	2	311	**13.** $9x^2 - 42xy + 49y^2$
14	4.4	3	312	**14.** $9x^2 - 25y^2$
15	4.5	1	320	**15.** $x^3 + 5x^2 + 8x + 4$
16	4.5	2	321	**16.** $2x^3 + 14x^2 + 20x$
17	4.5	3	322	**17.** $x^3 + 18x^2 + 108x + 216$
18	4.5	4	322	**18.** $4x^4 - 2x^2 + \dfrac{1}{4}$
19	4.5	5	322	**19.** $4x^4 - 49$
20	4.6	1, 2, 3	328, 330	**20.** $2x^2 + 6x - 1$ R 2

From the *Guinness Book of World Records.*

OBJECTIVES

REVIEW

Before starting this section you should know:

1. How to evaluate an expression (Section 1.6).
2. How to add, subtract, and multiply expressions (Sections 2.6, 2.7).

OBJECTIVES

You should be able to:

A. Classify a given polynomial as a monomial, binomial, or trinomial.
B. Find the degree of a polynomial in one variable.
C. Write a polynomial in descending order.
D. Evaluate a polynomial in one variable.

The diver in the photograph has jumped from a height of 118 ft. Do you know how many feet above the water the diver will be after t sec? Scientists have determined a formula for finding the answer:

$$-16t^2 + 118 \quad \text{(feet)}$$

above the water. The expression $-16t^2 + 118$ is an example of a *polynomial*. Here are some other polynomials:

$$5x, \quad 9x - 2, \quad -5t^2 + 18t - 4, \quad \text{and} \quad y^5 - 2y^2 + \tfrac{4}{5}y - 6$$

We construct these polynomials by adding or subtracting products of numbers and a variable raised to whole-number exponents. Of course, if we use any other operations, the result may not be a polynomial. For example,

$$x^2 - \frac{3}{x} \quad \text{and} \quad x^{-7} + 4x$$

are *not* polynomials (we divided by the variable x in the first one and used negative exponents in the second one). Thus, a **polynomial** is an algebraic expression formed by using the operations of addition and subtraction on products of a number and a variable raised to whole-number exponents.

The parts of a polynomial separated by plus ($+$) signs are called the **terms** of the polynomial. If there are subtraction signs, we can rewrite the polynomial using addition signs, since $a - b = a + (-b)$. For example,

$5x$ has one term: $\quad\quad 5x$

$9x - 2$ has two terms: $\quad 9x$ and -2 $\quad$ Recall that
$$9x - 2 = 9x + (-2).$$

$-5t^2 + 18t - 4$ has
three terms: $\quad\quad\quad -5t^2, 18t,$ and -4 $\quad -5t^2 + 18t - 4 =$
$$-5t^2 + 18t + (-4)$$

A. MONOMIALS, BINOMIALS, AND TRINOMIALS

Polynomials can be classified according to the number of terms they have. Thus:

$5x$ has *one* term; it is called a *monomial*.   *mono* means one
$9x - 2$ has *two* terms; it is called a *binomial*.   *bi*   means two
$-5t^2 + 18t - 4$ has *three* terms; it is called   *tri*   means three
a *trinomial*.

Clearly, a polynomial of one term is a **monomial,** two terms, a **binomial,** and three, a **trinomial.**

EXAMPLE 1 Classify each of the following polynomials as a monomial, binomial, or trinomial.
a. $6x - 1$ **b.** -8 **c.** $-4 + 3y - y^2$

Solution

a. $6x - 1$ has two terms; it is a binomial.
b. -8 has only one term; it is a monomial.
c. $-4 + 3y - y^2$ has three terms; it is a trinomial.

Problem 1 Classify as a monomial, binomial, or trinomial.
a. $-4t + t^2 - 8$ **b.** $-5y$
c. $278 + 6x$

B. THE DEGREE OF A POLYNOMIAL

All the polynomials we have seen contain only one variable and are called *polynomials in one variable*. Polynomials in one variable, such as $x^2 + 3x - 7$, can also be classified according to the highest exponent of the variable. The highest exponent of the variable is called the **degree** of the polynomial. To find the degree of a polynomial, you simply examine each term and find the *highest* exponent of the variable. Thus, the degree of $3x^2 + 5x^4 - 2$ is found by looking at the exponent of the variable in each of the terms.

The exponent in $3x^2$ is 2.

The exponent in $5x^4$ is 4.

The exponent in -2 is 0 because $-2 = -2x^0$. (Recall that $x^0 = 1$.)

Thus the degree of $3x^2 + 5x^4 - 2$ is 4, the highest exponent of the variable in the polynomial. Similarly, the degree of $4y^3 - 3y^5 + 9y^2$ is 5, since 5 is the highest exponent of the variable present in the polynomial. By convention, a number such as -4 or 7 is called a **polynomial of degree 0**, because if $a \neq 0$, $a = ax^0$. Thus, $-4 = -4x^0$ and $7 = 7x^0$ are polynomials of degree 0. The number 0 itself is called the **zero polynomial** and is *not* assigned a degree. (Note that $0 \cdot x^1 = 0$; $0 \cdot x^2 = 0$, $0 \cdot x^3 = 0$, and so on, so zero polynomials cannot have a degree.)

EXAMPLE 2 Find the degree.

a. $-2t^2 + 7t - 2 + 9t^3$ **b.** 8
c. $-3x + 7$ **d.** 0

Solution

a. The highest exponent of the variable t in the polynomial $-2t^2 + 7t - 2 + 9t^3$ is 3; thus the degree of the polynomial is 3.

b. The degree of 8 is, by convention, 0.

Problem 2 Find the degree.
a. $-5y - 3$ **b.** $4x^2 - 5x^3 + x^8$
c. -9 **d.** 0

c. Since $x = x^1$, $-3x + 7$ can be written as $-3x^1 + 7$, making the degree of the polynomial 1.

d. 0 is the zero polynomial; it does not have a degree.

C. DESCENDING ORDER

The degree of a polynomial is easier to find if we agree to write the polynomial in **descending** order, that is, the term with the *highest* exponent is written *first,* the *second* highest is *next,* and so on. Fortunately, the associative and commutative laws of addition permit us to do this rearranging! Thus, instead of writing $3x^2 - 5x^3 + 4x - 2$, we rearrange the terms and write $-5x^3 + 3x^2 + 4x - 2$ with exponents in the terms arranged in *descending* order. Similarly, to write $-3x^3 + 7 + 5x^4 - 2x$ in descending order, we use the associative and commutative laws and write $5x^4 - 3x^3 - 2x + 7$. Of course, it would not be incorrect to write this polynomial in ascending order (or with no order at all); it is just that we *agree* to write polynomials in descending order for uniformity and convenience.

Note that it is easier to find the *degree* of a polynomial if the polynomial is written in descending order. You simply look at the exponent in the *first* term.

EXAMPLE 3   Write in descending order.

a. $-9x + x^2 - 17$ **b.** $-5x^3 + 3x - 4x^2 + 8$

Solution

a. $-9x + x^2 - 17 = x^2 - 9x - 17$ when written in descending order.
b. $-5x^3 + 3x - 4x^2 + 8 = -5x^3 - 4x^2 + 3x + 8$

Problem 3   Write in descending order.
a. $-2x^4 + 5x - 3x^3 + 9$
b. $-8 + 5x^2 - 3x$

D. EVALUATION OF POLYNOMIALS

Now, let us go back to our diver. You may be wondering why the height above the water after t sec was $-16t^2 + 118$ ft. This expression does not even look like a number! But polynomials represent numbers when they are *evaluated*. For example, our diver is $-16t^2 + 118$ ft above the water after t sec. After **1** sec, that is, when $t = 1$, our diver will be

$$-16(1)^2 + 118 = -16 + 118 = 102 \text{ ft}$$

above the water.

After **2** sec, that is, when $t = 2$, the height will be

$$-16(2)^2 + 118 = -16 \cdot 4 + 118 = 54 \text{ ft}$$

above the water.

We usually say that

At $t = 1$, $-16t^2 + 118 = 102$
At $t = 2$, $-16t^2 + 118 = 54$

and so on.

EXAMPLE 4   When $t = 3$, what is the value of $-16t^2 + 118$?

Solution   When $t = 3$,

$$\begin{aligned} -16t^2 + 118 &= -16(3)^2 + 118 \\ &= -16(9) + 118 \\ &= -144 + 118 \\ &= -26 \end{aligned}$$

Problem 4   When $t = 2.5$, what is the value of $-16t^2 + 118$?

▲

Note that in this case, the answer is *negative,* which means that the diver should be *below* the water surface. However, you do not continue to free-fall after hitting the water. Thus, it took the diver between 2 and 3 seconds to hit the water.

EXAMPLE 5   Evaluate $3x^2 - 5x + 8$ when $x = 2$.

Solution   When $x = 2$,

$$3x^2 - 5x + 8 = 3(2^2) - 5(2) + 8$$

$$\begin{aligned} &= 3(4) - 10 + 8 &&\text{Multiplying } 2 \cdot 2 = 2^2 \\ &= 12 - 10 + 8 &&\text{Multiplying } 3 \cdot 4 \\ &= 2 + 8 &&\text{Subtracting } 12 - 10 \\ &= 10 &&\text{Adding } 2 + 8 \end{aligned}$$

Note that to evaluate this polynomial, we followed the order of operations (PMDAS) given on page 66.

Problem 5   Evaluate $2x^2 - 3x + 10$ when $x = 2$.

NAME

CLASS

SECTION

A, B. Classify as a monomial (M), binomial (B), or trinomial (T) and give the degree.

ANSWERS

1. $-5x + 7$
2. $8 + 9x^3$

3. $7x$
4. $-3x^4$

5. $-2x + 7x^2 + 9$
6. $-x + x^3 - 2x^2$

7. 18
8. 0

9. $9x^3 - 2x$
10. $-7x + 8x^6$

1. _____
2. _____
3. _____
4. _____
5. _____
6. _____
7. _____
8. _____
9. _____
10. _____

B, C. Write in descending order and give the degree of each polynomial.

11. $-3x + 8x^3$
12. $7 - 2x^3$

13. $4x - 7 + 8x^2$
14. $9 - 3x + x^3$

15. $5x + x^2$
16. $-3x - 7x^3$

17. $3 + x^3 - x^2$
18. $-3x^2 + 8 - 2x$

19. $4x^5 + 2x^2 - 3x^3$
20. $4 - 3x^3 + 2x^2 + x$

11. _____
12. _____
13. _____
14. _____
15. _____
16. _____
17. _____
18. _____
19. _____
20. _____

D. Find the value of the polynomial when (a) $x = 2$ and (b) $x = -2$.

21. $3x - 2$
22. $x^2 - 3$

23. $2x^2 - 1$
24. $x^3 - 1$

21. a. _____
    b. _____
22. a. _____
    b. _____
23. a. _____
    b. _____
24. a. _____
    b. _____

25. $3x^2 + x - 1$ **26.** $2x^2 + 2x + 1$

25. a. _____
   b. _____

26. a. _____
   b. _____

27. $3x - 1 + x^2$ **28.** $2x - 3 - x^2$

27. a. _____
   b. _____

28. a. _____
   b. _____

29. $-3 + x + x^2$ **30.** $-x - 4 - x^2$

29. a. _____
   b. _____

30. a. _____
   b. _____

E. Applications

31. If an object drops from an altitude of k feet, its height above ground after t seconds is given by $-16t^2 + k$ feet. If the object is dropped from an altitude of 150 ft, what would be the height of the object after each time?
  a. t seconds
  b. 1 second
  c. 2 seconds

31. a. _____
   b. _____
   c. _____

32. After t seconds have passed, the velocity of an object dropped from a height of 96 ft is $-32t$ feet/second. What would be the velocity of the object after each time?
  a. 1 second
  b. 2 seconds

32. a. _____
   b. _____

33. If an object drops from an altitude of k meters, its height above the ground after t seconds is given by $-4.9t^2 + k$ meters. If the object is dropped from an altitude of 200 m, what would be the height of the object after each time?
  a. t seconds
  b. 1 second
  c. 2 seconds

33. a. _____
   b. _____
   c. _____

34. After t seconds have passed, the velocity of an object dropped from a height of 300 m is $-9.8t$ meters/second. What would be the velocity of the object after each time?
  a. 1 second
  b. 2 seconds

34. a. _____
   b. _____

✓ **SKILL CHECKER**

Find.

35. $-3ab + (-4ab)$ **36.** $-8a^2b + (-5a^2b)$

35. _____

36. _____

ANSWER (to problem on page 288)
5. 12

37. $-3x^2y + 8x^2y - 2x^2y$ **38.** $-2xy^2 + 7xy^2 - 9xy^2$

39. $5xy^2 - (-3xy^2)$ **40.** $7x^2y - (-8x^2y)$

4.1   USING YOUR KNOWLEDGE

We have already stated that if an object is simply *dropped* from a certain height, its velocity after t seconds is given by $-32t$. What will happen if we actually *throw* the object down with an initial velocity, say v_0? Since the velocity $-32t$ is being helped by the velocity v_0, the new final velocity will be given by $-32t + v_0$ (v_0 is *negative* if the object is *thrown downward*).

1. Find the velocity of a ball thrown downward with an initial velocity of 10 ft/sec after t seconds have elapsed.

2. What will be the velocity of the ball in Exercise 1 after each time?
   a. 1 second
   b. 2 seconds

3. In the metric system, the velocity after t seconds of an object thrown downward with an initial velocity v_0 is given by

   $-9.8t + v_0$ (meters)

   What would be the velocity of a ball thrown downward with an initial velocity of 2 m/sec after each time?
   a. 1 second
   b. 2 seconds

4. The height of an object after t seconds have elapsed depends on two factors: the initial velocity v_0 and the height s_0 from which the object is thrown. The polynomial giving this height is:

   $-16t^2 + v_0 t + s_0$ (feet)

   where v_0 is the initial velocity and s_0 is the height from which the object is thrown. What would be the height of a ball thrown downward from a 300-ft tower with an initial velocity of 10 ft/sec after each time?
   a. 1 second
   b. 2 seconds

1. _____

2. a. _____
   b. _____

3. a. _____
   b. _____

4. a. _____
   b. _____

CALCULATOR CORNER

The evaluation of polynomials is easily done with a calculator. For example, suppose we wish to evaluate the polynomial of Problem 26 when $x = 2$. If you have a square key $\boxed{x^2}$, the keystrokes will be

$\boxed{2}\ \boxed{\times}\ \boxed{2}\ \boxed{x^2}\ \boxed{+}\ \boxed{2}\ \boxed{\times}\ \boxed{2}\ \boxed{+}\ \boxed{1}\ \boxed{=}$

The correct answer, 13, will then be displayed. When $x = -2$, we must use the $\boxed{+/-}$ key (even though $(-2)^2$ and $(2)^2$ will yield the same result). The keystrokes for $x = -2$ are as follows:

$$\boxed{2}\ \boxed{\times}\ \boxed{2}\ \boxed{+/-}\ \boxed{x^2}\ \boxed{+}\ \boxed{2}\ \boxed{\times}\ \boxed{2}\ \boxed{+/-}\ \boxed{+}\ \boxed{1}\ \boxed{=}$$

The correct result is 5. We will work with polynomials of higher degree than 2 later in the book. If you are allowed, use your calculator to work Problems 21 through 30.

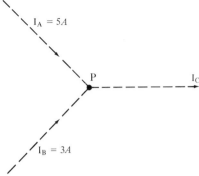

The diagram illustrates Kirchhoff's law (the sum of the currents *into* P must be equal to the sum of the currents *out* of P).

OBJECTIVES

REVIEW

Before starting this section you should know:

1. How to add and subtract integers.
2. How to remove parentheses in expressions preceded by a minus (−) sign.

OBJECTIVES

You should be able to:

A. Add polynomials.
B. Subtract polynomials.
C. Find areas using addition of polynomials.

How much current is going into P?

$$5A + 3A = 8A$$

We found the answer by *combining like terms*. As you recall from Section 2.6 $5A$ and $3A$ are *like* terms.

A. ADDING POLYNOMIALS

The addition of monomials and polynomials is also a matter of combining like terms. For example, suppose we wish to add $3x^2 + 7x - 3$ and $5x^2 - 2x + 9$; that is, we wish to find

$$(3x^2 + 7x - 3) + (5x^2 - 2x + 9)$$

Using the commutative, associative, and distributive properties, we write

$$\begin{aligned}
(3x^2 + 7x - 3) + (5x^2 - 2x + 9) &= (3x^2 + 5x^2) + (7x - 2x) + (-3 + 9) \\
&= (3 + 5)x^2 + (7 - 2)x + (-3 + 9) \\
&= 8x^2 + 5x + 6
\end{aligned}$$

Similarly, the sum of $4x^3 + \frac{3}{7}x^2 - 2x + 3$ and $6x^3 - \frac{1}{7}x^2 + 9$ is written as

$$\begin{aligned}
\left(4x^3 + \frac{3}{7}x^2 - 2x + 3\right) &+ \left(6x^3 - \frac{1}{7}x^2 + 9\right) \\
&= (4x^3 + 6x^3) + \left(\frac{3}{7}x^2 - \frac{1}{7}x^2\right) + (-2x) + (3 + 9) \\
&= 10x^3 + \frac{2}{7}x^2 - 2x + 12
\end{aligned}$$

In both examples the polynomials have been written in descending order for convenience in combining like terms.

EXAMPLE 1 Add $3x + 7x^2 - 7$ and $-4x^2 + 9 - 3x$.

Solution We first write both polynomials in descending order and then combine like terms, obtaining

$$\begin{aligned}
(7x^2 + 3x - 7) + (-4x^2 - 3x + 9) &= (7x^2 - 4x^2) + (3x - 3x) + (-7 + 9) \\
&= 3x^2 + 0 + 2 \\
&= 3x^2 + 2
\end{aligned}$$

▲

Problem 1 Add $2x + 3x^2 - 6$ and $-5x^2 + 10 - x$.

As in arithmetic, the addition of polynomials can be done by writing the polynomials in descending order and then placing like terms in a column. In arithmetic, you add 345 and 678 by writing the numbers in a column as shown:

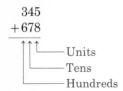

Thus, to add $4x^3 + 3x - 7$ and $7x - 3x^3 + x^2 + 9$, we first write both polynomials in descending order with like terms in the same column, leaving space for any missing terms. We then simply add the terms in each of the columns, as shown:

The x^2 term is missing.

$$
\begin{array}{r}
4x^3 \qquad\; + 3x - 7 \\
-3x^3 + x^2 + 7x + 9 \\
\hline
x^3 + x^2 + 10x + 2
\end{array}
$$

EXAMPLE 2   Add $-3x + 7x^2 - 2$ and $-4x^2 - 3 + 5x$.

Solution   We first write both polynomials in descending order, place like terms in a column, and then add as shown:

$$
\begin{array}{r}
7x^2 - 3x - 2 \\
-4x^2 + 5x - 3 \\
\hline
3x^2 + 2x - 5
\end{array}
$$

Horizontally, we write:

$$
\begin{aligned}
&(7x^2 - 3x - 2) + (-4x^2 + 5x - 3) \\
=\;& (7x^2 - 4x^2) \;\; + (-3x + 5x) + (-2 - 3) \\
=\;& \qquad 3x^2 \qquad\; + 2x + (-5) \\
=\;& \qquad 3x^2 \qquad\; + 2x - 5
\end{aligned}
$$

Note that we first wrote each polynomial in descending order.

Problem 2   Add $-5x + 8x^2 - 3$ and $-3x^2 - 4 + 8x$.

B.   SUBTRACTING POLYNOMIALS

To subtract polynomials, we first recall that

$$a - (b + c) = a - b - c$$ To remove the parentheses from an expression preceded by a minus ($-$) sign, we must change the sign of each term *inside* the parentheses.

Thus,

$$
\begin{aligned}
(3x^2 - 2x + 1) - (4x^2 + 5x + 2) &= 3x^2 - 2x + 1 - 4x^2 - 5x - 2 \\
&= (3x^2 - 4x^2) + (-2x - 5x) + (1 - 2) \\
&= -x^2 - 7x + (-1) \\
&= -x^2 - 7x - 1
\end{aligned}
$$

Here is the way to do it using the columns:

$$\begin{array}{l} 3x^2 - 2x + 1 \\ (-)4x^2 + 5x + 2 \end{array} \xrightarrow{\text{is written}} \quad \begin{array}{l} 3x^2 - 2x + 1 \\ (+)\underline{-4x^2 - 5x - 2} \\ -x^2 - 7x - 1 \end{array}$$

Note that we changed the sign of *every* term in $4x^2 + 5x + 2$ and wrote $-4x^2 - 5x - 2$.

EXAMPLE 3 Subtract $4x - 3 + 7x^2$ from $5x^2 - 3x$.

Solution We first write the problem in a column, then change the sign and add as shown:

$$\begin{array}{l} 5x^2 - 3x \\ (-)\underline{7x^2 + 4x - 3} \end{array} \xrightarrow{\text{is written}} \quad \begin{array}{l} 5x^2 - 3x \\ (+)\underline{-7x^2 - 4x + 3} \\ -2x^2 - 7x + 3 \end{array}$$

Thus the answer is $-2x^2 - 7x + 3$.
If we do it horizontally, we write

$$(5x^2 - 3x) - (7x^2 + 4x - 3)$$

$$= 5x^2 - 3x - 7x^2 - 4x + 3 \qquad \text{Changing the sign of every term in } 7x^2 + 4x - 3$$

$$= (5x^2 - 7x^2) + (-3x - 4x) + 3 \quad \text{Using the commutative and associative laws}$$

$$= -2x^2 - 7x + 3 \qquad\qquad\qquad\qquad\qquad\qquad \blacktriangle$$

Just as in arithmetic, we can add or subtract more than two polynomials. For example, to add the polynomials $-7x + x^2 - 3$, $6x^2 - 8 + 2x$, and $3x - x^2 + 5$, we simply write each of the polynomials in descending order with like terms in the same column and add as shown:

$$\begin{array}{l} x^2 - 7x - 3 \\ 6x^2 + 2x - 8 \\ \underline{-x^2 + 3x + 5} \\ 6x^2 - 2x - 6 \end{array}$$

Horizontally, we write

$$(x^2 - 7x - 3) + (6x^2 + 2x - 8) + (-x^2 + 3x + 5)$$

$$= (x^2 + 6x^2 - x^2) + (-7x + 2x + 3x) + (-3 - 8 + 5)$$

$$= 6x^2 + (-2x) + (-6)$$

$$= 6x^2 - 2x - 6$$

EXAMPLE 4 Add $x^3 + 2x - 3x^2 - 5$, $-8 + 2x - 5x^2$, and $7x^3 - 4x + 9$.

Solution We first write all the polynomials in descending order with like terms in the same column and then add:

$$\begin{array}{l} x^3 - 3x^2 + 2x - 5 \\ - 5x^2 + 2x - 8 \\ \underline{7x^3 - 4x + 9} \\ 8x^3 - 8x^2 - 4 \end{array}$$

Horizontally, we have

$$(x^3 - 3x^2 + 2x - 5) + (-5x^2 + 2x - 8) + (7x^3 - 4x + 9)$$

$$= (x^3 + 7x^3) + (-3x^2 - 5x^2) + (2x + 2x - 4x) + (-5 - 8 + 9)$$

$$= 8x^3 + (-8x^2) + 0x + (-4)$$

$$= 8x^3 - 8x^2 - 4$$

Problem 3 Subtract $3 - 4x^2 + 5x$ from $9x^2 - 2x$.

Problem 4 Add $2x^3 + 3x - 5x^2 - 2$, $-6 + 5x - 2x^2$, and $6x^3 - 2x + 8$.

C. FINDING AREAS

Addition of polynomials can be used to find the sum of the areas of several rectangles. Thus, to find the total area of the shaded rectangles, add the individual areas. Since the area of a rectangle is the product of its length and its width, we have

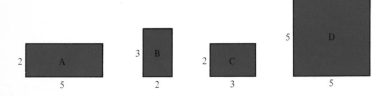

$\underbrace{\text{Area of } A}$ plus $\underbrace{\text{area of } B}$ plus $\underbrace{\text{area of } C}$ plus $\underbrace{\text{area of } D}$

$$5 \cdot 2 \quad + \quad 2 \cdot 3 \quad + \quad 3 \cdot 2 \quad + \quad 5 \cdot 5$$
$$= 10 \quad + \quad 6 \quad + \quad 6 \quad + \quad 25$$

Thus,

$$10 + 6 + 6 + 25 = 47 \quad \text{(square units)}$$

This same procedure can be used when some of the lengths are represented by variables, as shown in the next example.

EXAMPLE 5 Find the sum of the areas of the shaded rectangles.

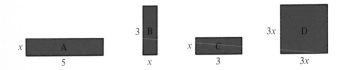

Solution The total area in square units is:

$\underbrace{\text{Area of } A}$ + $\underbrace{\text{area of } B}$ + $\underbrace{\text{area of } C}$ + $\underbrace{\text{area of } D}$

$$5x \quad + \quad 3x \quad + \quad 3x \quad + \quad (3x)^2$$
$$\underbrace{\qquad\qquad\qquad\qquad} \qquad\qquad$$
$$11x \qquad\qquad + \quad 9x^2$$

or

$$9x^2 + 11x \quad \text{In descending order}$$

Problem 5 Find the sum of the areas of the shaded rectangles.

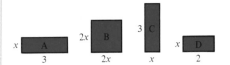

NAME

CLASS

SECTION

A. Add as indicated.

ANSWERS

1. $(5x^2 + 2x + 5) + (7x^2 + 3x + 1)$

2. $(3x^2 - 5x - 5) + (9x^2 + 2x + 1)$

3. $(-3x + 5x^2 - 1) + (-7 + 2x - 7x^2)$

4. $(3 - 2x^2 + 7x) + (-6 + 2x^2 - 5x)$

5. $(2x + 5x^2 - 2) + (-3 + 5x - 8x^2)$

6. $-3x - 2 + 3x^2$ and $-4 + 5x - 6x^2$

7. $-2 + 5x$ and $-3 - x^2 - 5x$

8. $-4x + 2 - 6x^2$ and $2 + 5x$

9. $x^3 - 2x + 3$ and $-2x^2 + x - 5$

10. $x^4 - 3 + 2x - 3x^3$ and $3x^4 - 2x^2 + 5 - x$

11. $-6x^3 + 2x^4 - x$ and $2x^2 + 5 + 2x - 2x^3$

12. $\frac{1}{2}x^3 + x^2 - \frac{1}{5}x$ and $\frac{3}{5}x + \frac{1}{2}x^3 - 3x^2$

13. $\frac{1}{3} - \frac{2}{5}x^2 + \frac{3}{4}x$ and $\frac{1}{4}x - \frac{1}{5}x^2 + \frac{2}{3}$

14. $0.3x - 0.1 - 0.4x^2$ and $0.1x^2 - 0.1x + 0.6$

15. $0.2x - 0.3 + 0.5x^2$ and $-\frac{1}{10} + \frac{1}{10}x - \frac{1}{10}x^2$

16. $\begin{array}{r} -x^2 + 5x + 2 \\ (+)\ \ 3x^2 - 7x - 2 \\ \hline \end{array}$

17. $\begin{array}{r} -3x^2 + 2x - 4 \\ (+)\ \ \ \ x^2 - 4x + 7 \\ \hline \end{array}$

18. $\begin{array}{r} -2x^4 \qquad + 2x - 1 \\ (+)\qquad - x^3 - 3x + 5 \\ \hline \end{array}$

19. $\begin{array}{r} 3x^4 \qquad - 3x + 4 \\ (+)\qquad x^3 - 2x - 5 \\ \hline \end{array}$

20. $\begin{array}{r} -3x^4 \qquad + 2x^2 - x + 5 \\ (+)\qquad - 2x^3 \qquad + 5x - 7 \\ \hline \end{array}$

21. $\begin{array}{r} -5x^4 \qquad - 5x^2 + 3 \\ (+)\qquad 5x^3 + 3x^2 - 5 \\ \hline \end{array}$

22. $\begin{array}{r} 3x^3 \qquad + x - 1 \\ x^2 - 2x + 5 \\ (+)\ 5x^3 \qquad - x \\ \hline \end{array}$

23. $\begin{array}{r} 5x^3 - x^2 \qquad - 3 \\ 5x + 9 \\ (+)\qquad - 3x^2 \qquad - 7 \\ \hline \end{array}$

24. $\begin{array}{r} -\dfrac{1}{3}x^3 \qquad\quad - \dfrac{1}{2}x + 5 \\[2mm] -\dfrac{1}{5}x^2 + \dfrac{1}{2}x - 1 \\[2mm] (+)\ \ \dfrac{2}{3}x^3 \qquad\quad + x - 2 \\ \hline \end{array}$

25. $\begin{array}{r} -\dfrac{2}{7}x^3 + \dfrac{1}{6}x^2 \qquad + 2 \\[2mm] \dfrac{1}{7}x^3 \qquad\quad + 5x - 3 \\[2mm] (+)\qquad - \dfrac{5}{6}x^2 \qquad + 1 \\ \hline \end{array}$

1. _____
2. _____
3. _____
4. _____
5. _____
6. _____
7. _____
8. _____
9. _____
10. _____
11. _____
12. _____
13. _____
14. _____
15. _____
16. _____
17. _____
18. _____
19. _____
20. _____
21. _____
22. _____
23. _____
24. _____
25. _____

26.
$$-\frac{1}{8}x^2 - \frac{1}{3}x + \frac{1}{5}$$
$$-x^3 + \frac{3}{8}x^2 \qquad - \frac{2}{5}$$
$$(+) \ -3x^3 \qquad\qquad - \frac{2}{3}x + \frac{4}{5}$$
_____

27.
$$-\frac{1}{7}x^3 \qquad\qquad\qquad + 2$$
$$-\frac{1}{9}x^2 - \ x - 3$$
$$(+) \ -\frac{2}{7}x^3 + \frac{2}{9}x^2 + 2x - 5$$
_____

28.
$$-2x^4 + 5x^3 - 2x^2 + 3x - 5$$
$$8x^3 \qquad\quad - 2x + 5$$
$$- \ x^4 \qquad\quad + 3x^2 - \ x - 2$$
$$(+) \qquad\quad 6x^3 \qquad\quad + 2x + 5$$
_____

29.
$$- \ 6x^3 + 2x^2 \qquad + 1$$
$$-x^4 + 3x^3 - 5x^2 + 3x$$
$$- \ x^3 \qquad\quad - 7x + 2$$
$$(+) \ -3x^4 \qquad\qquad\quad + 3x - 1$$
_____

30.
$$- \ 3x^4 \qquad\quad + 2x^2 \qquad\quad - 5$$
$$x^5 + \ x^4 - 2x^3 + 7x^2 + 5x$$
$$2x^4 \qquad\quad - 2x^2 \qquad\quad + 7$$
$$(+ \ ) \ 7x^5 \qquad\quad + 2x^3 \qquad\quad - 2x$$
_____

B. Find.

31. $(7x^2 + 2) - (3x^2 - 5)$

32. $(8x^2 - x) - (7x^2 + 3x)$

33. $(3x^2 - 2x - 1) - (4x^2 + 2x + 5)$

34. $(-3x + x^2 - 1) - (5x + 1 - 3x^2)$

35. $(-1 + 7x^2 - 2x) - (5x + 3x^2 - 7)$

36. $(7x^3 - x^2 + x - 1) - (2x^2 + 3x + 6)$

37. $(5x^2 - 2x + 5) - (3x^3 - x^2 + 5)$

38. $(3x^2 - x - 7) - (5x^3 + 5 - x^2 + 2x)$

39. $(6x^3 - 2x^2 - 3x + 1) - (-x^3 - x^2 - 5x + 7)$

40. $(x - 3x^2 + x^3 + 9) - (-8 + 7x - x^2 + x^3)$

41.
$$6x^2 - 3x + 5$$
$$(-) \ 3x^2 + 4x - 2$$
_____

42.
$$7x^2 + 4x - 5$$
$$(-) \ 9x^2 - 2x + 5$$
_____

43.
$$3x^2 - 2x - 1$$
$$(-) \ 3x^2 - 2x - 1$$
_____

44.
$$5x^2 \qquad\quad - 1$$
$$(-) \ 3x^2 - 2x + 1$$
_____

45.
$$4x^3 \qquad\quad - 2x + 5$$
$$(-) \qquad\quad 3x^2 + 5x - 1$$
_____

46.
$$- 3x^2 + 5x - 2$$
$$(-) \ x^3 - 2x^2 \qquad\quad + 5$$
_____

26. _____

27. _____

28. _____

29. _____

30. _____

31. _____

32. _____

33. _____

34. _____

35. _____

36. _____

37. _____

38. _____

39. _____

40. _____

41. _____

42. _____

43. _____

44. _____

45. _____

46. _____

ANSWER (to problem on page 296)

5. $4x^2 + 8x$

47.
$$3x^3 \qquad\quad - 2$$
$$(-) \qquad 2x^2 - x + 6$$
$$\overline{}$$

48.
$$\qquad\qquad x^2 - 2x + 1$$
$$(-) -3x^3 + x^2 + 5x - 2$$
$$\overline{}$$

49.
$$-5x^3 \qquad + x - 2$$
$$(-) \qquad 5x^2 - 3x + 7$$
$$\overline{}$$

50.
$$6x^3 \qquad\quad + 2x - 5$$
$$(-) \qquad -3x^2 - x$$
$$\overline{}$$

47. _____

48. _____

49. _____

50. _____

C. Find the sum of the areas of the shaded rectangles.

51.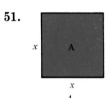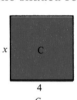

51. _____

52.

52. _____

53.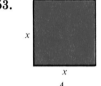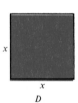

53. _____

54.

54. _____

55.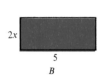

55. _____

✓ **SKILL CHECKER**

Simplify.

56. $(-5x^3) \cdot (2x^4)$

57. $(-2x^4) \cdot (3x^5)$

58. $5(x - 3)$

59. $6(y - 4)$

60. $-3(2y - 3)$

56. _____

57. _____

58. _____

59. _____

60. _____

4.2 USING YOUR KNOWLEDGE

Polynomials are sometimes used in business and economics. For example, the revenue R may be obtained by subtracting the cost C of the merchandise from its selling price S. In symbols:

$R = S - C$

Now, the cost C of the merchandise is made up of two parts: the *variable cost* per item and the *fixed cost*. For example, if you decide to manufacture Frisbees, you might spend $2 per Frisbee in materials, labor, and so forth. In addition, you might have $100 of fixed expenses. Then the cost for manufacturing x Frisbees is

$$C = 2x + 100$$

         ↗   ↑

The cost per Frisbee   The fixed expenses

If x Frisbees are then sold for $3 each, the total selling price S is $3x$, and the revenue R would be

$R = S - C$

$\quad = 3x - (2x + 100)$

$\quad = 3x - 2x - 100$

$\quad = x - 100$

Thus if the selling price S is $3 per Frisbee, the variable costs are $2 per Frisbee, and the fixed expenses are $100, the revenue after manufacturing x Frisbees is given by

$R = x - 100$

In Problems 1, 2, and 3, find the revenue R for the given cost C and selling price S:

1. $C = 3x + 50;\ S = 4x$

                                                1. _____

2. $C = 6x + 100;\ S = 8x$

    2. _____

3. $C = 7x;\ S = 9x$

    3. _____

4. In Problem 2, how many items were manufactured if the revenue was 0?

    4. _____

5. If the merchant of Problem 2 suffered a $40 loss ($-$$40 revenue), how many items were produced?

    5. _____

REVIEW

Before starting this section you should know:

1. How to multiply expressions (Section 2.7).
2. How to use the distributive law to remove parentheses in an expression (Sections 1.5 and 2.7).

OBJECTIVES

You should be able to:

A. Multiply two monomials.
B. Multiply a monomial and a binomial.
C. Multiply two binomials using the FOIL method.

How much does the beam bend (deflect) when a car or truck goes over the bridge? There is a formula that can tell us. For a certain beam of length L at a distance x from one end, the deflection is given by

$$(x - L)(x - 2L)$$

To multiply these two binomials, we first learn how to do several related types of multiplication.

A. MULTIPLYING TWO MONOMIALS

We have already multiplied two monomials in Section 2.7. The idea is to use the associative and commutative laws and the laws of exponents as shown in the next example.

EXAMPLE 1 Multiply $(-3x^2)$ by $(2x^3)$.

Solution

$$(-3x^2)(2x^3) = (-3 \cdot 2)(x^2 \cdot x^3) \quad \text{By the associative and commutative laws}$$
$$= -6x^{2+3} \quad \text{By the law of exponents}$$
$$= -6x^5$$

Problem 1 Multiply $(-7x^4)$ by $(5x^2)$.

B. MULTIPLYING A MONOMIAL AND A BINOMIAL

In Section 2.7, we also multiplied a monomial and a binomial. The procedure was based on the distributive laws, as shown next.

EXAMPLE 2 Remove parentheses (simplify).

a. $5(x - 2y)$ b. $(x^2 + 2x)3x^4$

Solution

a. $5(x - 2y) = 5x - 5 \cdot 2y$
$$= 5x - 10y$$

Problem 2 Remove parentheses (simplify).
a. $6(x - 3y)$ b. $(x^3 + 5x)4x^5$

b. $(x^2 + 2x)3x^4 = x^2 \cdot 3x^4 + 2x \cdot 3x^4$
$$= 3 \cdot x^2 \cdot x^4 + 2 \cdot 3 \cdot x \cdot x^4$$
$$= 3x^6 + 6x^5$$

C. MULTIPLYING TWO BINOMIALS

If you successfully completed Example 2b, you probably noticed that the $3x^4$ is used in the same manner as c in the distributive law; that is,

$$(a + b)c = ac + bc$$
$$(x^2 + 2x)3x^4 = x^2 \cdot 3x^4 + 2x \cdot 3x^4$$

The same idea can be used to multiply $(x + 2)$ by $(x + 3)$. Here is how. Think of $(x + 3)$ as c in the distributive law. Of course, this makes x like a and 2 like b. This is the way it is done:

$$(a + b)c = ac + bc$$
$$(x + 2)(x + 3) = x(x + 3) + 2(x + 3)$$
$$= x^2 + 3x + 2x + 6 \quad \text{Recall that } x(x + 3) = x^2 + 3x \text{ and}$$
$$= x^2 + 5x + 6 \qquad 2(x + 3) = 2x + 6.$$

Similarly,

$$(x - 3)(x + 5) = x(x + 5) - 3(x + 5)$$
$$= x^2 + 5x - 3x - 15$$
$$= x^2 + 2x - 15$$

Can you see a pattern developing? Look at the answers:

$$(x + 2)(x + 3) = x^2 + 5x + 6$$
$$(x - 3)(x + 5) = x^2 + 2x - 15$$

It seems that the *first* term in each answer (x^2) is obtained by multiplying the *first* terms in the factors (x and x). Similarly, the *last* terms (6 and -15) are obtained by multiplying the *last* terms ($2 \cdot 3$ and $-3 \cdot 5$). Here is how it works so far:

Need the
middle term

$x \cdot x$
$(x + 2)(x + 3) = x^2 + \underline{\quad\quad} + 6$
$2 \cdot 3$

Need the
middle term

$x \cdot x$
$(x - 3)(x + 5) = x^2 + \underline{\quad\quad} - 15$
$-3 \cdot 5$

But what about the *middle* terms? In $(x + 2)(x + 3)$, the *middle* term was obtained by adding $3x$ and $2x$, which were the results we got when multiplying the *outer* terms (x and 3) and adding the product of the *inner* terms (2 and x). Here is the diagram showing how the middle term is obtained:

$$(x + 2)(x + 3) = x^2 + 3x + 2x + 6$$

Outer terms

Inner terms

$x \cdot 3$

$$(x - 3)(x + 5) = x^2 + 5x - 3x - 15$$

Outer terms

Inner terms

$x \cdot 5$

Do you see how it works now? Here is a summary of the procedure:

First terms are multiplied first.

Outer terms are multiplied second.

Inner terms are multiplied third.

Last terms are multiplied last.

Of course, we call this method the **FOIL** method. We shall do one more example, step by step, so you can have more practice.

F	$(x + 7)(x - 4) \rightarrow x^2$
O	$(x + 7)(x - 4) \rightarrow x^2 - 4x$
I	$(x + 7)(x - 4) \rightarrow x^2 - 4x + 7x$
L	$(x + 7)(x - 4) = x^2 - 4x + 7x - 28$
	$(x + 7)(x - 4) = x^2 + 3x - 28$

EXAMPLE 3 Find.

a. $(x + 5)(x - 2)$ **b.** $(x - 4)(x + 3)$

Solution

$$\begin{array}{ccccccc} & (\text{First}) & (\text{Outer}) & (\text{Inner}) & (\text{Last}) \\ & F & O & I & L \\ \textbf{a.}\ (x + 5)(x - 2) = & x \cdot x & - & 2x & + & 5x & - & 5 \cdot 2 \\ = & x^2 & + & & 3x & & - & 10 \\ \textbf{b.}\ (x - 4)(x + 3) = & x \cdot x & + & 3x & - & 4x & - & 4 \cdot 3 \\ = & x^2 & - & & x & & - & 12 \end{array}$$

▲

As in the case of arithmetic, we use the ideas we have just discussed to do more complicated problems. Thus we use the FOIL technique to multiply expressions such as $(2x + 5)$ and $(3x - 4)$. The procedure is the same as before; just remember your laws of exponents and the **FOIL** sequence.

EXAMPLE 4 Find.

a. $(2x + 5)(3x - 4)$ **b.** $(3x - 2)(5x - 1)$

Solution

$$\begin{array}{ccccccc} & (\text{First}) & (\text{Outer}) & (\text{Inner}) & (\text{Last}) \\ & F & O & I & L \\ \textbf{a.}\ (2x + 5)(3x - 4) = & (2x)(3x) & + & (2x)(-4) & + & 5(3x) & + & (5)(-4) \\ = & 6x^2 & - & 8x & + & 15x & - & 20 \\ = & 6x^2 & + & & 7x & & - & 20 \end{array}$$

Problem 3 Find.
a. $(x + 7)(x - 3)$
b. $(x - 2)(x + 8)$

Problem 4 Find.
a. $(3x + 4)(5x - 2)$
b. $(4x - 3)(3x - 1)$

$$
\begin{array}{cccccc}
& & \text{F} & \text{O} & \text{I} & \text{L} \\
\textbf{b.} \ (3x-2)(5x-1) & = & (3x)(5x) + 3x(-1) - 2(5x) - 2(-1) \\
& = & 15x^2 & - & 3x & - & 10x & + & 2 \\
& = & 15x^2 & - & & 13x & + & 2 & \blacktriangle
\end{array}
$$

Would the same technique work when the binomials to be multiplied contain more than one variable? Fortunately, yes. Again, just remember your FOIL method and the laws of exponents. For example, to multiply $(3x + 2y)$ by $(2x + 5y)$, we use the FOIL method to obtain:

$$
\begin{array}{cccccc}
& & \text{F} & \text{O} & \text{I} & \text{L} \\
(2x+5y)(3x+2y) & = & (2x)(3x) + (2x)(2y) + (5y)(3x) + (5y)(2y) \\
& = & 6x^2 & + & 4xy & + & 15xy & + & 10y^2 \\
& = & 6x^2 & + & & 19xy & & + & 10y^2
\end{array}
$$

EXAMPLE 5 Find.
a. $(5x + 2y)(2x + 3y)$ **b.** $(3x - y)(4x - 3y)$

Solution

$$
\begin{array}{cccccc}
& & \text{F} & \text{O} & \text{I} & \text{L} \\
\textbf{a.} \ (5x+2y)(2x+3y) & = & (5x)(2x) + (5x)(3y) + (2y)(2x) + (2y)(3y) \\
& = & 10x^2 & + & 15xy & + & 4xy & + & 6y^2 \\
& = & 10x^2 & + & & 19xy & & + & 6y^2
\end{array}
$$

$$
\begin{array}{cccccc}
& & \text{F} & \text{O} & \text{I} & \text{L} \\
\textbf{b.} \ (3x-y)(4x-3y) & = & (3x)(4x) + (3x)(-3y) + (-y)(4x) + (-y)(-3y) \\
& = & 12x^2 & - & 9xy & - & 4xy & + & 3y^2 \\
& = & 12x^2 & - & & 13xy & & + & 3y^2 & \blacktriangle
\end{array}
$$

Now, one more thing. How do we multiply the expression

$(x - L)(x - 2L)$

We do it in the next example.

EXAMPLE 6 Perform the indicated operation.

$(x - L)(x - 2L)$

Solution

$$
\begin{array}{cccccc}
& & \text{F} & \text{O} & \text{I} & \text{L} \\
(x-L)(x-2L) & = & x \cdot x + (x)(-2L) + (-L)(x) + (-L)(-2L) \\
& = & x^2 & - & 2xL & - & xL & + & 2L^2 \\
& = & x^2 & - & & 3xL & & + & 2L^2
\end{array}
$$

NAME

CLASS

SECTION

A. Find.

1. $(5x^3)(9x^2)$

2. $(8x^4)(9x^3)$

3. $(-2x)(5x^2)$

4. $(-3y^2)(4y^3)$

5. $(-2y^2)(-3y)$

6. $(-5z)(-3z)$

B. Remove parentheses (simplify).

7. $3(x + y)$

8. $5(2x + y)$

9. $5(2x - y)$

10. $4(3x - 4y)$

11. $4x(2x - 3)$

12. $6x(5x - 3)$

13. $(x^2 + 4x)x^3$

14. $(x^2 + 2x)x^2$

15. $(x - x^2)4x$

16. $(x - 3x^2)5x$

17. $(x + y)3x$

18. $(x + 2y)5x^2$

19. $(2x - 3y)4y^2$

20. $(3x^2 - 4y)5y^3$

C. Use the FOIL method to perform the indicated operation.

21. $(x + 1)(x + 2)$

22. $(y + 3)(y + 8)$

23. $(y + 4)(y - 9)$

24. $(y + 6)(y - 5)$

25. $(x - 7)(x + 2)$

26. $(z - 2)(z + 9)$

27. $(x - 3)(x - 9)$

28. $(x - 2)(x - 11)$

ANSWERS

1. _____

2. _____

3. _____

4. _____

5. _____

6. _____

7. _____

8. _____

9. _____

10. _____

11. _____

12. _____

13. _____

14. _____

15. _____

16. _____

17. _____

18. _____

19. _____

20. _____

21. _____

22. _____

23. _____

24. _____

25. _____

26. _____

27. _____

28. _____

29. $(y - 3)(y - 3)$

30. $(y + 4)(y + 4)$

31. $(2x + 1)(3x + 2)$

32. $(4x + 3)(3x + 5)$

33. $(3y + 5)(2y - 3)$

34. $(4y - 1)(3y + 4)$

35. $(5z - 1)(2z + 9)$

36. $(2z - 7)(3z + 1)$

37. $(2x - 4)(3x - 11)$

38. $(5x - 1)(2x - 1)$

39. $(4z + 1)(4z + 1)$

40. $(3z - 2)(3z - 2)$

41. $(3x + y)(2x + 3y)$

42. $(4x + z)(3x + 2z)$

43. $(2x + 3y)(x - y)$

44. $(3x + 2y)(x - 5y)$

45. $(5z - y)(2z + 3y)$

46. $(2z - 5y)(3z + 2y)$

47. $(3x - 2z)(4x - z)$

48. $(2x - 3z)(5x - z)$

49. $(2x - 3y)(2x - 3y)$

50. $(3x + 5y)(3x + 5y)$

29. _____

30. _____

31. _____

32. _____

33. _____

34. _____

35. _____

36. _____

37. _____

38. _____

39. _____

40. _____

41. _____

42. _____

43. _____

44. _____

45. _____

46. _____

47. _____

48. _____

49. _____

50. _____

D. Applications

51. The area A of a rectangle is obtained by multiplying its length L by its width W, that is, $A = LW$. Find the area of the rectangle in the margin.

51. _____

52. Use the formula in Problem 51 to find the area of a rectangle of width $x - 4$ and length $x + 3$.

52. _____

53. The height reached by an object t sec after being thrown upward with a velocity of 96 ft/sec is given by $16t(6 - t)$. Use the distributive law to simplify this expression.

53. _____

ANSWERS   (to problems on page 304)

5. a. $12x^2 + 17xy + 6y^2$

   b. $10x^2 - 17xy + 3y^2$

6. $x^2 - 4xL + 3L^2$

54. The resistance R of a resistor varies with the temperature T according to the equation $R = (T + 100)(T + 20)$. Use the distributive law to simplify this expression.

54. _____

55. In chemistry, when V is the volume and P is the pressure of a certain gas, we find the expression $(V_2 - V_1)(CP + PR)$, C and R constants. Use the distributive law to simplify this expression.

55. _____

✓ **SKILL CHECKER**

Find.

56. $(4y)^2$ 57. $(3x)^2$

56. _____

57. _____

58. $(-A)^2$ 59. $(-3A)^2$

58. _____

59. _____

60. $A(-A)$

60. _____

4.3 USING YOUR KNOWLEDGE

Do you know how much money (profit) you may make in a certain business? The profit P is the difference between the revenue R and the expense E of doing business. In symbols:

$P = R - E$

Of course, R depends on the number n of items sold and their price p (in dollars); that is,

$R = np$

Thus,

$P = np - E$

1. If $n = -3p + 60$ and $E = -5p + 100$, find P.

1. _____

2. If $n = -2p + 50$ and $E = -3p + 300$, find P.

2. _____

3. In Problem 1, if the price was \$2, what was the profit P?

3. _____

4. In Problem 2, if the price was \$10, what was the profit P?

4. _____

4.4 SPECIAL PRODUCTS OF POLYNOMIALS

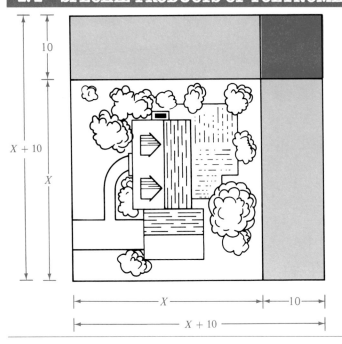

OBJECTIVES

REVIEW

Before starting this section, you should know:

1. How to use the FOIL method to multiply two binomials.
2. How to multiply expressions (Section 2.7).

OBJECTIVES

You should be able to multiply (expand) binomials of the form:

A. $(X + A)^2$
B. $(X - A)^2$
C. $(X + A)(X - A)$

Do you know the area of the square piece of land shown in the picture? Since the land is a square, the area is

$$(X + 10)(X + 10) = (X + 10)^2$$

The expression $(X + 10)^2$ is the square of the binomial $X + 10$. This type of expression is so common in algebra that we have *special products* or formulas to multiply them.

The expression $(X + A)(X + B)$ is also a special product. As you recall from the preceding section,

$$
\begin{aligned}
(X + A)(X + B) &\overset{\text{F\quad O\quad I\quad L}}{=} X^2 + BX + AX + AB \\
&= X^2 + (B + A)X + AB \\
&= X^2 + (A + B)X + AB
\end{aligned}
$$

Thus our first special product is given by the following:

Special Product (1)

PRODUCT OF TWO BINOMIALS
$(X + A)(X + B) = X^2 + (A + B)X + AB$

A. SQUARING SUMS

We can use this special product to find $(X + 10)^2$. If we let $A = B$ in (1), we have

$$(X + A)(X + A) = X^2 + (A + A)X + A \cdot A \qquad \text{Note that } (A + A)X = 2AX.$$

Since $A = B$, we substitute A for B.

Thus,

$$(X + A)^2 = X^2 + 2AX + A^2 \qquad \begin{array}{l}\text{Note that}\\ (X + A)(X + A) = (X + A)^2.\end{array}$$

Now we have our second special product:

Special
Product
(2)

THE SQUARE OF A BINOMIAL SUM

$$(X + A)^2 = X^2 + 2AX + A^2$$

Note that
$(X+A)^2 \neq X^2 + A^2$.
(See 4.4 *Using
Your Knowledge*.)

Here is the pattern used in Equation (2).

$$(\underset{\substack{\text{First} \\ \text{term}}}{X} + \underset{\substack{\text{Second} \\ \text{term}}}{A})^2 = \underset{\substack{\text{Square the} \\ \text{first term}}}{X^2} + \underset{\substack{\text{Multiply} \\ \text{the terms} \\ \text{and double}}}{2AX} + \underset{\substack{\text{Square} \\ \text{the last} \\ \text{term}}}{A^2}$$

Note that you can use the FOIL method to multiply $(X + A)(X + A)$, obtaining

$$
\begin{aligned}
(X + A)(X + A) &= \overset{F}{X^2} + \overset{O}{AX} + \overset{I}{AX} + \overset{L}{A \cdot A} \\
&= X^2 + 2AX + A^2
\end{aligned}
$$

as before. We are now ready to multiply (sometimes we call it *expand*) $(X + 10)^2$:

$$
\begin{aligned}
(X + 10)^2 &= \underset{\substack{\text{Square the} \\ \text{first term}}}{X^2} + \underset{\substack{\text{Multiply } X \\ \text{by 10,} \\ \text{double it}}}{2 \cdot 10 \cdot X} + \underset{\substack{\text{Square the} \\ \text{last term}}}{10^2} \\
&= X^2 + 20X + 100
\end{aligned}
$$

Similarly,

$$
\begin{aligned}
(x + 7)^2 &= x^2 + 2 \cdot 7 \cdot x + 7^2 \\
&= x^2 + 14x + 49
\end{aligned}
$$

EXAMPLE 1 Find (expand).

a. $(x + 9)^2$ **b.** $(2x + 3)^2$ **c.** $(3x + 4y)^2$

Solution

a. $(x + 9)^2 = x^2 + 2 \cdot 9 \cdot x + 9^2$ Letting $A = 9$ in Equation (2)

$\quad\quad\quad\quad = x^2 + 18x + 81$

b. $(2x + 3)^2 = (2x)^2 + 2 \cdot 3 \cdot 2x + 3^2$ Letting $X = 2x$

$\quad\quad\quad\quad\quad = 4x^2 + 12x + 9$ and $A = 3$ in Equation (2)

c. $(3x + 4y)^2 = (3x)^2 + 2(4y)(3x) + (4y)^2$ Letting $X = 3x$

$\quad\quad\quad\quad\quad = 9x^2 + 24xy + 16y^2$ and $A = 4y$ in Equation (2)

B. SQUARING DIFFERENCES

Can we also expand $(X - 10)^2$? Of course! But we first have to learn how to expand $(X - A)^2$. To do this, we simply write $-A$ instead of A in Equation (2), obtaining

$$
\begin{aligned}
(X - A)^2 &= X^2 + 2(-A)X + (-A)^2 \\
&= X^2 - 2AX + A^2
\end{aligned}
$$

Problem 1 Find (expand).
a. $(x + 5)^2$ **b.** $(3x + 2)^2$
c. $(2x + 3y)^2$

This is the special product we need and we call it Equation (3), the square of a binomial difference.

Special Product (3)

THE SQUARE OF A BINOMIAL DIFFERENCE
$(X - A)^2 = X^2 - 2AX + A^2$

Note that $(X - A)^2 \neq X^2 - A^2$, so the only difference between the square of a sum and the square of a difference is the sign preceding $2AX$. Here is the comparison:

(2) $\quad (X + A)^2 = X^2 + 2AX + A^2$

(3) $\quad (X - A)^2 = X^2 - 2AX + A^2$

Keeping this in mind,

$$(X - 10)^2 = X^2 - 2 \cdot 10 \cdot X + 10^2$$
$$= X^2 - 20X + 100$$

Similarly,

$$(x - 3)^2 = x^2 - 2 \cdot 3 \cdot x + 3^2$$
$$= x^2 - 6x + 9$$

EXAMPLE 2 Find.

a. $(x - 5)^2$ **b.** $(3x - 2)^2$ **c.** $(2x - 3y)^2$

Problem 2 Find.
a. $(x - 8)^2$ **b.** $(2x - 3)^2$
c. $(3x - 2y)^2$

Solution

a. $(x - 5)^2 = x^2 - 2 \cdot 5 \cdot x + 5^2$ Letting $A = 5$ in Equation (3)
           $= x^2 - 10x + 25$

b. $(3x - 2)^2 = (3x)^2 - 2 \cdot 2 \cdot 3x + 2^2$ Letting $X = 3x$,
           $= 9x^2 - 12x + 4$ $A = 2$ in Equation (3)

c. $(2x - 3y)^2 = (2x)^2 - 2 \cdot 3y \cdot 2x + (3y)^2$ Letting $X = 2x$,
           $= 4x^2 - 12xy + 9y^2$ $A = 3y$ in Equation (3)

C. MULTIPLYING SUMS AND DIFFERENCES

We have one more special product, and this one is really special. Suppose we multiply the sum of two terms by the difference of the same two terms; that is, suppose we wish to multiply

$$(X + A)(X - A)$$

If we substitute $-A$ for B in Equation (1),

$$(X + A)(X + B) = X^2 + (A + B)X + AB$$

becomes

$$(X + A)(X - A) = X^2 + (A - A)X + A(-A)$$
$$= X^2 + 0X - A^2$$
$$= X^2 - A^2$$

This gives us our last very special product:

Special
Product
(4)

THE PRODUCT OF THE SUM AND DIFFERENCE OF TWO TERMS

$$(X + A)(X - A) = X^2 - A^2$$

Note that $(X + A)(X - A) = (X - A)(X + A)$; so, $(X - A)(X + A) = X^2 - A^2$. Thus, to multiply the sum and difference of two terms, we simply square the first term and then subtract from this the square of the last term. If we do the multiplication using the FOIL method, we have

$$(X + A)(X - A) = X^2 \underbrace{- AX + AX}_{0} - A^2$$

$$= X^2 - A^2$$

Note that the middle term is 0. Thus

$$(x + 3)(x - 3) = x^2 - 3^2 = x^2 - 9$$
$$(x + 6)(x - 6) = x^2 - 6^2 = x^2 - 36$$

Similarly, by the commutative law, we also have

$$(x - 3)(x + 3) = x^2 - 3^2 = x^2 - 9$$
$$(x - 6)(x + 6) = x^2 - 6^2 = x^2 - 36$$

EXAMPLE 3 Find.

a. $(x + 10)(x - 10)$ **b.** $(2x + y)(2x - y)$ **c.** $(3x - 5y)(3x + 5y)$

Solution

a. $(x + 10)(x - 10) = x^2 - 10^2$ Letting $A = 10$ in Equation (4)
$$= x^2 - 100$$

b. $(2x + y)(2x - y) = (2x)^2 - y^2$ Letting $X = 2x$ and
$$= 4x^2 - y^2$$ $A = y$ in Equation (4)

c. $(3x - 5y)(3x + 5y) = (3x)^2 - (5y)^2$
$$= 9x^2 - 25y^2$$ ▲

Note that in this last example,

$$(3x - 5y)(3x + 5y) = (3x + 5y)(3x - 5y)$$

(by the commutative law), so Equation (4) still applies.
    A final word of advice before you do the exercises. Equation (1) (using the FOIL method) is very important and should be thoroughly understood before attempting the problems. This is because *all* the special products are *derived* from this formula. As a matter of fact, you can successfully complete all the problems in Exercise 4.4 if you fully understand the result given in Equation (1).

Problem 3 Find.
a. $(x + 5)(x - 5)$
b. $(3x + y)(3x - y)$
c. $(5x - 3y)(5x + 3y)$

Now, let us look at the results given in Equations (1)–(4) using FOIL.

$$
\begin{array}{cccc}
& \text{F} & \text{O} & \text{I} & \text{L} \\
\end{array}
$$

(1) $(X + A)(X + B) = X \cdot X + BX + AX + A \cdot B$

$\qquad\qquad\qquad = X^2 + AX + BX + AB \qquad$ Using the commutative law

$\qquad\qquad\qquad = X^2 + (A + B)X + AB \qquad$ Using the distributive law

$$
\begin{array}{cccc}
& \text{F} & \text{O} & \text{I} & \text{L} \\
\end{array}
$$

(2) $(X + A)(X + A) = X \cdot X + AX + AX + A \cdot A$

$\qquad\qquad\qquad = X^2 + 2AX + A^2$

$$
\begin{array}{cccc}
& \text{F} & \text{O} & \text{I} & \text{L} \\
\end{array}
$$

(3) $(X - A)(X - A) = X \cdot X - AX - AX + A \cdot A$

$\qquad\qquad\qquad = X^2 - 2AX + A^2$

(4) $(X + A)(X - A) = X \cdot X - AX + AX - A \cdot A$

$\qquad\qquad\qquad = X^2 - A^2$

Now you are ready for the exercises.

A. Expand.

1. $(x + 1)^2$

2. $(x + 6)^2$

3. $(x + 8)^2$

4. $(x + 11)^2$

5. $(2x + 1)^2$

6. $(2x + 3)^2$

7. $(3x + 5)^2$

8. $(4x + 7)^2$

9. $(3x + 2y)^2$

10. $(4x + 5y)^2$

11. $(5x + y)^2$

12. $(6x + 5y)^2$

B. Expand.

13. $(x - 1)^2$

14. $(x - 2)^2$

15. $(x - 4)^2$

16. $(x - 11)^2$

17. $(2x - 1)^2$

18. $(3x - 4)^2$

19. $(5x - 6)^2$

20. $(2x - 9)^2$

21. $(3x - y)^2$

22. $(4x - y)^2$

23. $(6x - 5y)^2$

24. $(4x - 3y)^2$

25. $(2x - 7y)^2$

26. $(3x - 5y)^2$

C. Find.

27. $(x + 2)(x - 2)$

28. $(x + 1)(x - 1)$

ANSWERS

1. _____
2. _____
3. _____
4. _____
5. _____
6. _____
7. _____
8. _____
9. _____
10. _____
11. _____
12. _____

13. _____
14. _____
15. _____
16. _____
17. _____
18. _____
19. _____
20. _____
21. _____
22. _____
23. _____
24. _____
25. _____
26. _____

27. _____
28. _____

29. $(x + 4)(x - 4)$ **30.** $(2x + y)(2x - y)$

31. $(3x + 2y)(3x - 2y)$ **32.** $(2x + 5y)(2x - 5y)$

33. $(x - 6)(x + 6)$ **34.** $(x - 11)(x + 11)$

35. $(x - 12)(x + 12)$ **36.** $(x - 9)(x + 9)$

37. $(3x - y)(3x + y)$ **38.** $(5x - 6y)(5x + 6y)$

39. $(2x - 7y)(2x + 7y)$ **40.** $(5x - 8y)(5x + 8y)$

Use the special products to multiply.

41. $(x^2 + 2)(x^2 + 5)$ **42.** $(x^2 - 3)(x^2 + 2)$

43. $(x^2 + y)^2$ **44.** $(2x^2 + y)^2$

45. $(3x^2 - 2y^2)^2$ **46.** $(4x^3 - 5y^3)^2$

47. $(x^2 - 2y^2)(x^2 + 2y^2)$ **48.** $(x^2 - 3y^2)(x^2 + 3y^2)$

49. $(2x + 4y^2)(2x - 4y^2)$ **50.** $(5x^2 + 2y)(5x^2 - 2y)$

29. _____
30. _____
31. _____
32. _____
33. _____
34. _____
35. _____
36. _____
37. _____
38. _____
39. _____
40. _____

41. _____
42. _____
43. _____
44. _____
45. _____
46. _____
47. _____
48. _____
49. _____
50. _____

✓ **SKILL CHECKER**

Use the distributive law to multiply.

51. $5(x^2 + x - 2)$ **52.** $x(x^2 + x - 2)$

53. $-3(x^2 - 2x - 4)$ **54.** $x(x^2 - 2x - 4)$

51. _____
52. _____
53. _____
54. _____

4.4   USING YOUR KNOWLEDGE

A common fallacy (*mistake*) when multiplying binomials is to *assume* that

$$(x + y)^2 = x^2 + y^2$$

Here are some arguments that should convince you this is *not* true.

1. Let $x = 1$, $y = 2$.
   a. What is $(x + y)^2$?
   b. What is $x^2 + y^2$?
   c. Is $(x + y)^2 = x^2 + y^2$?

2. Let $x = 2$, $y = 1$.
   a. What is $(x - y)^2$?
   b. What is $x^2 - y^2$?
   c. Is $(x - y)^2 = x^2 - y^2$?

3. Look at the large square in the margin. Its area is $(x + y)^2$. The square is divided into four smaller areas numbered 1, 2, 3, and 4.
   a. What is the area of square 1?
   b. What is the area of rectangle 2?
   c. What is the area of square 3?
   d. What is the area of rectangle 4?

1. a. _____
   b. _____
   c. _____

2. a. _____
   b. _____
   c. _____

3. a. _____
   b. _____
   c. _____
   d. _____

4. The total area of the square is $(x + y)^2$. It is also the sum of the four areas numbered 1, 2, 3, and 4. What is the sum of these four areas? (Simplify your answer.)

4. _____

5. From your answer to Problem 4, what can you say about $x^2 + 2xy + y^2$ and $(x + y)^2$?

5. _____

6. $x^2 + y^2$ is the sum of the areas of the squares numbered 1 and 3. Is $x^2 + y^2 = (x + y)^2$?

6. _____

4.5 MORE SPECIAL PRODUCTS

OBJECTIVES

REVIEW

Before starting this section, you should know:

1. How to use the distributive law to simplify expressions.
2. How to use the FOIL method to multiply polynomials.

OBJECTIVES

You should be able to:

A. Multiply a binomial by a trinomial.
B. Use the special products or the FOIL method to multiply polynomials in general.

The woman in the photo is baking cookies. What does that have to do with polynomials? Scientists have determined that the temperature of the oven (in degrees Fahrenheit) after t minutes is given by

$(5t - 12)(2t^2 - 27t + 109)$ This formula works only when t is between 3 and 7 min, inclusive.

In the preceding section we used the FOIL method to multiply binomials. Can we use this method to multiply any two polynomials? Unfortunately, no. But wait; if algebra is a generalized arithmetic, we should be able to multiply a binomial by a trinomial using the same techniques we employed to multiply 23 by 342, for example.

First, let us see how we multiplied 23 by 342. Here are the steps:

STEP 1	STEP 2	STEP 3
$\begin{array}{r} 342 \\ \times\ \ 23 \\ \hline 1026 \end{array}$	$\begin{array}{r} 342 \\ \times\ \ 23 \\ \hline 1026 \\ 6840 \\ \hline \end{array}$	$\begin{array}{r} 342 \\ \times\ \ 23 \\ \hline 1026 \\ 6840 \\ \hline 7866 \end{array}$
$3 \times 342 = 1026$	$20 \times 342 = 6840$	$1026 + 6840 = 7866$

A. MULTIPLYING A BINOMIAL BY A TRINOMIAL

Now let us use this same technique to multiply the two polynomials $(x + 5)$ and $(x^2 + x - 2)$.

STEP 1	STEP 2	STEP 3
$x^2 + x - 2$ $\underline{x + 5}$ $5x^2 + 5x - 10$	$x^2 + x - 2$ $\underline{x + 5}$ $5x^2 + 5x - 10$ $\underline{x^3 + x^2 - 2x}$	$x^2 + x - 2$ $\underline{x + 5}$ $5x^2 + 5x - 10$ $\underline{x^3 + x^2 - 2x}$ $x^3 + 6x^2 + 3x - 10$
$5(x^2 + x - 2)$ $= 5x^2 + 5x - 10$	$x(x^2 + x - 2)$ $= x^3 + x^2 - 2x$	$(5x^2 + 5x - 10)$ $\underline{(x^3 + x^2 - 2x)}$ $x^3 + 6x^2 + 3x - 10$

Note that in Step 3, all *like* terms are placed in the same column so they can be combined.

For obvious reasons this method is called the **vertical scheme** and can be used when one of the polynomials to be multiplied has *three or more terms*. Of course, we could have obtained the same result by using the distributive law:

$$(a + b)c = ac + bc$$

The result would look like this:

$$
\begin{aligned}
(a + b) \quad \cdot \quad c \quad &= a \quad \cdot \quad c \quad + b \quad \cdot \quad c \\
(x + 5)(x^2 + x - 2) &= x(x^2 + x - 2) + 5(x^2 + x - 2) \\
&= x^3 + x^2 - 2x + 5x^2 + 5x - 10 \\
&= x^3 + (x^2 + 5x^2) + (-2x + 5x) - 10 \\
&= x^3 + 6x^2 + 3x - 10
\end{aligned}
$$

EXAMPLE 1   Find $(x - 3)(x^2 - 2x - 4)$.

Solution   Using the vertical scheme:

$$
\begin{array}{l}
x^2 - 2x - 4 \\
\underline{x - 3} \\
-3x^2 + 6x + 12 \quad \text{Multiplying } x^2 - 2x - 4 \text{ by } -3 \\
\underline{x^3 - 2x^2 - 4x} \quad \text{Multiplying } x^2 - 2x - 4 \text{ by } x \\
x^3 - 5x^2 + 2x + 12 \quad \text{Adding like terms}
\end{array}
$$

Thus the result of multiplying $(x^2 - 2x - 4)$ by $(x - 3)$ is $x^3 - 5x^2 + 2x + 12$. You can also do this problem by using the distributive law, $(a + b)c = ac + bc$. The result would be

$$
\begin{aligned}
(a - b) \quad \cdot \quad c \quad &= a \quad \cdot \quad c \quad - b \quad \cdot \quad c \\
(x - 3)(x^2 - 2x - 4) &= x(x^2 - 2x - 4) - 3(x^2 - 2x - 4) \\
&= x^3 - 2x^2 - 4x - 3x^2 + 6x + 12 \\
&= x^3 + (-2x^2 - 3x^2) + (-4x + 6x) + 12 \\
&= x^3 - 5x^2 + 2x + 12
\end{aligned}
$$

▲

Problem 1   Find $(x - 2)(x^2 - 4x - 3)$.

Note that the same result is obtained in both cases. Here is the idea we used:

RULE TO MULTIPLY *ANY* TWO POLYNOMIALS

To multiply two polynomials, multiply each term of one by every term of the other and add the results.

B.   USING THE SPECIAL PRODUCTS

Now that you have learned all the basic techniques used to multiply polynomials, you should be able to tackle any polynomial multiplication. To do this, you must *first* decide what special product (if any) is involved. Here are the special products we have studied.

MULTIPLYING ANY TWO POLYNOMIALS
(TERM-BY-TERM MULTIPLICATION)

(1)   FOIL $(X + A)(X + B) = X^2 + (A + B)X + AB$
(2) $(X + A)(X + A) = (X + A)^2 = X^2 + 2AX + A^2$
(3) $(X - A)(X - A) = (X - A)^2 = X^2 - 2AX + A^2$
(4) $(X + A)(X - A) = X^2 - A^2$

Of course, the FOIL method works for the last three types, but you should try to recognize these last three special products.

EXAMPLE 2   Find $3x(x + 5)(x + 6)$.

Solution   You can use FOIL first as follows:

$$
\begin{aligned}
3x(x + 5)(x + 6) &= 3x(x^2 + 6x + 5x + 30) \\
&= 3x(x^2 + 11x + 30) \\
&= 3x^3 + 33x^2 + 90x
\end{aligned}
$$

where the top line shows F, O, I, L over the terms.

or use the distributive law to multiply $(x + 5)$ by $3x$ and then FOIL, as shown:

$$
\begin{aligned}
3x(x + 5)(x + 6) &= (3x \cdot x + 3x \cdot 5)(x + 6) \\
&= (3x^2 + 15x)(x + 6) && \text{Simplifying} \\
&= 3x^2 \cdot x + 3x^2 \cdot 6 + 15x \cdot x + 15x \cdot 6 && \text{By FOIL} \\
&= 3x^3 + 18x^2 + 15x^2 + 90x && \text{Simplifying} \\
&= 3x^3 + 33x^2 + 90x && \text{Collecting like terms}
\end{aligned}
$$

Of course, the result is the same for both methods.

Problem 2   Find $2x(x + 2)(x + 3)$.

EXAMPLE 3 Find $(x + 3)^3$.

Solution Recall that $(x + 3)^3 = (x + 3)(x + 3)(x + 3)$. Thus, we can square the sum $(x + 3)$ first and then use the distributive law. It goes like this:

$$(x + 3)(x + 3)(x + 3) = (x + 3)(x^2 + 6x + 9)$$
$$= x(x^2 + 6x + 9) + 3(x^2 + 6x + 9)$$
$$= x \cdot x^2 + x \cdot 6x + x \cdot 9 + 3 \cdot x^2 + 3 \cdot 6x + 3 \cdot 9$$
$$= x^3 + 9x^2 + 27x + 27$$

EXAMPLE 4 Find $\left(5t^2 - \dfrac{1}{2}\right)^2$.

Solution Using Equation (3) we obtain

$$\left(5t^2 - \frac{1}{2}\right)^2 = (5t^2)^2 - 2 \cdot \frac{1}{2} \cdot (5t^2) + \left(\frac{1}{2}\right)^2$$
$$= 25t^4 - 5t^2 + \frac{1}{4}$$

EXAMPLE 5 Find $(2x^2 + 5)(2x^2 - 5)$.

Solution This time we use Equation (4). We have:

$$(2x^2 + 5)(2x^2 - 5) = (2x^2)^2 - (5)^2$$
$$= 4x^4 - 25 \qquad \blacktriangle$$

Which method is best? A general one to keep in mind is the FOIL method, but before attempting a problem ask yourself these questions:

TO MULTIPLY TWO POLYNOMIALS

1. Is the product the square of a binomial? If so, use Equation (2) or (3):

   (2) $(X + A)^2 = (X + A)(X + A) = X^2 + 2AX + A^2$
   (3) $(X - A)^2 = (X - A)(X - A) = X^2 - 2AX + A^2$

   Note that both answers have *three* terms.

2. Is the product the sum and difference of the same two terms? If so, use Equation (4):

   (4) $(X + A)(X - A) = X^2 - A^2$

   The answer has *two* terms.

3. Is the product different from those in 1 and 2? If so, use FOIL. The answer will have *three* or *four* terms.

4. Is the product different from all the ones mentioned in 1, 2, and 3? If so, multiply every term of the first polynomial by every term of the second and collect like terms. The answer will have *more than two* terms.

Try using these ideas when working the exercises.

Problem 3 Find $(x + 2)^3$.

Problem 4 Find $\left(4t^2 - \dfrac{1}{4}\right)^2$.

Problem 5 Find $(3x^2 + 2)(3x^2 - 2)$.

ANSWERS (to problems on pages 320–321)
1. $x^3 - 6x^2 + 5x + 6$
2. $2x^3 + 10x^2 + 12x$

ANSWERS

A. Find.

1. $(x + 3)(x^2 + x + 5)$

2. $(x + 2)(x^2 + 5x + 6)$

3. $(x + 4)(x^2 - x + 3)$

4. $(x + 5)(x^2 - x + 2)$

5. $(x + 3)(x^2 - x - 2)$

6. $(x + 4)(x^2 - x - 3)$

7. $(x - 2)(x^2 + 2x + 4)$

8. $(x - 3)(x^2 + x + 1)$

9. $(x - 1)(x^2 - x + 2)$

10. $(x - 2)(x^2 - 2x + 1)$

11. $(x - 4)(x^2 - 4x - 1)$

12. $(x - 3)(x^2 - 2x - 2)$

B. Find.

13. $2x(x + 1)(x + 2)$

14. $3x(x + 2)(x + 5)$

15. $3x(x - 1)(x + 2)$

16. $4x(x - 2)(x + 3)$

17. $4x(x - 1)(x - 2)$

18. $2x(x - 3)(x - 1)$

19. $5x(x + 1)(x - 5)$

20. $6x(x + 2)(x - 4)$

21. $(x + 5)^3$

22. $(x + 4)^3$

23. $(2x + 3)^3$

24. $(3x + 2)^3$

25. $(2x + 3y)^3$

26. $(3x + 2y)^3$

27. $(4t^2 + 3)^2$

28. $(5t^2 + 1)^2$

1. _____
2. _____
3. _____
4. _____
5. _____
6. _____
7. _____
8. _____
9. _____
10. _____
11. _____
12. _____

13. _____
14. _____
15. _____
16. _____
17. _____
18. _____
19. _____
20. _____
21. _____
22. _____
23. _____
24. _____
25. _____
26. _____
27. _____
28. _____

29. $(4t^2 + 3u)^2$

30. $(5t^2 + u)^2$

31. $\left(3t^2 - \dfrac{1}{3}\right)^2$

32. $\left(4t^2 - \dfrac{1}{2}\right)^2$

33. $\left(3t^2 - \dfrac{1}{3}u\right)^2$

34. $\left(4t^2 - \dfrac{1}{2}u\right)^2$

35. $(3x^2 + 5)(3x^2 - 5)$

36. $(4x^2 + 3)(4x^2 - 3)$

37. $(3x^2 + 5y^2)(3x^2 - 5y^2)$

38. $(4x^2 + 3y^2)(4x^2 - 3y^2)$

39. $(4x^3 - 5y^3)(4x^3 + 5y^3)$

40. $(2x^4 - 3y^3)(2x^4 + 3y^3)$

29. _____

30. _____

31. _____

32. _____

33. _____

34. _____

35. _____

36. _____

37. _____

38. _____

39. _____

40. _____

C. Applications

41. The drain current I_D for a certain gate source is

$$I_D = 0.004\left(1 + \frac{V}{2}\right)^2$$

where V is the voltage at the gate source. Expand this expression.

41. _____

42. The equation of a transconductance curve is

$$I_D = I_{\text{DSS}}\left[1 - \frac{V_1}{V_2}\right]^2$$

Expand this expression. (Here I_{DSS} is the current from drain to source with shorted gate, V_1 is the voltage at the gate source, and V_2 is the voltage when the gate source is off.)

42. _____

43. The heat transmission between two objects of temperature T_2 and T_1 involves the expression

$$(T_1^2 + T_2^2)(T_1^2 - T_2^2)$$

Multiply this expression.

43. _____

44. The deflection of a certain beam involves the expression $w(l^2 - x^2)^2$. Expand this expression.

44. _____

45. The heat output from a natural draught convector is given by $K(t_n - t_a)^2$. Expand this expression.

45. _____

ANSWERS (to problems on page 322)
3. $x^3 + 6x^2 + 12x + 8$
4. $16t^4 - 2t^2 + \dfrac{1}{16}$ **5.** $9x^4 - 4$

✓ **SKILL CHECKER**

Perform the indicated operations.

46. $\frac{1}{5}(x + 10)$ **47.** $\frac{1}{7}(y + 14)$

48. $\frac{1}{2}(4x + 8)$ **49.** $\frac{1}{3}(6x + 9)$

50. $\frac{4x^3}{2x}$ **51.** $\frac{8x^3}{2x}$

46. _____

47. _____

48. _____

49. _____

50. _____

51. _____

4.5 USING YOUR KNOWLEDGE

At the beginning of this section, we mentioned that the temperature of the oven (in degrees Fahrenheit) after t minutes was

$$(5t - 12)(2t^2 - 27t + 109)$$

1. Multiply (expand) this expression.

1. _____

CALCULATOR CORNER

We have mentioned that the temperature of the oven (in degrees Fahrenheit) after t minutes is given by

$$(5t - 12)(2t^2 - 27t + 109)$$

For $t = 3$, this temperature will be

$$(5 \cdot 3 - 12)(2 \cdot 3^2 - 27 \cdot 3 + 109)$$

If your calculator has a set of parentheses, key in

| (| 5 | × | 3 | − | 1 | 2 |) | × | (| 2 | × | 3 |
| × | 3 | − | 2 | 7 | × | 3 | + | 1 | 0 | 9 |) | = |

to obtain a temperature of 138° F.

1. Find the temperature after 4 min, that is, when $t = 4$.

2. Find the temperature when $t = 5$.

3. Find the temperature when $t = 6$.

4. Find the temperature when $t = 7$.

5. Find the temperature when $t = 8$.

Most ovens heat only to 500° F. Do you see why the formula works only up to 7 min?

1. _____

2. _____

3. _____

4. _____

5. _____

4.6 DIVISION OF POLYNOMIALS

OBJECTIVES

REVIEW

Before starting this section you should know:

1. How to use the distributive law to simplify expressions.
2. How to use the FOIL method to multiply polynomials.

OBJECTIVES

You should be able to:

A. Divide a polynomial by a monomial.
B. Divide one polynomial by another polynomial.

How efficient is your refrigerator or your heat pump? Their efficiency E is given by

$$E = \frac{T_1 - T_2}{T_1}$$

where T_1 and T_2 are the initial and final temperatures between which they operate. Can you do the indicated division in

$$\frac{T_1 - T_2}{T_1}$$

Follow the steps in the procedure.

$$\frac{T_1 - T_2}{T_1} = (T_1 - T_2) \div T_1$$

$$= (T_1 - T_2)\left(\frac{1}{T_1}\right) \quad \text{Division by } T_1 \text{ is the same as multiplication by } \frac{1}{T_1}$$

$$= T_1\left(\frac{1}{T_1}\right) - T_2\left(\frac{1}{T_1}\right) \quad \text{Using the distributive law}$$

$$= \frac{T_1}{T_1} - \frac{T_2}{T_1} \quad \text{Multiplying}$$

$$= 1 - \frac{T_2}{T_1} \quad \text{Since } \frac{T_1}{T_1} = 1$$

A. DIVIDING A POLYNOMIAL BY A MONOMIAL

To divide the binomial $4x^3 - 8x^2$ by $2x$, we proceed similarly.

$$\frac{4x^3 - 8x^2}{2x} = (4x^3 - 8x^2) \div 2x$$

$$= (4x^3 - 8x^2)\left(\frac{1}{2x}\right)$$

$$= 4x^3\left(\frac{1}{2x}\right) - 8x^2\left(\frac{1}{2x}\right)$$

$$= \frac{4x^3}{2x} - \frac{8x^2}{2x}$$

$$= \frac{\overset{2x^2}{\cancel{4x^3}}}{\cancel{2x}} - \frac{\overset{4x}{\cancel{8x^2}}}{\cancel{2x}}$$

$$= 2x^2 - 4x$$

We then have

$$\frac{4x^3 - 8x^2}{2x} = \frac{4x^3}{2x} - \frac{8x^2}{2x} = 2x^2 - 4x$$

These examples suggest the following rule.

RULE TO DIVIDE A POLYNOMIAL BY A MONOMIAL

To divide a *polynomial* by a *monomial,* divide *each* term in the polynomial by the *monomial.*

EXAMPLE 1 Find.

a. $\dfrac{28x^4 - 14x^3}{7x^2}$ **b.** $\dfrac{20x^3 - 5x^2 + 10x}{10x^2}$

Solution

a. $\dfrac{28x^4 - 14x^3}{7x^2} = \dfrac{28x^4}{7x^2} - \dfrac{14x^3}{7x^2}$

$\qquad\qquad = 4x^2 - 2x$

b. $\dfrac{20x^3 - 5x^2 + 10x}{10x^2} = \dfrac{20x^3}{10x^2} - \dfrac{5x^2}{10x^2} + \dfrac{10x}{10x^2}$

$\qquad\qquad = 2x - \dfrac{1}{2} + \dfrac{1}{x}$

Problem 1 Find.

a. $\dfrac{24x^4 - 18x^3}{6x^2}$

b. $\dfrac{16x^4 - 4x^2 + 8x}{8x^2}$

B. DIVIDING ONE POLYNOMIAL BY ANOTHER POLYNOMIAL

If we wish to divide a polynomial, called the **dividend,** by another polynomial, called the **divisor,** we proceed very much as we did in long division in arithmetic. To show you that this is so, we are performing the division of 337 by 16 and $(x^2 + 3x + 3)$ by $(x + 1)$ side by side.

1. $16\overline{)337}$ with 2 above
Divide 33 by 16. It goes twice. Write **2** over the 33.

$x + 1\overline{)x^2 + 3x + 3}$ with x above
Divide x^2 by x. It goes x times. Write x over the $3x$.

2. $16\overline{)337}$ with 2 above, -32, 1
Multiply 2 by 16 and subtract the product 32 from 33, obtaining 1.

$x + 1\overline{)x^2 + 3x + 3}$ with x above, $(-)\,x^2 + x$, $0 + 2x$
Multiply x by $x + 1$ and subtract the product $x^2 + x$ from $x^2 + 3x$ obtaining $0 + 2x$.

3. $16\overline{)337}$ with 21 above, -32, 17
Bring down the **7**. Now, divide 17 by 16. It goes once. Write **1** after the 2.

$x + 1\overline{)x^2 + 3x + 3}$ with $x + 2$ above, $(-)\,x^2 + x$, $0 + 2x + 3$
Bring down the **3**. Now, divide $2x + 3$ by $x + 1$. It goes **2** times. Write **+2** after the x.

4. $16\overline{)337}$ with 21 above, -32, 17, -16, 1
Multiply 1 by 16 and subtract the result from 17. The remainder is **1**.

$x + 1\overline{)x^2 + 3x + 3}$ with $x + 2$ above, $(-)\,x^2 + x$, $0 + 2x + 3$, $(-)\,2x + 2$, 1
Multiply 2 by $x + 1$ obtaining $2x + 2$. Subtract this result from $2x + 3$. The remainder is 1.

5. The answer (**quotient**) can be written as 21 R 1 (read "21 remainder 1") or as $21 + \frac{1}{16}$, which is $21\frac{1}{16}$.

The answer (**quotient**) can be written as $x + 2$ R 1 (read "$x + 2$ remainder 1") or as

$$x + 2 + \frac{1}{x + 1}$$

6. You can check this answer by multiplying 21 by 16 (336) and adding the remainder 1 to obtain 337, the dividend.

You can check the answer by multiplying $(x + 2)(x + 1)$ obtaining $x^2 + 3x + 2$ and then adding the remainder 1 to get $x^2 + 3x + 3$, the dividend.

EXAMPLE 2 Divide $x^2 + 2x - 17$ by $x - 3$.

Solution

$$
\begin{array}{r}
x + 5 \\
x - 3 \overline{\smash{\big)}\ x^2 + 2x - 17} \\
(-)\ x^2 - 3x \\
\hline
5x - 17 \\
(-)\ 5x - 15 \\
\hline
2
\end{array}
$$

x^2 divided by x is x.

$5x$ divided by x is 5.

$x(x - 3) = x^2 - 3x$

$5(x - 3) = 5x - 15$

Remainder

Thus, $(x^2 + 2x - 17) \div (x - 3) = x - 5\ \text{R}\ 2$. ▲

If there are missing terms in the polynomial being divided, we insert zero coefficients as shown in the next example.

EXAMPLE 3 Divide $2x^3 - 2 - 4x$ by $2 + 2x$.

Solution We write the polynomials in *descending* order, inserting $\mathbf{0x^2}$ in the dividend, since the x^2 term is missing. We then have:

$$
\begin{array}{r}
x^2 - x - 1 \\
2x + 2 \overline{\smash{\big)}\ 2x^3 + 0x^2 - 4x - 2} \\
(-)\ 2x^3 + 2x^2 \\
\hline
0 - 2x^2 - 4x \\
(-)\ -2x^2 - 2x \\
\hline
-2x - 2 \\
(-)\ -2x - 2 \\
\hline
0
\end{array}
$$

$2x^3$ divided by $2x$ is x^2.

$x^2(2x + 2) = 2x^3 + 2x^2$

$-2x^2$ divided by $2x$ is $-x$.
$-x(2x + 2) = -2x^2 - 2x$
$-2x$ divided by $2x$ is -1.
$-1(2x + 2) = -2x - 2$
There is no remainder.

Thus, $(2x^3 - 4x - 2) \div (2x + 2) = x^2 - x - 1$.

EXAMPLE 4 Divide $x^4 + x^3 - 3x^2 + 1$ by $x^2 - 3$.

Solution We write the polynomials in *descending* order, inserting $0x$ for the missing term in the dividend. We then have:

$$
\begin{array}{r}
x^2 + x \\
x^2 - 3 \overline{\smash{\big)}\ x^4 + x^3 - 3x^2 + 0x + 1} \\
(-)\ x^4 \quad\ \ - 3x^2 \\
\hline
0 + x^3 \qquad\quad + 1 \\
(-)\ x^3 \qquad - 3x \\
\hline
3x + 1
\end{array}
$$

x^4 divided by x^2 is x^2.
$x^2(x^2 - 3) = x^4 - 3x^2$
x^3 divided by x^2 is x.
$x(x^2 - 3) = x^3 - 3x$
We cannot divide $3x + 1$ by x^2, so we stop. The remainder is $3x + 1$.

Thus, $(x^4 + x^3 - 3x^2 + 1) \div (x^2 - 3) = (x^2 + x)\ \text{R}\ (3x + 1)$. You can also write the answer as

$$
x^2 + x + \frac{3x + 1}{x^2 - 3}
$$

Problem 2 Divide $x^2 + 4x - 15$ by $x - 2$.

Problem 3 Divide $2x^3 + x - 3$ by $x - 1$.

Problem 4 Divide $x^4 + x^3 - 2x^2 + 1$ by $x^2 - 2$.

ANSWER (to problem on page 328)

1. a. $4x^2 - 3x$ **b.** $2x^2 - \dfrac{1}{2} + \dfrac{1}{x}$

A. Divide.

1. $\dfrac{3x + 9y}{3}$

2. $\dfrac{6x + 8y}{2}$

3. $\dfrac{10x - 5y}{5}$

4. $\dfrac{24x - 12y}{6}$

5. $\dfrac{8y - 32}{-4}$

6. $\dfrac{9y - 45}{-3}$

7. $10x^2 + 8x$ by x

8. $12x^2 + 18x$ by x

9. $15x^3 - 10x^2$ by $5x^2$

10. $18x^4 - 24x^2$ by $3x^2$

B. Divide.

11. $x^2 + 5x + 6$ by $x + 3$

12. $x^2 + 9x + 20$ by $x + 4$

13. $y^2 + 3y - 10$ by $y + 5$

14. $y^2 + 2y - 15$ by $y + 5$

15. $2x + x^2 - 24$ by $x - 4$

16. $4x + x^2 - 21$ by $x - 3$

17. $-8 + 2x + 3x^2$ by $2 + x$

18. $-6 + x + 2x^2$ by $2 + x$

19. $2y^2 + 9y - 36$ by $7 + y$

20. $3y^2 + 13y - 32$ by $6 + y$

21. $2x^3 - 4x - 2$ by $2x + 2$

22. $3x^3 - 9x - 6$ by $3x + 3$

23. $y^4 - y^2 - 2y - 1$ by $y^2 + y + 1$

24. $y^4 - y^2 - 4y - 4$ by $y^2 + y + 2$

25. $8x^3 - 6x^2 + 5x - 9$ by $2x - 3$

26. $2x^4 - x^3 + 7x - 2$ by $2x + 3$

27. $x^3 + 8$ by $x + 2$

28. $x^3 + 64$ by $x + 4$

1. _____

2. _____

3. _____

4. _____

5. _____

6. _____

7. _____

8. _____

9. _____

10. _____

11. _____

12. _____

13. _____

14. _____

15. _____

16. _____

17. _____

18. _____

19. _____

20. _____

21. _____

22. _____

23. _____

24. _____

25. _____

26. _____

27. _____

28. _____

29. $8y^3 - 64$ by $2y - 4$ **30.** $27x^3 - 8$ by $3x - 2$

29. _____

30. _____

31. $x^4 - x^2 - 2x - 1$ by $x^2 - x - 1$

31. _____

32. $y^4 - y^3 - 3y - 9$ by $y^2 - y - 3$

32. _____

33. $x^5 - x^4 + 6x^2 - 5x + 3$ by $x^2 - 2x + 3$

33. _____

34. $y^6 - y^5 + 6y^3 - 5y^2 + 3y$ by $y^2 - 2y + 3$

34. _____

35. $m^4 - 11m^2 + 34$ by $m^2 - 3$

35. _____

36. $n^3 - n^2 - 6n$ by $n^2 + 3n$

36. _____

37. $\dfrac{x^3 - y^3}{x - y}$ **38.** $\dfrac{x^3 + 8}{x + 2}$

37. _____

38. _____

39. $\dfrac{x^3 + 8}{x - 2}$ **40.** $\dfrac{x^5 + 32}{x - 2}$

39. _____

40. _____

✓ **SKILL CHECKER**

Find the LCM.

41. 20 and 18   **42.** 30 and 16

41. _____

42. _____

43. 40 and 12   **44.** 10, 18, and 12

43. _____

44. _____

45. 20, 30, and 18

45. _____

4.6 USING YOUR KNOWLEDGE

In economics, the average cost per unit of a product, denoted by $\overline{C(x)}$ is defined by

$$\overline{C(x)} = \frac{C(x)}{x}$$

where $C(x)$ is a polynomial in the variable x and x is the number of units produced.

1. If $C(x) = 3x^2 + 5x$, find $\overline{C(x)}$

1. _____

2. If $C(x) = 30 + 3x^2$, find $\overline{C(x)}$

2. _____

The average profit $\overline{P(x)}$ is

$$\overline{P(x)} = \frac{P(x)}{x}$$

where $P(x)$ is a polynomial in the variable x and x is the number of units sold in a certain period of time.

ANSWERS   (to problems on page 330)

2. $x + 6$ R $- 3$ **3.** $2x^2 + 2x + 3$

4. $x^2 + x + \dfrac{2x + 1}{x^2 - 2}$

3. If $P(x) = 50x + x^2 - 7000$ (dollars), find the average profit.

3. _____

4. If in Problem 3, 100 units are sold in a period of 1 week, what is the average profit?

4. _____

SUMMARY

SECTION	ITEM	MEANING	EXAMPLE
4.1	Polynomial	An algebraic expression formed by using the operations of addition and subtraction on products of a number and a variable raised to whole number exponents	$x^2 + 3x - 5$, $2x + 8 - x^3$ and $9x^7 - 3x^3 + 4x^8 - 10$ are polynomials but $\sqrt{x} - 3$, $\dfrac{x^2 - 2x + 3}{x}$, and $x^{3/2} - x$ are not.
4.1	Terms	The parts of a polynomial separated by plus $(+)$ signs are the terms.	The terms of $x^2 - 2x + 3$ are x^2, $-2x$, and 3.
4.1A	Monomial	A polynomial with one term	$3x$, $7x^2$, and $-3x^{10}$ are monomials.
4.2A	Binomial	A polynomial with two terms	$3x + x^2$, $7x - 8$, and $x^3 - 8x^7$ are binomials.
4.2A	Trinomial	A polynomial with three terms	$-8 + 3x + x^2$, $7x - 8 + x^4$, and $x^3 - 8x^7 + 9$ are trinomials.
4.2B	Degree	The degree of a polynomial is the highest exponent of the variable.	The degree of $8 + 3x + x^2$ is 2 and the degree of $7x - 8 + x^4$ is 4.
4.3C	FOIL Method	To multiply two binomials such as $(x + a)(x + b)$ multiply the **F**irst terms, the **O**uter terms, the **I**nner terms, and the **L**ast terms and add.	$\overset{\text{F}\quad\text{O}\quad\text{I}\quad\text{L}}{(x + 2)(x + 3) = x^2 + 3x + 2x + 6}$
4.4A	The square of a binomial sum	$(X + A)^2 = X^2 + 2AX + A^2$	$(x + 5)^2 = x^2 + 2 \cdot 5 \cdot x + 5^2$ $= x^2 + 10x + 25$
4.4B	The square of a binomial difference	$(X - A)^2 = X^2 - 2AX + A^2$	$(x - 5)^2 = x^2 - 2 \cdot 5 \cdot x + 5^2$ $= x^2 - 10x + 25$
4.4C	The product of the sum and difference of two terms	$(X + A)(X - A) = X^2 - A^2$	$(x + 7)(x - 7) = x^2 - 7^2$ $= x^2 - 49$

(If you need help with these exercises, look in the section indicated in brackets.)

ANSWERS

1. [4.1A] Classify as a monomial (M), binomial (B), or trinomial (T).
   a. $9x^2 - 9 + 7x$
   b. $7x^2$
   c. $3x - 1$

1. a. _____
   b. _____
   c. _____

2. [4.1B] Find the degree of the given polynomial.
   a. $3x^2 - 7x + 8x^4$
   b. $-4x + 2x^2 - 3$
   c. $8 + 3x - 4x^2$

2. a. _____
   b. _____
   c. _____

3. [4.1C] Write the given polynomial in descending order of exponents.
   a. $4x^2 - 8x + 9x^4$
   b. $-3x + 4x^2 - 3$
   c. $8 + 3x - 4x^2$

3. a. _____
   b. _____
   c. _____

4. [4.1D] Find the value of $-16t^2 + 300$ for each value of t.
   a. $t = 1$
   b. $t = 3$
   c. $t = 5$

4. a. _____
   b. _____
   c. _____

5. [4.2A] Add the given polynomials.
   a. $-5x + 7x^2 - 3$ and $-2x^2 - 7 + 4x$
   b. $-3x^2 + 8x - 1$ and $3 + 7x - 2x^2$
   c. $-4 + 3x^2 - 5x$ and $6x^2 - 2x + 5$

5. a. _____
   b. _____
   c. _____

6. [4.2B] Subtract the first polynomial from the second.
   a. $3x - 4 + 7x^2$ from $6x^2 - 4x$
   b. $5x - 3 + 2x^2$ from $9x^2 - 2x$
   c. $6 - 2x + 5x^2$ from $2x - 5$

6. a. _____
   b. _____
   c. _____

7. [4.3A] Find.
   a. $(-6x^2)(3x^5)$
   b. $(-8x^3)(5x^6)$
   c. $(-9x^4)(3x^7)$

7. a. _____
   b. _____
   c. _____

8. [4.3B] Remove parentheses (simplify).
   a. $-2x^2(x + 2y)$
   b. $-3x^3(2x + 3y)$
   c. $-4x^3(5x + 7y)$

8. a. _____
   b. _____
   c. _____

9. [4.3C] Find.
   a. $(x + 6)(x + 9)$
   b. $(x + 2)(x + 3)$
   c. $(x + 7)(x + 9)$

9. a. _____
   b. _____
   c. _____

10. [4.3C] Find.
    a. $(x + 7)(x - 3)$
    b. $(x + 6)(x - 2)$
    c. $(x + 5)(x - 1)$

10. a. _____
    b. _____
    c. _____

11. [4.3C] Find.
    a. $(x + 3)(x - 7)$
    b. $(x + 2)(x - 6)$
    c. $(x + 1)(x - 5)$

11. a. _____
    b. _____
    c. _____

12. [4.3C] Find.
    a. $(3x - 2y)(2x - 3y)$
    b. $(5x - 3y)(4x - 3y)$
    c. $(4x - 3y)(2x - 5y)$

12. a. _____
    b. _____
    c. _____

13. [4.4A] Expand.
    a. $(2x + 3y)^2$
    b. $(3x + 4y)^2$
    c. $(4x + 5y)^2$

13. a. _____
    b. _____
    c. _____

14. [4.4B] Expand.
    a. $(2x - 3y)^2$
    b. $(3x - 2y)^2$
    c. $(5x - 2y)^2$

14. a. _____
    b. _____
    c. _____

15. [4.4C] Find.
    a. $(3x - 5y)(3x + 5y)$
    b. $(3x - 2y)(3x + 2y)$
    c. $(3x - 4y)(3x + 4y)$

15. a. _____
    b. _____
    c. _____

16. [4.5A] Find.
    a. $(x + 1)(x^2 + 3x + 2)$
    b. $(x + 2)(x^2 + 3x + 2)$
    c. $(x + 3)(x^2 + 3x + 2)$

16. a. _____
    b. _____
    c. _____

17. [4.5B] Find.
    a. $3x(x + 1)(x + 2)$
    b. $4x(x + 1)(x + 2)$
    c. $5x(x + 1)(x + 2)$

17. a. _____
    b. _____
    c. _____

18. [4.5B] Expand.
    a. $(x + 2)^3$
    b. $(x + 3)^3$
    c. $(x + 4)^3$

18. a. _____
    b. _____
    c. _____

19. [4.5B] Expand.

    a. $\left(5x^2 - \dfrac{1}{2}\right)^2$

    b. $\left(7x^2 - \dfrac{1}{2}\right)^2$

    c. $\left(9x^2 - \dfrac{1}{2}\right)^2$

19. a. _____
    b. _____
    c. _____

20. [4.5B] Expand.
    a. $(3x^2 + 2)(3x^2 - 2)$
    b. $(3x^2 + 4)(3x^2 - 4)$
    c. $(3x^2 + 5)(3x^2 - 5)$

21. [4.6A] Find.

    a. $\dfrac{18x^3 - 9x^2}{9x}$

    b. $\dfrac{20x^3 - 10x^2}{5x}$

    c. $\dfrac{24x^3 - 12x^2}{6x}$

22. [4.6A] Divide.
    a. $x^2 + 4x - 12$ by $x - 2$
    b. $x^2 + 4x - 21$ by $x - 3$
    c. $x^2 + 4x - 32$ by $x - 4$

23. [4.6B] Divide.
    a. $8x^3 - 16x - 8$ by $2 + 2x$
    b. $12x^3 - 24x - 12$ by $2 + 2x$
    c. $4x^3 - 8x - 4$ by $2 + 2x$

24. [4.6B] Divide.
    a. $2x^3 - 20x + 8$ by $x - 3$
    b. $2x^3 - 21x + 12$ by $x - 3$
    c. $3x^3 - 4x + 5$ by $x - 1$

25. [4.6B] Divide.
    a. $x^4 + x^3 - 4x^2 + 1$ by $x^2 - 4$
    b. $x^4 + x^3 - 5x^2 + 1$ by $x^2 - 5$
    c. $x^4 + x^3 - 6x^2 + 1$ by $x^2 - 6$

20. a. _____
    b. _____
    c. _____

21. a. _____
    b. _____
    c. _____

22. a. _____
    b. _____
    c. _____

23. a. _____
    b. _____
    c. _____

24. a. _____
    b. _____
    c. _____

25. a. _____
    b. _____
    c. _____

(*Answers on page* 340)

NAME

CLASS

SECTION

ANSWERS

1. Classify as a monomial (M), binomial (B), or trinomial (T):
   a. $3x - 5$
   b. $5x^3$
   c. $8x^2 - 2 + 5x$

1. a. _____
   b. _____
   c. _____

2. Find the degree of the polynomial $5x - 3x^2 + 7$.

2. _____

3. Write the polynomial $-3x + 7 + 8x^2$ in descending order of exponents.

3. _____

4. Find the value of $-16t^2 + 100$ when $t = 2$.

4. _____

5. Add $-4x + 8x^2 - 3$ and $-5x^2 - 4 + 2x$.

5. _____

6. Subtract $5x - 2 + 8x^2$ from $3x^2 - 2x$.

6. _____

7. Remove parentheses (simplify): $-2x^2(x + 3y)$.

7. _____

8. Find $(x + 5)(x + 7)$.

8. _____

9. Find $(x + 8)(x - 3)$.

9. _____

10. Find $(x + 4)(x - 6)$.

10. _____

11. Find $(5x - 2y)(4x - 3y)$.

11. _____

12. Expand $(3x + 5y)^2$.

12. _____

13. Expand $(2x - 7y)^2$.

13. _____

14. Find $(2x - 5y)(2x + 5y)$.

14. _____

15. Find $(x + 2)(x^2 + 5x + 3)$.

15. _____

16. Find $3x(x + 2)(x + 5)$.

16. _____

17. Expand $(x + 7)^3$.

17. _____

18. Expand $\left(3x^2 - \dfrac{1}{2}\right)^2$.

18. _____

19. Find $(3x^2 + 7)(3x^2 - 7)$.

19. _____

20. Divide $2x^3 - 9x + 5$ by $x - 2$.

20. _____

IF YOU MISSED QUESTION	SECTION	EXAMPLES	PAGE	ANSWERS
1	4.1	1	286	1. a. B b. M c. T
2	4.1	2	286	2. 2
3	4.1	3	287	3. $8x^2 - 3x + 7$
4	4.1	4	287	4. 36
5	4.2	1, 2	293, 294	5. $3x^2 - 2x - 7$
6	4.2	3	295	6. $-5x^2 - 7x + 2$
7	4.3	1, 2	301	7. $-2x^3 - 6x^2y$
8	4.3	3	303	8. $x^2 + 12x + 35$
9	4.3	3, 4	303	9. $x^2 + 5x - 24$
10	4.3	3, 4	303	10. $x^2 - 2x - 24$
11	4.3	4, 5	303, 304	11. $20x^2 - 23xy + 6y^2$
12	4.4	1	310	12. $9x^2 + 30xy + 25y^2$
13	4.4	2	311	13. $4x^2 - 28xy + 49y^2$
14	4.4	3	312	14. $4x^2 - 25y^2$
15	4.5	1	320	15. $x^3 + 7x^2 + 13x + 6$
16	4.5	2	321	16. $3x^3 + 21x^2 + 30x$
17	4.5	3	322	17. $x^3 + 21x^2 + 147x + 343$
18	4.5	4	322	18. $9x^4 - 3x^2 + \dfrac{1}{4}$
19	4.5	5	322	19. $9x^4 - 49$
20	4.6	1, 2, 3	328, 330	20. $2x^2 + 4x - 1$ R 3

CHAPTER 5

FACTORING

We learned how to multiply polynomials in the preceding chapter. We now learn how to "undo" these multiplications by using a method called factoring. When we factor a polynomial, we write it as a product. There are several techniques that allow the factorization of polynomials, and we discuss these in the first four sections. We end the chapter by giving an important application of factoring: the solution of quadratic equations.

NAME

CLASS

SECTION

ANSWERS

(*Answers on page* 344)

1. Factor $12x^3 - 24x^5$.

2. Factor $\dfrac{3}{7}x^6 - \dfrac{4}{7}x^5 + \dfrac{2}{7}x^4 + \dfrac{1}{7}x^2$.

3. Factor $3x^3 + 6x^2 + x + 2$.

4. Factor $x^2 + 8x + 12$.

5. Factor $x^2 - 9x + 18$.

6. Factor $6x^2 + 13x - 5$.

7. Factor $6x^2 - 17xy + 5y^2$.

8. Factor $x^2 + 18x + 81$.

9. Factor $9x^2 + 12xy + 4y^2$.

10. Factor $x^2 - 18x + 81$.

11. Factor $4x^2 - 12xy + 9y^2$.

12. Factor $x^2 - 81$.

13. Factor $9x^2 - 25$.

14. Factor $2x^3 - 2x^2 - 8x$.

15. Factor $3x^3 - 12x^2 - 15x$.

16. Factor $2x^2 + 4x + x + 2$.

17. Factor $9kx^2 + 12kx + 4k$.

18. Factor $-9x^4 + 9x^2$.

19. Solve $x^2 - 2x - 15 = 0$.

20. Solve $3x^2 - 4x = 4$.

1. _____

2. _____

3. _____

4. _____

5. _____

6. _____

7. _____

8. _____

9. _____

10. _____

11. _____

12. _____

13. _____

14. _____

15. _____

16. _____

17. _____

18. _____

19. _____

20. _____

IF YOU MISSED QUESTION	SECTION	EXAMPLES	PAGE	ANSWERS
1	5.1	1	347	1. $12x^3(1 - 2x^2)$
2	5.1	2, 3	347	2. $\frac{1}{7}x^2(3x^4 - 4x^3 + 2x^2 + 1)$
3	5.1	4, 5	348, 349	3. $(x + 2)(3x^2 + 1)$
4	5.2	1	356	4. $(x + 6)(x + 2)$
5	5.2	1	356	5. $(x - 3)(x - 6)$
6	5.2	2, 3	359, 360	6. $(3x - 1)(2x + 5)$
7	5.2	4	360	7. $(3x - y)(2x - 5y)$
8	5.3	2	371	8. $(x + 9)^2$
9	5.3	2	371	9. $(3x + 2y)^2$
10	5.3	3	371	10. $(x - 9)^2$
11	5.3	3	371	11. $(2x - 3y)^2$
12	5.3	4	372	12. $(x + 9)(x - 9)$
13	5.3	4	372	13. $(3x + 5)(3x - 5)$
14	5.4	1	378	14. $2x(x^2 - x - 4)$
15	5.4	1	378	15. $3x(x + 1)(x - 5)$
16	5.4	2	378	16. $(x + 2)(2x + 1)$
17	5.4	3	379	17. $k(3x + 2)^2$
18	5.4	4, 5	379	18. $-9x^2(x + 1)(x - 1)$
19	5.5	1	387	19. $x = 5$ or $x = -3$
20	5.5	2, 3	388, 389	20. $x = 2$ or $x = -\frac{2}{3}$

5.1 FACTORING POLYNOMIALS

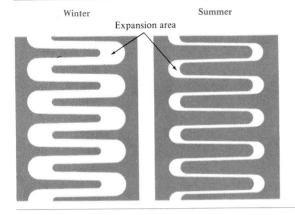

Winter   Summer

Expansion area

OBJECTIVES

REVIEW

Before starting this section you should know:

1. How to write a polynomial in descending order.
2. The distributive laws (Section 2.4).

OBJECTIVES

You should be able to:

A. Factor a common term from a polynomial.
B. Factor a four-term expression by grouping.

The drawing shows an expansion joint used in bridges and buildings. The linear expansion of a solid is:

$$e = \alpha L t_2 - \alpha L t_1$$ α is the coefficient of linear expansion, L is the length of the material, and t_2 and t_1 are the temperatures in degrees Celsius.

The expression on the right-hand side of the equation can be written in a simpler way if we *factor* it.

To *factor* an expression is to write the expression as a product of its factors. Does that sound complicated? It is not! Suppose we give you the numbers 3 and 5 and tell you to multiply them. You will probably write

$$3 \times 5 = 15$$

In the reverse process, we give you the product 15 and tell you to **factor** it! (Factoring is the *reverse* of multiplying.) You will then write

Product⌐   ⌐Factors

$$15 = \overbrace{3 \times 5}$$

Why? Because you have multiplied 3 by 5 many times before and gotten 15. What about factoring the number 20? Here you may write

$20 = 4 \times 5$

or

$20 = 2 \times 10$

$20 = 4 \times 5$ and $20 = 2 \times 10$ are not *completely* factored because they contain factors that are *not* prime numbers, numbers only divisible by themselves and 1. The first few prime numbers are 2, 3, 5, 7, 11, and so on.

Neither of these factorizations is *complete*. Note that in the first factorization $4 = (\mathbf{2 \times 2})$, so

$$20 = (2 \times 2) \times 5$$

whereas in the second factorization $10 = (\mathbf{2 \times 5})$. Thus,

$$20 = 2 \times (2 \times 5)$$

In either case, the complete factorization for 20 is

$$20 = 2 \times 2 \times 5$$

From now on, when we say factor a number, we mean *factor completely.* Moreover, the numerical factors are assumed to be prime numbers. Thus we do not factor 20 as

$$20 = \frac{1}{4} \times 80$$

With all these preliminaries out of the way, we can go on to factor algebraic expressions. There are different factoring techniques used in different situations. We will study most of them in this chapter. To start, compare the multiplications with the factors and see if you discover a pattern.

FINDING THE PRODUCT	FINDING THE FACTORS
$4(x + y) = 4x + 4y$	$4x + 4y = 4(x + y)$
$5(a - 2b) = 5a - 10b$	$5a - 10b = 5(x - 2b)$
$2x(x + 3) = 2x^2 + 6x$	$2x^2 + 6x = 2x(x + 3)$

A. FACTORING OUT COMMON FACTORS

What do all these operations have in common? They use the *distributive law.* When multiplying, we have

$$a(b + c) = ab + ac$$

When factoring,

$ab + ac = a(b + c)$ We are factoring a monomial (a) from a binomial ($ab + ac$).

Of course, we are now more interested in the latter. It tells us that to factor a binomial, we must find a factor (a in this case) common to all terms. The trick in having a *completely factored* expression is to select the common factor with the *largest coefficient* and the *largest exponent.* This factor is called the **greatest common factor** (GCF). Thus, to factor $6x^3 + 18x^2$, we could write

$6x^3 + 18x^2 = 3x(2x^2 + 6x)$ $2x^2 + 6x$ can be factored further.

but this is *not completely* factored. Here the *largest* number dividing 6 and 18 is 6, and the *highest* power of x dividing x^2 and x^3 is x^2. Thus the complete factorization is

$6x^3 + 18x^2 = 6x^2(x + 3)$ $6x^2$ is the GCF.

Of course, it might actually help your accuracy and understanding if you write an intermediate step indicating the common factor present in each term. Thus, to factor $6x^3 + 18x^2$, you may write

$6x^3 + 18x^2 = 6x^2 \cdot x + 6x^2 \cdot 3$

$\qquad\qquad\quad = 6x^2(x + 3)$ Note that since the 6 is a coefficient, it is *not* written in factored form; that is, we write $6x^2(x + 3)$ and *not* $2 \cdot 3x^2(x + 3)$.

When factoring polynomials, *do not* factor the final coefficients or the final constant terms. Similarly, to factor $4x - 28$, you may write

$$4x - 28 = 4 \cdot x - 4 \cdot 7$$
$$= 4(x - 7)$$

One more thing. When an expression such as $-3x + 12$ is to be factored, we have two possible factorizations:

$$-3(x - 4) \quad \text{or} \quad 3(-x + 4)$$

The first one is the *preferred* one, since in that case the first term of the binomial $x - 4$ has a positive sign.

EXAMPLE 1 Factor.

a. $8x + 24$ **b.** $-6y + 12$ **c.** $10x^2 - 25x^3$

Solution

a. $8x + 24 = 8 \cdot x + 8 \cdot 3$
$$= 8(x + 3)$$
b. $-6y + 12 = -6 \cdot y - 6(-2)$
$$= -6(y - 2)$$
c. $10x^2 - 25x^3 = 5x^2 \cdot 2 - 5x^2 \cdot 5x$
$$= 5x^2(2 - 5x)$$ ▲

We can also factor polynomials with more than two terms, as shown next.

EXAMPLE 2 Factor.

a. $6x^3 + 12x^2 + 18x$ **b.** $10x^6 - 15x^5 + 20x^4 + 30x^2$

Solution

a. $6x^3 + 12x^2 + 18x = 6x \cdot x^2 + 6x \cdot 2x + 6x \cdot 3$ $6x$ is the GCF.
$$= 6x(x^2 + 2x + 3)$$
b. $10x^6 - 15x^5 + 20x^4 + 30x^2 = 5x^2 \cdot 2x^4 - 5x^2 \cdot 3x^3 + 5x^2 \cdot 4x^2 + 5x^2 \cdot 6$
$$= 5x^2(2x^4 - 3x^3 + 4x^2 + 6) \quad 5x^2 \text{ is the GCF.}$$

EXAMPLE 3 Factor.

$$\frac{3}{4}x^2 - \frac{1}{4}x + \frac{5}{4}$$

Solution   Here, we can see that the greatest common factor is $\frac{1}{4}$. Thus,

$$\frac{3}{4}x^2 - \frac{1}{4}x + \frac{5}{4} = \frac{1}{4} \cdot 3x^2 - \frac{1}{4} \cdot x + \frac{1}{4} \cdot 5$$

$$= \frac{1}{4}(3x^2 - x + 5)$$

Problem 1 Factor.
a. $6x + 48$ **b.** $-3y + 21$
c. $4x^2 - 32x^3$

Problem 2 Factor.
a. $7x^3 + 14x^2 - 49x$
b. $3x^6 - 6x^5 + 12x^4 + 27x^2$

Problem 3 Factor.
$$\frac{2}{5}x^2 - \frac{3}{5}x - \frac{1}{5}$$

B. FACTORING BY GROUPING

Can we factor $x^3 + 2x^2 + 3x + 6$? It seems that there is no common factor except 1. However, we can group and factor the first two terms and also the last two terms and then use the distributive law, as shown next:

STEP 1. Group terms with common factors using the associative law.

$$x^3 + 2x^2 + 3x + 6 = (x^3 + 2x^2) + (3x + 6)$$

STEP 2. Factor each resulting binomial.

$$= x^2(x + 2) + 3(x + 2)$$

STEP 3. Factor out the GCF, $(x + 2)$.

$$= (x + 2)(x^2 + 3)$$

Thus,

$$x^3 + 2x^2 + 3x + 6 = (x + 2)(x^2 + 3)$$

Note that $x^2(x + 2) + 3(x + 2)$ can also be written as $(x^2 + 3)(x + 2)$, since $ac + bc = (a + b)c$. Hence

$$x^2 + 2x^2 + 3x + 6 = (x^2 + 3)(x + 2)$$

Either factorization is correct.

EXAMPLE 4 Factor.

a. $3x^3 + 6x^2 + 2x + 4$ **b.** $6x^3 - 3x^2 - 4x + 2$

Solution

a. We proceed by steps.

STEP 1. Group terms with common factors using the associative law.

$$3x^3 + 6x^2 + 2x + 4 = (3x^3 + 6x^2) + (2x + 4)$$

STEP 2. Factor each resulting binomial.

$$= 3x^2(x + 2) + 2(x + 2)$$

STEP 3. Factor out the GCF, $(x + 2)$, using the distributive law.

$$= (x + 2)(3x^2 + 2)$$

Note that if you write $3x^3 + 2x + 6x^2 + 4$ in Step 1, your answer would be $(3x^2 + 2)(x + 2)$. Since $(3x^2 + 2)(x + 2) = (x + 2)(3x^2 + 2)$, both answers are correct. We will factor our problems by first writing the polynomials in *descending* order.

Problem 4 Factor.
a. $2x^3 + 2x^2 + 3x + 3$
b. $6x^3 - 9x^2 - 2x + 3$

b. STEP 1. Group terms with common factors using the associative law.

$$6x^3 - 3x^2 - 4x + 2 = (6x^3 - 3x^2) - (4x - 2)$$

Note that
$-(4x - 2) = -4x + 2.$

STEP 2. Factor each resulting binomial.

$$= 3x^2(2x - 1) - 2(2x - 1)$$

STEP 3. Factor out the GCF, **$(2x - 1)$.**

$$= (2x - 1)(3x^2 - 2)$$

Thus, $6x^3 - 3x^2 - 4x + 2 = (2x - 1)(3x^2 - 2)$. Note that $2x - 1$ and $3x^2 - 2$ *cannot* be factored any further, so the polynomial is *completely* factored.

EXAMPLE 5 Factor.

a. $2x^3 - 4x^2 - x + 2$ **b.** $6x^4 - 9x^2 + 4x^2 - 6$

Solution

a. STEP 1. $2x^3 - 4x^2 - x + 2 = (2x^3 - 4x^2) - (x - 2)$

STEP 2. $\qquad\qquad = 2x^2(x - 2) - 1(x - 2)$

STEP 3. $\qquad\qquad = (x - 2)(2x^2 - 1)$

b. STEP 1. $6x^4 - 9x^2 + 4x^2 - 6 = (6x^4 - 9x^2) + (4x^2 - 6)$

STEP 2. $\qquad\qquad\quad = 3x^2(2x^2 - 3) + 2(2x^2 - 3)$

STEP 3. $\qquad\qquad\quad = (2x^2 - 3)(3x^2 + 2)$

Problem 5 Factor.
a. $3x^3 - 6x^2 - x + 2$
b. $6x^4 - 9x^2 + 2x^2 - 3$

4. a. $(x + 1)(2x^2 + 3)$ **b.** $(2x - 3)(3x^2 - 1)$
5. a. $(x - 2)(3x^2 - 1)$ **b.** $(2x^2 - 3)(3x^2 + 1)$

_____
_____
_____
_____
_____

NAME

CLASS

SECTION

ANSWERS

A. Factor.

1. $3x + 15$

2. $5x + 45$

3. $9y - 18$

4. $11y - 33$

5. $-5y + 20$

6. $-4y + 28$

7. $-3x - 27$

8. $-6x - 36$

9. $4x^2 + 32x$

10. $5x^3 + 20x$

11. $6x - 42x^2$

12. $7x - 14x^3$

13. $-5x^2 - 25x^4$

14. $-3x^3 - 18x^6$

15. $3x^3 + 6x^2 + 9x$

16. $8x^3 + 4x^2 - 16x$

17. $9y^3 - 18y^2 + 27y$

18. $10y^3 - 5y^2 + 10y$

19. $6x^6 + 12x^5 - 18x^4 + 30x^2$

20. $5x^7 - 15x^6 + 10x^3 - 20x^2$

21. $8y^8 + 16y^5 - 24y^4 + 8y^3$

22. $12y^9 - 4y^6 + 6y^5 + 8y^4$

23. $\dfrac{4}{7}x^3 + \dfrac{3}{7}x^2 - \dfrac{9}{7}x + \dfrac{3}{7}$

24. $\dfrac{2}{5}x^3 + \dfrac{3}{5}x^2 - \dfrac{2}{5}x + \dfrac{4}{5}$

25. $\dfrac{7}{8}y^9 + \dfrac{3}{8}y^6 - \dfrac{5}{8}y^4 + \dfrac{5}{8}y^2$

26. $\dfrac{4}{3}y^7 - \dfrac{1}{3}y^5 + \dfrac{2}{3}y^4 - \dfrac{5}{3}y^3$

B. Factor by grouping.

27. $x^3 + 2x^2 + x + 2$

28. $x^3 + 3x^2 + x + 3$

1. _____
2. _____
3. _____
4. _____
5. _____
6. _____
7. _____
8. _____
9. _____
10. _____
11. _____
12. _____
13. _____
14. _____
15. _____
16. _____
17. _____
18. _____
19. _____
20. _____
21. _____
22. _____
23. _____
24. _____
25. _____
26. _____
27. _____
28. _____

29. $y^3 - 3y^2 + y - 3$ **30.** $y^3 - 5y^2 + y - 5$

31. $4x^3 + 6x^2 + 2x + 3$ **32.** $6x^3 + 3x^2 + 2x + 1$

33. $6x^3 - 2x^2 + 3x - 1$ **34.** $6x^3 - 9x^2 + 2x - 3$

35. $4y^3 + 8y^2 + y + 2$ **36.** $2y^3 - 6y^2 - y + 3$

37. $2a^3 + 3a^2 + 2a + 3$ **38.** $3a^3 + 2a^2 + 3a + 2$

39. $3x^4 + 12x^2 + x^2 + 4$ **40.** $2x^4 + 2x^2 + x^2 + 1$

41. $6y^4 + 9y^2 + 2y^2 + 3$ **42.** $12y^4 + 8y^2 + 3y^2 + 2$

43. $4y^4 + 12y^2 + y^2 + 3$ **44.** $2y^4 + 2y^2 + y^2 + 1$

45. $3a^4 - 6a^2 - 2a^2 + 4$ **46.** $4a^4 - 12a^2 - 3a^2 + 9$

29. _____
30. _____
31. _____
32. _____
33. _____
34. _____
35. _____
36. _____
37. _____
38. _____
39. _____
40. _____
41. _____
42. _____
43. _____
44. _____
45. _____
46. _____

C. Applications

47. Factor $\alpha L t_2 - \alpha L t_1$, where α is the coefficient of linear expansion, L the length of the material, and t_2 and t_1 the temperature in degrees Celsius.

47. _____

48. Factor the expression $-kx - kl$ (k is a constant), which represents the restoring force of a spring stretched an amount l to its equilibrium position and then an additional x units.

48. _____

49. When solving for the equivalent resistance of two circuits we have to factor the expression $R^2 - R - R + 1$. Factor this expression by grouping.

49. _____

50. The bending moment of a cantilever beam of length L, at x inches from its support, involves the expression $L^2 - Lx - Lx + x^2$. Factor this expression by grouping.

50. _____

✓ **SKILL CHECKER**

Find.

51. $(x + 5)(x + 3)$ **52.** $(x - 5)(x + 2)$

53. $(2x - 1)(x - 4)$ **54.** $(5x + 2)(x + 1)$

55. $(2x + 1)(2x - 3)$

51. _____
52. _____
53. _____
54. _____
55. _____

5.1 USING YOUR KNOWLEDGE

There are many formulas that can be simplified by factoring. Here are a few.

1. The vertical shear at any section of a cantilever beam of uniform cross section is

    $-w\ell + wz$

    Factor this expression.

2. The bending moment of any section of a cantilever beam of uniform cross section is

    $-P\ell + Px$

    Factor this expression.

3. The surface area of a square pyramid is

    $a^2 + 2as$

    Factor this expression.

4. The energy of a moving object is given by

    $900\,m - mv^2$

    Factor this expression.

5. The height of a rock thrown from the roof of a certain building is given by

    $-16t^2 + 80t + 240$

    Factor this expression. (*Hint:* -16 is a common factor.)

1. _____

2. _____

3. _____

4. _____

5. _____

The factoring techniques studied in this section can be used to evaluate higher-degree polynomials. Moreover, the evaluation can be done with a calculator performing only the four basic operations of addition, subtraction, multiplication, and division. We illustrate the five-step procedure next for the polynomial $2a^3 + 3a^2 + 4a + 5$

STEP 1. Given. $\qquad\qquad\qquad\qquad 2a^3 + 3a^2 + 4a + 5$

STEP 2. Group the terms involving a, and factor out $\boldsymbol{a}$. $\qquad (2a^2 + 3a + 4)a + 5$

STEP 3. Repeat this process for the expression in parentheses. $\qquad [(2a + 3)a + 4]a + 5$

STEP 4. Repeat the process within the innermost grouping symbol. $\qquad \{[(2)a + 3]a + 4\}a + 5$

STEP 5. You can stop when the innermost expression (the 2) is a constant.

The keystroke sequence to evaluate this polynomial for any number $\boldsymbol{a}$ would then be:

$\boxed{2}\;\boxed{\times}\;\boxed{a}\;\boxed{+}\;\boxed{3}\;\boxed{=}\;\boxed{\times}\;\boxed{a}\;\boxed{+}\;\boxed{4}\;\boxed{=}\;\boxed{\times}\;\boxed{a}\;\boxed{+}\;\boxed{5}\;\boxed{=}$

What is so great about that? Well, notice that after you enter the first 2, you simply repeat over and over the keystrokes

$\boxed{\times}\ \boxed{a}\ \boxed{+}\ \boxed{\text{a number}}\ \boxed{=}$

For example, when $a = 3$, we obtain

$\boxed{2}\ \boxed{\times}\ \boxed{3}\ \boxed{+}\ \boxed{3}\ \boxed{=}\ \boxed{\times}\ \boxed{3}\ \boxed{+}\ \boxed{4}\ \boxed{=}\ \boxed{\times}\ \boxed{3}\ \boxed{+}\ \boxed{5}\ \boxed{=}$

The result would be 90. In case some terms are "missing" in a given polynomial, they are inserted with a zero coefficient. Thus, to write $3a^3 + 5a + 6$ using this procedure, we first insert the "missing" a^2 term and write

$3a^3 + 0a^2 + 5a + 6 = \{[(3)a + 0]a + 5\}a + 6$

Use this procedure to evaluate the polynomials in Problems 15 through 18 for x (or y) equals 2.

4-B   THE BRANDON TRIBUNE, Wednesday, April 11

OBJECTIVES

REVIEW

Before starting this section you should know:

1. How to expand $(X + A)(X + B)$.
2. How to multiply integers.

OBJECTIVES

You should be able to:

A. Factor a trinomial of the form $x^2 + bx + c$.
B. Factor a trinomial of the form $ax^2 + bx + c$ using the ac method.
C. Factor a trinomial of the form $ax^2 + bx + c$ using trial and error.

Ralph Castellano

About Business

FREEZE DAMAGES ORANGE CROP PRICES TO RISE

Why do prices go up? One of the reasons is the supply of the product involved. The supply function for a certain product might be

$p^2 + 3p - 70$ Where p is the price of the product

The expression $p^2 + 3p - 70$ is factorable. How do we factor it? By using *reverse* multiplication.

A. FACTORING TRINOMIALS OF THE FORM $x^2 + bx + c$

Since factoring is the *reverse* of multiplying, we use our special products again. Of course, we rewrite them to suit our purposes, that is, so that we can use them for *factoring*. For example, the most important of the special products was given in Equation (1) (Section 4.4). We rewrite this product as

(F1)
$$X^2 + (A + B)X + AB = (X + A)(X + B)$$

Thus, to factor $x^2 + bx + c$, we need to find two binomials whose product is $x^2 + bx + c$. Now, suppose we wish to factor the polynomial

$x^2 + 8x + 15$

To do this, we could use (F1). Here is how the two compare:

$X^2 + (A + B)X + AB$
$x^2 + 8x + 15$

As you can see, 15 is used instead of AB and 8 instead of $A + B$, that is, we have two numbers A and B such that $AB = 15$ and $A + B = 8$. We write the possible factors and sums in the table.

FACTORS	SUM
15, 1	16
5, 3	8

The numbers should be $A = 5$ and $B = 3$. Now,

$$X^2 + \underbrace{(A + B)}X + \underbrace{AB} = (X + A)(X + B)$$

$$x^2 + \quad 8x \quad + 15 \; = \; (x + 5)(x + 3)$$

By the commutative law, we can also write

$$x^2 + 8x + 15 = (x + 3)(x + 5)$$

as our answer. Do you see what the idea is? To factor a trinomial of the form $x^2 + bx + c$, we must find two numbers A and B so that $A + B = b$ and $AB = c$. Then

$$x^2 + bx + c = x^2 + (A + B)x + AB = (x + A)(x + B)$$

Thus, to factor

$$x^2 - 3x - 10$$

we need two numbers whose product is -10 and whose sum is -3. Here is a table showing the possibilities.

FACTORS	SUM
$-10, 1$	-9
$10, -1$	9
$5, -2$	3
$-5, 2$	-3

The numbers are -5 and $+2$. Thus,

$$x^2 - 3x - 10 = (x - 5)(x + 2)$$

Note that the answer $(x + 2)(x - 5)$ is also correct, by the commutative law.

Similarly, to factor $x^2 + 5x - 14$ we need two numbers whose product is -14 and whose sum is 5. The numbers are -2 and $+7$ ($-2 \cdot 7 = -14$ and $-2 + 7 = 5$). Thus,

$$x^2 + 5x - 14 = (x - 2)(x + 7)$$

We can also write $x^2 + 5x - 14 = (x + 7)(x - 2)$.

EXAMPLE 1 Factor.

a. $x^2 + 5x + 6$ **b.** $x^2 - 6x + 5$ **c.** $p^2 + 3p - 70$

Solution

a. To factor $x^2 + 5x + 6$, we need two numbers with product 6 and sum 5. The numbers are 3 and 2; thus,

$$x^2 + 5x + 6 = (x + 3)(x + 2)$$

b. Here we need two numbers with product 5 and sum -6. In order to obtain the positive product 5, both numbers must be negative, so the desired numbers are -5 and -1. Hence

$$x^2 - 6x + 5 = (x - 5)(x - 1)$$

c. Now we need two numbers with product -70 and sum 3. Since -70 is negative, the numbers must have *opposite* signs. Moreover, since the sum of the two numbers is 3, the *larger* number must be *positive*. Here are the possibilities:

Problem 1 Factor.
a. $x^2 + 7x + 12$ **b.** $x^2 - 5x + 6$
c. $p^2 + 2p - 80$

FACTORS	SUM
70, −1	69
35, −2	33
14, −5	9
10, −7	3

The numbers we need are 10 and −7. Thus,

$$p^2 + 3p - 70 = (p + 10)(p - 7) \qquad \blacktriangle$$

Unfortunately, not all trinomials are of the form $x^2 + bx + c$, where the x^2 has a coefficient of 1 ($x^2 = 1x^2$). Some trinomials are of the form $ax^2 + bx + c$, with a as the coefficient of x^2. There is a way to find out if a trinomial of the form $ax^2 + bx + c$ is factorable. Here is the way it works:

---

ac TEST

$ax^2 + bx + c$ is factorable if there are two integers with product **ac** and sum **b.**

Here is a diagram to help.

---

ac TEST

We need two numbers whose product is ac:

$$\overset{\downarrow \qquad \downarrow}{ax^2 + bx + c}$$

The sum of the numbers must be b.

Naturally, we call this the **ac test,** where a and c are the first and last numbers in $ax^2 + bx + c$ and b is the coefficient of x. The diagram may help you visualize the procedure. Now, suppose you want to know if $6x^2 + 7x + 2$ is factorable. You can do it in three steps:

STEP 1. Multiply 6 by 2 ($6 \times 2 = 12$).

STEP 2. Find two integers whose product is 12 and whose sum is 7.

We need two numbers whose product is $6 \cdot 2 = 12$.

$$6x^2 + 7x + 2$$

The sum of the two numbers must be 7.

STEP 3. A little searching will produce 4 and 3.

CHECK. $\underbrace{4 \times 3}_{\text{Product}} = 12, \qquad \underbrace{4 + 3}_{\text{Sum}} = 7$

Thus, $6x^2 + 7x + 2$ is factorable.

On the other hand, consider the trinomial $2x^2 + 5x + 4$. Here is the ac test for this trinomial.

STEP 1. Multiply 2 by 4 ($2 \times 4 = 8$).

STEP 2. Find two integers whose product is 8 and whose sum is 5.

STEP 3. The factors of 8 are 4 and 2, and 8 and 1. Neither pair adds up to 5 ($4 + 2 = 6$, $8 + 1 = 9$). Thus the trinomial $2x^2 + 5x + 4$ is *not* factorable.

B. FACTORING TRINOMIALS OF THE FORM $ax^2 + bx + c$

At this point, you should be convinced that the ac test really tells you if a trinomial of the form $ax^2 + bx + c$ is factorable; however, we still do not *know* how to do the actual factorization. But we are in luck; the number ac still plays an important part in factoring the trinomial $ax^2 + bx + c$. In fact, the number ac is so important that we shall call it the **key number** in the factorization of $ax^2 + bx + c$. To get a little practice, we have found and circled the key numbers of a few trinomials.

	a	c	ac
$6x^2 + 8x + 5$	6	5	$\boxed{30}$
$2x^2 - 7x - 4$	2	-4	$\boxed{-8}$
$-3x^2 + 2x + 5$	-3	5	$\boxed{-15}$

As before, by examining these key numbers and the coefficient of the middle term, you should be able to determine if the trinomial is factorable. For example, the key number of the trinomial $6x^2 + 8x + 5$ is 30. But since there are *no* integers with sum 8 whose product is 30, this trinomial is *not* factorable. (The factors of 30 are 6 and 5, 10 and 3, 15 and 2, and 30 and 1. None of these pairs has a sum of 8.)

On the other hand, the key number for $2x^2 - 7x - 4$ is -8 and -8 has two factors (-8 and 1) whose product is -8 and whose sum is the coefficient of the middle term, that is, -7. Thus, $2x^2 - 7x - 4$ is factorable; we shall factor it now. Here are the steps.

1. Find the key number [$2 \cdot (-4) = -8$]. $2x^2 - 7x - 4$ $\boxed{-8}$

2. Find the factors of the key number and use the appropriate ones to rewrite the middle term.

$2x^2 \underline{- 8x + 1x} - 4$ $-8, 1$

$-7x = -8x + 1x$ $-8(1) = -8$

                             $-8 + 1 = -7$

3. Group the terms into pairs (as we did in Section 5.1). $(2x^2 - 8x) + (1x - 4)$

4. Factor each pair. $2x(x - 4) + 1(x - 4)$

5. Note that $(x - 4)$ is the greatest common factor. $(x - 4)(2x + 1)$

Thus, $2x^2 - 7x - 4 = (x - 4)(2x + 1)$. You should check that this is the correct factorization by multiplying $(x - 4)$ by $(2x + 1)$. Now, a word of warning. You can write the factorization of $ax^2 + bx + c$ in *two* ways. Suppose you wish to factor the trinomial $5x^2 + 7x + 2$. Here is one way of doing it.

1. Find the key number $[5 \cdot 2 = 10]$. $5x^2 + \underline{7x} + 2$ ⑩

2. Find the factors of the key number; use them to rewrite the middle term. $5x^2 + \underline{5x + 2x} + 2$ $5, 2$

3. Group the terms into pairs. $(5x^2 + 5x) + (2x + 2)$

4. Factor each pair. $5x(x + 1) + 2(x + 1)$

5. Note that $(x + 1)$ is the greatest common factor. $(x + 1)(5x + 2)$

Thus, $5x^2 + 7x + 2 = (x + 1)(5x + 2)$.
Another way of proceeding is as follows:

1. Find the key number $[5 \cdot 2 = 10]$. $5x^2 + \underline{7x} + 2$ ⑩

2. Find the factors of the key number and use them to rewrite the middle term. $5x^2 + \underline{2x + 5x} + 2$ $2, 5$

3. Group the terms into pairs. $(5x^2 + 2x) + (5x + 2)$

4. Factor each pair. $x(5x + 2) + 1(5x + 2)$

5. Note that $(5x + 2)$ is the greatest common factor. $(5x + 2)(x + 1)$

In this case, we found that

$$5x^2 + 7x + 2 = (5x + 2)(x + 1)$$

Is the correct factorization $(x + 1)(5x + 2)$ or $(5x + 2)(x + 1)$? The answer is that *both* factorizations are correct! This is because the multiplication of real numbers is commutative and the variable x, as well as the trinomials involved, also represent real numbers; thus the *order* in which the product is written, according to the commutative law of multiplication, *makes no difference in the final answer*.

EXAMPLE 2   Factor.

a. $6x^2 - 3x + 4$ **b.** $4x^2 - 3 - 4x$

Solution

a. We proceed by steps:

1. Find the key number $[6 \cdot 4 = 24]$. $6x^2 - 3x + 4$ ㉔

2. Find the factors of the key number and use them to rewrite the middle term. Unfortunately, it is impossible to find two numbers with product 24 and sum -3. This trinomial is *not* factorable.

b. We first rewrite the polynomial *(in descending order)* as $4x^2 - 4x - 3$, and then proceed by steps.

1. Find the key number $[4 \cdot (-3) = -12]$. $4x^2 - 4x - 3$ ⌊-12⌋

2. Find the factors of the key number and use them to rewrite the middle term. $4x^2 \underline{- 6x + 2x} - 3$ $-6, 2$

3. Group the terms into pairs. $(4x^2 - 6x) + (2x - 3)$

4. Factor each pair. $2x(2x - 3) + 1(2x - 3)$

5. Note that $(2x - 3)$ is the greatest common factor. $(2x - 3)(2x + 1)$

Thus, $4x^2 - 4x - 3 = (2x - 3)(2x + 1)$, as can easily be verified by multiplication.   ▲

Problem 2   Factor.
a. $5x^2 - 2x + 2$ **b.** $3x^2 - 4 - 4x$

Factoring problems in which the third term in Step 2 contains a negative number as a coefficient require special care with the signs. Thus to factor the trinomial $4x^2 - 5x + 1$, we proceed as usual.

1. Find the key number $[4 \cdot 1 = 4]$. $4x^2 - 5x + 1$ ④

2. Find the factors of the key number and use them to rewrite the middle term. $4x^2 \underline{- 4x - 1x} + 1$ $-4, -1$

   Note that the third term has a negative coefficient, -1.

3. Group the terms into pairs. $(4x^2 - 4x) + (-1x + 1)$

4. Factor each pair. $4x(x - 1) - 1(x - 1)$

   Recall that $-1(x - 1) = -x + 1$.

5. Note that $(x - 1)$ is the greatest common factor. $(x - 1)(4x - 1)$

   If the first pair has $(x - 1)$ as a factor, the second pair will also have $(x - 1)$ as a factor.

Thus, $4x^2 - 5x + 1 = (x - 1)(4x - 1)$.

EXAMPLE 3   Factor $5x^2 - 11x + 2$.

Solution

1. Find the key number $[5 \cdot 2 = 10]$. $5x^2 - 11x + 2$ ⑩

2. Find the factors of the key number and use them to rewrite the middle term. $5x^2 \underline{- 10x - 1x} + 2$ $-10, -1$

3. Group the terms into pairs. $(5x^2 - 10x) + (-1x + 2)$

4. Factor each pair. $5x(x - 2) - 1(x - 2)$

5. Note that $(x - 2)$ is the greatest common factor. $(x - 2)(5x - 1)$

Thus the factorization of $5x^2 - 11x + 2$ is $(x - 2)(5x - 1)$.   ▲

So far, we have factored trinomials in one variable only. A procedure similar to the one used for factoring a trinomial of the form $ax^2 + bx + c$ can be used to factor certain trinomials in two variables. We illustrate the procedure in the next example.

EXAMPLE 4   Factor $6x^2 - xy - 2y^2$.

Solution

1. Find the key number $[6 \cdot (-2) = -12]$. $6x^2 - xy - 2y^2$ ⑫ (−12)

2. Find the factors of the key number and use them to rewrite the middle term. $6x^2 \underline{- 4xy + 3xy} - 2y^2$ $-4, 3$

3. Group the terms into pairs. $(6x^2 - 4xy) + (3xy - 2y^2)$

4. Factor each pair. $2x(3x - 2y) + y(3x - 2y)$

5. Note that $(3x - 2y)$ is the greatest common factor. $(3x - 2y)(2x + y)$

Thus, $6x^2 - xy - 2y^2 = (3x - 2y)(2x + y)$.

Problem 3   Factor $2x^2 - 11x + 5$.

Problem 4   Factor $2x^2 - xy - 6y^2$.

ANSWER   (to problem on page 359)
2. **a.** Not factorable
   b. $(3x + 2)(x - 2)$ or $(x - 2)(3x + 2)$

C. FACTORING $ax^2 + bx + c$ BY TRIAL AND ERROR

Sometimes, it is easier to factor a polynomial of the form $ax^2 + bx + c$ by *trial and error*. This is especially so when the a or c is a prime number like 2, 3, 5, 7, 11, etc. or when ac is large. We do this by writing:

$$ax^2 + bx + c = (\underline{}x + \underline{})(\underline{}x + \underline{})$$

Product must be c

Product must be a

Now,

1. The product of the numbers in the *first* blanks must be a.
2. The coefficients of the *outside* products and the *inside* products must add up to b.
3. The products of the numbers in the *last* blanks must be c.

For example, to factor $2x^2 + 5x + 3$, we write:

$$2x^2 + 5x + 3 = (\underline{}x + \underline{})(\underline{}x + \underline{})$$

③

②

We now look for two numbers whose product is 2. These numbers are 2, 1 or -2, -1. We have these possibilities:

$$(2x + \underline{})(x + \underline{}) \quad \text{or} \quad (-2x + \underline{})(-x + \underline{})$$

Now, we look for numbers whose product is 3. These numbers are 3, 1 or -3, -1, which we substitute into the blanks, obtaining:

$(2x + 3)(x + 1)$	$(-2x + 3)(-x + 1)$
$(2x + 1)(x + 3)$	$(-2x + 1)(-x + 3)$
$(2x - 3)(x - 1)$	$(-2x - 3)(-x - 1)$
$(2x - 1)(x - 3)$	$(-2x - 1)(-x - 3)$

Since the final results must be $2x^2 + 5x + 3$, the two shaded expressions yield the desired factorization. We choose the one in which the *first* coefficients are *positive*. Thus,

$$2x^2 + 5x + 3 = (2x + 3)(x + 1)$$

EXAMPLE 5 Factor $3x^2 + 7x + 2$.

Solution Since we want the first coefficients in the factorization to be positive, the only two factors of 3 we consider are 3 and 1. We then have:

$$3x^2 + 7x + 2 = (3x + \underline{})(x + \underline{})$$

②

Problem 5 Factor $3x^2 + 5x + 2$.

We now need to fill in the blanks with factors whose product is 2. The factors are **2** and **1,** and the possibilities are:

$(3x + 2)(x + 1)$ or $(3x + 1)(x + 2)$

Since the second of these gives the desired result,

$3x^2 + 7x + 2 = (3x + 1)(x + 2)$ ▲

Note that the trial-and-error method is based on FOIL. Thus to multiply $(2x + 3)(3x + 4)$ using FOIL, we write

$$
\begin{array}{cccc}
 & \text{F} & \text{O} \quad \text{I} & \text{L} \\
(2x + 3)(3x + 4) = 6x^2 & + & 8x + 9x & + \ 12 \\
= 6x^2 & + & 17x & + \ 12 \\
 & \uparrow & \uparrow & \uparrow \\
 & \text{F} & \text{O} + \text{I} & \text{L} \\
 & 2 \cdot 3 & 2 \cdot 4 + 3 \cdot 3 & 3 \cdot 4
\end{array}
$$

Now, to factor $6x^2 + 17x + 12$, we do the reverse, using trial and error. Since the factors of 6 are 6, 1 or 3, 2 (we will not use -6, -1 and -3, -2 because then the first coefficients will be negative), the possible combinations are

$(6x + \underline{})(x + \underline{})$ $(3x + \underline{})(2x + \underline{})$

Since the product of the last two numbers is 12, the possible factors are **12, 1, 6, 2,** and **3, 4.** The possibilities are

$(6x + 12)(x + 1)^*$ $(6x + 1)(x + 12)$
$(6x + 6)(x + 2)^*$ $(6x + 2)(x + 6)^*$ Note that the
$(6x + 3)(x + 4)^*$ $(6x + 4)(x + 3)^*$ starred items
$(3x + 12)(2x + 1)^*$ $(3x + 1)(2x + 12)^*$ are not completely
$(3x + 6)(2x + 2)^*$ $(3x + 2)(2x + 6)^*$ factored.
$(3x + 3)(2x + 4)^*$ $(3x + 4)(2x + 3)$

Thus, $6x^2 + 17x + 12 = (3x + 4)(2x + 3)$.

Note that if there is a *common* factor, we must factor it out first. Thus, to factor $12x^2 + 2x - 2$, we must *first* factor out the common factor 2, as illustrated in Example 6.

EXAMPLE 6   Factor $12x^2 + 2x - 2$.

Solution   Since 2 is a common factor, we first factor it out, obtaining

$12x^2 + 2x - 6 = 2 \cdot 6x^2 + 2 \cdot x - 2 \cdot 1$
$= 2(6x^2 + x - 1)$

Now, we factor $6x^2 + x - 1$. The factors of 6 are **6, 1** or **3, 2.** Thus,

$6x^2 + x - 3 = (6x + \underline{})(x + \underline{})$

or

$6x^2 + x - 3 = (3x + \underline{})(2x + \underline{})$

The product of the last two terms must be -1. The possible factors are $-1, 1$. The possibilities are

$(6x - 1)(x + 1)$ $(6x + 1)(x - 1)$
$(3x - 1)(2x + 1)$ $(3x + 1)(2x - 1)$

Problem 6   Factor $16x^2 + 4x - 2$.

The only product that yields $6x^2 + x - 1$ is $(3x - 1)(2x + 1)$. Thus,

$$12x^2 + 2x - 6 = 2(6x^2 + x - 1)$$
$$= 2(3x - 1)(2x + 1)$$

Don't forget to write
the common factor.

A. Factor.

ANSWERS

1. $y^2 + 6y + 8$

2. $y^2 + 10y + 21$

3. $x^2 + 7x + 10$

4. $x^2 + 13x + 22$

5. $y^2 + 3y - 10$

6. $y^2 + 5y - 24$

7. $x^2 + 5x - 14$

8. $x^2 + 5x - 36$

9. $x^2 - 6x - 7$

10. $x^2 - 7x - 8$

11. $y^2 - 5y - 14$

12. $y^2 - 4y - 12$

13. $y^2 - 3y + 2$

14. $y^2 - 11y + 30$

15. $x^2 - 5x + 4$

16. $x^2 - 12x + 27$

1. _____

2. _____

3. _____

4. _____

5. _____

6. _____

7. _____

8. _____

9. _____

10. _____

11. _____

12. _____

13. _____

14. _____

15. _____

16. _____

B. Factor (if possible). Use trial and error if you wish.

17. $2x^2 + 5x + 3$

18. $2x^2 + 7x + 3$

19. $6x^2 + 11x + 3$

20. $6x^2 + 17x + 5$

21. $6x^2 + 11x + 4$

22. $5x^2 + 2x + 1$

23. $2x^2 + 3x - 2$

24. $2x^2 + x - 3$

25. $3x^2 + 16x - 12$

26. $6x^2 + x - 12$

27. $4y^2 - 11y + 6$

28. $3y^2 - 17y + 10$

17. _____

18. _____

19. _____

20. _____

21. _____

22. _____

23. _____

24. _____

25. _____

26. _____

27. _____

28. _____

29. $4y^2 - 8y + 6$

30. $3y^2 - 11y + 6$

29. _____

30. _____

31. $6y^2 - 10y - 4$

32. $12y^2 - 10y - 12$

31. _____

32. _____

33. $12y^2 - y - 6$

34. $3y^2 - y - 1$

33. _____

34. _____

35. $18y^2 - 21y - 9$

36. $36y^2 - 12y - 15$

35. _____

36. _____

37. $3x^2 + 2 + 7x$

38. $2x^2 + 2 + 5x$

37. _____

38. _____

39. $5x^2 + 2 + 11x$

40. $5x^2 + 3 + 12x$

39. _____

40. _____

41. $6x^2 - 5 + 15x$

42. $5x^2 - 8 + 6x$

41. _____

42. _____

43. $3x^2 - 2 - 5x$

44. $5x^2 - 8 - 6x$

43. _____

44. _____

45. $15x^2 - 2 + x$

46. $8x^2 + 15 - 14x$

45. _____

46. _____

47. $8x^2 + 20xy + 8y^2$

48. $12x^2 + 28xy + 8y^2$

47. _____

48. _____

49. $6x^2 + 7xy - 3y^2$

50. $3x^2 + 13xy - 10y^2$

49. _____

50. _____

51. $7x^2 - 10xy + 3y^2$

52. $6x^2 - 17xy + 5y^2$

51. _____

52. _____

53. $15x^2 - xy - 2y^2$

54. $5x^2 - 6xy - 8y^2$

53. _____

54. _____

55. $15x^2 - 2xy - 2y^2$

56. $4x^2 - 13xy - 3y^2$

55. _____

56. _____

C. Applications

57. To find the flow g (in hundreds of gallons per minute) in 100 ft of $2\frac{1}{2}$-in. rubber-lined hose when the friction loss is 36 lb/in.^{2}, we need to evaluate the expression

$$2g^2 + g - 36$$

Factor this expression.

57. _____

58. To find the flow g (in hundreds of gallons per minute) in 100 ft of $2\frac{1}{2}$-in. rubber-lined hose when the friction loss is 55 lb/in.^{2}, we must evaluate the expression

$2g^2 + g - 55$

Factor this expression.

58. _____

59. When solving for the equivalent resistance R of two electric circuits we find the expression

$2R^2 - 3R + 1$

Factor this expression.

59. _____

60. To find the time t at which an object thrown upward at 12 m/sec will be 4 meters above the ground, we must evaluate the expression

$5t^2 - 12t + 4$

Factor this expression.

60. _____

✓ **SKILL CHECKER**

Expand.

61. $(x + 4)^2$

62. $(x + 6)^2$

61. _____

62. _____

63. $(x - 3)^2$

64. $(x - 5)^2$

63. _____

64. _____

65. $(3x + 2y)^2$

66. $(4x + 3y)^2$

65. _____

66. _____

67. $(5x - 2y)^2$

68. $(3x - 4y)^2$

67. _____

68. _____

69. $(3x + 4y)(3x - 4y)$

70. $(5x + 2y)(5x - 2y)$

69. _____

70. _____

5.2   USING YOUR KNOWLEDGE

The ideas presented in this section are very important in many other fields. Use your knowledge to factor the following problems.

1. To find the deflection of a beam of length L at a distance of 3 ft from its end, we must evaluate the expression

$2L^2 - 9L + 9$

Factor this expression.

1. _____

2. In Exercise 1, if the distance from the end is x ft, then we must use the expression

$2L^2 - 3xL + x^2$

Factor this expression.

2. _____

3. The distance traveled in t seconds by an object thrown upward at 12 m/sec is

$$-5t^2 + 12t$$

To determine the time at which the object will be 7 m above ground, we must solve the equation

$$5t^2 - 12t + 7 = 0$$

Factor the trinomial on the left of this equation.

3. _____

5.3 FACTORING SQUARES OF BINOMIALS

OBJECTIVES

REVIEW

Before starting this section you should know:

1. How to expand a binomial sum or difference (Section 4.4).
2. How to find the product of the sum and difference of two terms.

OBJECTIVES

You should be able to:

A. Recognize the square of a binomial (a perfect square trinomial).
B. Factor the square of a binomial (a perfect square trinomial).
C. Factor the difference of two squares.

What is the bending moment on the above crane? At x feet from its support, the moment involves the expression

$$\frac{w}{2}\left(x^2 - 20x + 100\right)$$

where w is the weight of the crane in pounds per foot. The expression $x^2 - 20x + 100$ is the result of *squaring a binomial* and is called a **perfect square trinomial** because $x^2 - 20x + 100 = (x - 10)^2$. Similarly, $x^2 + 12x + 36 = (x + 6)^2$ is the square of a binomial. We can factor these two expressions by using the formulas studied in Section 4.4 in reverse. For example, $x^2 - 20x + 100$ has the same form as Equation (3), whereas $x^2 + 12x + 36$ looks like Equation (2). We rewrite these two equations here for your convenience.

(F2) $X^2 + 2AX + A^2 = (X + A)^2$ Note that $X^2 + A^2 \neq (X + A)^2$.
(F3) $X^2 - 2AX + A^2 = (X - A)^2$ Note that $X^2 - A^2 \neq (X - A)^2$.

A. SQUARES OF BINOMIALS

Note that to be the **square of a binomial** (a *perfect square trinomial*), three things must happen to the trinomial:

1. The first and last terms (X^2 and A^2) must be perfect squares.
2. There must be no minus signs before A^2 or X^2.
3. The middle term is twice the product of the square roots of the first and last terms, ($2AX$) or its additive inverse ($-2AX$). Note the X and A are the terms of the binomial being squared.

EXAMPLE 1 Determine if the given expression is the square of a binomial.

a. $x^2 + 8x + 16$ **b.** $x^2 + 6x - 9$
c. $x^2 + 4x + 16$ **d.** $4x^2 - 12xy + 9y^2$

Problem 1 Determine if the given expression is the square of a binomial.
a. $x^2 + 6x - 9$
b. $x^2 + 6x + 9$
c. $x^2 + 8x + 64$
d. $4x^2 - 20xy + 25y^2$

Solution

a. We check the three conditions necessary for having a perfect square trinomial.

  1. x^2 and $16 = 4^2$ are perfect squares.
  2. There are no minus signs before x^2 or 16.
  3. The middle term is twice the product of the square roots of the first and last terms; that is, the middle term is $2 \cdot (x \cdot 4) = 8x$. Thus, $x^2 + 8x + 16$ is a perfect square trinomial (the square of a binomial).

b. 1. x^2 and $9 = 3^2$ are perfect squares.
  2. However, there is a minus sign before the 9. Thus, $x^2 + 6x - 9$ is *not* the square of a binomial.

c. 1. x^2 and $16 = 42$ are perfect squares. The middle term should be $2 \cdot (x \cdot 4) = 8x$, but instead it is $4x$. Thus, $x^2 + 4x + 16$ is *not* a perfect square trinomial.

d. 1. $4x^2 = (2x)^2$ and $9y^2 = (3y)^2$ are perfect squares.
  2. There are no minus signs before $4x^2$ or $9y^2$.
  3. The middle term is the additive inverse of twice the product of the square roots of the first and last terms; that is, $-2 \cdot (2x \cdot 3y) = -12xy$. Thus, $4x^2 - 12xy + 9y^2$ *is* a perfect square trinomial.

B.   FACTORING SQUARES OF BINOMIALS

The formula given in Equation (F2) can be used to factor any trinomials that are perfect squares. For example, the trinomial $9x^2 + 12x + 4$ can be factored using Equation (F2) if we first notice that

1. $9x^2$ and 4 are perfect squares, since $9x^2 = (3x)^2$ and $4 = 2^2$.
2. There are no minus signs before $9x^2$ or 4.
3. $12x = 2 \cdot (2 \cdot 3x)$. (Twice the product of 2 and $3x$, the square roots of the first and last terms.

We then write

$$9x^2 + 12x + 4 = (3x)^2 + 2 \cdot (2 \cdot 3x) + 2^2 \quad \text{We are letting } X = 3x,$$
$$= (3x + 2)^2 \qquad\qquad A = 2 \text{ in (F2)}$$

Here are some other examples of this form:

$$9x^2 + 6x + 1 = (3x)^2 + 2 \cdot (1 \cdot 3x) + 1^2 = (3x + 1)^2$$
$$\text{Letting } X = 3x, A = 1 \text{ in (F2)}$$
$$16x^2 + 24x + 9 = (4x)^2 + 2 \cdot (3 \cdot 4x) + 3^2 = (4x + 3)^2$$
$$\text{Here } X = 4x, A = 3.$$
$$4x^2 + 12xy + 9y^2 = (2x)^2 + 2 \cdot (3y \cdot 2x) + (3y)^2 = (2x + 3y)^2$$
$$\text{Here } X = 2x, A = 3y.$$

Study these examples carefully before going on. Note that the key for factoring these triomials is to recognize that the first and last terms are *perfect squares*. (Of course, you have to check the middle term too.)

EXAMPLE 2 Factor.
a. $x^2 + 16x + 64$ b. $25x^2 + 20x + 4$ c. $9x^2 + 12xy + 4y^2$

Solution

a. We first write the trinomial in the form $X^2 + 2AX + A^2$. Thus

$$x^2 + 16x + 64 = x^2 + 2 \cdot (8 \cdot x) + 8^2 = (x + 8)^2$$

b. $25x^2 + 20x + 4 = (5x)^2 + 2 \cdot (2 \cdot 5x) + 2^2 = (5x + 2)^2$
c. $9x^2 + 12xy + 4y^2 = (3x)^2 + 2 \cdot (2y \cdot 3x) + (2y)^2 = (3x + 2y)^2$ ▲

A similar technique can be used to factor $x^2 - 16x + 64$ or $25x^2 - 20x + 4$. All we need is the equivalent of Equation (3) in Section 4.4:

(F3) $\boxed{X^2 - 2AX + A^2 = (X - A)^2}$ Perfect square trinomial

Since

$$x^2 - 16x + 64 = x^2 - 2 \cdot (8 \cdot x) + 8^2$$
$$x^2 - 16x + 64 = (x - 8)^2$$

Similarly, since

$$25x^2 - 20x + 4 = (5x)^2 - 2 \cdot (2 \cdot 5x) + 2^2$$
$$25x^2 - 20x + 4 = (5x - 2)^2$$

EXAMPLE 3 Factor.

a. $x^2 - 10x + 25$ b. $4x^2 - 12x + 9$ c. $4x^2 - 20xy + 25y^2$

Solution

a. $x^2 - 10x + 25 = x^2 - 2 \cdot (5 \cdot x) + 5^2 = (x - 5)^2$
b. $4x^2 - 12x + 9 = (2x)^2 - 2 \cdot (3 \cdot 2x) + 3^2 = (2x - 3)^2$
c. $4x^2 - 20xy + 25y^2 = (2x)^2 - 2 \cdot (5y \cdot 2x) + (5y)^2 = (2x - 5y)^2$

C. FACTORING THE DIFFERENCE OF TWO SQUARES

Can we factor $x^2 - 9$ as a product of two binomials? Note that $x^2 - 9$ has no middle term. The only special product with no middle term we have studied is the product of the sum and difference of two terms. Here is the corresponding factoring formula:

(F4) $\boxed{X^2 - A^2 = (X + A)(X - A)}$

Problem 2 Factor.
a. $x^2 + 4x + 4$
b. $9x^2 + 30x + 25$
c. $16x^2 + 24xy + 9y^2$

Problem 3 Factor.
a. $x^2 - 6x + 9$
b. $9x^2 - 12x + 4$
c. $9x^2 - 24xy + 16y^2$

We can now factor binomials of the form $x^2 - 16$ and $9x^2 - 25y^2$. To do this, we proceed as follows:

$$x^2 - 16 = (x)^2 - (4)^2 = (x + 4)(x - 4)$$

and

$$9x^2 - 25y^2 = (3x)^2 - (5y)^2 = (3x + 5y)(3x - 5y)$$

Check your answers by using FOIL.

Note that $x^2 + A^2$ *cannot* be factored!

EXAMPLE 4 Factor.

a. $x^2 - 4$ **b.** $25x^2 - 9$ **c.** $16x^2 - 9y^2$ **d.** $x^4 - 16$

Solution

a. $x^2 - 4 = (x)^2 - (2)^2 = (x + 2)(x - 2)$
b. $25x^2 - 9 = (5x)^2 - (3)^2 = (5x + 3)(5x - 3)$
c. $16x^2 - 9y^2 = (4x)^2 - (3y)^2 = (4x + 3y)(4x - 3y)$
d. $x^4 - 16 = (x^2)^2 - (4)^2 = (x^2 + 4)(x^2 - 4)$
    But $(x^2 - 4)$ is factorable, so $(x^2 + 4)(x^2 - 4) = (x^2 + 4)(x + 2)(x - 2)$.
    Thus

$$x^4 - 16 = \underbrace{(x^2 + 4)}(x + 2)(x - 2)$$
$$\uparrow$$
Not factorable

Problem 4 Factor.
a. $x^2 - 1$ **b.** $9x^2 - 16$
c. $9x^2 - 25y^2$ **d.** $x^4 - 81$

EXERCISE 5.3

NAME

CLASS

SECTION

ANSWERS

A. Determine if the given expression is a perfect square trinomial (the square of a binomial).

1. $x^2 + 14x + 49$ **2.** $x^2 + 18x + 81$

3. $25x^2 + 10x - 1$ **4.** $9x^2 + 12x - 4$

5. $25x^2 + 10x + 1$ **6.** $9x^2 + 12x + 4$

7. $y^2 - 4y - 4$ **8.** $y^2 - 20y - 100$

9. $16y^2 - 40yz + 25z^2$ **10.** $49y^2 - 56yz + 16z^2$

B. Factor completely.

11. $x^2 + 2x + 1$ **12.** $x^2 + 6x + 9$

13. $3x^2 + 30x + 75$ **14.** $2x^2 + 28x + 98$

15. $9x^2 + 6x + 1$ **16.** $16x^2 + 8x + 1$

17. $9x^2 + 12x + 4$ **18.** $25x^2 + 10x + 1$

19. $16x^2 + 40xy + 25y^2$ **20.** $9x^2 + 30xy + 25y^2$

21. $25x^2 + 20xy + 4y^2$ **22.** $36x^2 + 60xy + 25y^2$

23. $y^2 - 2y + 1$ **24.** $y^2 - 4y + 4$

25. $3y^2 - 24y + 48$ **26.** $2y^2 - 40y + 200$

27. $9x^2 - 6x + 1$ **28.** $4x^2 - 20x + 25$

Answers list: 1–28.

29. $16x^2 - 56x + 49$ **30.** $25x^2 - 30x + 9$

31. $9x^2 - 12xy + 4y^2$ **32.** $16x^2 - 40xy + 25y^2$

33. $25x^2 - 10xy + y^2$ **34.** $49x^2 - 56xy + 16y^2$

C. Factor completely.

35. $x^2 - 49$ **36.** $x^2 - 121$

37. $9x^2 - 49$ **38.** $16x^2 - 81$

39. $25x^2 - 81y^2$ **40.** $81x^2 - 25y^2$

41. $x^4 - 1$ **42.** $x^4 - 256$

43. $16x^4 - 1$ **44.** $16x^4 - 81$

✓ SKILL CHECKER

Find.

45. $(R + r)(R - r)$ **46.** $(P + q)(P - q)$

Factor.

47. $6x^2 - 18x - 24$

48. $4x^4 + 12x^3 + 40x^2$

49. $2x^2 - 18$

50. $3x^2 - 27$

5.3 USING YOUR KNOWLEDGE

Many of the ideas in business are made precise by using expressions called **functions**. These expressions are often given in factored form. Use your knowledge to factor the given expressions (functions).

1. When x units of an item are demanded by consumers, the price per unit is given by the **demand** function $D(x)$ (read "D of x")

$$D(x) = x^2 - 14x + 49$$

Factor this expression.

29. _____
30. _____
31. _____
32. _____
33. _____
34. _____

35. _____
36. _____
37. _____
38. _____
39. _____
40. _____
41. _____
42. _____
43. _____
44. _____

45. _____
46. _____

47. _____
48. _____
49. _____
50. _____

1. _____

ANSWER (to problem on page 372)
4. a. $(x + 1)(x - 1)$ **b.** $(3x + 4)(3x - 4)$
  c. $(3x + 5y)(3x - 5y)$
  d. $(x^2 + 9)(x + 3)(x - 3)$

2. When x units are supplied by sellers, the price per unit of an item is given by the **supply** function $S(x)$

$S(x) = x^2 + 4x + 4$

Factor this expression.

2. _____

3. When x units are produced, the cost function $C(x)$ for a certain item is given by

$C(x) = x^2 + 12x + 36$

Factor this expression.

3. _____

4. When the market price is p dollars, the supply function $S(p)$ for a certain commodity is given by

$S(p) = p^2 - 6p + 9$

Factor this expression.

4. _____

Courtesy Bill Davis and Eldra Solomon.

OBJECTIVES

REVIEW

Before starting this section you should know how to factor polynomials using Equations F1–F4, pages 355, 369, 371.

OBJECTIVES

You should be able to:

A. Factor a given polynomial completely by using the general strategy given.

B. Factor expressions whose leading coefficient is preceded by a minus $(-)$ sign.

The photo shows the cross section of an artery. The speed (in centimeters per second) of the blood is given by

$$CR^2 - Cr^2$$

To factor this expression we proceed as follows:

STEP 1. Factor out any common factors (in this case, C).

$$CR^2 - Cr^2 = C(R^2 - r^2)$$

STEP 2. Look at the terms inside the parentheses. We have the difference of two terms in the expression: $R^2 - r^2$, that is the difference of two squares; we factor it.

$$= C(R + r)(R - r)$$

STEP 3. Make sure the expression is *completely* factored.

Note that in factoring this expression, we use two of the techniques we learned in the previous sections. We now give you a general strategy involving all the techniques we have studied.

A. A GENERAL FACTORING STRATEGY

To make this procedure easier for you, we write the steps in our general factoring strategy. Remember that when we say *factor,* we mean *completely* factor.

1. Factor out all common factors (the GCF).

2. Look at the number of terms inside the parentheses (or in the original polynomial):

   Four terms: Factor by grouping.

   Three terms: See if the expression is a perfect square trinomial; if so, factor it. Otherwise, use the *ac* method to factor.

   Two terms: Make sure you have the difference of two squares $(X^2 - A^2)$. Note that $X^2 + A^2$ is *not* factorable.

EXAMPLE 1   Factor.

a. $6x^2 - 18x - 24$ **b.** $4x^4 - 12x^3 + 40x^2$

Solution

a. We follow the three steps in our general factoring strategy.

   1. $6x^2 - 18x - 24 = 6(x^2 - 3x - 4)$ Factoring out the GCF
   2. $x^2 - 3x - 4$ has three terms and it is factored by finding two numbers whose product is -4, and whose sum is -3. These numbers are 1 and -4. Thus,

   $x^2 - 3x - 4 = (x + 1)(x - 4)$

   We then have

   $6x^2 - 18x - 24 = 6(x + 1)(x - 4)$

   3. This expression cannot be factored any further.

b. 1. Here the greatest common factor is $4x^2$. Thus,

   $4x^4 + 12x^3 + 40x^2 = 4x^2(x^2 + 3x + 10)$

   2. The trinomial $x^2 + 3x + 10$ is *not* factorable, since there are no numbers whose product is 10 with a sum of 3.
   3. The complete factorization is simply

   $4x^4 + 12x^3 + 40x^2 = 4x^2(x^2 + 3x + 10)$

Problem 1   Factor.
a. $8x^2 - 16x - 24$
b. $5x^4 - 10x^3 + 20x^2$

EXAMPLE 2   Factor $3x^3 + 9x^2 + x + 3$.

Solution

1. There are no common monomial factors.
2. Since the expression has four terms, we factor by grouping:

$$3x^3 + 9x^2 + x + 3 = (3x^3 + 9x^2) + (x + 3)$$
$$= 3x^2(x + 3) + 1 \cdot (x + 3)$$
$$= (x + 3)(3x^2 + 1)$$

3. This result cannot be factored any further, so the factorization is complete.

Problem 2   Factor $3x^3 + 12x^2 + x + 4$.

EXAMPLE 3 The heat output from a natural draught convector is $kt_n^2 - 2kt_nt_a + kt_a^2$. ($t_n^2$ is read as "t sub n squared." The n is called a **subscript**.) Factor this expression.

Problem 3 Factor $kt_n^2 + 2kt_nt_a + kt_a^2$.

Solution As usual, we proceed by steps.

1. The greatest common factor is k. Hence,

$$kt_n^2 - 2kt_nt_a + kt_a^2 = k(t_n^2 - 2t_nt_a + t_a^2)$$

2. $t_n^2 - 2t_nt_a + t_a^2$ is a trinomial square, which factors into $(t_n - t_a)^2$. Thus,

$$kt_n^2 - 2kt_nt_a + kt_a^2 = k(t_n - t_a)^2$$

3. This expression cannot be factored further.

EXAMPLE 4 Factor $D^4 - d^4$.

Problem 4 Factor $x^4 - 81$.

Solution

1. There are no common monomial factors.
2. The expression has *two* terms separated by a minus sign. It is the difference of *two* squares. Thus,

$$D^4 - d^4 = (D^2)^2 - (d^2)^2$$
$$= (D^2 + d^2)(D^2 - d^2)$$

3. Note that the expression $D^2 - d^2$ is also the difference of two squares, which can be factored into $(D + d)(D - d)$. Thus,

$$D^4 - d^4 = (D^2 + d^2)\underbrace{(D^2 - d^2)}$$
$$= (D^2 + d^2)\underbrace{(D + d)(D - d)}$$

Note that $D^2 + d^2$, which is the *sum* of two squares, *cannot* be factored.

B. USING -1 AS A FACTOR

In the preceding problems, we have *not* factored expressions in which the leading coefficient is preceded by a minus $(-)$ sign. Can some of these expressions be factored? The answer is yes, but we must first factor -1 from each term. Thus, to factor $-x^2 + 6x - 9$, we first write

$$-x^2 + 6x - 9 = -1 \cdot (x^2 - 6x + 9)$$
$$= -1 \cdot (x - 3)^2 \qquad \text{Note that } x^2 - 6x + 9 = (x - 3)^2$$
$$= -(x - 3)^2 \qquad \text{Since } -1 \cdot a = -a,$$
$$-1 \cdot (x - 3)^2 = -(x - 3)^2$$

EXAMPLE 5 Factor, if possible.

a. $-x^2 - 8x - 16$ **b.** $-4x^2 + 12xy - 9y^2$
c. $-9x^2 - 12xy + 4y^2$ **d.** $-4x^4 + 25x^2$

Problem 5 Factor, if possible.
a. $-x^2 - 10x - 25$
b. $-9x^2 - 30xy + 25y^2$
c. $-9x^2 + 30xy - 25y^2$
d. $-9x^4 + 4x^2$

Solution

a. We first factor out -1, obtaining

$$-x^2 - 8x - 16 = -1 \cdot (x^2 + 8x + 16)$$
$$= -1 \cdot (x + 4)^2 \qquad x^2 + 8x + 16 = (x + 4)^2$$
$$= -(x + 4)^2$$

b. $-4x^2 + 12xy - 9y^2 = -1 \cdot (4x^2 - 12xy + 9y^2)$

$\qquad\qquad\qquad\quad = -1 \cdot (2x - 3y)^2 \qquad 4x^2 - 12xy + 9y^2 = (2x - 3y)^2$

$\qquad\qquad\qquad\quad = -(2x - 3y)^2$

c. $-9x^2 - 12xy + 4y^2 = -1 \cdot (9x^2 + 12xy - 4y^2)$

But $9x^2 + 12xy - 4y^2$ has a minus $(-)$ before the last term $4y^2$ and is *not* a perfect square trinomial; thus, $-9x^2 - 12xy + 4y^2$ is not factorable.

d. $-4x^4 + 25x^2 = -x^2 \cdot (4x^2 - 25)$ $\qquad$ The GCF is $-x^2$.

$\qquad\qquad\quad = -x^2 \cdot (2x + 5)(2x - 5) \quad 4x^2 - 25 = (2x + 5)(2x - y)$

$\qquad\qquad\quad = -x^2(2x + 5)(2x - 5)$

NAME

CLASS

SECTION

A. Factor.

1. $3x^2 - 3x - 18$

2. $4x^2 - 12x - 16$

3. $5x^2 + 10x - 40$

4. $6x^2 + 18x - 60$

5. $3x^3 + 6x^2 + 21x$

6. $6x^3 + 18x^2 + 12x$

7. $2x^4 - 4x^3 - 10x^2$

8. $3x^4 - 12x^3 - 9x^2$

9. $4x^4 + 12x^3 + 18x^2$

10. $5x^4 + 25x^3 + 30x^2$

11. $3x^3 + 6x^2 + x + 2$

12. $2x^3 + 8x^2 + x + 4$

13. $3x^3 + 3x^2 + 2x + 2$

14. $4x^3 + 8x^2 + 3x + 6$

15. $2x^3 + 2x^2 - x - 1$

16. $3x^3 + 6x^2 - x - 2$

17. $3x^2 + 24x + 48$

18. $2x^2 + 12x + 18$

19. $kx^2 + 4kx + 4k$

20. $kx^2 + 10kx + 25k$

21. $4x^2 - 24x + 36$

22. $5x^2 - 20x + 20$

23. $kx^2 - 12kx + 36$

24. $kx^2 - 10kx + 25k$

25. $3x^3 + 12x^2 + 12x$

26. $2x^3 + 16x^2 + 32x$

27. $18x^3 + 12x^2 + 2x$

28. $12x^3 + 12x^2 + 3x$

29. $12x^4 - 36x^3 + 27x^2$

30. $18x^4 - 24x^3 + 8x^2$

ANSWERS

1. _____
2. _____
3. _____
4. _____
5. _____
6. _____
7. _____
8. _____
9. _____
10. _____
11. _____
12. _____
13. _____
14. _____
15. _____
16. _____
17. _____
18. _____
19. _____
20. _____
21. _____
22. _____
23. _____
24. _____
25. _____
26. _____
27. _____
28. _____
29. _____
30. _____

31. $x^4 - 1$

32. $x^4 - 16$

33. $x^4 - y^4$

34. $x^4 - z^4$

35. $x^4 - 16y^4$

36. $x^4 - 81y^4$

37. $-x^2 - 6x - 9$

38. $-x^2 - 10x - 25$

39. $-x^2 - 4x - 4$

40. $-x^2 - 6x - 9$

41. $-4x^2 - 4xy - y^2$

42. $-9x^2 - 6xy - y^2$

43. $-9x^2 - 12xy - 4y^2$

44. $-4x^2 - 12xy - 9y^2$

45. $-4x^2 + 12xy - 9y^2$

46. $-9x^2 + 12xy - 4y^2$

47. $-18x^3 - 24x^2y - 8xy^2$

48. $-12x^3 - 36x^2y - 27xy^2$

49. $-18x^3 - 60x^2y - 50xy^2$

50. $-12x^3 - 60x^2y - 75xy^2$

51. $-x^3 + x$

52. $-x^3 + 9x$

53. $-x^4 + 4x^2$

54. $-x^4 + 16x^2$

55. $-4x^4 + 9x^2$

56. $-9x^4 + 4x^2$

57. $-8x^3 + 18x$

58. $-12x^3 + 3x$

59. $-18x^4 + 8x^2$

60. $-12x^4 + 27x^2$

✓ **SKILL CHECKER**

Use the *ac* method to factor (if possible).

61. $10x^2 + 13x - 3$

62. $3x^2 - 5x - 1$

63. $2x^2 - 5x - 3$

64. $2x^2 - 3x - 2$

NAME

31. _____
32. _____
33. _____
34. _____
35. _____
36. _____
37. _____
38. _____
39. _____
40. _____
41. _____
42. _____
43. _____
44. _____
45. _____
46. _____
47. _____
48. _____
49. _____
50. _____
51. _____
52. _____
53. _____
54. _____
55. _____
56. _____
57. _____
58. _____
59. _____
60. _____

61. _____
62. _____
63. _____
64. _____

5.4 USING YOUR KNOWLEDGE

Many of the ideas presented in this section are used by engineers and technicians. Use your knowledge to solve the following problems.

1. The bend allowance needed to bend a piece of metal of thickness t through an angle A when the inside radius of the bend is IR is given by the expression

    $$\frac{2\pi A}{360} IR + \frac{2\pi A}{360} Kt \qquad \text{K is a constant}$$

    Factor this expression.

2. The change in kinetic energy of a moving object of mass m with initial velocity v_1 and terminal velocity v_2 is given by

    $$\frac{1}{2}mv_1^2 - \frac{1}{2}mv_2^2$$

    Factor this expression.

3. The parabolic distribution of shear stress on the cross section of a certain beam is given by

    $$\frac{3Sd^2}{2bd^3} - \frac{12Sz^2}{2bd^3}$$

    Factor this expression.

4. The polar moment of inertia J of a hollow round shaft of inner diameter d_1 and outer diameter d is given by

    $$\frac{\pi d^4}{32} - \frac{\pi d_1^4}{32}$$

    Factor this expression.

1. _____

2. _____

3. _____

4. _____

5.5 SOLVING QUADRATIC EQUATIONS BY FACTORING

OBJECTIVES

REVIEW

Before starting this section, you should know:

1. How to factor an expression of the form $ax^2 + bx + c$.
2. How to solve a linear equation of the form $x + a = b$.

OBJECTIVE

A. You should be able to solve quadratic equations by factoring.

The girl has thrown the ball with an initial velocity of 4 m/sec. Since she released the ball 1 m above the ground, the *height* of the ball after t seconds is

$$-5t^2 + 4t + 1$$

To find out how long it takes the ball to hit the ground, we must set this expression equal to 0. The reason for this is that when the ball *is* on the ground, the height is 0. Thus, we write

$$-5t^2 + 4t + 1 = 0 \quad \text{Remember that the girl's hand was 1 m above the ground.}$$

This is the height after t seconds

This is the height when the ball hits the ground

$-5t^2 + 4t + 1 = 0$ is a *quadratic equation,* an equation in which the greatest exponent of the variable is 2. We will learn how to solve these equations next.

A. SOLVING QUADRATIC EQUATIONS BY FACTORING

How can we solve this equation? For starters, we make the leading coefficient positive by multiplying each side of the equation by -1, obtaining

$$5t^2 - 4t - 1 = 0 \quad -1(-5t^2 + 4t + 1) = 5t^2 - 4t - 1 \text{ and } -1 \cdot 0 = 0$$

Since the *ac* number is -5,

$$5t^2 - 5t + 1t - 1 = 0$$
$$5t(t - 1) + 1(t - 1) = 0$$
$$(t - 1)(5t + 1) = 0$$

(You can also use trial and error.) At this point, we note that the product of two expressions $(t - 1)$ and $(5t + 1)$ gives us a result of 0. What does this mean?

We know that if we have two numbers and at least one of them is 0, their product is 0. For example,

$$-5 \cdot 0 = 0$$
$$0 \cdot 8 = 0$$
$$\frac{3}{2} \cdot 0 = 0$$
$$0 \cdot x = 0$$
$$x \cdot 0 = 0$$

and

$$0 \cdot 0 = 0$$

As you can see, in all cases in which the answer is 0, at least one of the factors is 0. In general, it can be shown that if the product of the two factors is 0, at least one of the factors must be 0. We shall call this idea the **principle of zero product.**

PRINCIPLE OF ZERO PRODUCT

If $A \cdot B = 0$, then $A = 0$ or $B = 0$.

Now, let us go back to our original equation. We can think of $(t - 1)$ as A and $(5t + 1)$ as B. Then our equation

$$(t - 1)(5t + 1) = 0$$

becomes

$$A \cdot B = 0$$

By the principle of zero product, if $A \cdot B = 0$,

$$A = 0 \quad \text{or} \quad B = 0$$

Thus,

$$t - 1 = 0 \quad \text{or} \quad 5t + 1 = 0$$
$$t = 1 \quad \text{or} \quad 5t = -1 \quad \text{We added 1 in the first equation}$$
$$\text{and subtracted 1 in the second.}$$
$$t = 1 \quad \text{or} \quad t = \frac{-1}{5}$$

Thus the ball reaches the ground after 1 sec or after $\frac{-1}{5}$ sec. The answer $\frac{-1}{5}$ is negative, which is impossible because we start counting when $t = 0$; so we can see that the ball thrown by the girl at 4 m/sec will reach the ground after $t = 1$ sec. You can check this by letting $t = 1$:

$$-5t^2 + 4t + 1 \stackrel{?}{=} 0$$
$$-5(1)^2 + 4(1) + 1 \stackrel{?}{=} 0$$
$$-5 + 4 + 1 \stackrel{?}{=} 0$$
$$0 = 0$$

Similarly, if we want to find how long a ball thrown from level ground at 10 m/sec takes to return to the ground, we need to solve the equation

$$5t^2 - 10t = 0$$

Factoring:

$$5t(t - 2) = 0$$

By the principle of zero product:

$$5t = 0 \quad \text{or} \quad t - 2 = 0$$
$$t = 0 \quad \text{or} \qquad t = 2 \quad \text{If } 5t = 0,\ t = 0.$$

Thus, the ball returns to the ground after 2 sec. (The other possible answer, $t = 0$, indicates that the ball was on the ground when $t = 0$, which is true.)

EXAMPLE 1 Solve.

a. $3x^2 + 11x - 4 = 0$ **b.** $6x^2 - x - 2 = 0$

Solution

a. To solve this equation, we must first factor the left-hand side. Here is the work.

The key number is -12.

$$3x^2 + 11x - 4 = 0$$
$$12(-1) = -12 \text{ and } 12 + (-1) = 11$$

Rewrite the middle term.
$$3x^2 + 12x - 1x - 4 = 0$$

Factor each pair.
$$3x(x + 4) - 1(x + 4) = 0$$

Factor out the GCF, $(x + 4)$.
$$(x + 4)(3x - 1) = 0$$

Use the principle of zero product.
$$x + 4 = 0 \quad \text{or} \quad 3x - 1 = 0$$

Solve each equation.
$$x = -4 \quad \text{or} \qquad 3x = 1$$
$$x = -4 \quad \text{or} \qquad x = \frac{1}{3}$$

Thus, the possible solutions are $x = -4$ and $x = \frac{1}{3}$. To verify that -4 is a correct solution, we substitute -4 in the original equation, obtaining

$$3(-4)^2 + 11(-4) - 4 = 3(16) - 44 - 4$$
$$= 48 - 44 - 4 = 0$$

You can verify that $\frac{1}{3}$ is also a solution.

b. As before, we must factor the left-hand side.

The key number is -12.
$$6x^2 - x - 2 = 0$$

Rewrite the middle term.
$$6x^2 - 4x + 3x - 2 = 0$$

Factor each pair.
$$2x(3x - 2) + 1(3x - 2) = 0$$

Factor out the GCF, $(3x - 2)$.
$$(3x - 2)(2x + 1) = 0$$

Use the principle of zero product.
$$3x - 2 = 0 \quad \text{or} \quad 2x + 1 = 0$$

Solve each equation.
$$3x = 2 \quad \text{or} \qquad 2x = -1$$
$$x = \frac{2}{3} \quad \text{or} \qquad x = -\frac{1}{2}$$

Thus, the solutions are $\frac{2}{3}$ and $-\frac{1}{2}$. You can check this by substituting these values in the original equation. ▲

Problem 1 Solve.
a. $5x^2 + 9x - 2 = 0$
b. $3x^2 - 2x - 5 = 0$

In the preceding discussion, we solved equations in which the highest exponent of the variable is 2. These equations are called *quadratic equations*. A quadratic equation is an equation in which the greatest exponent of the variable is 2. Moreover, to solve these quadratic equations, we must have one of the sides of the equation equaling 0. These two ideas are summarized next.

If a, b, and c are real numbers ($a \neq 0$),

$$ax^2 + bx + c = 0$$

is a **quadratic equation** in **standard form.**

To solve a quadratic equation by the method of factoring, the equation must be in *standard form.*

EXAMPLE 2 Solve $10x^2 + 13x = 3$.

Problem 2 Solve $10x^2 - 13x = 3$.

Solution This equation is *not* in standard form, so we cannot solve it as written. However, if we subtract 3 from each side of the equation, we have a quadratic equation.

$$10x^2 + 13x - 3 = 0$$

which is in standard form. We can now solve by factoring using trial and error or the *ac* method. To use the *ac* method, we write:

The key number is -30.	$10x^2 + \underline{13x} - 3 = 0$
Rewrite the middle term.	$10x^2 + \underline{15x - 2x} - 3 = 0$
Factor each pair.	$5x(2x + 3) - 1(2x + 3) = 0$
Factor out the GCF, $(2x + 3)$.	$(2x + 3)(5x - 1) = 0$
Use the principle of zero product.	$2x + 3 = 0$ or $5x - 1 = 0$
Solve each equation.	$2x = -3 \qquad 5x = 1$
	$x = -\dfrac{3}{2} \qquad x = \dfrac{1}{5}$

Thus the solutions of the equation are

$$x = -\frac{3}{2} \quad \text{or} \quad x = \frac{1}{5}$$ ▲

Sometimes we need to do some simplification before writing the equation in standard form. Thus, to solve the equation

$$(3x + 1)(x - 1) = 3(x + 1) - 2$$

we need to remove parentheses by multiplying the factors involved. Here is the work.

Given.	$(3x + 1)(x - 1) = 3(x + 1) - 2$
Multiply.	$3x^2 - 2x - 1 = 3x + 3 - 2$
Simplify on the right.	$3x^2 - 2x - 1 = 3x + 1$
Subtract $3x$ and 1 from each side.	$3x^2 - 5x - 2 = 0$

To solve this equation, we factor. (You can also use trial and error.)

The key number is -6.	$3x^2 - 6x + 1x - 2 = 0$
Factor each pair.	$3x(x - 2) + 1(x - 2) = 0$
Factor out the GCF, $(x - 2)$.	$(x - 2)(3x + 1) = 0$
Use the principle of zero product.	$x - 2 = 0 \quad \text{or} \quad 3x + 1 = 0$
Solve each equation.	$x = 2 \qquad\qquad 3x = -1$
	$x = 2 \quad \text{or} \qquad x = -\dfrac{1}{3}$

EXAMPLE 3   Solve $(2x + 1)(x - 2) = 2(x - 1) + 3$.

Solution

Multiply.	$2x^2 - 3x - 2 = 2x - 2 + 3$
Simplify.	$2x^2 - 3x - 2 = 2x + 1$
Subtract $2x$.	$2x^2 - 5x - 2 = 1$
Subtract 1.	$2x^2 - 5x - 3 = 0$
Rewrite the middle term.	$2x^2 - 6x + 1x - 3 = 0$
Factor each pair.	$2x(x - 3) + 1(x - 3) = 0$
Factor out the GCF, $(x - 3)$.	$(x - 3)(2x + 1) = 0$
Use the principle of zero product.	$x - 3 = 0 \quad \text{or} \quad 2x + 1 = 0$
Solve each equation.	$x = 3 \qquad\qquad x = -\dfrac{1}{2}$ ▲

Problem 3   Solve
$(3x - 2)(x - 1) = 2(x + 3) + 2.$

Finally, it is possible that we have to use the special formulas developed in Section 4.4 to solve certain equations. We illustrate this possibility in Example 4.

EXAMPLE 4   Solve $(4x - 1)(x - 1) = 2(x + 2) - 3x - 4$.

Solution

Multiply.	$4x^2 - 5x + 1 = 2x + 4 - 3x - 4$
Simplify.	$4x^2 - 5x + 1 = -x$
Add x.	$4x^2 - 4x + 1 = 0$
Rewrite the middle term.	$4x^2 - 2x - 2x + 1 = 0$
Factor each pair.	$2x(2x - 1) - 1(2x - 1) = 0$
Factor the GCF, $(2x - 1)$.	$(2x - 1)(2x - 1) = 0$
Use the principle of zero product.	$2x - 1 = 0 \quad \text{or} \quad 2x - 1 = 0$
Solve the equations.	$x = \dfrac{1}{2} \quad \text{or} \qquad x = \dfrac{1}{2}$

Problem 4   Solve
$(9x - 2)(x - 1) = 2(x + 1) - 7x - 1.$

Note that in this case there is really only *one* solution, $x = \frac{1}{2}$. A lot of work could be avoided if you notice that the expression $4x^2 - 4x + 1 = 0$ is a perfect square trinomial (F3) that can be factored as $(2x - 1)^2$ or, equivalently, $(2x - 1)(2x - 1)$. To avoid extra work, look for perfect trinomial squares before you factor!

ANSWERS

A. Solve.

1. $2x^2 + 7x + 3 = 0$

2. $2x^2 + 5x + 3 = 0$

3. $2x^2 + x - 3 = 0$

4. $6x^2 + x - 12 = 0$

5. $3y^2 - 11y + 6 = 0$

6. $4y^2 - 11y + 6 = 0$

7. $3y^2 - 2y - 1 = 0$

8. $12y^2 - y - 6 = 0$

9. $6x^2 + 11x = -4$

10. $5x^2 + 6x = -1$

11. $3x^2 - 5x = 2$

12. $12x^2 - x = 6$

13. $5x^2 + 6x = 8$

14. $6x^2 + 13x = 5$

15. $5x^2 - 13x = -8$

16. $3x^2 + 5x = -2$

17. $3y^2 = 17y - 10$

18. $3y^2 = 2y + 1$

19. $2y^2 = -5y - 2$

20. $5y^2 = -6y + 8$

Solve [you may use Equations (F2) and (F3)].

21. $9x^2 + 6x + 1 = 0$

22. $x^2 + 14x + 49 = 0$

23. $y^2 - 8y = -16$

24. $y^2 - 20y = -100$

25. $9x^2 + 12x = -4$

26. $25x^2 + 10x = -1$

27. $4y^2 - 20y = -25$

28. $16y^2 - 56y = -49$

29. $x^2 = -10x - 25$

30. $x^2 = -16x - 64$

1. _____
2. _____
3. _____
4. _____
5. _____
6. _____
7. _____
8. _____
9. _____
10. _____
11. _____
12. _____
13. _____
14. _____
15. _____
16. _____
17. _____
18. _____
19. _____
20. _____
21. _____
22. _____
23. _____
24. _____
25. _____
26. _____
27. _____
28. _____
29. _____
30. _____

31. $(2x - 1)(x - 3) = 3x - 5$

32. $(3x + 1)(x - 2) = x + 7$

33. $(2x + 3)(x + 4) = 2(x - 1) + 4$

34. $(5x - 2)(x + 2) = 3(x + 1) - 7$

35. $(2x - 1)(x - 1) = x - 1$

36. $(3x - 2)(3x - 1) = 1 - 3x$

31. _____

32. _____

33. _____

34. _____

35. _____

36. _____

In Problems 37 through 40 use

$$h = 5t^2 + V_0 t$$

where h is the distance traveled after t seconds by an object thrown downward with an initial velocity V_0.

37. An object is thrown downward at 5 m/sec from a height of 10 m. How long would it take the object to hit the ground?

38. An object is thrown downward from a height of 28 m with an initial velocity of 4 m/sec. How long would it take the object to hit the ground?

39. An object is thrown down from a building 15 m high at 10 m/sec. How long would it take the object to hit the ground?

40. How long would it take a package thrown downward from a plane at 10 m/sec to hit the ground 175 m below?

37. _____

38. _____

39. _____

40. _____

✓ **SKILL CHECKER**

Solve.

41. $7 = 3x - 2$

42. $11 = 4x + 3$

43. $8 = 2x - 4$

44. $17 = 3x - 7$

41. _____

42. _____

43. _____

44. _____

5.5 USING YOUR KNOWLEDGE

1. Solve the equation in Problem 1 of the 5.3 *Using Your Knowledge* (page 375) when $D(x) = 0$.

2. Do the same for Problem 4 (page 375) when $S(p) = 0$.

1. _____

2. _____

CALCULATOR CORNER

A calculator is ideal to check the solutions of quadratic equations. This time we will use the memory key denoted by $\boxed{\text{STO}}$ (or $\boxed{\text{M+}}$) and the memory recall key $\boxed{\text{RCL}}$ (or $\boxed{\text{MR}}$). The $\boxed{\text{STO}}$ key will enable you to store the solution and then recall it by using the $\boxed{\text{RCL}}$ key. Thus, to check the solution $-\frac{3}{2}$ for the equation $10x^2 + 13x = 3$ of Example 2, store the solution $-\frac{3}{2}$ by using the keystrokes

$\boxed{3}\ \boxed{+/-}\ \boxed{\div}\ \boxed{2}\ =\ \boxed{\text{STO}}$

EXERCISE 5.5

Then check the answer by pressing

$$\boxed{1}\,\boxed{0}\,\boxed{\times}\,\boxed{\text{RCL}}\,\boxed{x^2}\,\boxed{+}\,\boxed{1}\,\boxed{3}\,\boxed{\times}\,\boxed{\text{RCL}}\,\boxed{=}$$

The answer should be 3. To check the other answer $(\frac{1}{5})$, enter

$$\boxed{1}\,\boxed{\div}\,\boxed{5}\,\boxed{=}\,\boxed{\text{STO}}$$

and then repeat the keystrokes used to check $-\frac{3}{2}$.

If you are allowed to use calculators, check your answers to Problems 1 through 28.

SUMMARY

SECTION	ITEM	MEANING	EXAMPLE
5.3A	Factoring squares of binomials	$X^2 + 2AX + A^2 = (X + A)^2$ $X^2 - 2AX + A^2 = (X - A)^2$	$x^2 + 10x + 25 = (x + 5)^2$ $x^2 - 10x + 25 = (x - 5)^2$
5.3C	Factoring the difference of two squares	$X^2 - A^2 = (X + A)(X - A)$	$x^2 - 36 = (x + 6)(x - 6)$

5.4A	General factoring strategy	1. Factor out the GCF.
		 2. Look at the number of terms inside the parentheses or in the original polynomial.
		    Four terms: Grouping
		    Three terms: Perfect square trinomial or ac method
		    Two terms: Difference of two squares	

5.5	Quadratic equation	An equation in which the greatest exponent of the variable is 2	$x^2 + 3x - 7 = 0$ is a quadratic equation.
5.5A	Principle of zero products	If $A \cdot B = 0$, then $A = 0$ or $B = 0$.	If $(x + 1)(x + 2) = 0$, then $x + 1 = 0$ or $x + 2 = 0$.

(If you need help with these exercises, look in the section indicated in brackets.)

ANSWERS

1. [5.1A] Factor.
   a. $20x^3 - 45x^5$
   b. $14x^4 - 35x^6$
   c. $16x^7 - 40x^9$

   1. a. _____
      b. _____
      c. _____

2. [5.1A] Factor.

   a. $\dfrac{3}{7}x^6 - \dfrac{5}{7}x^5 + \dfrac{2}{7}x^4 - \dfrac{1}{7}x^2$

   b. $\dfrac{4}{9}x^7 - \dfrac{2}{9}x^6 + \dfrac{2}{9}x^5 - \dfrac{1}{9}x^3$

   c. $\dfrac{3}{8}x^9 - \dfrac{7}{8}x^8 + \dfrac{3}{8}x^7 - \dfrac{1}{8}x^5$

   2. a. _____
      b. _____
      c. _____

3. [5.1B] Factor.
   a. $3x^3 + 21x^2 + x + 7$
   b. $3x^3 + 18x^2 + x + 6$
   c. $4x^3 + 8x^2 + x + 2$

   3. a. _____
      b. _____
      c. _____

4. [5.2A] Factor.
   a. $x^2 + 8x + 7$
   b. $x^2 + 10x + 9$
   c. $x^2 + 6x + 5$

   4. a. _____
      b. _____
      c. _____

5. [5.2A] Factor.
   a. $x^2 - 7x + 10$
   b. $x^2 - 9x + 14$
   c. $x^2 - 6x + 8$

   5. a. _____
      b. _____
      c. _____

6. [5.2B] Factor.
   a. $6x^2 - 6 + 5x$
   b. $6x^2 - 1 + x$
   c. $6x^2 - 5 + 13x$

   6. a. _____
      b. _____
      c. _____

7. [5.2B] Factor.
   a. $6x^2 - 17xy + 5y^2$
   b. $6x^2 - 7xy + 2y^2$
   c. $6x^2 - 11xy + 4y^2$

   7. a. _____
      b. _____
      c. _____

8. [5.3B] Factor.
   a. $x^2 + 4x + 4$
   b. $x^2 + 6x + 9$
   c. $x^2 + 8x + 16$

   8. a. _____
      b. _____
      c. _____

9. [5.3B] Factor.
   a. $9x^2 + 12xy + 4y^2$
   b. $9x^2 + 30xy + 25y^2$
   c. $9x^2 + 24xy + 16y^2$

   9. a. _____
      b. _____
      c. _____

10. [5.3B] Factor.
   a. $x^2 - 4x + 4$
   b. $x^2 - 6x + 9$
   c. $x^2 - 8x + 16$

10. a. _____
    b. _____
    c. _____

11. [5.3B] Factor.
   a. $4x^2 - 12xy + 9y^2$
   b. $4x^2 - 20xy + 25y^2$
   c. $4x^2 - 28xy + 49y^2$

11. a. _____
    b. _____
    c. _____

12. [5.3B] Factor.
   a. $x^2 - 36$
   b. $x^2 - 49$
   c. $x^2 - 81$

12. a. _____
    b. _____
    c. _____

13. [5.3B] Factor.
   a. $16x^2 - 81y^2$
   b. $25x^2 - 64y^2$
   c. $9x^2 - 100y^2$

13. a. _____
    b. _____
    c. _____

14. [5.4A] Factor.
   a. $3x^3 - 6x^2 + 27x$
   b. $3x^3 - 6x^2 + 30x$
   c. $4x^3 - 8x^2 + 32x$

14. a. _____
    b. _____
    c. _____

15. [5.4A] Factor.
   a. $2x^3 - 2x^2 - 4x$
   b. $3x^3 - 6x^2 - 9x$
   c. $4x^3 - 12x^2 - 16x$

15. a. _____
    b. _____
    c. _____

16. [5.4A] Factor.
   a. $2x^3 + 8x^2 + x + 4$
   b. $2x^3 + 10x^2 + x + 5$
   c. $2x^3 + 12x^2 + x + 6$

16. a. _____
    b. _____
    c. _____

17. [5.4A] Factor.
   a. $9kx^2 + 12kx + 4k$
   b. $9kx^2 + 30kx + 25k$
   c. $4kx^2 + 20kx + 25k$

17. a. _____
    b. _____
    c. _____

18. [5.4B] Factor.
   a. $-3x^4 + 27x^2$
   b. $-4x^4 + 64x^2$
   c. $-5x^4 + 20x^2$

18. a. _____
    b. _____
    c. _____

19. [5.5A] Solve.
   a. $x^2 - 4x - 5 = 0$
   b. $x^2 - 5x - 6 = 0$
   c. $x^2 - 6x - 7 = 0$

19. a. _____
    b. _____
    c. _____

20. [5.5A] Solve.
   a. $2x^2 + x = 10$
   b. $2x^2 + 3x = 5$
   c. $2x^2 + x = 3$

20. a. _____
    b. _____
    c. _____

ANSWERS

(Answers on page 398)

1. Factor $10x^3 - 35x^5$.

1. _____

2. Factor $\dfrac{4}{5}x^6 - \dfrac{3}{5}x^5 + \dfrac{2}{5}x^4 - \dfrac{1}{5}x^2$.

2. _____

3. Factor $2x^3 + 6x^2 + x + 3$.

3. _____

4. Factor $x^2 + 7x + 6$.

4. _____

5. Factor $x^2 - 8x + 12$.

5. _____

6. Factor $6x^2 - 3 + 7x$.

6. _____

7. Factor $6x^2 - 11xy + 3y^2$.

7. _____

8. Factor $x^2 + 12x + 36$.

8. _____

9. Factor $4x^2 + 12xy + 9y^2$.

9. _____

10. Factor $x^2 - 14x + 49$.

10. _____

11. Factor $9x^2 - 12xy + 4y^2$.

11. _____

12. Factor $x^2 - 100$.

12. _____

13. Factor $16x^2 - 25y^2$.

13. _____

14. Factor $3x^3 - 6x^2 + 24x$.

14. _____

15. Factor $2x^3 - 8x^2 - 10x$.

15. _____

16. Factor $2x^2 + 6x + x + 3$.

16. _____

17. Factor $4kx^2 + 12kx + 9k$.

17. _____

18. Factor $-9x^4 + 36x^2$.

18. _____

19. Solve $x^2 - 3x - 10 = 0$.

19. _____

20. Solve $2x^2 - x = 15$.

20. _____

IF YOU MISSED QUESTION	SECTION	EXAMPLES	PAGE	ANSWERS
1	5.1	1	347	1. $5x^3(2 - 7x^2)$
2	5.1	2, 3	347	2. $\frac{1}{5}x^2(4x^4 - 3x^3 + 2x^2 - 1)$
3	5.1	4, 5	348, 349	3. $(x + 3)(2x^2 + 1)$
4	5.2	1	356	4. $(x + 6)(x + 1)$
5	5.2	1	356	5. $(x - 2)(x - 6)$
6	5.2	2, 3	359, 360	6. $(3x - 1)(2x + 3)$
7	5.2	4	360	7. $(3x - y)(2x - 3y)$
8	5.3	2	371	8. $(x + 6)^2$
9	5.3	2	371	9. $(2x + 3y)^2$
10	5.3	3	371	10. $(x - 7)^2$
11	5.3	3	371	11. $(3x - 2y)^2$
12	5.3	4	372	12. $(x + 10)(x - 10)$
13	5.3	4	372	13. $(4x + 5y)(4x - 5y)$
14	5.4	1	378	14. $3x(x^2 - 2x + 8)$
15	5.4	1	378	15. $2x(x + 1)(x - 5)$
16	5.4	2	378	16. $(x + 3)(2x + 1)$
17	5.4	3	379	17. $k(2x + 3)^2$
18	5.4	4, 5	379	18. $-9x^2(x + 2)(x - 2)$
19	5.5	1	387	19. $x = 5$ or $x = -2$
20	5.5	2, 3	388, 389	20. $x = 3$ or $x = -\dfrac{5}{2}$

CHAPTER 6

GRAPHS, LINEAR EQUATIONS, AND SLOPES

In this chapter we introduce the idea of an ordered pair of numbers, its location on the Cartesian coordinate plane, and the graph of an equation using these ordered pairs. We then introduce the idea of inequalities and their graphs on the Cartesian plane. We also discuss the idea of slope and the different forms in which the equation of a straight line can be written.

NAME

CLASS

SECTION

(*Answers on pages* 404–405)

ANSWERS

1. Graph the point $(2, -3)$.

1.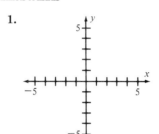

2. What are the coordinates of point A?

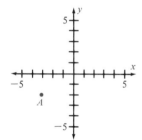

2. _____

3. Determine if the ordered pair $(1, -2)$ is a solution of $2x - y = 0$.

3. _____

4. Find x in the ordered pair $(x, 8)$ so that the ordered pair satisfies the equation $3x - y = 10$.

4. _____

5. Graph $2x + y = 4$.

5.

6. Graph $y = -3$.

6.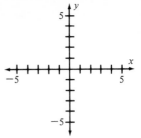

7. Graph $2x - 6 = 0$.

7.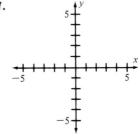

8. What is the slope of the line $2y - 4x = 8$?

8. _____

9. What is the y-intercept of the line $2y - 4x = 8$?

9. _____

10. Graph the inequality $2x - 3y < -6$.

10.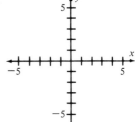

11. Graph the inequality $-y \leq -2x + 2$.

11.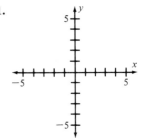

12. Find the slope of the line going through the points $(1, -8)$ and $(-1, -2)$.

12. _____

13. Find an equation of the line that goes through the point $(2, -4)$ and has slope $m = -3$.

13. _____

14. Find an equation of the line having slope 4 and y intercept -3.

14. _____

15. Find the pair of parallel lines.

15. _____

    (1) $4y = x + 5$

    (2) $2x - 8y = 5$

    (3) $8y + 2x = 5$

REVIEW

IF YOU MISSED QUESTION	SECTION	EXAMPLES	PAGE	ANSWERS
1	6.1	1	408	1.
2	6.1	2	409	2. $(-3, -2)$
3	6.1	3	410	3. No
4	6.1	4	412	4. $x = 6$
5	6.2	1	420	5.
6	6.2	2	424	6.
7	6.2	2	424	7.
8	6.2	3	426	8. 2
9	6.2	3	426	9. 4
10	6.3	1	439	10.

IF YOU MISSED QUESTION	SECTION	EXAMPLES	PAGE	ANSWERS
11	6.3	2	440	**11.**
12	6.4	1, 2	445, 446	**12.** -3
13	6.4	3	447	**13.** $y - (-4) = -3(x - 2)$ or $y = -3x + 2$
14	6.4	4	447	**14.** $y = 4x - 3$
15	6.4	5	448	**15.** (1) and (2)

6.1 ORDERED PAIRS

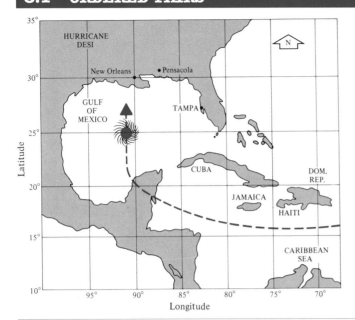

REVIEW

Before starting this section, you should know how to evaluate an expression.

OBJECTIVES

You should be able to:

A. Graph an ordered pair of numbers in a Cartesian coordinate system.
B. Determine the coordinates of a point in a Cartesian coordinate system or on a graph.
C. Determine if an ordered pair satisfies a given equation.
D. Find the missing coordinate of an ordered pair so that the result satisfies a given equation.

The map at the beginning of this section shows the position of Hurricane Desi. The hurricane is near the intersection of the vertical line indicating 90° of longitude and the horizontal line indicating 25° of latitude. This point can be identified by assigning to it an **ordered pair** of numbers, called **coordinates,** showing the longitude first and the latitude second. Thus, the hurricane would have the coordinates

$$(90, 25)$$

This is the longitude.   This is the latitude.

The method we have described is used to give the position of cities, islands, ships, airplanes, and so on. For example, the coordinates of New Orleans on the map are (90, 30), whereas those of Pensacola are approximately (87, 31). In mathematics, we use a system very similar to this one to locate points in a plane.

We first draw a number line and label the points in the usual manner (see Fig. 1). We then draw another number line perpendicular to the first one and crossing it at 0 (see Fig. 2). Now, every point in the plane determined by these lines can be associated with an ordered pair of numbers. For example, the point P in Fig. 3 is associated with the ordered pair (2, 3), whereas the point Q is associated with the ordered pair $(-1, 2)$. In the language of algebra the *horizontal* number line is called the **x-axis** and is labeled with the letter x; the *vertical* number line is called the **y-axis** and is labeled with the letter y. The two axes divide the plane into four regions called **quadrants.** These quadrants are numbered in counterclockwise order using Roman numerals and starting in the upper right-hand region. The whole arrangement is called a **Cartesian coordinate system,** a **rectangular coordinate system,** or simply a **coordinate plane.** Moreover, the point at which the two number lines cross is called the **origin** and has coordinates (0, 0) (see Fig. 4). Note that in Fig. 3 the point P with coordinates (2, 3) corresponds to the point at which $x = 2$ and $y = 3$. For this reason, we

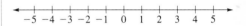

FIG. 1

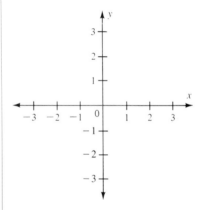

FIG. 2

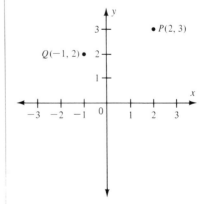

FIG. 3

say that the x-coordinate (or **abscissa**) of $(2, 3)$ is 2 and the y-coordinate (or **ordinate**) is 3.

A. GRAPHING ORDERED PAIRS

In general, if a point P has coordinates (x, y) we can always locate or graph the point in the coordinate plane. We start at the origin and go x units to the *right* if x is *positive;* to the *left* if x is *negative*. We then go y units *up* if y is *positive, down* if y is *negative*.

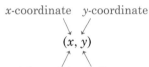

Tells us to go right or left   Tells us to go up or down

For example, to graph the point $R(3, -2)$, we start at the origin and go 3 units right (since the x-coordinate 3 is positive) and 2 units down (since the y-coordinate -2 is negative). The point is graphed in Fig. 5.

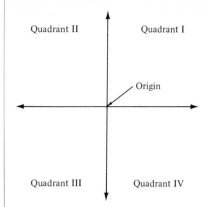

FIG. 4

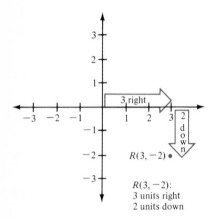

FIG. 5

EXAMPLE 1   Graph the points.

a. $A(1, 3)$ **b.** $B(2, -1)$ **c.** $C(-4, 1)$ **d.** $D(-2, -4)$

Solution

a. We start at the origin. To reach point $(1, 3)$, go 1 unit to the right and 3 units up. The graph of A is shown in Fig. 6.

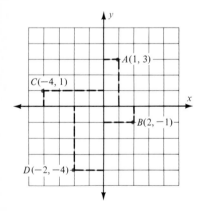

FIG. 6

Problem 1   Graph the points.
a. $A(4, 2)$ **b.** $B(3, -2)$
c. $C(-2, 1)$ **d.** $D(-1, -3)$

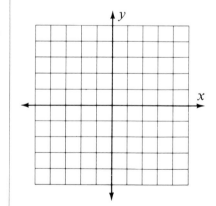

b. To graph $(2, -1)$, we start at the origin, go 2 units right and 1 unit down. The graph of B is shown in Fig. 6.

c. As usual, start at the origin. $(-4, 1)$ means to go 4 units left and 1 unit up as shown in Fig. 6.

d. The point D has both coordinates negative. Thus from the origin we go 2 units left and 4 units down; see Fig. 6.

B. FINDING COORDINATES OF POINTS

The graph shows the sales of cordless phones (in millions of units). The x-coordinate gives the *year,* whereas the y-coordinate gives the *number* of units sold. Thus, in 1985 the number of units sold was 4 million and the ordered pair giving this information is (1985, 40)

$$(x, y)$$

This is the year.   This is the number of units sold.

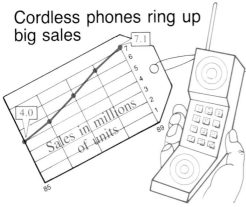

Cordless phones ring up big sales

Source: Electronic Industries Association; estimated figures

Adapted from USA Today, Mar. 8, 1989, p. 1Ð.

EXAMPLE 2

a. Find an ordered pair giving the number of phones sold in 1986.

b. Find an ordered pair giving the year in which sales reached 7.1 million units.

Solution

a. In 1986 the number of units sold was about 4.5 million. Thus, the corresponding ordered pair is (1986, 4.5).

b. The graph shows that sales reached 7.1 million in 1989. Thus, the corresponding ordered pair is (1989, 7.1).

C. ORDERED PAIRS AS SOLUTIONS OF EQUATIONS

How is the graph showing the cordless phone sales constructed? First, the x-axis is labeled with the years from 1985 to 1989, whereas the y-axis shows the number of units sold (in millions). The ordered pairs (x, y) representing the year and the sales are then graphed and the resulting points are joined with a line. Does this line have any relation to algebra? The answer is *yes,* especially if the line is *approximately* a

Problem 2   Find an ordered pair giving

a. The number of phones sold in 1988

b. The year in which sales reached 5.5 million

straight line. Thus, if we label the x-axis with the years 0, 1, 2, 3, and 4, corresponding to 1985 to 1989, respectively, the line representing the number of millions of units sold can be approximated by the equation

$$y = \frac{3}{4}x + 4$$

Note that in 1985 (when $x = \mathbf{0}$),

$$y = \frac{3}{4} \cdot 0 + 4 = 4$$

If we write $x = 0$ and $y = 4$ as the ordered pair $(0, 4)$, then we say that the ordered pair $(0, 4)$ **satisfies,** or is a **solution** of, the equation $y = \frac{3}{4}x + 4$. What about 1987? Since 1987 is 2 yr after 1985,

$$x = 2 \quad \text{and} \quad y = \frac{3}{4} \cdot 2 + 4 = \frac{6}{4} + 4 = \frac{3}{2} + 4 = 5\frac{1}{2} = 5.5$$

Hence the ordered pair $(2, 5.5)$ also satisfies the equation $y = \frac{3}{4}x + 4$, since $5.5 = \frac{3}{4} \cdot 2 + 4$. Note that the equation $y = \frac{3}{4}x + 4$ is different from all other equations we have studied because it has two variables, x and y. Because of this, such equations are called **equations in two variables.**

As you can see, to determine if an ordered pair is a solution of an equation in two variables, we simply substitute the x-coordinate for x and the y-coordinate for y in the given equation. If the resulting statement is *true,* then the ordered pair is a solution of, or *satisfies,* the equation. Thus, $(2, 5)$ is a solution of $y = x + 3$, since in the ordered pair $(2, 5)$, $x = 2$, $y = 5$ and

$$y = x + 3$$

becomes

$$5 = 2 + 3$$

which is a true statement.

EXAMPLE 3 Determine if the given ordered pairs are solutions of $2x + 3y = 10$.

a. $(2, 2)$ **b.** $(-3, 4)$ **c.** $(-4, 6)$

Solution

a. In the ordered pair $(2, 2)$, $x = 2$ and $y = 2$. Substituting in

$$2x + 3y = 10$$

we get

$$2(2) + 3(2) = 10$$

or

$$4 + 6 = 10$$

which is true. Thus, $(2, 2)$ is a solution of

$$2x + 3y = 10$$

Problem 3 Determine if the given ordered pairs are solutions of $3x + 2y = 10$.
a. $(3, 3)$ **b.** $(-2, 8)$ **c.** $(-4, 11)$

ANSWER (to problem on page 408)

1.

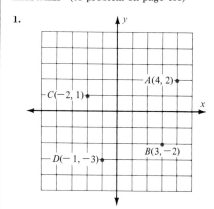

b. In the ordered pair $(-3, 4)$, $x = -3$ and $y = 4$. Substituting, we get

$$2(-3) + 3(4) = 10$$
$$-6 + 12 = 10$$

which is *not* true. Thus, $(-3, 4)$ is not a solution of the given equation.

c. In the ordered pair $(-4, 6)$, the x-coordinate is -4 and the y-coordinate is 6. Substituting these numbers for x and y, respectively, we have

$$2x + 3y = 10$$
$$2(-4) + 3(6) = 10$$

or

$$-8 + 18 = 10$$

which is a true statement. Thus, $(-4, 6)$ satisfies, or is a solution of, the given equation.

D. FINDING MISSING COORDINATES

In some cases (see the next section), rather than verifying that a certain ordered pair *satisfies* an equation, we actually have to *find* ordered pairs that are solutions of the given equation. For example, we might have the equation

$$y = 2x + 5$$

and would like to find several ordered pairs that satisfy this equation. To do this we substitute any number for x in the equation and then find the corresponding y value. An easy number to choose for x is **0.** In this case,

$$y = 2 \cdot 0 + 5 = 5 \quad \text{0 is easy to work with because only the } y$$
$$\text{variable is left when 0 is substituted for } x.$$

Thus, the ordered pair $(0, 5)$ satisfies the equation $y = 2x + 5$. For $x = \mathbf{1}$, $y = 2 \cdot \mathbf{1} + 5 = 7$; hence $(1, 7)$ also satisfies the equation. Clearly, we can let x be any number in the equation and then find the corresponding y value. Conversely, we can let y be any number in the given equation and then find the x value. For example, if we are given the equation

$$y = 3x - 2$$

and we are asked to find the value of x in the ordered pair $(x, 7)$, we simply let y be 7, obtaining

$$7 = 3x - 2$$

We then solve for x by rewriting the equation as

$$3x - 2 = 7$$
$$3x = 9 \quad \text{Adding 2}$$
$$x = 3 \quad \text{Dividing by 3}$$

Thus, $x = 3$ and the ordered pair satisfying $y = 3x - 2$ is $(3, 7)$, as can easily be verified since $7 = 3(3) - 2$.

ANSWER (to problem on page 409)
2. a. $(1988, 6.5)$ **b.** $(1987, 5.5)$

EXAMPLE 4 Complete the given ordered pairs so that they satisfy the equation $y = 4x + 3$.

a. $(x, 11)$ **b.** $(-2, y)$

Solution

a. In the ordered pair $(x, 11)$, y is 11. Substituting 11 for y in the given equation, we have

$$11 = 4x + 3$$
$$4x + 3 = 11 \qquad \text{Rewriting}$$
$$4x = 8 \qquad \text{Subtracting 3}$$
$$x = 2 \qquad \text{Dividing by 4}$$

Thus, $x = 2$ and the ordered pair is $(2, 11)$.

b. Here $x = -2$. Substituting this value in the given equation yields

$$y = 4(-2) + 3 = -8 + 3 = -5$$

Thus $y = -5$ and the ordered pair is $(-2, -5)$.

Problem 4 Complete the given ordered pairs so that they satisfy the equation $y = 3x + 4$.
a. $(x, 10)$ **b.** $(-2, y)$

A. Graph on the given coordinate system.

ANSWERS

1. **a.** $A(1, 2)$
 b. $B(-2, 3)$
 c. $C(-3, 1)$
 d. $D(-4, -1)$

1.

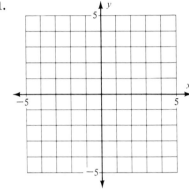

2. **a.** $A\left(-2\frac{1}{2}, 3\right)$

 b. $B\left(-1, 3\frac{1}{2}\right)$

 c. $C\left(-\frac{1}{2}, -4\frac{1}{2}\right)$

 d. $D\left(\frac{1}{3}, 4\right)$

2.

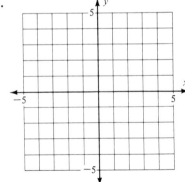

3. **a.** $A(0, 2)$

 b. $B(-3, 0)$

 c. $C\left(3\frac{1}{2}, 0\right)$

 d. $D\left(0, -1\frac{1}{4}\right)$

3.

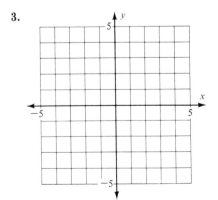

4. **a.** $A(20, 20)$
   b. $B(-10, 20)$
   c. $C(35, -15)$
   d. $D(-25, -45)$

4.

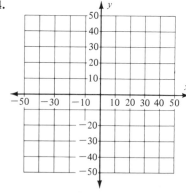

5. **a.** $A(0, 40)$
   b. $B(-35, 0)$
   c. $C(-40, -15)$
   d. $D(0, -25)$

5.

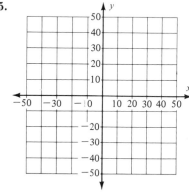

B. In Problems 6 through 10 give the coordinates of the points.

6.

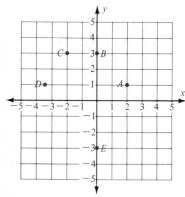

7.

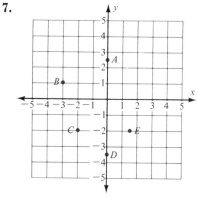

6. *A.* _____
   B. _____
   C. _____
   D. _____
   E. _____

7. *A.* _____
   B. _____
   C. _____
   D. _____
   E. _____

8.

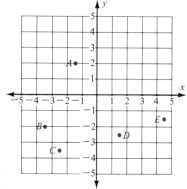

8. *A.* _____
   B. _____
   C. _____
   D. _____
   E. _____

ANSWER (to problem on page 412)
4. a. $(2, 10)$ **b.** $(-2, -2)$

9.

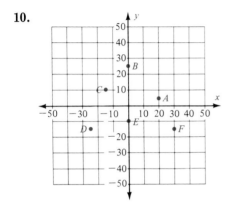

9. A. _____
   B. _____
   C. _____
   D. _____
   E. _____

10.

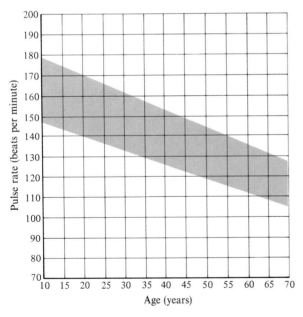

10. A. _____
    B. _____
    C. _____
    D. _____
    E. _____
    F. _____

The accompanying graph will be used in Problems 11 through 14.

Are you exercising hard? Your target zone can tell you. It works like this. Take your pulse after exercising, find your age on the x-axis and follow a vertical line up to the lower edge of the shaded area, then go across to the number on the y-axis at the left. That pulse rate is the lower limit for your target zone. To find the upper limit, continue vertically on your age line to the top of the shaded area and then go across to the y-axis to locate your pulse rate.

11. What is the lower limit pulse rate for a 20-year-old? Write the answer as an ordered pair.

11. _____

12. What is the upper limit pulse rate for a 20-year-old? Write the answer as an ordered pair.

12. _____

13. What is the upper limit pulse rate for a 45-year-old? Write the answer as an ordered pair.

13. _____

14. What is the lower limit pulse rate for a 50-year-old? Write the answer as an ordered pair.

14. _____

C. Determine if the ordered pair is a solution of the equation.

15. $(3, 2)$; $x + 2y = 7$

15. _____

16. $(4, 2)$; $x - 3y = 2$

16. _____

17. $(5, 3)$; $2x - 5y = -5$

17. _____

18. $(-2, 1)$; $-3x = 5y + 1$

18. _____

19. $(2, 3)$; $-5x = 2y + 4$

19. _____

20. $(-1, 1)$; $4y = -2x + 2$

20. _____

D. Find the missing coordinate.

21. $(3, \underline{\quad})$ is a solution of $2x - y = 6$.

21. _____

22. $(-2, \underline{\quad})$ is a solution of $-3x + y = 8$.

22. _____

23. $(\underline{\quad}, 2)$ is a solution of $3x + 2y = -2$.

23. _____

24. $(\underline{\quad}, -5)$ is a solution of $x - y = 0$.

24. _____

25. $(0, \underline{\quad})$ is a solution of $3x - y = 3$.

25. _____

26. $(0, \underline{\quad})$ is a solution of $x - 2y = 8$.

26. _____

27. $(\underline{\quad}, 0)$ is a solution of $2x - y = 6$.

27. _____

28. $(\underline{\quad}, 0)$ is a solution of $-2x - y = 10$.

28. _____

29. $(-3, \underline{\quad})$ is a solution of $-2x + y = 8$.

29. _____

30. $(-5, \underline{\quad})$ is a solution of $-3x - 2y = 9$.

30. _____

E. Applications

The following equation will be used in Problems 31 through 33.

Gross national product
    $\downarrow$
    x

Net national product
    $\downarrow$
    y

Depreciation
    $\downarrow$
    D

$$x = y + D$$

31. In 1929 the gross national product was $103 billion and the depreciation was $8 billion. Find the corresponding ordered pair (x, y) satisfying the given equation.

31. _____

32. If the net national product is $408 billion and the depreciation is $39 billion, find the corresponding pair (x, y) satisfying the given equation.

32. _____

33. If the net national product is $956 billion and the depreciation is $94 billion, find the corresponding ordered pair (x, y) satisfying the given equation.

33. _____

34. The height h (in centimeters) of a female is related to the length f of her femur bone by the equation $h = 2f + 73$. If the height of a female is 151 cm, find the corresponding ordered pair (h, f) satisfying the equation.

34. _____

35. The height (in centimeters) of a male is related to the length H of his humerus bone by the equation $h = 3H + 71$. If the height of a male is 176 cm, find the corresponding ordered pair (h, H) satisfying the equation.

35. _____

✓ SKILL CHECKER

Solve.

36. $C = 0.10m + 10$ when $m = 30$

36. _____

37. $3x + y = 9$ when $x = 0$

37. _____

38. $y = 50x + 450$ when $x = 2$

38. _____

39. $3x + y = 6$ when $x = 1$

39. _____

40. $3x + y = 6$ when $y = 0$

40. _____

6.1 USING YOUR KNOWLEDGE

The ideas presented in this section are vital for understanding graphs. For example, have you been exercising in the summer? To determine the risk of exercising in the heat, you must know how to read the graph. This can be done by first finding the temperature on the y-axis and then reading across from it to the right, stopping at the vertical line representing the relative humidity. Thus, on a 90° F day, if the humidity is less than 30%, the weather is in the safe zone.

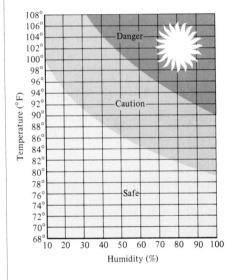

1. If the humidity is 50%, how high can the temperature go and still be in the safe zone for exercising? (Answer to the nearest degree.)

1. _____

2. If the humidity is 70%, at what temperature will the danger zone start?

2. _____

3. If the temperature is 100° F, what does the humidity have to be so that it is safe to exercise?

3. _____

4. Between what temperatures should you use caution when exercising if the humidity is 80%?

4. _____

5. Suppose the temperature is 86° F and the humidity is 60%. How many degrees can the temperature rise before you get to the danger zone?

5. _____

Your calculator can be used to determine if a given ordered pair is a solution of an equation. For example, to find out if $(-3, 4)$ is a solution of $2x + 3y = 10$, key in

| 2 | × | 3 | +/− | + | 3 | × | 4 | = |

if your calculator uses algebraic logic. The result is 6 (not 10 as in the equation). Thus, $(-3, 4)$ is not a solution of $2x + 3y = 10$. If you do not have algebraic logic in your calculator, you can use the parentheses keys to check if $(-3, 4)$ satisfies the equation $2x + 3y = 10$. The keystrokes will be

| 2 | × | 3 | +/− | + | (| 3 | × | 4 |) | = |

If your instructor allows it, solve Problems 15 through 20 with your calculator.

DAVIS RENT-A-CAR

TYPE	MILEAGE RATE			FLAT RATE	
	DAY	WEEK	MILE	DAY	WEEK
Economy	$9.00	$49.00	$0.10	$15.00	$79.00
Subcompact	10.00	55.00	0.10	17.00	89.00
Compact	11.00	60.00	0.11	17.00	89.00
Intermediate sedan	12.00	65.00	0.12	19.00	99.00
Sedans	13.00	70.00	0.13	17.95	99.00
Premium sedans	15.00	85.00	0.15	20.00	109.00
Station wagon	18.00	90.00	0.18	25.00	135.00

Adapted from *Advertising Age*, Jan. 9, 1989.

OBJECTIVES

REVIEW

Before starting this section, you should know:

1. How to evaluate an expression.
2. How to solve a linear equation in one variable.

OBJECTIVES

You should be able to:

A. Graph a linear equation of the form $Ax + By = C$, A, B, C constants (numbers), x, y variables.
B. Graph vertical ($x = k$) and horizontal ($y = k$) lines.
C. Write a given linear equation in the slope-intercept form

$$y = mx + b$$

and graph it.

In the preceding section, we mentioned the Cartesian coordinate system as a very important tool in mathematics. Do you wonder why we call it the Cartesian system? The reason is that the system was invented by the French mathematician René Des*cartes*. Descartes worked in the area known as **analytic geometry.** This area deals with two types of problems:

1. Given an equation, find its graph.
2. Given the graph of a geometric figure, find its equation.

A. GRAPHING LINEAR EQUATIONS

In this section we shall deal with the first type of problem, that is, given an equation we shall try to find its graph. We shall concentrate on obtaining graphs of lines. Let us look at the ad at the beginning of this section.

The equation that gives the daily cost C of renting a subcompact when m miles are traveled is

$$C = 0.10m + 10$$

Ten cents per mile   Ten dollars each day

Now, remember that a solution of this equation must be an ordered pair of numbers of the form (m, C). For example, if you travel 10 miles, $m = 10$ and the cost is

$$C = 0.10(10) + 10 = 1 + 10 = \$11$$

Thus **(10**, 11) is an ordered pair satisfying the equation; that is, (10, 11) is a *solution* of the equation. If we go **20** miles, $m = 20$ and

$$C = 0.10(20) + 10 = 2 + 10 = \$12$$

Hence (20, 12) is also a solution. As you see, we can go on forever finding solutions. It is much better to organize our work and list these two solutions and some others and then graph the points obtained in a Cartesian coordinate system. In this system the number of miles m will appear on the horizontal axis, and the cost C on the vertical axis. The table and the corresponding points appear in the following figure.

m	C
0	10
10	11
20	12
30	13

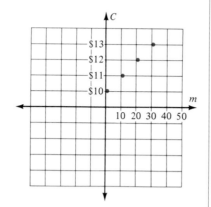

Note that m must be *positive* or *zero*. We have selected values for m that make the cost easy to compute—namely, 0, 10, 20, and 30.

It seems that if we join the points appearing in the graph, we obtain a straight line. Of course, if we knew this for sure, we could have saved time! Why? Because if we knew that the graph of an equation is a straight line, we could simply find *two* solutions of the equation, graph the two points, and then join them with a straight line. As it turns out, the graph of $C = 0.10m + 10$ is a straight line. Here is the rule that tells us when a given equation will give us a straight line for its graph:

The graph of an equation of the form

$Ax + By = C$ **A, B, C** constants (**A** and **B** not both 0)
                x, y variables

is a *straight line,* and every straight line has an equation that can be written in this form.

Technically, we should say that the graph of the *solution* set of an equation of the form $Ax + By = C$ is a straight line.

EXAMPLE 1   Graph the equation $3x + y = 6$.

Solution   The equation is of the form $Ax + By = C$, and thus the graph is a straight line. We shall graph two points and join them with a straight line, the graph of the equation. Two easy points to use occur if we let $x = 0$ and then find y and if we let $y = 0$ and find x. For $x = 0$,

$3x + y = 6$

becomes

$3 \cdot 0 + y = 6$ or $y = 6$

Thus, $(0, 6)$ is on the graph.
When $y = 0$,

$3x + y = 6$

becomes

$3x + 0 = 6$

Problem 1   Graph the equation $x + 2y = 4$.

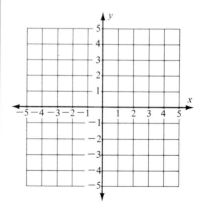

or

$3x = 6$

That is,

$x = 2$

Hence (2, 0) is also on the graph. It is a good idea to pick a third point as a check. For example, if we let $x = 1$,

$3x + y = 6$

becomes

$3 \cdot 1 + y = 6$

or

$3 + y = 6$

that is,

$y = 3$

Now we have our third point (1, 3), as shown in the table of the following figure. The points (0, 6), (2, 0), and (1, 3), as well as the completed graph of the line, are shown in the figure. The y-coordinate of the point where the line crosses the y-axis, 6, is called the **y-intercept,** and the x-coordinate of the point at which the line crosses the x-axis, 2, is called the **x-intercept.**

x	y
0	6
2	0
1	3

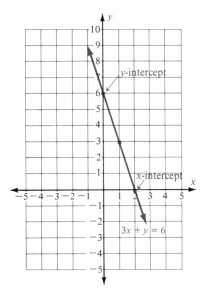

The figure cannot show the entire line, which extends indefinitely in both directions; the graph shown is simply a part of the line that continues without end in both directions as indicated by the arrows in the figure. We chose $x = 0$ and $y = 0$ because the calculations are easy. There are infinitely many points satisfying the equation $3x + y = 6$, but the points (0, 6) and (2, 0)—the y- and x-intercepts—are easier to find. Note that to find the graph of $x + 5y = 0$, for example, you can start by letting $x = 0$, obtaining $0 + 5y = 0$, or $y = 0$. This means that (0, 0) is on the line $x + 5y = 0$. But if you now let $y = 0$ you get $x = 0$. This line goes through the origin (0, 0), and you need to find another point. An easy one would be when $x = 5$. We obtain $5 + 5y = 0$, or $y = -1$. Thus, another point is $(5, -1)$, and the graph is as follows.

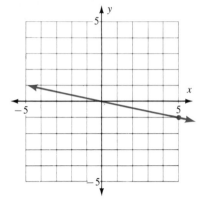

x	y
0	0
5	−1

B. GRAPHING HORIZONTAL AND VERTICAL LINES

Not all linear equations are written in the form $Ax + By = C$. For example, consider the equation $y = 3$. It seems that this equation is not in the form $Ax + By = C$. However, we can write the equation $y = 3$ as

$$0 \cdot x + y = 3$$

which is an equation written in the desired form. How do we graph the equation $y = 3$? Since it does not matter what value we give x, the result is always $y = 3$. Here is a table with three pairs satisfying the equation $y = 3$.

x	y	
0	3	In the equation $0 \cdot x + y = 3$, x can be any number; you always get
1	3	$y = 3$.
2	3	

These ordered pairs, as well as the completed graph, appear in the following figure.

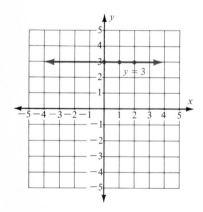

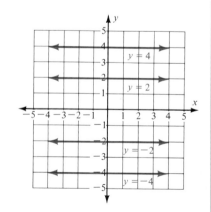

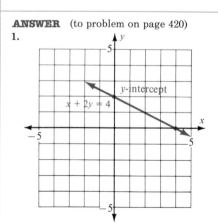

Note that the equation $y = 3$ has for its graph a *horizontal* line crossing the y-axis at $y = 3$. The graphs of some other horizontal lines, all of which have equations of the form $y = k$ (k a constant), appear in the figure on the right. If the graph of any equation $y = k$ is a horizontal line, what would the graph of the equation $x = 3$ be? A vertical line, of course! We first note that the equation $x = 3$ can be written as

$$x + 0 \cdot y = 3$$

Thus, the equation is of the form $Ax + By = C$, so that its graph is a straight line. Now, for any value of y, the value of x remains 3. Three values of x and the corresponding y values appear in the table. These three points, as well as the completed graph, are shown in the figure. We also give the graphs of some other vertical lines.

x	y
3	0
3	1
3	2

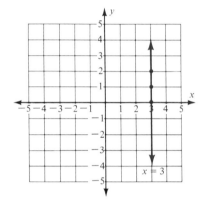
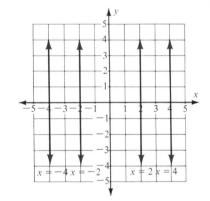

The graph on the right shows the vertical lines $x = -4$, $x = -2$, $x = 2$, and $x = 4$. This discussion can be summarized by the following rules.

The graph of any equation of the form

$$y = k, \quad \boldsymbol{k} \text{ a constant}$$

is a horizontal line crossing the y-axis at k.

The graph of any equation of the form

$$x = k, \quad \boldsymbol{k} \text{ a constant}$$

is a vertical line crossing the x-axis at k.

EXAMPLE 2   Graph the line.
a. $2x - 4 = 0$ **b.** $3 + 3y = 0$

Solution

a. We first solve for x.

Given: $2x - 4 = 0$

Add 4: $2x = 4$

Divide by 2: $x = 2$

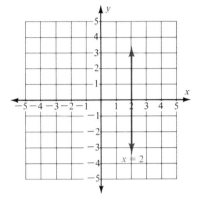

The graph of $x = 2$ is a vertical line crossing the x-axis at 2, as shown.

b. In this case, we solve for y.

Given: $3 + 3y = 0$

Subtract 3: $3y = -3$

Divide by 3: $y = -1$

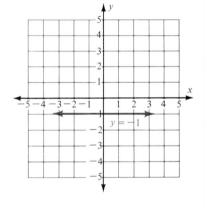

The graph of $y = -1$ is a horizontal line crossing the y-axis at -1, as shown.

C. EQUATIONS OF THE FORM $y = mx + b$

Are there any other forms of equations whose graphs are straight lines? As we mentioned before, any equation of the form $Ax + By = C$ has a straight line for its graph. Let us look at the equation of the line approximately representing the average cost for a 30-sec TV ad during the Super Bowl. The equation is $y = 50x + 450$, where y represents the cost of the ad (in thousands) and x shows the years after 1984. This equation can be written as $-50x + y = 450$—that is, in the form $Ax + By = C$, where $A = -50$, $B = 1$, and $C = 450$. Thus the graph of the equation $y = 50x + 450$ is a straight line. Note that the cost of an ad in 1984 (when $x = 0$) was \$450 (thousand). This is the point at which the line intersects the y-axis. Also note that costs go up about \$50 (thousand) each year between 1984 and 1989. Can you see the relationship between \$450 (thousand), \$50 (thousand), and the equation $y = 50x + 450$?

Problem 2   Graph the line.
a. $3x - 6 = 0$ **b.** $4 + 2y = 0$

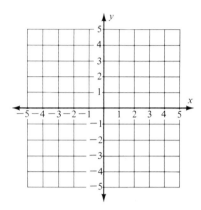

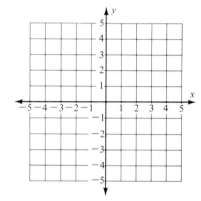

COST OF SPOTS SOARS

Average cost for a Super Bowl
30-second spot shown in thousands
of dollars

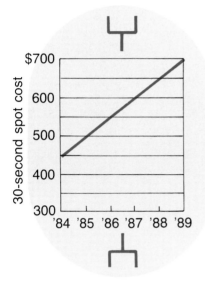

Nielsen ratings for 1989 are estimated.
Source: Nielsen Media Research.

It is easy to see that 50 gives the rate of change for the cost, whereas 450 is the starting point for the graph on the y-axis. In mathematics, we call 50 the *slope* of the line and 450 the *y-intercept*. In general, we define the slope of a line to be the *ratio of the vertical change to the corresponding horizontal change* between any two points on the line. If the two points are (x_1, y_1) and (x_2, y_2), the slope m is given by

$$m = \frac{\text{vertical change}}{\text{horizontal change}} = \frac{y_2 - y_1}{x_2 - x_1}$$

Thus, in the preceding discussion,

$$\text{Slope} = \frac{700 - 450}{1989 - 1984} = \frac{250}{5} = 50$$

Vertical change

Horizontal change

For the equation $y = mx + b$, we have the following rule:

SLOPE-INTERCEPT FORM

The graph of an equation of the form

$y = mx + b$ m, b constants
x, y variables

is a straight line. m is called the **slope** and b is the y-intercept.

Note that $\frac{250}{5}$, the slope, gives the 50 in the equation $y = 50x + 450$.

The figures show several lines with different slopes.

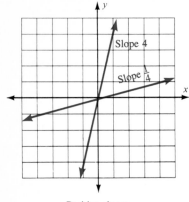

Positive slopes

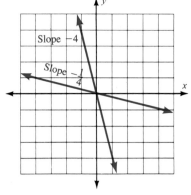

Negative slopes

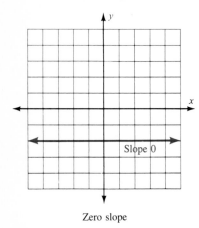

Zero slope

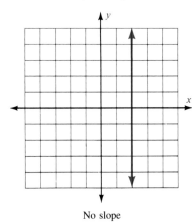

No slope

On these graphs, a horizontal line has slope 0, a line slightly inclined has slope $\frac{1}{4}$, but one that is steeper has a slope of 4. Thus a *positive* slope means that the line is *tilted upward to the right*. On the other hand, a line that is slightly tilted downward to the right has a small negative slope; that is, a *negative* slope means that the line is *tilted downward to the right*.

EXAMPLE 3   Consider the equation $3y - 6x = 9$.

a. Write the equation in slope-intercept form.
b. Find the slope.
c. Find the y-intercept.
d. Graph the equation.

Solution

a. We have to write the equation in the form $y = mx + b$, that is, we have to solve for y.

Given: $\qquad\qquad\qquad\qquad 3y - 6x = 9$

Add **6x:** $\qquad\qquad\qquad 3y - 6x + 6x = 6x + 9$

Simplify: $\qquad\qquad\qquad\qquad 3y = 6x + 9$

Divide each term by 3: $\qquad\qquad \dfrac{3y}{3} = \dfrac{6x}{3} + \dfrac{9}{3}$

Simplify: $\qquad\qquad\qquad\qquad y = 2x + 3$

Now the equation is in the form $y = mx + b$, with $m = 2$ and $b = 3$.

Problem 3   Consider the equation $4x - 2y = 8$.

a. Write the equation in slope-intercept form.
b. Find the slope.
c. Find the y-intercept.
d. Graph the equation.

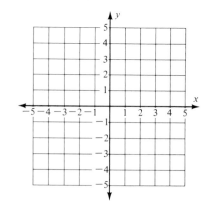

b. When an equation is in the form $y = mx + b$, m is the slope of the line. From part (a), $y = 2x + 3$; thus, **2** is the slope.

c. A line with an equation of the form $y = mx + b$ has y-intercept b. Thus the line $y = 2x + 3$ has y-intercept **3**.

d. We graph the equation $3y - 6x = 9$—or, equivalently and more easily, $y = 2x + 3$—by using the y-intercept 3 and the slope 2. These tell us that the line crosses the y-axis at $(0, 3)$, and starting at $(0, 3)$ a second point on the line can be obtained by going any distance to the right and twice this distance up. For instance, the points $(1, 3 + 2) = (1, 5)$, $(2, 3 + 4) = (2, 7)$, and so on, are all on the line. The completed graph appears in the margin. ▲

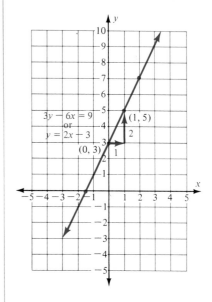

As the preceding examples indicate, it is easy to see that the graph of a first-degree equation in two variables that can be written in the form

$$Ax + By = C \qquad (1)$$

is a straight line.

$$Ax + By = C$$

is called a *linear equation* because its graph is a straight line. If $B \neq 0$, we can solve for y and write the equation in the form

$$y = mx + b$$

where $m = -\dfrac{A}{B}$ and $b = \dfrac{C}{B}$. If $B = 0$, we can solve for x and write the equation in the form $x = k$. In either case, the result is a *straight line*. For this reason, any equation that can be written in the form (1) is called a **linear equation** in two variables. Here is a summary of the different ways in which we can graph a linear equation.

TO GRAPH A LINEAR EQUATION

1. If the equation has *one* variable only, the graph will be parallel to an axis. ($x = k$ will be a vertical line and $y = k$ will be a horizontal line.)

2. If the equation has *two* variables, find the intercepts. Graph the resulting equation using these intercepts when possible.

3. If the intercept is $(0, 0)$, the line goes through the origin. You need only find one more point and then graph the line using $(0, 0)$ and the point you found.

4. If you find the x- and y-intercepts, find a third point to use as a check; then graph the equation.

ANSWER (to problem on page 424)

2. a.

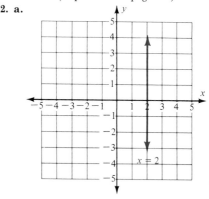

b.

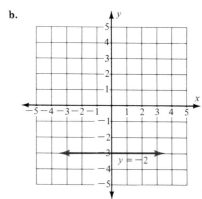

ANSWER (to problem on page 426)
3. **a.** $y = 2x - 4$ **b.** Slope 2
   c. y-intercept -4
   d.

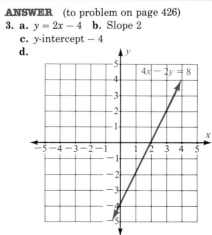

NAME

CLASS

SECTION

A. Graph.

1. $2x + y = 4$

2. $y + 3x = 3$

ANSWERS

1. _____

2. _____

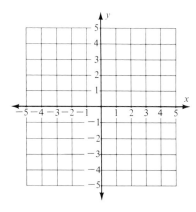

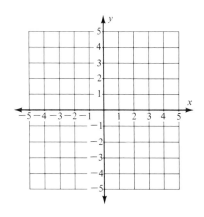

3. $-2x - 5y = -10$

4. $-3x - 2y = -6$

3. _____

4. _____

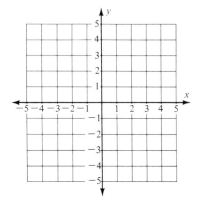

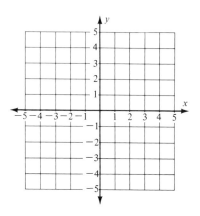

5. $y = 3x - 3$

6. $y = -2x + 4$

5. _____

6. _____

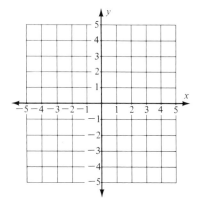

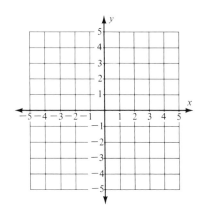

7. $6 = 3x - 6y$

8. $6 = 2x - 3y$

7. _____

8. _____

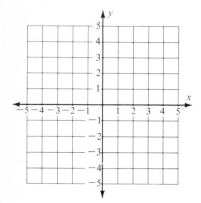

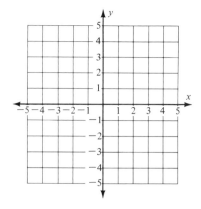

9. $-3y = 4x + 12$

9. _____

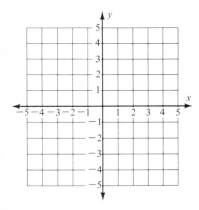

B. Graph.

10. $-2x = 5y + 10$

11. $y = -4$

10. _____

11. _____

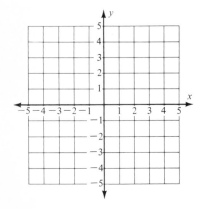

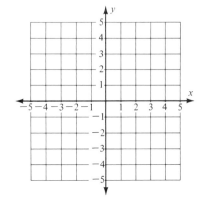

12. $y = \dfrac{-3}{2}$

13. $2y + 6 = 0$

12. _____

13. _____

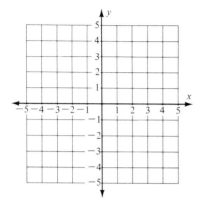

14. $-3y + 9 = 0$

15. $x = \dfrac{-5}{2}$

14. _____

15. _____

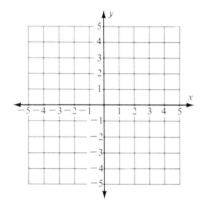

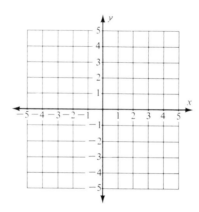

16. $x = \dfrac{7}{2}$

17. $2x + 4 = 0$

16. _____

17. _____

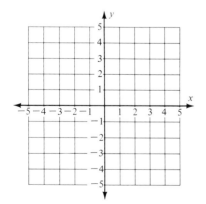

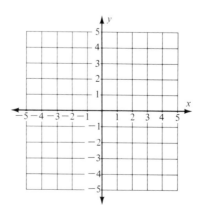

18. $3x - 12 = 0$

19. $2x - 9 = 0$

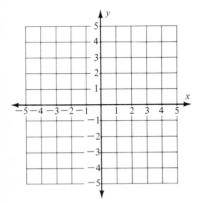

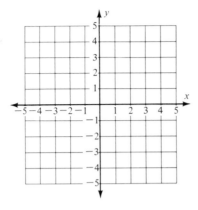

20. $-2x + 7 = 0$

20. _____

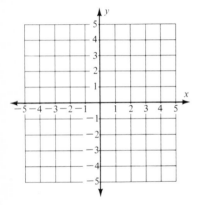

C. In Problems 21 through 30:

a. Write the equation in slope-intercept form.
b. Find the slope.
c. Find the y-intercept.
d. Graph the equation.

21. $2y - 4x = 8$

21. **a.** _____
   b. _____
   c. _____

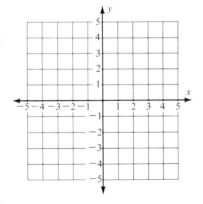

22. $3x - 6y = 9$

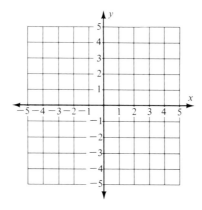

22. **a.** _____
    b. _____
    c. _____

23. $2y - 5x = 10$

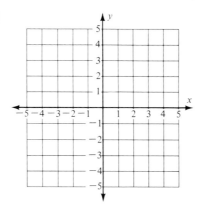

23. **a.** _____
    b. _____
    c. _____

24. $3y = 6x$

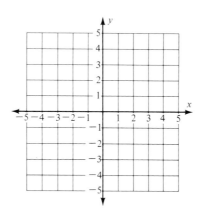

24. **a.** _____
    b. _____
    c. _____

25. $-2y = x$

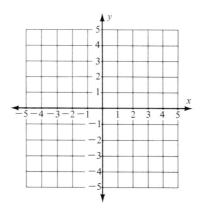

25. **a.** _____
    b. _____
    c. _____

26. $-2x = 4y + 6$

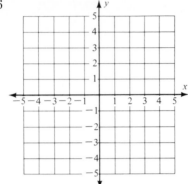

26. a. _____

b. _____

c. _____

27. $-2y = x + 4$

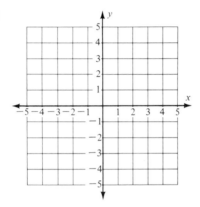

27. a. _____

b. _____

c. _____

28. $-3x = 3y + 9$

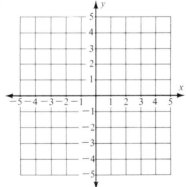

28. a. _____

b. _____

c. _____

29. $2y + 2x = 3x + 2$

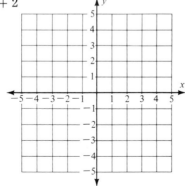

29. a. _____

b. _____

c. _____

30. $y - x = x - y$

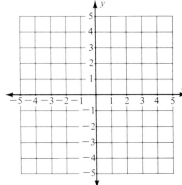

30. a. _____
    b. _____
    c. _____

✓ **SKILL CHECKER**

Fill in the blank with < or > so that the result is a true statement.

31. 0 _____ -8

32. 0 _____ 6

33. 10 _____ 0

34. -5 _____ 0

35. -1 _____ 0

31. _____

32. _____

33. _____

34. _____

35. _____

6.2 USING YOUR KNOWLEDGE

In economics and business, the slope m and the y-intercept of an equation play an important part. Let us see how.

Suppose you wish to go into the business of manufacturing fancy candles. First, you have to buy some ingredients such as wax, paint, and so on. Assume all these ingredients cost you $100. This is the *fixed cost*. Now, suppose it costs $2 to manufacture each candle. (This is the *marginal cost*.) What would be the total cost y if the marginal cost is $2, x units are produced, and the fixed cost is $100? The answer is

$$y \;=\; \underset{\substack{\uparrow \\ \text{Cost for} \\ x \text{ units}}}{2x} \;+\; \underset{\substack{\uparrow \\ \text{Fixed} \\ \text{cost}}}{100}$$

$\underset{\substack{\uparrow \\ \text{Total} \\ \text{cost}}}{y}$

In general, an equation of the form

$$y = mx + b$$

gives the total cost y of producing x units, when m is the cost of producing 1 unit and b is the fixed cost.

1. Find the total cost y of producing x units of a product costing $2 per unit if the fixed cost is $50.

2. Find the total cost y of producing x units of a product whose production cost is $7 per unit if the fixed cost is $300.

1. _____

2. _____

3. The total cost y of producing x units of a certain product is given by

$$y = 2x + 75$$

a. What is the production cost for each unit?
b. What is the fixed cost?

3. a. _____
   b. _____

DAVIS RENT-A-CAR					
TYPE	**FLAT RATE**		**1** MILEAGE RATE		
	DAY	**WEEK**	**MILE**	**DAY**	**WEEK**
Economy	$9.00	$49.00	$0.10	$15.00	$79.00
Subcompact	10.00	55.00	0.10	17.00	89.00
Compact	11.00	60.00	0.11	17.00	89.00
Intermediate sedan	12.00	65.00	0.12	19.00	99.00
Sedans	13.00	70.00	0.13	17.95	99.00
Premium sedans	15.00	85.00	0.15	20.00	109.00
Station wagon	18.00	90.00	0.18	25.00	135.00

REVIEW

Before starting this section you should know:

1. How to graph a line.
2. How to evaluate an expression.

OBJECTIVES

A. You should be able to graph a linear inequality in two variables.

In the preceding section, we learned how to graph a linear equation in two variables. A natural question at this point is: Do we have linear *inequalities* in two variables? The answer is *yes*. Just as the inequality $x < 2$ can be graphed by shading on a number line all the numbers that satisfy it (the graph of $x < 2$ consists of all points to the left of $x = 2$; see Section 3.4), an inequality of the form $y < mx + b$ can also be graphed by shading the region containing the points that satisfy it. Of course, in the case of the inequality $y < mx + b$, the solutions are *ordered pairs* of numbers. For example, suppose you wish to rent a subcompact costing $10 a day and 10 cents per mile. The total cost T depends on the number of days x the car is rented and on the number of miles y traveled. The equation is

$$T \; = \; 10x \; + \; 0.10y$$

Total cost   Number of   Number of
              days   miles

Suppose, moreover, that you want the cost T to be exactly $50. Then

$$10x + 0.10y = 50 \qquad\qquad (1)$$

We can graph this equation by letting $x = 0$, 1, and 2 and then finding the corresponding y-values. The information is then entered into a table and graphed.

x	y
0	500
1	400
2	300

When $x = 0$, $0.10y = 50$ or $y = 500$

When $x = 1$, $10 + 0.10y = 50$, $0.10y = 40$, $y = 400$

When $x = 2$, $20 + 0.10y = 50$, $0.10y = 30$, $y = 300$

The completed graph is shown in the figure.

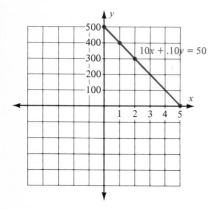

Now, suppose we wish the total cost T to be less than \$50. We then have

$$10x + 0.10y < 50 \tag{2}$$

We know that all the points $10x + 0.10y = 50$ are on the line. Where are the points for which $10x + 0.10y < 50$? Clearly, the line $10x + 0.10y = 50$ divides the plane into three parts:

1. The points *below* the line.
2. The points *on* the line.
3. The points *above* the line.

In more advanced courses, it is shown that if any point on one side of the line $Ax + By = C$ satisfies the inequality $Ax + By < C$, then *all* the points on that side also satisfy the inequality and no point on the other side of the line does.

We see that $(0, 0)$ is *below* the line, and it obviously satisfies inequality (2) because $0 + 0 < 50$. Therefore, all the other points below the line satisfy the inequality also. Because of this, the graph of $10x + 0.10y < 50$ consists of all the points *below* the line $10x + 0.10y = 50$, as shown shaded in the figure.

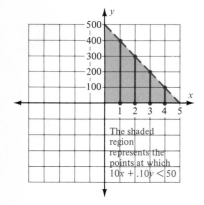

$(0, 0)$ satisfies $10x + 0.10y < 50$ because when $x = 0$ and $y = 0$, then $10x + 0.10y = 10 \cdot 0 + 0.10 \cdot 0 < 50$.

As a check, we verify that the point $(5, 500)$, which is above the line, does not satisfy the inequality $10x + 0.10y < 50$. (This is easy to see since when $x = 5$ and $y = 500$, $10x + 0.10y = 10(5) + 0.10(500)$ which is *not* less than 50.) Note that the line $10x + 0.10y = 50$ is *not* part of the graph and, for this reason, it is shown dashed in the figure. Now, to apply this result to our rental car problem, we must make use of the fact that x must be an integer. (You are not allowed to rent a car for $\frac{2}{5}$ d!) Hence the solution to our problem consists of the points on the heavy line segments at $x = 1, 2, 3,$ and 4 in the preceding figure. The graph shows, for instance, that you can rent the car for 2 days and drive up to 300 mi for less than $50.

EXAMPLE 1   Graph $2x - 4y < -8$.

Solution   We first graph the line $2x - 4y = -8$.

x	y
0	2
-4	0

When $x = 0$, $-4y = -8$,   and $y = 2$

When $y = 0$, $2x = -8$,   and $x = -4$

The graph is shown below. We select an easy test point and see if it satisfies the inequality. If it does, the solution lies on the same side of the line as the test point, otherwise the solution is on the other side of the line.

An easy point is $(0, 0)$, which is *below* the line. If we substitute $x = \mathbf{0}$ and $y = \mathbf{0}$ in the inequality $2x - 4y < -8$, we obtain

$$2 \cdot 0 - 4 \cdot 0 < -8$$

or

$$0 < -8$$

which is false. Thus the point $(0, 0)$ is not part of the solution. Because of this, the solution consists of the points *above* (on the other side of) the line $2x - 4y = -8$ as shown shaded in the figure below. Note that the line itself is shown dashed to indicate that it is not part of the solution.

Problem 1   Graph $3x - 2y < -6$.

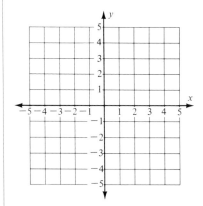

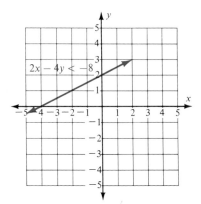

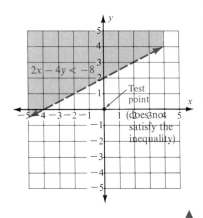

In the preceding problem, the test point was not part of the solution for the given inequality. Next, we give an example in which the test point is part of the solution for the inequality.

EXAMPLE 2 Graph $y \le -2x + 6$.

Solution As usual, we first graph the line $y = -2x + 6$.

When $x = 0$, $y = 6$
When $y = 0$, $0 = -2x + 6$ or $x = 3$

x	y
0	6
3	0

The graph of the line is shown in the figure at the left below. Now, we select the point $(0, 0)$ as a test point. When $x = 0$ and $y = 0$, the inequality

$$y \le -2x + 6$$

becomes

$$0 \le -2 \cdot 0 + 6$$

or

$$0 \le 6$$

which is *true*. Thus all the points on the same side of the line as $(0, 0)$, that is, the points *below* the line, are solutions of $y \le -2x + 6$. These solutions are shown shaded in the figure at the right below. This time, the line is shown *solid* because it is part of the solution, since $y \le -2x + 6$ allows $y = -2x + 6$. (For example, the point $(3, 0)$ satisfies the inequality $y \le -2x + 6$ because $0 \le -2 \cdot 3 + 6$ yields $0 \le 0$, which is true.)

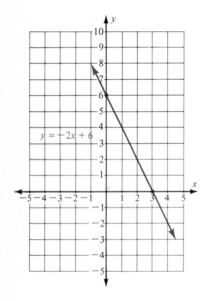

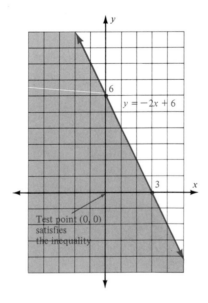

Problem 2 Graph $y \le -4x + 8$.

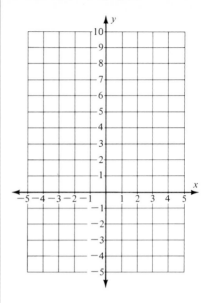

ANSWER (to problem on page 439)
1.

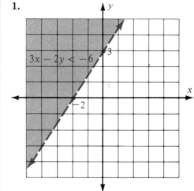

NAME

CLASS

SECTION

A. Graph.

ANSWERS

1. $2x + y > 4$

2. $y + 3x > 3$

1. _____

2. _____

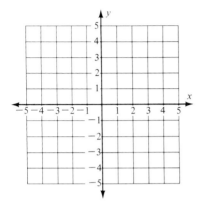

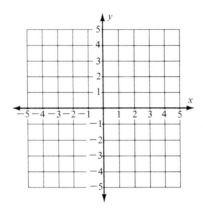

3. $-2x - 5y \leq -10$

4. $-3x - 2y \leq -6$

3. _____

4. _____

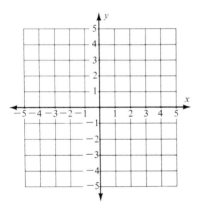

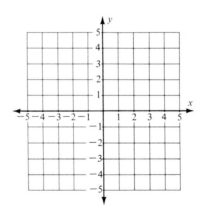

5. $y \geq 3x - 3$

6. $y \geq -2x + 4$

5. _____

6. _____

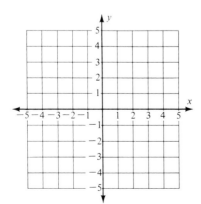

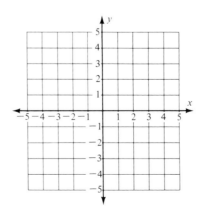

7. $6 < 3x - 6y$

8. $6 < 2x - 3y$

7. _____

8. _____

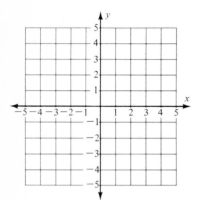

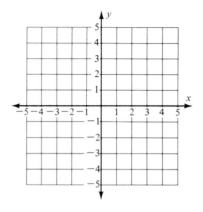

9. $3x + 4y \geq 12$

10. $-3y \geq 6x + 6$

9. _____

10. _____

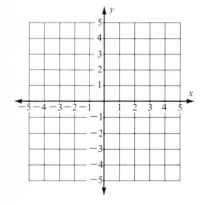

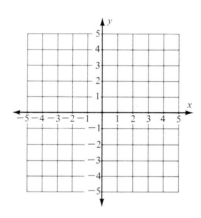

11. $10 < -2x + 5y$

12. $4 < x - y$

11. _____

12. _____

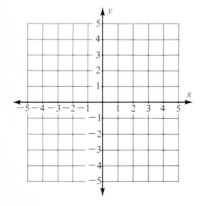

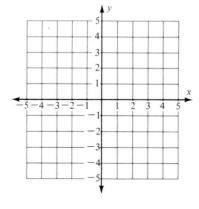

13. $x \geq 2y - 4$

14. $2x \geq 4y + 2$

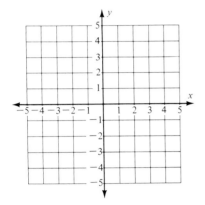

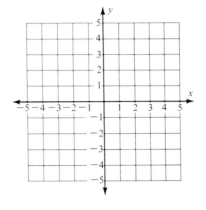

13. _____

14. _____

15. $y < -x + 5$

16. $2y < 4x - 8$

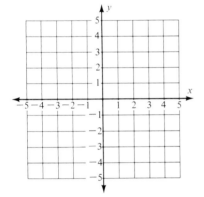

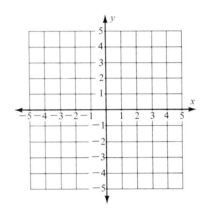

15. _____

16. _____

✓ **SKILL CHECKER**

Find.

17. $3 - (-6)$

18. $5 - (-7)$

19. $-6 - 3$

20. $-5 - 7$

17. _____

18. _____

19. _____

20. _____

6.3 USING YOUR KNOWLEDGE

The ideas discussed in this section can save you money when renting a car. Here is how.

Suppose you have the choice of renting a car from company A or from company B. The rates for these companies are as follows:

Company A: $8.50 a day plus 13¢ per mile

Company B: $5 a day plus 20¢ per mile

If x is the number of miles traveled in a day and y represents the cost for that day, the equations representing the cost for each company are:

Company A: $y = 0.13x + 8.50$
Company B: $y = 0.20x + 5$

1. Graph the equation representing the cost for A.

1, 2.

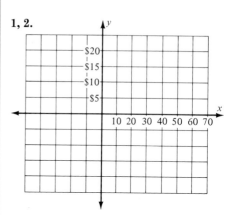

2. On the same coordinate axes, graph the equation representing the cost for B.

3. When is the cost the same for both companies?

3. _____

4. When is the cost less for company A?

4. _____

5. When is the cost less for company B?

5. _____

ANSWER (to problem on page 440)

2.

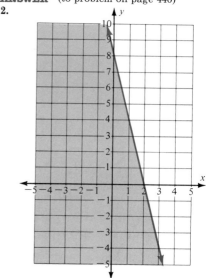

6.4 SLOPES AND EQUATIONS OF LINES

How cellular phone costs have dropped

If you bought a cellular telephone in 1983, you probably paid about $3,000. Today, you could buy nearly five of them for that amount.

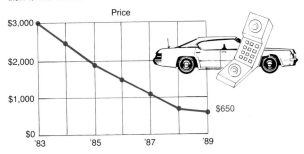

Price

Source: *Cellular Marketing Magazine*
Adapted from *USA Today*, Aug. 18, 1989, p. 1B.

The graph shows the costs of cellular phones from 1983 to 1989. This graph indicates that the drop in prices between 1983 and 1988 was about the same each year ($500). From 1988 to 1989, however, the drop was slower. That is, if a line segment were drawn from 1988 to 1989, it would be less steep than a segment drawn from 1983 to 1988.

A. SLOPE

The steepness of a line can be measured by using the ratio of the **vertical rise** (or **fall**) to the corresponding **horizontal run.** This ratio is called the *slope*. For example, a staircase that rises 3 ft in a horizontal distance of 4 ft is said to have a slope of $\frac{3}{4}$. As before, the definition of slope is as follows:

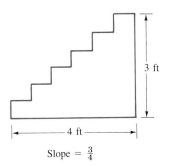

Slope = $\frac{3}{4}$

The slope m of the line going through the points (x_1, y_1) and (x_2, y_2), where $x_1 \neq x_2$, is given by

$$m = \frac{y_2 - y_1}{x_2 - x_1}$$

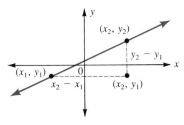

We do not define the slope for a vertical line. The slope of a horizontal line is 0 because all points on such a line have the same y values (try it for $y = 7$).

EXAMPLE 1 Find the slope of the line going through the points $(0, -6)$ and $(3, 3)$.

Solution The two given points are shown in the figure. Suppose we choose $(x_1, y_1) = (0, -6)$ and $(x_2, y_2) = (3, 3)$. Then we get

$$m = \frac{3 - (-6)}{3 - 0} = \frac{9}{3} = 3$$

Problem 1 Find the slope of the line going through the points $(0, -4)$ and $(2, 2)$.

If we choose $(x_1, y_1) = (3, 3)$ and $(x_2, y_2) = (0, -6)$, then

$$m = \frac{-6 - 3}{0 - 3} = \frac{-9}{-3} = 3$$

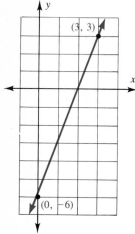

Line with positive slope

As you can see, it makes no difference which point is labeled (x_1, y_1) and which is labeled (x_2, y_2). Since an interchange of the two points simply changes the sign of both the numerator and the denominator in the slope formula, the result is the same in both cases.

EXAMPLE 2 Find the slope of the line that goes through the two points $(3, -4)$ and $(-2, 3)$. See the figure.

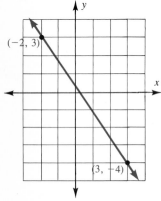

Line with negative slope

Solution We take $(x_1, y_1) = (-2, 3)$ so that $(x_2, y_2) = (3, -4)$. Then

$$m = \frac{-4 - 3}{3 - (-2)} = -\frac{7}{5}$$

▲

Examples 1 and 2 are illustrations of the fact that a line that rises from left to right has a *positive slope* and one that falls from left to right has a *negative slope*.

Problem 2 Find the slope of the line that goes through the two points $(5, -6)$ and $(-3, 4)$.

B. THE POINT-SLOPE EQUATION OF A LINE

You can use the slope of a line to obtain the equation of the line provided you are given one point on the line. Thus, suppose a line has slope m and passes through the point (x_1, y_1). If we let (x, y) be a second point on the line, the slope of the line is given by

$$\frac{y - y_1}{x - x_1} = m$$

Multiplying both sides by $(x - x_1)$, we get the *point-slope equation* of the line.

> $$y - y_1 = m(x - x_1) \tag{1}$$
>
> is the **point-slope** equation of the line going through (x_1, y_1) and having slope m.

EXAMPLE 3   Find an equation of the line that goes through the point $(2, -3)$ and has slope $m = -4$.

Solution   Using the point-slope equation (1), we get

$$y - (-3) = -4(x - 2)$$
$$y + 3 = -4x + 8$$
$$y = -4x + 5$$

Problem 3   Find an equation of the line that goes through the point $(2, -4)$ and has slope $m = -3$.

C. THE SLOPE-INTERCEPT EQUATION OF A LINE

An important special case of equation (1) is that in which the given point is the point where the line intersects the y-axis. Let this point be denoted by $(0, b)$. Then b is called the *y-intercept* of the line. Using equation (1), we obtain

$$y - b = m(x - 0)$$

or

$$y - b = mx$$

By adding b to both sides, we get the *slope-intercept form* of the equation of the line.

> $y = mx + b$ is the **slope-intercept** form of the equation of the line having slope m and y-intercept b.

EXAMPLE 4   Find the equation of the line having slope 5 and y-intercept -4.

Solution   In this case $m = 5$ and $b = -4$. Substituting in the slope-intercept form, we obtain:

$$y = 5x + (-4)$$

or

$$y = 5x - 4$$

Problem 4   Find the equation of the line with slope 3 and y-intercept -6.

D. PARALLEL LINES

Since the slope of a line determines its direction, *two lines with the same slope and different y-intercepts must be parallel lines*. The next example uses this idea.

EXAMPLE 5 Show that $3y = x + 2$ and $2x - 6y = 7$ are parallel lines.

Solution We must show that both lines have the same slope but different y-intercepts. To do this, we write each equation in slope-intercept form by solving for y. For the first equation, we have:

Given. $\qquad\qquad\qquad 3y = 1x + 2$

Divide each term by 3. $\qquad y = \dfrac{1}{3}x + \dfrac{2}{3}$

For the second equation,

$$2x - 6y = 7$$

Add $6y$ on both sides. $\qquad 2x = 7 + 6y$

Subtract 7. $\qquad\qquad\quad 2x - 7 = 6y$

Divide each term by 6. $\qquad \dfrac{2}{6}x - \dfrac{7}{3} = y$

$$\dfrac{1}{3}x - \dfrac{7}{3} = y$$

As you can see, both equations have a slope of $\frac{1}{3}$, but they have different y-intercepts; thus the two lines are parallel. ▲

At this point, many students ask, "Which formula should we use in the problems?" The table tells you which formula to use, depending on what information is given. Study this table before you attempt the problems in Exercise 6.4.

TO FIND THE EQUATION OF A LINE, GIVEN:	USE
Two points (x_1, y_1) and (x_2, y_2), $x_1 \neq x_2$	Two-point form: $y - y_1 = m(x - x_1)$, where $m = (y_2 - y_1)/(x_2 - x_1)$
A point (x_1, y_1) and the slope m	Point-slope form: $y - y_1 = m(x - x_1)$
The slope m and the y intercept b	Slope-intercept form: $y = mx + b$

The resulting equation can always be written in the general form: $Ax + By = C$.

Problem 5 Show that $4y = x + 6$ and $2x - 8y = 5$ are parallel lines.

NAME

CLASS

SECTION

A. In Problems 1 through 10, find the slope of the line that goes through the two given points.

ANSWERS

1. $(1, 2)$ and $(3, 4)$

2. $(1, -2)$ and $(-3, -4)$

3. $(0, 5)$ and $(5, 0)$

4. $(3, -6)$ and $(5, -6)$

5. $(-1, -3)$ and $(7, -4)$

6. $(-2, -5)$ and $(-1, -6)$

7. $(0, 0)$ and $(12, 3)$

8. $(-1, -1)$ and $(-10, -10)$

9. $(3, 5)$ and $(-2, 5)$

10. $(4, -3)$ and $(2, -3)$

1. _____

2. _____

3. _____

4. _____

5. _____

6. _____

7. _____

8. _____

9. _____

10. _____

B. In Problems 11 through 16, find the slope-intercept form (if possible) of the equation of the line that has the given properties (m is the slope).

11. Goes through $(1, 2)$; $m = \dfrac{1}{2}$

12. Goes through $(-1, -2)$; $m = -2$

13. Goes through $(2, 4)$; $m = -1$

14. Goes through $(-3, 1)$; $m = \dfrac{3}{2}$

15. Goes through $(4, 5)$; $m = 0$

16. Goes through $(3, 2)$; slope is not defined (does not exist)

11. _____

12. _____

13. _____

14. _____

15. _____

16. _____

C. In Problems 17 through 26 find an equation of the line with the given slope and intercept.

17. Slope, 2; y-intercept, -3

18. Slope, 3; y-intercept, -5

19. Slope, -4; y-intercept, 6

20. Slope, -6; y-intercept, -7

21. Slope, $\dfrac{3}{4}$; y-intercept, $\dfrac{7}{8}$

22. Slope, $\dfrac{7}{8}$; y-intercept, $\dfrac{3}{8}$

17. _____

18. _____

19. _____

20. _____

21. _____

22. _____

23. Slope, 2.5; y-intercept, -4.7

24. Slope, 2.8; y-intercept, -3.2

25. Slope, -3.5; y-intercept, 5.9

26. Slope, -2.5; y-intercept, 6.4

23. _____

24. _____

25. _____

26. _____

D. In Problems 27 through 32, determine whether the given lines are parallel.

27. $y = 2x + 5$; $4x - 2y = 7$

28. $y = 4 - 5x$; $15x + 3y = 3$

29. $2x + 5y = 8$; $5x - 2y = -9$

30. $3x + 4y = 4$; $2x - 6y = 7$

31. $x + 7y = 7$; $2x + 14y = 21$

32. $y = 5x - 12$; $y = 3x - 8$

27. _____

28. _____

29. _____

30. _____

31. _____

32. _____

✓ **SKILL CHECKER**

Graph.

33. $2x + y = 4$

34. $x + y = 4$

33. _____

34. _____

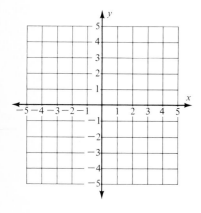

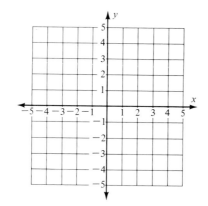

35. $x + 2y = 4$

35. _____

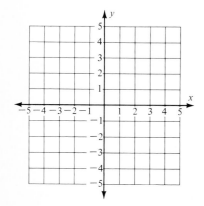

ANSWER (to problem on page 448)

5. Both lines have the same slope $\left(\dfrac{1}{4}\right)$ but different y-intercepts.

6.4 USING YOUR KNOWLEDGE

Look at the graph at the beginning of this section. Can we find an equation for the drop in prices between 1983 and 1987? For simplicity, let 1983 correspond to 0, 1984 to 1, 1985 to 2, 1986 to 3, and 1987 to 4. The points on the graph are: (0, 3000), (1, 2500), (2, 2000), (3, 1500), and (4, 1000).

1. Find the slope of the line passing through (0, 3000) and (4, 1000).

2. Use the slope you obtained in (1) and the point (0, 3000) to write the equation of the line in the *point-slope form*.

3. Look at the graph. What is the *y*-intercept?

4. Use the slope you found in (1) and the *y*-intercept from (3) to write the equation of the line in *slope-intercept form*.

1. _____

2. _____

3. _____

4. _____

SUMMARY

SECTION	ITEM	MEANING	EXAMPLE
6.1	*x*-axis	A horizontal number line on a coordinate plane	
6.1	*y*-axis	A vertical number line on a coordinate plane	
6.1	Quadrant	One of the four regions into which the axes divide the plane	
6.1	Abscissa	The first coordinate in an ordered pair	The abscissa in the ordered pair (3, 4) is 3.
6.1	Ordinate	The second coordinate in an ordered pair	The ordinate in the ordered pair (3, 4) is 4.
6.2A	Graph of a linear equation	The graph of a linear equation of the form $Ax + By = C$ is a straight line, and every straight line has an equation of this form.	The linear equation $3x + 6y = 12$ has a straight line for its graph.
6.2A	*x*-intercept	The *x*-coordinate of the point at which a line crosses the *x*-axis.	The *x*-intercept of the line $3x + 6y = 12$ is 4.
6.2A	*y*-intercept	The *y*-coordinate of the point at which a line crosses the *y*-axis.	The *y*-intercept of the line $3x + 6y = 12$ is 2.
6.2B	Horizontal line	A line whose equation can be written in the form $y = k$	The line $y = 3$ is a horizontal line.
6.2B	Vertical line	A line whose equation can be written in the form $x = k$	The line $x = -5$ is a vertical line.
6.2C	Slope of a line	The ratio of the vertical change to the horizontal change of a line.	
6.2C	Slope of a line through (x_1, y_1) and (x_2, y_2), $x_2 \neq x_1$	$m = \dfrac{y_2 - y_1}{x_2 - x_1}$	The slope of the line through (3, 5) and (6, 8) is $m = \dfrac{8 - 5}{6 - 3} = 1$.
6.2C	Slope-intercept form	The slope-intercept form of a line with slope m and *y*-intercept b is $y = mx + b$.	The slope-intercept form of the line with slope -3 and *y*-intercept -5 is $y = -3x - 5$.

(Continued)

SUMMARY (*Continued*)

SECTION	ITEM	MEANING	EXAMPLE
6.3	Linear inequality	An inequality that can be written in the form $Ax + By < C$	$3x + 6y < 12$ is a linear inequality.
6.4B	Point-slope form	The point-slope form of the line passing through (x_1, y_1) and having slope m is $y - y_1 = m(x - x_1)$.	The point-slope form of the line passing through $(3, -4)$ and having slope 6 is $y - (-4) = 6(x - 3)$.
6.4D	Parallel lines	Two lines are parallel if their slopes are equal and they have different y-intercepts.	The lines $y = 2x + 5$ and $3y - 6x = 8$ are parallel.

_____
_____
_____
_____
_____
_____

NAME

CLASS

SECTION

(If you need help with these exercises, look in the section indicated in brackets.)

ANSWERS

1. [6.1A] In the accompanying coordinate system, graph the points.
   a. $(-1, 2)$
   b. $(-2, 1)$
   c. $(-3, 3)$

1.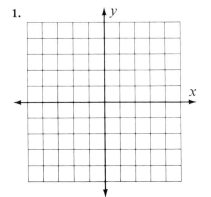

2. [6.1B] Find the coordinates of the point.
   a. A
   b. B
   c. C

2. a. _____
   b. _____
   c. _____

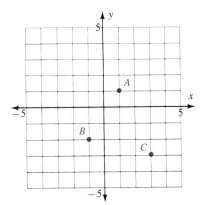

3. [6.1C] Determine if the given point is a solution of $x - 2y = -3$.
   a. $(1, -2)$
   b. $(2, -1)$
   c. $(-1, -1)$

3. a. _____
   b. _____
   c. _____

4. [6.1D] Find x in the given ordered pair so that the pair satisfies the equation, $2x - y = 4$.
   a. $(x, 2)$
   b. $(x, 4)$
   c. $(x, 0)$

4. a. _____
   b. _____
   c. _____

5. [6.2A] Graph.
    a. $x + y = 4$
    b. $x + y = 2$
    c. $x + 2y = 2$

5.

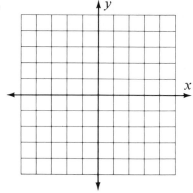

6. [6.2B] Graph.
    a. $y = -1$
    b. $y = -3$
    c. $y = -4$

6.

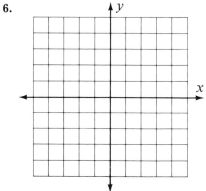

7. [6.2B] Graph.
    a. $2x - 6 = 0$
    b. $2x - 2 = 0$
    c. $2x - 4 = 0$

7.

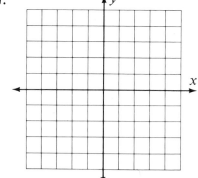

8. [6.2C] Find the slope of the line.
    a. $2y - 6x = 6$
    b. $3y - 6x = 12$
    c. $2y - 8x = 8$

8. a. _____
   b. _____
   c. _____

9. [6.2C] Find the y-intercept of the line.
    a. $4y - 2x = 12$
    b. $3y - 6x = 12$
    c. $3y - 4x = 12$

9. a. _____
   b. _____
   c. _____

10. [6.3] Graph.

   a. $2x - 4y < -8$

10. a.

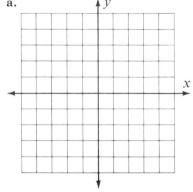

   b. $4x - 6y < -12$

b.

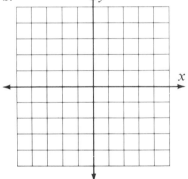

   c. $4x - 2y < -8$

c.

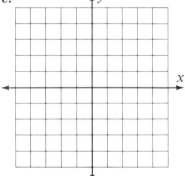

11. [6.3] Graph.

   a. $-y \le -2x + 2$

11. a.

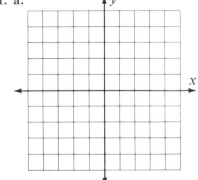

b. $-y \le -2x + 4$

b.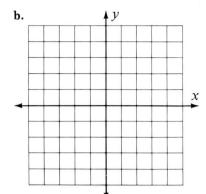

c. $-y \le -x + 3$

c.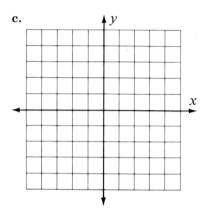

12. [6.4A] Find the slope of the line going through the given points.
    a. $(1, -4)$ and $(2, -3)$
    b. $(5, -2)$ and $(8, -5)$
    c. $(3, -4)$ and $(4, -8)$

12. **a.** _____
    b. _____
    c. _____

13. [6.4B] Find an equation of the line that goes through the point $(3, -5)$ and has the given slope.
    a. $m = -2$
    b. $m = -3$
    c. $m = -4$

13. **a.** _____
    b. _____
    c. _____

14. [6.4C] Find an equation of the line with the following slope and y-intercept.
    a. Slope 5, y-intercept -2
    b. Slope 4, y-intercept -3
    c. Slope 6, y-intercept -4

14. **a.** _____
    b. _____
    c. _____

15. [6.4D] Which line is parallel to $2y = x + 8$?
    a. $2x + y = 8$ or $2x - 4y = 8$
    b. $4x + 2y = 8$ or $2x - 4y = 8$
    c. $4y = 8x + 8$ or $8x = 4y + 8$

15. **a.** _____
    b. _____
    c. _____

NAME

CLASS

SECTION

(Answers on pages 460–461)

ANSWERS

1. Graph the point $(-2, 3)$.

1.
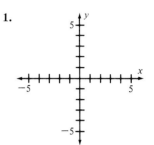

2. What are the coordinates of point A?

2. _____

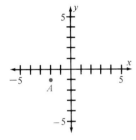

3. Determine if the ordered pair $(1, -2)$ is a solution of $2x - y = -2$.

3. _____

4. Find x in the ordered pair $(x, 2)$ so that the ordered pair satisfies the equation $3x - y = 10$.

4. _____

5. Graph $x + 2y = 4$.

5.
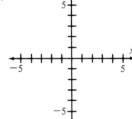

6. Graph $y = -2$.

6.
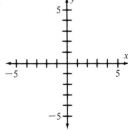

7. Graph $3x - 6 = 0$.

7.

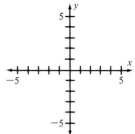

8. What is the slope of the line $2y - 4x = 12$?

8. _____

9. What is the y-intercept of the line $2y - 4x = 12$?

9. _____

10. Graph the inequality $3x - 2y < -6$.

10.

11. Graph the inequality $-y \leq -3x + 3$.

11.

12. Find the slope of the line going through the points $(2, -8)$ and $(-4, -2)$

12. _____

13. Find an equation of the line that goes through the point $(2, -6)$ and has slope $m = -5$.

13. _____

14. Find an equation of the line having slope 5 and y-intercept -4.

14. _____

15. Find the pair of parallel lines.

   (1) $3y = x + 5$
   (2) $2x - 6y = 5$
   (3) $6y - 3x = 5$

15. _____

IF YOU MISSED QUESTION	SECTION	EXAMPLES	PAGE	ANSWERS
1	6.1	1	408	1. _____
				2. $(-2, -1)$
				3. No
				4. $x = 4$
2	6.1	2	409	
3	6.1	3	410	
4	6.1	4	412	
5	6.2	1	420	5. _____
6	6.2	2	424	6. _____
7	6.2	2	424	7. _____
				8. $\dfrac{2}{}$
				9. 6
8	6.2	3	426	
9	6.2	3	426	
10	6.3	1	439	10. _____

IF YOU MISSED QUESTION	SECTION	EXAMPLES	PAGE	ANSWERS
11	6.3	2	440	**11.**
12	6.4	1, 2	445, 446	**12.** -1
13	6.4	3	447	**13.** $y - (-6) = -5(x - 2)$ $y = -5x + 4$
14	6.4	4	447	**14.** $y = 5x - 4$
15	6.4	5	448	**15.** (1) and (2)

CHAPTER 7

SYSTEMS OF LINEAR EQUATIONS

In this chapter we learn how to solve systems of linear equations by three different methods: graphical, substitution, and elimination. We then use these ideas to solve applications dealing with coin problems, distance problems, and other general word problems.

ANSWERS

(*Answers on page* 464)

1. Use the graphical method to solve the system

    $x + 2y = 6$

    $4y - x = 0$

1.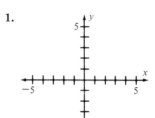

2. Use the graphical method to solve the system

    $y - 2x = 2$

    $3y - 6x = 12$

2.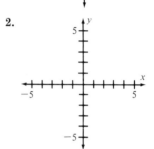

3. Use the method of substitution to solve the system (if possible).

    $x + 3y = 6$

    $2x + 6y = 10$

3. _____

4. Use the method of substitution to solve the system (if possible).

    $x + 3y = 6$

    $2x + 6y = 12$

4. _____

5. Solve the system (if possible).

    $2x + 3y = -7$

    $3x + y = -7$

5. _____

6. Solve the system (if possible).

    $3x - 2y = 6$

    $-9x + 6y = -8$

6. _____

7. Solve the system (if possible).

    $2y + 3x = -13$

    $6x + 4y = -26$

7. _____

8. Eva has \$3 in nickels and dimes. She has twice as many dimes as nickels. How many nickels and how many dimes does she have?

8. _____

9. The sum of two numbers is 140. Their difference is 80. What are the numbers?

9. _____

10. A plane flies 600 mi with a tail wind in 2 hr. It takes the same plane 3 hr to fly the 600 mi when flying against the wind. What is the plane's speed in still air?

10. _____

IF YOU MISSED QUESTION	SECTION	EXAMPLES	PAGE	ANSWERS
1	7.1	1	466	1. The solution is (4, 1).
2	7.1	2, 3	467, 468	2. The lines are parallel; there is no solution
3	7.2	1, 2	480, 481	3. No solution
4	7.2	3	481	4. Dependent (infinitely many solutions)
5	7.3	1	488	5. $(-2, -1)$
6	7.3	1, 2	488, 489	6. No solution
7	7.3	3	490	7. Dependent (infinitely many solutions)
8	7.4	1	497	8. 12 nickels 24 dimes
9	7.4	2	497	9. 110 and 30
10	7.4	3	498	10. 250 mi/hr

7.1 SOLVING SYSTEMS OF EQUATIONS BY GRAPHING

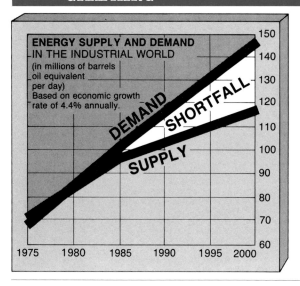

OBJECTIVES

REVIEW

Before starting this section, you should know:

1. How to graph the equation of a line.
2. How to determine if an ordered pair is a solution of an equation (Section 6.1).

OBJECTIVES

You should be able to:

A. Solve a system of two equations in two variables by using the graphical method.
B. Determine if a system of equations in two variables is consistent (one solution) inconsistent (no solution) or dependent (infinitely many solutions).

Look at the above graph. Can you tell when the energy supply and the demand were about the same? This happened around 1980. At this point, according to economists, the demand and the supply were equal and prices reached equilibrium. If we denote the year by x and the millions of barrels of oil per day by y, that point at which the demand was the same as the supply—that is, the point at which these lines intersect—is (1980, 85). Since we have learned how to graph linear equations, we can graph a pair of linear equations and find an ordered pair such as (1980, 85) which is a solution of both equations by finding the point (if there is one) at which the lines *intersect*. The point (1980, 85) is the point of intersection of the lines representing supply and demand. The coordinates of this point will be the *common solution* to both equations. For example, the solution of the system

$$x + 2y = 4$$
$$2y - x = 0$$

is (2, 1). This can be easily checked by letting $x = 2$ and $y = 1$ in both equations.

$x + 2y = 4$	$2y - x = 0$
$2 + 2(1) = 4$	$2(1) - 2 = 0$
$2 + 2 = 4$	$2 - 2 = 0$
$4 = 4$	$0 = 0$

Clearly, a true statement results in both cases. In the language of algebra, a system of two linear equations is called a **system** of simultaneous **equations**. To solve one of these systems we need to find (if possible) all ordered pairs of numbers that satisfy both equations.

A. SOLVING SYSTEMS BY THE GRAPHICAL METHOD

To solve the system

$x + 2y = 4$ This system is called a system of simultaneous equations
$2y - x = 0$ because we have to find a solution that satisfies both
equations simultaneously.

we graph each of the equations in the usual way. To graph $x + 2y = 4$, we make the following table.

x	y
0	2
4	0

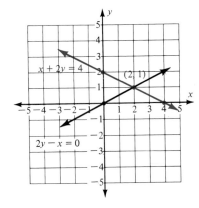

These two points, as well as the completed graph, are shown in color in the figure. The equation $2y - x = 0$ is graphed similarly using the following table.

x	y
0	0

Note that we need another point in the table. We can pick any x we want and then find y. If we pick $x = 4$, $2y - 4 = 0$, or $y = 2$, giving us the point (4, 2). The graph of the equation $2y - x = 0$ is shown in black. Since the lines intersect at (2, 1), the point **(2, 1)** is the solution of the system of equations. (You can check this by substituting 2 for x and 1 for y in each equation.)

EXAMPLE 1   Use the graphical method to find the solution of the system.

$2x + y = 4$
$y - 2x = 0$

Solution   We first graph the equation $2x + y = 4$ using the following table.

x	y
0	4
2	0

The two points and the complete graph are shown in color. We then graph $y - 2x = 0$ using the following table.

x	y
0	0
2	4

Problem 1   Use the graphical method to find the solution of the system.

$x + 2y = 4$
$2x - 4y = 0$

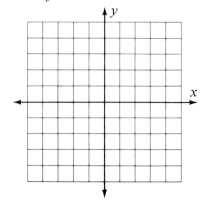

The graph for $y - 2x = 0$ is shown in black. Since the lines intersect at **(1, 2)**, the point **(1, 2)** is the solution of the system of equations. (Check this!)

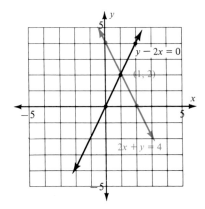

EXAMPLE 2 Use the graphical method to find the solution of the system.

$$y - 2x = 4$$
$$2y - 4x = 12$$

Solution We first graph the equation $y - 2x = 4$ using the following table.

x	y
0	4
-2	0

The two points, as well as the completed graph, are shown in color in the figure. We then graph $2y - 4x = 12$ using the accompanying table.

x	y
0	6
-3	0

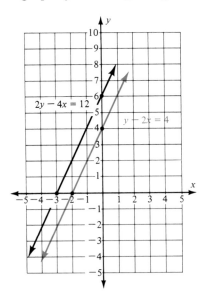

The graph of $2y - 4x = 12$ is shown in black in the figure. The two lines appear to be parallel and not intersecting. If we examine the equations more carefully, we see that by dividing the second equation by 2, we get $y - 2x = 6$. Thus one equation says $y - 2x = 4$ and the

Problem 2 Use the graphical method to find the solution of the system.

$$y - 3x = 3$$
$$2y - 6x = 12$$

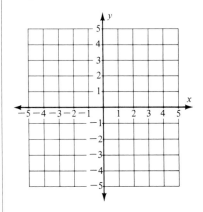

other says that $y - 2x = 6$. Obviously, both equations cannot be true at the same time and their graphs cannot intersect. Thus there is *no solution* for this system, since the two lines do not have any points in common; the system is said to be *inconsistent*. Note that the slopes of the two lines are the same (2) but their y-intercepts are different.

EXAMPLE 3   Use the graphical method to solve the system.

$$2x + y = 4$$
$$2y + 4x = 8$$

Solution   We use the table

x	y
0	4
2	0

to graph $2x + y = 4$ shown in color on the graph. To graph $2y + 4x = 8$, we first let $x = 0$ obtaining $2y = 8$ or $y = 4$. For $y = 0$, $4x = 8$ or $x = 2$. Thus, the two points on our table will be

x	y
0	4
2	0

But these points are exactly the same as those obtained in the preceding table. What does this mean? It means that the graphs of the lines $2x + y = 4$ and $2y + 4x = 8$ **coincide** (are the same). Thus, a solution of one equation is *automatically* a solution for the other. In fact, there are *infinitely many* solutions. Such a system is called *dependent*. In a dependent system, one of the equations is a **multiple** of the other. (If you multiply both sides of the first equation by 2, you get the second equation.)

B.   CONSISTENT, INCONSISTENT, AND DEPENDENT EQUATIONS

As you can see from the examples we have given, a system of equations can have *one solution* (when the lines *intersect,* as in Example 1), *no solution* (when the lines are *parallel,* as in Example 2), and *infinitely many* solutions (when the graphs of the two lines are *identical,* as in Example 3). These examples illustrate that there are three possibilities when solving a system of simultaneous equations:

1. **Consistent and independent equations:** The graphs of the equations intersect at one point, whose coordinates give the solution of the system.

2. **Inconsistent equations:** The graphs of the equations are *parallel* lines; there is *no solution* for the system.

3. **Dependent equations:** The graphs of the equations *coincide* (are the same). There are *infinitely many* solutions for this system.

Problem 3   Use the graphical method to solve the system.

$$x + 2y = 4$$
$$4y + 2x = 8$$

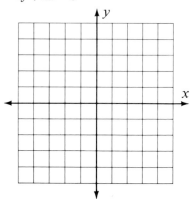

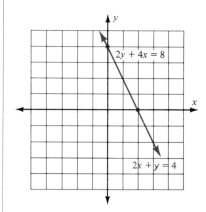

$2y + 4x = 8$

$2x + y = 4$

ANSWER   (to problem on page 466)
1.

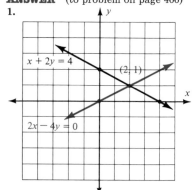

$x + 2y = 4$

$(2, 1)$

$2x - 4y = 0$

The solution is (2, 1).

Here is a diagram that might help you further.

TYPE OF LINES	SLOPES	y-INTERCEPT	NUMBER OF SOLUTIONS	TYPE OF SYSTEM
Intersecting	Different	Same or different	One	Consistent
Parallel	Same	Different	None	Inconsistent
Coinciding	Same	Same	Infinite	Dependent

EXAMPLE 4 Use the graphical method to solve the given system of equations. Classify each system as consistent (one solution), inconsistent (no solution), or dependent (infinitely many solutions).

a. $x + y = 4$
$2y - x = -1$

b. $x + 2y = 4$
$2x + 4y = 6$

c. $x + 2y = 4$
$4y + 2x = 8$

Solution

a. The table for $x + y = 4$ is   The table for $2y - x = -1$ is

x	y
0	4
4	0

x	y
0	$\dfrac{-1}{2}$
1	0

The graphs of these two lines are shown. As you can see, the solution is **(3, 1)**. (Check this!) The system is *consistent*.

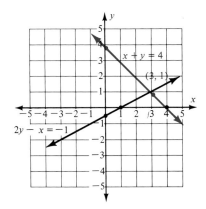

b. The table for $x + 2y = 4$ is   The table for $2x + 4y = 6$ is

x	y
0	2
4	0

x	y
0	$\dfrac{3}{2}$
3	0

Problem 4 Use the graphical method to solve the given system. Classify each system as consistent (one solution), inconsistent (no solution), or dependent (infinitely many solutions).

a. $x + y = 4$
$2y - x = 2$

b. $2x + y = 4$
$2y + 4x = 6$

c. $2x + y = 4$
$2y + 4x = 8$

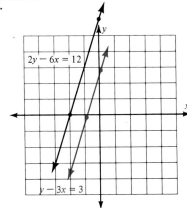

The graphs of the two lines are shown. There is no solution because the lines are parallel. The system is *inconsistent*.

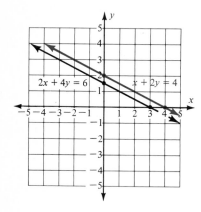

c. The table for $x + 2y = 4$ is   The table for $4y + 2x = 8$ is

x	y
0	2
4	0

x	y
0	2
4	0

which is identical to the previous table. There are infinitely many solutions because the lines coincide (see the graph). The system is *dependent*.

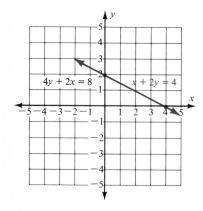

▲

Note that in a *dependent* system, one equation is a *multiple* of the other. Thus,

$$x + 2y = 4$$

and

$$4y + 2x = 8$$

are dependent because $4y + 2x = 8$ is a *multiple* of $x + 2y = 4$. Note that

$$2(x + 2y = 4)$$

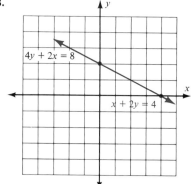

becomes

$$2x + 4y = 8$$

or

$$4y + 2x = 8$$

the second equation.

ANSWER   (to problem on page 469)

4. a.

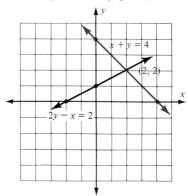

The solution is (2, 2)
The system is consistent.

b.

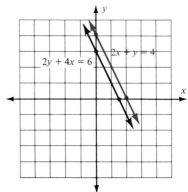

The lines are parallel.
There are no solutions.
The system is inconsistent.

c.

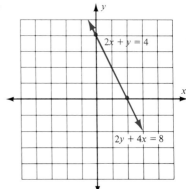

The lines coincide.
There are infinitely many solutions.
The system is dependent.

NAME

CLASS

SECTION

A, B. Solve by graphing. Label each system as consistent (one solution), inconsistent (no solution), or dependent (infinitely many solutions).

ANSWERS

1. $x + y = 4$
$x - y = -2$

2. $x + y = 3$
$x - y = -5$

1. _____

2. _____

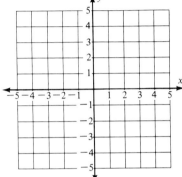

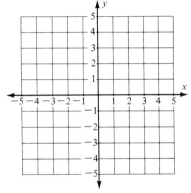

3. $x + 2y = 0$
$x - y = -3$

4. $y + 2x = -3$
$y - x = 3$

3. _____

4. _____

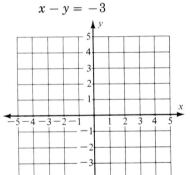

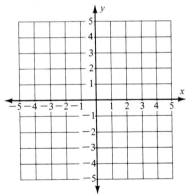

5. $3x - 2y = 6$
$6x - 4y = 12$

6. $2x + y = -2$
$8x + 4y = 8$

5. _____

6. _____

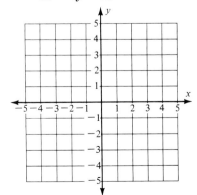

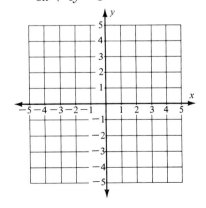

7. $3x - y = -3$
$y - 3x = 3$

8. $4x - 2y = 8$
$y - 2x = -4$

7. _____

8. _____

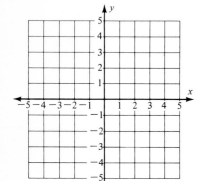

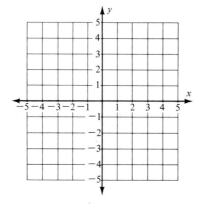

9. $2x - y = -2$
$y = 2x + 4$

10. $2x + y = -2$
$y = -2x + 4$

9. _____

10. _____

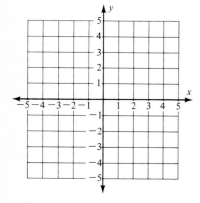

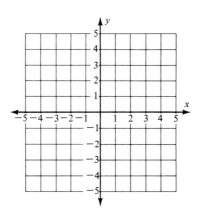

11. $y = -2$
$2y = x + 2$

12. $3y = 6 - x$
$y = 3$

11. _____

12. _____

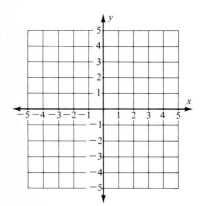

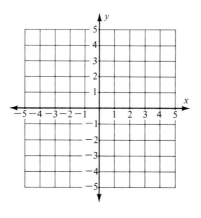

13. $x = 3$
$y = 2x - 4$

14. $y = -x + 2$
$x = -1$

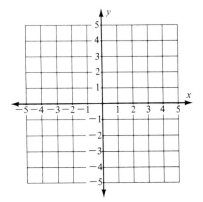

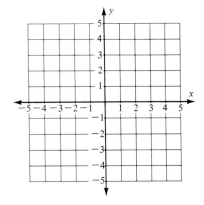

15. $x + y = 3$
$2x - y = 0$

16. $x + y = 5$
$x - 4y = 0$

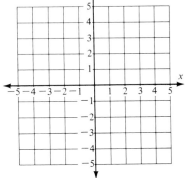

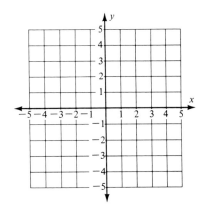

17. $5x + y = 5$
$5x = 15 - 3y$

18. $2x - y = -4$
$4x = 4 + 2y$

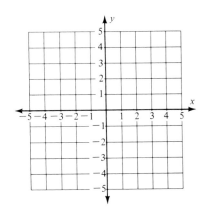

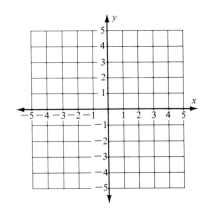

NAME

19. $3x + 4y = 12$
$8y = 24 - 6x$

20. $2x - 3y = 6$
$6x = 18 + 9y$

19. _____

20. _____

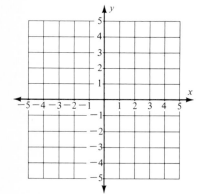
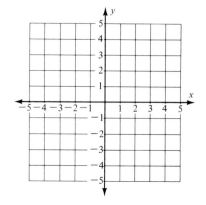

✓ SKILL CHECKER

Solve.

21. $3x + 72 = 2.5x + 74$

22. $5x + 20 = 3.5x + 26$

21. _____

22. _____

Determine if the given point is a solution of the equation.

23. $(3, 5);$ $2x + y = 11$

24. $(-1, 4);$ $2x - y = -6$

25. $(-1, 2);$ $2x - y = 0$

26. $(-2, 6);$ $3x - y = 0$

23. _____

24. _____

25. _____

26. _____

7.1 USING YOUR KNOWLEDGE

Do you want to save money when renting a car? Suppose you have the choice of renting one of those subcompacts we mentioned (see page 437). The cost y of renting a subcompact from company A is $10 a day and 10¢ per mile. If x is the number of miles traveled, the equation for the daily cost is

$y = 0.10x + 10$

Now, suppose Hurt's rentals has a car whose cost is $5 a day and 15¢ per mile.

1. If y is the daily cost for the Hurt's car and x is the number of miles traveled, write an equation that will represent the daily cost y.

1. _____

2. How many miles do you have to travel so that the cost for both companies is the same?

2. _____

3. Graph the cost for the subcompact and the Hurt's car in the accompanying coordinate system.

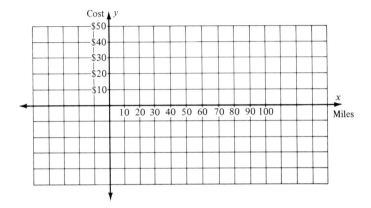

4. Based on the graph, which car would you rent if you were going to travel 60 mi on that day?

5. Based on the graph, which car would you rent if you were going to travel 40 mi on that day?

7.2 SOLVING SYSTEMS OF EQUATIONS BY SUBSTITUTION

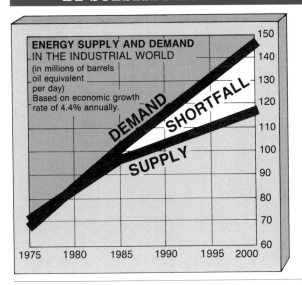

ENERGY SUPPLY AND DEMAND IN THE INDUSTRIAL WORLD

(in millions of barrels oil equivalent per day) Based on economic growth rate of 4.4% annually.

REVIEW

Before starting this section, you should know:

1. How to solve linear equations.
2. How to determine if an ordered pair satisfies an equation.

OBJECTIVES

You should be able to:

A. Solve a system of two equations in two variables by using the substitution method.

B. Determine if a system of two equations in two variables is consistent (one solution), inconsistent (no solution), or dependent (infinitely many solutions).

As you can see from the graph, the supply and the demand were about the same in the year 1980. However, this solution is only an *approximate* one because the x-scale representing the years is numbered at 5-yr intervals, and it is hard to pinpoint the exact point at which the lines intersect. Suppose we have the equations for the supply and the demand between 1975 and 1985. These equations are

Supply: $y = 2.5x + 72$ x = the number of years elapsed after 1975

Demand: $y = 3x + 70$

Can we now tell *exactly* where the lines meet? Not graphically! For one thing, if we let $x = 0$ in the first equation, we obtain $y = 2.5 \cdot 0 + 72$, or $y = 72$. Thus, we either need a piece of graph paper with 72 units, or else we have to make each division on the graph paper 10 units, thus *losing accuracy*. But there is a way out. We can use an *algebraic* method rather than a *graphical* one. Since we are looking for the point at which the supply (y) is the same as the demand (y), we may *substitute* the expression for y in the demand equation, that is, $3x + 70$, into the supply equation. Thus, we have

$3x + 70 = 2.5x + 72$

$3x = 2.5x + 2$ Subtracting 70

$0.5x = 2$ Subtracting $2.5x$

$\dfrac{0.5x}{0.5} = \dfrac{2}{0.5}$ Dividing by 0.5

$x = 4$ Simplifying

Thus, 4 yr after 1975 (or in 1979) the supply would equal the demand. At this time the demand would be

$y = 3(4) + 70 = 12 + 70 = 82$ million barrels

A. THE SUBSTITUTION METHOD

Here is a summary of the substitution method we have just used.

TO SOLVE A SYSTEM OF EQUATIONS BY THE SUBSTITUTION METHOD

1. Solve one of the equations for x or y.

2. Substitute the resulting expression into the other equation. (Now you have an equation in one variable.)

3. Solve the new equation for the variable.

4. Substitute the value of the variable into one of the original equations and solve this equation to get the value for the second variable.

5. Check the solution by substituting the numerical values of the variables in both equations.

EXAMPLE 1   Use the given five-step procedure to solve the system.

$$x + y = 8$$
$$2x - 3y = -9$$

Solution

1. Solve one of the equations for x or y (we solve the first equation for y). $y = 8 - x$

2. Substitute $8 - x$ for y into $2x - 3y = -9$. $2x - 3(8 - x) = -9$

3. Solve the new equation for the variable. $2x - 3(8 - x) = -9$

   Simplify. $2x - 24 + 3x = -9$

   Combine like terms. $5x - 24 = -9$

   Add 24 to both sides. $5x = 15$

   Divide by 5. $x = 3$

4. Substitute the value of the variable $x = 3$ into one of the original equations. (We substitute $x = 3$ in the equation $x + y = 8$.) Then solve for the second variable. $3 + y = 8$
   $y = 5$

5. Our solution is the ordered pair (3, 5).

Check: When $x = 3$, $y = 5$,

$$x + y = 8$$

becomes

$$3 + 5 = 8$$
$$8 = 8$$

which is true.

$$2x - 3y = -9$$

Problem 1   Use the given five-step procedure to solve the system.

$$x + y = 5$$
$$2x - 3y = -5$$

becomes

$$2(3) - 3(5) = -9$$
$$6 - 15 = -9$$
$$-9 = -9$$

which is true.

Thus, our solution $(3, 5)$ is correct.

EXAMPLE 2   Solve the system.

$$x + 2y = 4$$
$$2x + 4y = 6$$

Solution   We use the five-step procedure.

1. Solve one of the equations for one of the variables (we solve the first equation for x).

   $x = 4 - 2y$

2. Substitute $x = 4 - 2y$ into $2x + 4y = 6$.

   $2(4 - 2y) + 4y = 6$

3. Solve the new equation for the variable. Note that we have obtained the result $8 = 6$, which is never true (a contradiction). Since our procedure is correct, we conclude that the given system has *no solution*; it is *inconsistent*.

   $8 - 4y + 4y = 6$
   $8 = 6$

Note that if you divide the second equation by 2, you get $x + 2y = 3$ which *contradicts* the first equation, $x + 2y = 4$.

EXAMPLE 3   Solve the system.

$$x + 2y = 4$$
$$4y + 2x = 8$$

Solution   As before, we use the five-step procedure.

1. Solve the first equation for x.

   $x = 4 - 2y$

2. Substitute $x = 4 - 2y$ into $4y + 2x = 8$.

   $4y + 2(4 - 2y) = 8$

3. Solve the new equation for the variable. Note that in this case we have obtained a true statement, regardless of the value we assign to x or to y.

   $4y + 8 - 4y = 8$
   $8 = 8$

   Thus, the equations are dependent, that is, there are infinitely many solutions. For example, if we let $x = 0$ in the equation $x + 2y = 4$, we obtain $2y = 4$, or $y = 2$. Similarly, if we let $x = 0$ in the equation $4y + 2x = 8$, we obtain $4y = 8$, or $y = 2$. Thus $(0, 2)$ is a solution for both equations. It can also be shown that $x = 2$ and $y = 1$ are solutions for both equations. Therefore $(2, 1)$ is another solution, and so on. Note that if you divide the second equation by 2 and rearrange, you get $x + 2y = 4$, which is identical to the first equation. Thus, any solution of the first equation is also a solution of the second equation.

Problem 2   Solve the system.

$$x - 3y = 6$$
$$2x - 6y = 8$$

Problem 3   Solve the system.

$$x - 3y = 6$$
$$6y - 2x = -12$$

B. CONSISTENT, INCONSISTENT, AND DEPENDENT SYSTEMS

WHEN WE USE THE SUBSTITUTION METHOD, ONE OF THREE THINGS CAN OCCUR:

1. The equations are *consistent;* there is only *one* solution (x, y).

2. The equations are *inconsistent;* you will get a contradictory (false) statement and there will be *no solution.*

3. The equations are *dependent;* you will get a statement that is true for all values of the remaining variable and there will be infinitely *many* solutions.

Keep this in mind when doing the exercises, and be very careful with your arithmetic!

ANSWERS   (to problems on pages 480–481)
1. (2, 3)
2. There is no solution. The system is inconsistent.
3. There are infinitely many solutions. The equations are dependent. Here are some solutions: $(0, -2)$, $(6, 0)$, $(3, -1)$.

A, B. Solve by the substitution method. Label each system as consistent (one solution), inconsistent (no solution), or dependent (infinitely many solutions).

1. $y = 2x - 4$
    $-2x = y - 4$

2. $y = 2x + 2$
    $-x = y + 1$

3. $x + y = 5$
    $3x + y = 9$

4. $x + y = 5$
    $3x + y = 3$

5. $y - 4 = 2x$
    $y = 2x + 2$

6. $y + 5 = 4x$
    $y = 4x + 7$

7. $x = 8 - 2y$
    $x + 2y = 4$

8. $x = 4 - 2y$
    $x - 2y = 0$

9. $x + 2y = 4$
    $x = -2y + 4$

10. $x + 3y = 6$
    $x = -3y + 6$

11. $x = 2y + 1$
    $y = 2x + 1$

12. $y = 3x + 2$
    $x = 3y + 2$

13. $2x - y = -4$
    $4x = 4 + 2y$

14. $5x + y = 5$
    $5x = 15 - 3y$

15. $x = 5 - y$
    $0 = x - 4y$

16. $x = 3 - y$
    $0 = 2x - y$

1. _____

2. _____

3. _____

4. _____

5. _____

6. _____

7. _____

8. _____

9. _____

10. _____

11. _____

12. _____

13. _____

14. _____

15. _____

16. _____

C. Applications

17. The supply y of a certain item is given by the equation $y = 2x + 8$, where x is the number of days elapsed. If the demand is given by $y = 4x$, in how many days will the supply equal the demand?

17. _____

18. The supply of a certain item is $y = 3x + 8$, where x is the number of days elapsed. If the demand is given by $y = 4x$, in how many days will the supply equal the demand?

18. _____

19. A company has 10 units of a certain item and can manufacture 5 items each day. If the demand for the item is $y = 7x$, where x is the number of days elapsed, in how many days will the demand equal the supply?

19. _____

20. Clonker Manufacturing has 12 clonkers in stock. They manufacture 3 other clonkers each day. If the clonker demand is 7 each day, in how many days will the supply equal the demand?

20. _____

✓ **SKILL CHECKER**

Solve.

21. $-0.5x = -4$

22. $-0.2y = -6$

23. $9x = -9$

24. $11y = -11$

21. _____

22. _____

23. _____

24. _____

7.2 USING YOUR KNOWLEDGE

The ideas presented in this section are very important in other fields. Use your knowledge to *solve* the following problems.

1. The total inductance of the inductors L_1 and L_2 in an oscillator must be 400 microhenrys. Thus,

$$L_1 = -L_2 + 400$$

To provide the correct regeneration for the oscillator circuit, $L_2/L_1 = 4$, that is,

$$L_2 = 4L_1$$

Solve the system

$$L_1 = -L_2 + 400$$
$$L_2 = 4L_1$$

by substitution.

2. The equations for the resistors in a voltage divider must be such that

$$R_1 = 3R_2$$
$$R_1 + R_2 = 400$$

Solve for R_1 and R_2 using the substitution method.

3. The total revenue R for a certain manufacturer is

$$R = 5x$$

the total cost C is

$$C = 4x + 500$$

where x represents the number of units produced and sold.

  a. Use your knowledge of the substitution method to write the equation that will result when

$$R = C$$

  b. The point at which $R = C$ is called the *break-even* point. Find the number of units the manufacturer must produce and sell in order to break even.

1. _____

2. _____

3. _____

CALCULATOR CORNER

If a system of equations has a solution, this solution can be substituted into the original equations and the result checked with a calculator.

Thus, in Example 3, the solution of the system is the ordered pair (3, 5). To check that this is indeed the case, we let $x = 3$ and $y = 5$ in the first equation using the keystrokes

$\boxed{3}\ \boxed{+}\ \boxed{5}\ \boxed{=}$

which yields 8, a true statement. The second equation is $2x - 3y = -9$. When $x = 3$ and $y = 5$, we key in

$\boxed{2}\ \boxed{\times}\ \boxed{3}\ \boxed{-}\ \boxed{3}\ \boxed{\times}\ \boxed{5}\ \boxed{=}$

obtaining -9. (If your calculator does not have algebraic logic, you have to use parentheses around the 3×5.) If you can use calculators, check the solutions you obtain in Exercises 7.2 and 7.3.

7.3 SOLVING SYSTEMS OF EQUATIONS BY ELIMINATION

The man in the photo sells coffee, ground to order. A customer wanted 10 lb of a mixture of coffee A costing $2 per pound, and coffee B costing $1.50 per pound. If the price for the purchase amounted to $18, how many pounds of each of the coffees did the customer get?

The solution of this problem uses two things we have learned: the RSTUV method and solving a system of equations. The information above can be summarized in a table like this:

	PRICE	POUNDS	TOTAL PRICE
Coffee A	2.00	a	$2a$
Coffee B	1.50	b	$1.50b$
Totals		$a + b$	$2a + 1.50b$

Since the customer bought a total of 10 lb of coffee, we know that

$$a + b = 10$$

Also, since the purchase came to $18, we know that

$$2a + 1.50b = 18$$

These two equations can be written as the system

$$\begin{aligned} a + b &= 10 \\ 2a + 1.50b &= 18 \end{aligned} \qquad (1)$$

A. SOLVING SYSTEMS BY ELIMINATION

To solve the problem, we need to solve this system of equations. We shall now use a method which consists of replacing the given system with an equivalent system, until we get a system with an obvious solution. To do this, we first write the equations in the form $Ax + By = C$. Recall that an *equivalent* system is one that has the *same* solution as the given one. Note that the equation

$A = B$

is equivalent to the equation

$kA = kB, \qquad k \neq 0$

Thus, the system

$A = B$

$C = D$

is equivalent to the system

$$A = B$$
$$k_1 A + k_2 C = k_1 B + k_2 D$$

Suppose that we multiply the first equation in the given system by -2; we then get the equivalent system

$$-2a - 2b \quad = -20$$

Add the equations. $\dfrac{2a + 1.50b = \quad 18}{0 - 0.50b = -2}$

$-0.50b = -2$

Multiply by -2 because we want the coefficients of A to be opposites (like -2 and 2).

Divide by -0.50. $\qquad b = 4$

Substitute 4 for b in $\qquad a + 4 = 10$
$a + b = 10$.

Solve for a. $\qquad a = 6$

Thus the customer bought 6 lb of coffee A and 4 lb of coffee B. This answer can be verified. If the customer bought 6 lb of coffee A and 4 of B, she did buy 10 lb. Her price for coffee A was $2 \cdot 6 = \$12$ and for coffee B, $1.50 \cdot 4 = \$6$. Thus the entire cost was $\$12 + \$6 = \$18$, as stated. What have we done? Well, the technique here depends on the fact that one (or both) of the equations in a system can be multiplied by a nonzero number to obtain two equivalent equations with opposite coefficients of x (or y). Here is the principle.

> You can multiply (or divide) one or both of the equations by any nonzero number you wish, but the idea is to obtain an equivalent system in which the coefficients of the x's (or of the y's) are opposites, thus *eliminating* x or y when the equations are added.

EXAMPLE 1 Solve the system.

$$2x + y = 1$$
$$3x - 2y = -9$$
(1)

Solution Remember the idea: to multiply one or both of the equations by a number or numbers that will cause either the coefficients of x or

Problem 1 Solve the system.

$$3x + y = 1$$
$$3x - 4y = 11$$

the coefficients of y to be opposites. This can be accomplished by multiplying the first equation by 2.

$$2x + y = 1 \xrightarrow{\text{Multiply by 2}} 4x + 2y = 2$$

$$3x - 2y = -9 \xrightarrow{\text{Leave as is}} \underline{3x - 2y = -9}$$

Add the equations.
$$7x + 0 = -7$$
$$7x = -7$$

Divide by 7.
$$x = -1$$

Substitute -1 for x in
$$2(-1) + y = 1$$

$2x + y = 1$.
$$-2 + y = 1$$

Add 2.
$$y = 3$$

Thus the solution of the system is $(-1, 3)$.

Check: When $x = -1$ and $y = 3$, $2x + y = 1$ becomes

$$2(-1) + 3 = 1$$
$$-2 + 3 = 1$$
$$1 = 1$$

a true statement.
$3x - 2y = -9$ becomes

$$3(-1) - 2(3) = -9$$
$$-3 - 6 = -9$$
$$-9 = -9$$

which is also true.

B. CONSISTENT, INCONSISTENT, AND DEPENDENT SYSTEMS

As we have seen, not all systems have solutions. How do we find out if they don't? Look at the next example, which shows a contradiction for a system that has no solution.

EXAMPLE 2 Solve the system.

$$2x + 3y = 3$$
$$4x + 6y = -6$$
$$(2)$$

Solution In this case, we are going to try to eliminate the variable x by multiplying the first equation by -2.

$$2x + 3y = 3 \xrightarrow{\text{Multiply by } -2} -4x - 6y = -6$$

$$4x + 6y = -6 \xrightarrow{\text{Leave as is}} \underline{4x + 6y = -6}$$
$$(2)$$
$$0 + 0 = -12 \quad \text{Adding}$$
$$0 = -12$$

Of course, this is a contradiction, so there is no solution; the system is *inconsistent*.

Problem 2 Solve.

$$3x + 2y = 1$$
$$6x + 4y = 12$$

EXAMPLE 3 Solve the system.

$$2x - 4y = 6$$
$$-x + 2y = -3 \qquad (3)$$

Solution We shall try to eliminate the variable x by multiplying the second equation by 2. We obtain

$$2x - 4y = 6 \quad \xrightarrow{\text{Leave as is}} \quad 2x - 4y = 6$$
$$-x + 2y = -3 \quad \xrightarrow{\text{Multiply by 2}} \quad \underline{-2x + 4y = -6} \qquad (3)$$
$$0 + 0 = 0 \quad \text{Adding}$$
$$0 = 0$$

Lo and behold, we have eliminated both variables! However, notice that if we had multiplied the second equation in the original system by -2, we would have obtained

$$2x - 4y = 6 \quad \xrightarrow{\text{Leave as is}} \quad 2x - 4y = 6$$
$$-x + 2y = -3 \quad \xrightarrow{\text{Multiply by } -2} \quad 2x - 4y = 6 \qquad (3)$$

This means that the first equation is a multiple of the second one, that is, they are equivalent equations. When a system of equations consists of two equivalent equations, the system is said to be *dependent,* and any solution of one equation is a solution of the other. Because of this, the system has infinitely many solutions. For example, if we let x be 0 in the first equation of Example 3, then y is $-\frac{3}{2}$, and $(0, -\frac{3}{2})$ is a solution of the system. Similarly, if we let y be 0 in the first equation, x is 3, and we obtain the solution $(3, 0)$. Many other solutions are possible; try to find some of them. ▲

Finally, in some cases we cannot multiply just one of the equations by an integer that will cause the coefficients of one of the variables to be opposites. For example, to solve the system

$$2x + 3y = 3$$
$$5x + 2y = 13 \qquad (4)$$

we must multiply both equations by integers chosen so that the coefficients of one of the variables are opposites. We can do this using either of the following methods.

METHOD 1. Multiply the first equation by 5 and the second one by -2 to obtain an equivalent system.

$$2x + 3y = 3 \quad \xrightarrow{\text{Multiply by 5}} \quad 10x + 15y = 15$$
$$5x + 2y = 13 \quad \xrightarrow{\text{Multiply by } -2} \quad \underline{-10x - 4y = -26} \qquad (4)$$

Add.
$$0 + 11y = -11$$
$$11y = -11$$

Divide by 11.
$$y = -1$$

Substitute -1 for y in
$$2x + 3(-1) = 3$$
$2x + 3y = 3$.

Simplify.
$$2x - 3 = 3$$

Add 3.
$$2x = 6$$

Divide by 2.
$$x = 3$$

Problem 3 Solve.

$$3x - 6y = 9$$
$$-x + 2y = -3$$

Thus the solution of the system is $(3, -1)$. This time we eliminated the x and solved for y. Alternatively we can eliminate the y first, as shown next.

METHOD 2. This time, we eliminate the y:

$$2x + 3y = 3 \xrightarrow{\text{Multiply by } -2} -4x - 6y = -6$$

$$5x + 2y = 13 \xrightarrow{\text{Multiply by } 3} \underline{15x + 6y = 39}$$

$$\text{Add.} \qquad \qquad \qquad 11x + 0 = 33 \tag{4}$$

$$11x = 33$$

Divide by 11. $x = 3$

Substitute 3 for x in $2(3) + 3y = 3$
$2x + 3y = 3$.

Simplify. $6 + 3y = 3$

Subtract 6. $3y = -3$

Divide by 3. $y = -1$

Thus, the solution is $(3, -1)$ as before.

Before going on, here are two important reminders:

1. There are *three* possibilities when solving simultaneous equations.
  a. *Consistent and independent* equations have *one solution.*
  b. *Inconsistent* equations have *no solution.* You can recognize them when you get a contradiction (a false statement) in your work, as we did in Example 2. (In Example 2, we got $0 = -12$, a contradiction.)
  c. *Dependent* equations have *infinitely many solutions.* You can recognize them when you get a true statement such as $0 = 0$ in Example 3. Remember that any solution of one of these equations is a solution of the other.

2. The second reminder has to do with the position of the variables in the equations. All the equations with which we have worked were written in the form

$$ax + by = c$$
$$dx + ey = f \tag{5}$$

Constant terms

y column

x column

If the equations are not in this form, rewrite them using this form. It helps to keep things straight! Practice this as in the next example.

EXAMPLE 4 Solve the system.

$$5y + 2x = 9$$
$$2y = 8 - 3x \tag{6}$$

Solution We first write the system in standard form as shown in (5), that is, the x's first, then the y's, and then the constants. The result is the equivalent system

$$2x + 5y = 9$$
$$3x + 2y = 8 \tag{7}$$

Problem 4 Solve.

$$5x + 4y = 6$$
$$3y = 4x - 11$$

This time we multiply the first equation by 3 and the second one by -2 so that, upon addition, the x's will be eliminated.

$$2x + 5y = 9 \xrightarrow{\text{Multiply by 3}} 6x + 15y = 27$$

$$3x + 2y = 8 \xrightarrow{\text{Multiply by } -2} \underline{-6x - 4y = -16}$$

(7)

Add.
$$0 + 11y = 11$$

$$11y = 11$$

Divide by 11.
$$y = 1$$

Substitute 1 for y in
$2x + 5y = 9$.
$$2x + 5(1) = 9$$

Simplify.
$$2x + 5 = 9$$

Subtract 5.
$$2x = 4$$

Divide by 2.
$$x = 2$$

Thus the solution is $(2, 1)$. You should verify this result to make sure it satisfies both equations.

ANSWERS   (to problems on pages 490–491)

3. Many solutions. The system is dependent.

4. $(2, -1)$

NAME

CLASS

SECTION

ANSWERS

A, B. Solve each system. If the system is inconsistent or dependent, say so.

1. $x + y = 3$
$x - y = -1$

2. $x + y = 5$
$x - y = 1$

3. $x + 3y = 6$
$x - 3y = -6$

4. $x + 2y = 4$
$x - 2y = 8$

5. $2x + y = 4$
$4x + 2y = 0$

6. $3x + 5y = 2$
$6x + 10y = 5$

7. $2x + 3y = 6$
$4x + 6y = 2$

8. $3x - 5y = 4$
$-6x + 10y = 0$

9. $x - 5y = 15$
$x + 5y = 5$

10. $-3x + 2y = 1$
$2x + y = 4$

11. $x + 2y = 2$
$2x + 3y = -10$

12. $3x - 2y = -1$
$x + 7y = -8$

13. $3x - 4y = 10$
$5x + 2y = 34$

14. $5x - 4y = 6$
$3x + 2y = 8$

15. $11x - 3y = 25$
$5x + 8y = 2$

16. $12x + 8y = 8$
$7x - 5y = 24$

17. $2x + 3y = 21$
$3x = y + 4$

18. $2x - 3y = 16$
$x = y + 7$

19. $x = 1 + 2y$
$-y = x + 5$

20. $3y = 1 - 2x$
$3x = -4y - 1$

21. $\dfrac{x}{4} + \dfrac{y}{3} = 4$

$\dfrac{x}{2} - \dfrac{y}{6} = 3$

(*Hint:* Multiply by the LCD first.)

22. $\dfrac{x}{5} + \dfrac{y}{6} = 5$

$\dfrac{2x}{5} + \dfrac{y}{3} = -2$

(*Hint:* Multiply by the LCD first.)

23. $\dfrac{1}{4}x - \dfrac{1}{3}y = -\dfrac{5}{12}$

$\dfrac{1}{5}x + \dfrac{2}{5}y = 1$

(*Hint:* Multiply by the LCD first.)

24. $\dfrac{x}{2} + \dfrac{y}{2} = \dfrac{5}{2}$

$\dfrac{x}{2} - \dfrac{y}{3} = \dfrac{5}{2}$

(*Hint:* Multiply by the LCD first.)

1. _____

2. _____

3. _____

4. _____

5. _____

6. _____

7. _____

8. _____

9. _____

10. _____

11. _____

12. _____

13. _____

14. _____

15. _____

16. _____

17. _____

18. _____

19. _____

20. _____

21. _____

22. _____

23. _____

24. _____

✓ SKILL CHECKER

Write an expression corresponding to the given sentence.

25. The sum of the nickels (n) and the dimes (d) equals 300.

25. _____

26. The difference of h and w is 922.

26. _____

27. The product of 4 and $(x - y)$ is 48.

27. _____

28. The quotient of x and y is 80.

28. _____

29. The number m is 3 less than the number n.

29. _____

30. The number m is 5 more than the number n.

30. _____

7.3 USING YOUR KNOWLEDGE

Have you read *Alice in Wonderland?* Do you know who the author of this book is? The answer is Lewis Carroll. Although better known as the author of *Alice in Wonderland,* Lewis Carroll was also a mathematician and logician. He also wrote another book called *Through the Looking Glass.* In this book, one of the characters, Tweedledee, is talking to Tweedledum. Here is the conversation.

Tweedledee: The sum of your weight and twice mine is 361 pounds.

Tweedledum: Contrariwise, the sum of your weight and twice mine is 360 pounds.

1. If Tweedledee weighs x pounds and Tweedledum y pounds, find their weights using the knowledge gained in this section.

1. _____

7.4 APPLICATIONS: WORD PROBLEMS

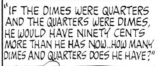

OBJECTIVES

REVIEW

Before starting this section, you should know:

1. The RSTUV procedure used to solve word problems (Section 3.5).
2. How to solve a system of two equations in two unknowns.

OBJECTIVES

You should be able to:

A. Solve word problems involving coins.
B. Solve general word problems.
C. Solve word problems using the distance formula.

A. COIN PROBLEMS

Patty is upset, she needs *help!* Why? She has not learned about systems of equations and the problem is beyond her, but we can do it! In the preceding sections, we studied systems of equations. We now use that knowledge to solve word problems involving two variables. As usual, we use the RSTUV method. If you have forgotten how that goes, go back and review the RSTUV method. Ready? Patty's problem is a type of word problem popular in algebra, the coin problem. Before tackling it, let us get down to nickels, dimes, and quarters. Suppose you are down to your last nickel; you have 5¢.

If you have **2** nickels, you have $5 \cdot 2 = 10$ cents.
If you have **3** nickels, you have $5 \cdot 3 = 15$ cents.
If you have **n** nickels, you have $5 \cdot n = 5n$ cents.

Follow the pattern:
$5 \cdot 1$
$5 \cdot 2$
$5 \cdot 3$
$5 \cdot n$

The same thing can be done with dimes.

If you have **1** dime, you have $10 \cdot 1 = 10$ cents.
If you have **2** dimes, you have $10 \cdot 2 = 20$ cents.
If you have **n** dimes, you have $10 \cdot n = 10n$ cents.

Follow the pattern:
$10 \cdot 1$
$10 \cdot 2$
$10 \cdot n$

Let us make a table that will work for nickels, dimes, and quarters and will help us summarize the information.

	VALUE (CENTS)	×	HOW MANY	=	TOTAL VALUE
Nickels	5		n		$5n$
Dimes	10		d		$10d$
Quarters	25		q		$25q$
Half-dollars	50		h		$50h$

Now we are ready to *help* poor Patty. We do it in five steps, RSTUV.

1. **R**ead the problem carefully. We are asked how many dimes and quarters the man has.
2. **S**elect d to stand for the number of dimes and q for the number of quarters.
3. **T**ranslate the problem. We do it by translating each of the sentences in (a) and (b) in the margin.

   a. A man has 20 coins consisting of dimes and quarters:

   $$20 = d + q$$

   b. This sentence seems hard to translate. Let us take the easy part first, how much money he has now. Since he has d dimes, the table tells us that he has $10d$ (cents). He also has q quarters, which are worth $25q$ (cents). Thus he has

   $(10d + 25q)$ cents

   What would happen if the dimes were quarters and the quarters dimes? We simply would change the amount the coins are worth, like this:

   He would have $(25d + 10q)$ cents

   Now let us translate the sentence:

If the dimes were quarters and the quarters were dimes	he would have	ninety cents more than he has now.
$\underbrace{25d + 10q}$	$\underbrace{=}$	$\underbrace{(10d + 25q) + 90}$

If we put together the information from (a) and (b), we have the system of equations:

$$\begin{aligned} d + q &= 20 \\ 25d + 10q &= (10d + 25q) + 90 \end{aligned} \qquad (1)$$

This system has to be written with all the variables and constants in the proper columns. We do this by subtracting $10d$ and $25q$ from both sides of the second equation, obtaining

$$\begin{aligned} d + q &= 20 \\ 15d - 15q &= 90 \end{aligned} \qquad (2)$$

We then divide each term in the second equation by 15 to get

$$\begin{aligned} d + q &= 20 \\ d - q &= 6 \end{aligned} \qquad (3)$$

4. **U**se the algebra we have learned to solve systems of equations.

$$\begin{aligned} d + q &= 20 \\ d - q &= 6 \end{aligned} \qquad (3)$$

Add.

$$2d = 26$$

Divide by 2.

$$d = 13 \quad \text{(number of dimes)}$$

Substitute 13 for d in $d + q = 20$.

$$13 + q = 20$$
$$q = 7 \quad \text{(number of quarters)}$$

Thus the man has 13 dimes ($1.30) and 7 quarters ($1.75), a total of $3.05.

5. **V**erify that the answer is correct. If the dimes were quarters and the quarters were dimes, the man would have 13 quarters ($3.25) and 7 dimes ($0.70), a total of $3.95, which is indeed $0.90 more than the $3.05 he now has. Patty, you got your help!

a.

b.

EXAMPLE 1 Jack has $3 in nickels and dimes. He has twice as many nickels as he has dimes. How many nickels and how many dimes does he have?

Solution As usual, we use the RSTUV method.

1. Read the problem. We are asking for the number of nickels and dimes.
2. Select n to be the number of nickels and d the number of dimes.
3. Translate the problem. Jack has $3 (300 cents) in nickels and dimes:

$$300 = 5n + 10d$$

He has twice as many nickels as he has dimes:

$$n = 2d$$

We then have the system

$$5n + 10d = 300$$
$$n = 2d \qquad (4)$$

4. Use algebra to solve this system. This time it is easy to use the substitution method.

$5n + 10d = 300 \xrightarrow{\text{Letting } n = 2d}$	$5(2d) + 10d = 300$
Simplify.	$10d + 10d = 300$
Combine like terms.	$20d = 300$
Divide by 20.	$d = 15$
Substitute 15 for d in $n = 2d$.	$n = 2(15) = 30$

Thus Jack has 15 dimes ($1.50) and 30 nickels ($1.50).

5. Verify this answer. Since Jack has $3 ($1.50 + $1.50) and he does have twice as many nickels as dimes, the answer is correct.

B. GENERAL PROBLEMS

We can also use systems of equations to solve other problems. Here is an interesting one.

EXAMPLE 2 The greatest weight difference recorded for a married couple is 922 lb (Mills Darden of North Carolina and his wife Mary). Their combined weight is 1118 lb. What is the weight of each of the Dardens? (He is the heavy one.)

Solution

1. We are asked for the weight of the Dardens.
2. Let h be the weight of the husband and w be the weight of the wife.
3. Translate the problem. The weight difference is 922 lb:

$$h - w = 922$$

Their combined weight is 1118 pounds:

$$h + w = 1118$$

We then have the system

$$h - w = 922$$
$$h + w = 1118 \qquad (5)$$

Problem 1 Jill has $2 in nickels and dimes. She has twice as many nickels as she has dimes. How many nickels and how many dimes does she have?

Problem 2 At birth, the Stimson twins weighed a total of 35 oz. If their weight difference was 3 oz, what was the weight of each of the twins?

4. Use algebra to solve system (5):

$$h - w = 922$$
$$\underline{h + w = 1118}$$
$$2h = 2040 \quad \text{Adding}$$
$$h = 1020 \quad \text{Dividing by 2}$$
$$1020 + w = 1118 \quad \text{Substituting } \mathbf{1020} \text{ for } h \text{ in } h + w = 1118$$
$$w = 98 \quad \text{Subtracting } 1020$$

Thus Mary weighs 98 lb and Mills weighs 1020 lb.

5. You can verify this in the *Guinness Book of Records!*

C. DISTANCE PROBLEMS

Remember the distance problems we solved earlier? They can also be done using two variables. The procedure is about the same! We write the given information in a chart labeled $R \times T = D$ and, as usual, use the RSTUV method.

EXAMPLE 3 The world's strongest current is the Saltstraumen in Norway. The current is so strong that a boat, which can go 48 mi downstream (with the current) in 1 hr, takes 4 hr to go the same 48 mi upstream (against the current). How fast is the current flowing?

Problem 3 A plane goes 1200 mi with a tail wind in 3 hr. It takes 4 hr to travel the same distance against the wind. Find the speed of the wind and the speed of the plane in still air.

Solution

1. Read the problem carefully.

2. Let x be the speed of the boat in still water and y be the speed of the current. Then $(x + y)$ is the speed of the boat going downstream; $(x - y)$ is the speed of the boat going upstream.

3. We enter this information in a chart:

	R	$\times$ T $=$	D	
Downstream:	$x + y$	1	$1 \cdot (x + y)$	$\rightarrow x + y = 48$
Upstream:	$x - y$	4	$4 \cdot (x - y)$	$\rightarrow 4(x - y) = 48$

4. Our system of equations is simplified as follows:

$$x + y = 48 \quad \xrightarrow{\text{Leave as is}} \quad x + y = 48$$

$$4(x - y) = 48 \quad \xrightarrow{\text{Divide by 4}} \quad x - y = 12$$

Add. $\phantom{4(x - y) = 48 \quad \text{Divide by 4}} \quad \overline{2x = 60}$

Divide by 2. $\phantom{4(x - y) = 48 \quad \text{Divide by 4}2x} \quad x = 30$

Substitute 30 for x in $ 30 + y = 48$
$x + y = 48$.

Subtract 30. $\phantom{4(x - y) = 48 \quad \text{Div}} y = 18$

Thus, the speed of the boat in still water is $x = 30$ mi/hr and the speed of the current is 18 mi/hr.

5. The verification is left for the student.

NAME

CLASS

SECTION

A. Solve.

ANSWERS

1. The sum of two numbers is 102. Their difference is 16. What are the numbers?

1. _____

2. The difference between two numbers is 28. Their sum is 82. What are the numbers?

2. _____

3. The sum of two integers is 126. If one of the integers is 5 times the other, what are the integers?

3. _____

4. The difference between two integers is 245. If one of the integers is 8 times the other, find the integers.

4. _____

5. The difference between two numbers is 16. One of the numbers exceeds the other by 4. What are the numbers?

5. _____

6. The sum of two numbers is 116. One of the numbers is 50 less than the other. What are the numbers?

6. _____

7. Longs Peak is 145 ft higher than Pikes Peak. If you were to put these two peaks on top of each other, you would still be 637 ft short of reaching the elevation of Mt. Everest, 29,002 ft. Find the elevations of Longs Peak and of Pikes Peak.

7. _____

B. Solve.

8. Two brothers had a total of $7500 in separate bank accounts. One of the brothers complained, and the other brother took $250 and put it in the complaining brother's account. They now had the same amount of money! How much did each of the brothers have in the bank before the transfer?

8. _____

9. Mida has $2.25 in nickels and dimes. She has four times as many dimes as nickels. How many dimes and how many nickels does she have?

9. _____

10. Dora has $5.50 in nickels and quarters. She has twice as many quarters as she has nickels. How many of each coin does she have?

10. _____

11. Mongo has 20 coins consisting of nickels and dimes. If the nickels were dimes and the dimes were nickels, he would have 50¢ more than he now has. How many nickels and how many dimes does he have?

11. _____

12. Desi has 10 coins consisting of pennies and nickels. Strangely enough, if the nickels were pennies and the pennies were nickels, she would have the same amount of money as she now has. How many pennies and nickels does she have?

12. _____

13. Don had $26 in his pocket. If he had only $1 bills and $5 bills, and he had a total of 10 bills, how many of each of the bills did he have?

13. _____

14. A person went to the bank to deposit $300. The money was in $10 and $20 bills, 25 bills in all. How many of each did the person have?

14. _____

C. Solve.

15. A plane flying from city A to city B at 300 mi/hr arrives $\frac{1}{2}$ hr later than scheduled. If the plane had flown at 350 mi/hr, it would have made the scheduled time. How far apart are A and B?

15. _____

16. A plane flies 540 mi/hr with a tail wind in $2\frac{1}{4}$ hr. The plane makes the return trip against the same wind and takes 3 hr. Find the speed of the plane in still air and the speed of the wind.

16. _____

17. A motor boat runs 45 mi downstream in $2\frac{1}{2}$ hr and 39 miles upstream in $3\frac{1}{4}$ hr. Find the speed of the boat in still water and the speed of the current.

17. _____

18. A small plane piloted by Balman travels 520 mi with the wind in 3 hr 20 min ($3\frac{1}{3}$ hr), the same time that it takes to travel 460 mi against the wind. What is the plane's speed in still air?

18. _____

19. If Bill drives from his home to his office at 40 mi/hr, he arrives 5 min early. If he drives at 30 mi/hr, he arrives 5 min late. How far is it from his home to his office?

19. _____

20. An unidentified plane approaching the U.S. coast is sighted on radar and determined to be 380 mi away and heading straight toward the coast at 600 mi/hr. 5 min ($\frac{1}{12}$ hr) later, a U.S. jet, flying at 720 mi/hr, scrambles from the coastline to meet the plane. How far from the coast does the interceptor meet the plane?

20. _____

✓ **SKILL CHECKER**

Find an equivalent fraction.

21. $\dfrac{3}{8} = \dfrac{?}{16}$

22. $\dfrac{5}{3} = \dfrac{?}{6}$

21. _____

22. _____

Reduce.

23. $\dfrac{6}{18}$

24. $\dfrac{7}{28}$

23. _____

24. _____

ANSWER (to problem on page 498)
3. Speed of the plane, 350 mi/hr; speed of the wind, 50 mi/hr

SUMMARY

SECTION	ITEM	MEANING	EXAMPLE
7.1A	System of equations	A set of equations that may have a common solution	$x + y = 2$ $x - y = 4$ is a system of equations.
7.1A	Inconsistent system	A system with no solution	$x + y = 2$ $x + y = 3$ is an inconsistent system.
7.1A	Dependent system	A system in which both equations are equivalent	$x + y = 2$ $2x + 2y = 4$ is a dependent system.
7.2A	Substitution method	A method used to solve systems of equations by solving one equation for one variable and substituting this result in the other equation	To solve the system $x + y = 2$ $2x - 3y = 6$ solve the first equation for $x = 2 - y$ and substitute in the other equation: $2(2 - y) + 3y = 6$ $4 - 2y + 3y = 6$ $y = 2$
7.3	Elimination method	A method used to solve systems of equations by multiplying by numbers that will cause the coefficients of one of the variables to be opposites	To solve the system $x + 2y = 5$ $x - y = -1$ multiply the second equation by 2 and add to the first equation.
7.4	RSTUV method	A method for solving word problems consisting of **R**eading, **S**electing the variables, **T**ranslating the problem, **U**sing algebra to solve, and **V**erifying the answer	

NAME

CLASS

SECTION

(If you need help with these exercises, look in the section indicated in brackets.)

ANSWERS

1. [7.1A] Use the graphical method to solve the system.

    a. $2x + y = 4$
    $y - 2x = 0$

1. a.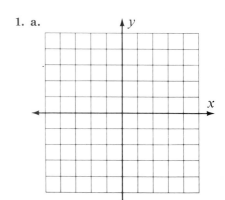

    b. $x + y = 4$
    $y - x = 0$

    b.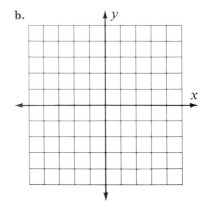

    c. $x + y = 4$
    $y - 3x = 0$

    c.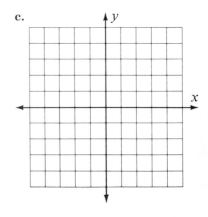

2. [7.1A,B] Use the graphical method to solve the system.

    a. $y - 3x = 3$
          $2y - 6x = 12$

2. a.

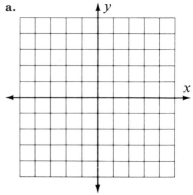

    b. $y - 2x = 2$
          $2y - 4x = 8$

b.

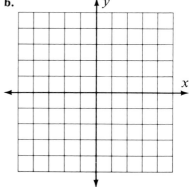

    c. $y - 3x = 6$
          $2y - 6x = 6$

c.

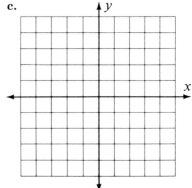

3. [7.2A] Use the substitution method to solve the system (if possible).

    a. $x + 4y = 5$
          $2x + 8y = 15$

    b. $x + 3y = 6$
          $3x + 9y = 12$

    c. $x + \ 4y = 5$
          $2x + 13y = 15$

3. a. $x = $ _____

      $y = $ _____

  b. $x = $ _____

      $y = $ _____

  c. $x = $ _____

      $y = $ _____

4. [7.2A,B] Use the substitution method to solve the system (if possible).

   a. $x + 4y = 5$
       $2x + 8y = 10$

   b. $x + 3y = 6$
       $3x + 9y = 18$

   c. $x + 4y = 5$
       $-2x - 8y = -10$

4. a. $x =$ _____
    $y =$ _____
  b. $x =$ _____
    $y =$ _____
  c. $x =$ _____
    $y =$ _____

5. [7.3A] Solve the system (if possible).

   a. $3x + 2y = 1$
       $2x + y = 0$

   b. $3x + 2y = 4$
       $2x + y = 3$

   c. $3x + 2y = -7$
       $2x + y = -4$

5. a. $x =$ _____
    $y =$ _____
  b. $x =$ _____
    $y =$ _____
  c. $x =$ _____
    $y =$ _____

6. [7.3A,B] Solve the system (if possible).

   a. $2x - 3y = 6$
       $-4x + 6y = -2$

   b. $3x - 2y = 8$
       $-9x + 6y = -4$

   c. $3x - 5y = 6$
       $-3x + 5y = -12$

6. a. $x =$ _____
    $y =$ _____
  b. $x =$ _____
    $y =$ _____
  c. $x =$ _____
    $y =$ _____

7. [7.3B] Solve the system (if possible).

   a. $3y + 2x = 1$
       $6y + 4x = 2$

   b. $2y + 3x = 1$
       $6x + 4y = 2$

   c. $3y + 4x = -11$
       $8x + 6y = -22$

7. a. $x =$ _____
    $y =$ _____
  b. $x =$ _____
    $y =$ _____
  c. $x =$ _____
    $y =$ _____

8. [7.4A] Desi has $3 in nickels and dimes. How many nickels and how many dimes does she have if
   a. She has the same number of nickels and dimes?
   b. She has 4 times as many nickels as she has dimes?
   c. She has 10 times as many nickels as she has dimes?

8. a. _____
  b. _____
  c. _____

9. [7.4B] The sum of two numbers is 180. What are the numbers if
   a. Their difference is 40?
   b. Their difference is 60?
   c. Their difference is 80?

9. a. _____
  b. _____
  c. _____

REVIEW EXERCISES CHAPTER 7

NAME

10. [7.4C] A plane flew 2400 mi with a tail wind in 3 hr. What is the plane's speed in still air if the return trip took:
    a. 8 hr?
    b. 10 hr?
    c. 12 hr?

10. a. ______
    b. ______
    c. ______

NAME

CLASS

SECTION

(Answers on page 508)

ANSWERS

1. Use the graphical method to solve the system

    $x + 2y = 4$

    $2y - x = 0$

    1.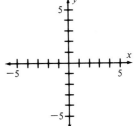

2. Use the graphical method to solve the system.

    $y - 2x = 2$

    $2y - 4x = 8$

    2.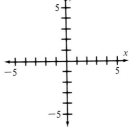

3. Use the method of substitution to solve the system (if possible).

    $x + 3y = 6$

    $2x + 6y = 8$

    3. _____

4. Use the method of substitution to solve the system (if possible).

    $x + 3y = 6$

    $3x + 9y = 18$

    4. _____

5. Solve the system (if possible).

    $2x + 3y = -8$

    $3x + y = -5$

    5. _____

6. Solve the system (if possible).

    $3x - 2y = 6$

    $-6x + 4y = -2$

    6. _____

7. Solve the system (if possible).

    $2y + 3x = -12$

    $6x + 4y = -24$

    7. _____

8. Eva has $2 in nickels and dimes. She has twice as many dimes as nickels. How many nickels and how many dimes does she have?

    8. _____

9. The sum of two numbers is 140. Their difference is 90. What are the numbers?

    9. _____

10. A plane flies 600 mi with a tail wind in 2 hr. It takes the same plane 3 hr to fly the 600 mi when flying against the wind. What is the plane's speed in still air?

    10. _____

IF YOU MISSED QUESTION	SECTION	EXAMPLES	PAGE	ANSWERS
1	7.1	1	466	1. The solution is (2, 1).
				The lines are parallel;
2	7.1	2, 3	467, 468	2. there is no solution.
3	7.2	1, 2	480, 481	3. No solution
4	7.2	3	481	Dependent (infinitely many 4. solutions)
5	7.3	1	488	5. $(-1, -2)$
6	7.3	1, 2	488, 489	6. No solution
7	7.3	3	490	Dependent (infinitely many 7. solutions)
8	7.4	1	497	8. 8 nickels, 16 dimes
9	7.4	2	497	9. 115 and 25
10	7.4	3	498	10. 250 mi/hr

CHAPTER 8

FRACTIONAL EXPRESSIONS AND EQUATIONS

This chapter integrates many of the ideas previously studied. We begin with numerical fractions and then generalize to algebraic fractions and their addition, subtraction, multiplication, and division. The idea of a complex fraction is also introduced. We then apply these concepts to solve fractional equations and end the chapter by discussing ratio, proportion, and word problems.

(*Answers on page* 512)

1. Write $\dfrac{3x}{7y}$ with a denominator of $28y^3$.

2. Reduce $\dfrac{-9(x^2 - y^2)}{3(x - y)}$ to lowest terms.

3. Simplify $\dfrac{-x}{x + x^2}$.

4. Reduce to lowest terms $\dfrac{x^2 + x - 12}{3 - x}$.

In Problems 5 through 12 perform the indicated operations and simplify.

5. Multiply $\dfrac{2y^2}{7} \cdot \dfrac{21x}{8y}$.

6. Multiply $(x - 1) \cdot \dfrac{x + 3}{x^2 - 1}$.

7. Divide $\dfrac{x^2 - 9}{x + 5} \div (x - 3)$.

8. Divide $\dfrac{x + 2}{x - 2} \div \dfrac{x^2 - 4}{2 - x}$.

9. Add $\dfrac{5}{2(x - 2)} + \dfrac{3}{2(x - 2)}$.

10. Subtract $\dfrac{10}{3(x + 1)} - \dfrac{1}{3(x + 1)}$.

11. Add $\dfrac{3}{x + 1} + \dfrac{1}{x - 1}$.

12. Subtract $\dfrac{x + 1}{x^2 + x - 2} - \dfrac{x + 2}{x^2 - 1}$.

13. Simplify $\dfrac{\dfrac{3}{4x} - \dfrac{1}{2x}}{\dfrac{1}{x} + \dfrac{2}{3x}}$.

14. Solve $\dfrac{4x}{x - 2} + 4 = \dfrac{6x}{x - 2}$.

15. Solve $\dfrac{x}{x^2 - 9} + \dfrac{3}{x - 3} = \dfrac{1}{x + 3}$.

16. Solve $\dfrac{x}{x + 6} - \dfrac{1}{6} = \dfrac{-6}{x + 6}$.

17. Solve $1 + \dfrac{2}{x - 3} = \dfrac{12}{x^2 - 9}$.

18. A car travels 150 mi on 9 gal of gas. How many gallons will it need to travel 300 mi?

19. Solve $\dfrac{x + 3}{7} = \dfrac{11}{6}$.

20. A woman can paint a house in 5 hr. Another one can do it in 6 hr. How long would it take both painters working together to finish the job?

IF YOU MISSED QUESTION	SECTION	EXAMPLES	PAGE	ANSWERS
1	8.1	1	516	1. $\dfrac{12xy^2}{28y^3}$
2	8.1	2	518	2. $-3(x+y)$
3	8.1	3	519	3. $\dfrac{-1}{1+x}$
4	8.1	4	521	4. $-(x+4)$
5	8.2	1, 2	528, 529	5. $\dfrac{3xy}{4}$
6	8.2	3	530	6. $\dfrac{x+3}{x+1}$
7	8.2	4	531	7. $\dfrac{x+3}{x+5}$
8	8.2	4, 5	531	8. $\dfrac{-1}{x-2}=\dfrac{1}{2-x}$
9	8.3	1a	538	9. $\dfrac{4}{x-2}$
10	8.3	1b	538	10. $\dfrac{3}{x+1}$
11	8.3	2	540	11. $\dfrac{4x-2}{x^2-1}$
12	8.3	3	542	12. $\dfrac{-2x-3}{(x+2)(x+1)(x-1)}$
13	8.4	1, 2	550	13. $\dfrac{3}{20}$
14	8.5	1, 2	556	14. $x=4$
15	8.5	3	557	15. $x=-4$
16	8.5	4	558	16. No solution
17	8.5	5	559	17. $x=-5$
18	8.6	1	566	18. 18
19	8.6	2	567	19. $x=\dfrac{59}{6}$
20	8.6	3	568	20. $2\dfrac{8}{11}$ hr

8.1 BUILDING AND REDUCING FRACTIONS

© 1966 United Features Syndicate.

OBJECTIVES

REVIEW

Before starting this section you should know:

1. How to write a fraction with a specified denominator.
2. How to simplify fractions (Section 1.1).

OBJECTIVES

You should be able to:

A. Write an algebraic fraction as an equivalent one with a specified denominator.
B. Reduce (simplify) an algebraic fraction to lowest terms.

In the very first section of this book, we mentioned that algebra was a generalized arithmetic. In arithmetic we study the natural numbers, the whole numbers, the integers, and the rational numbers. We have already studied polynomials, and you probably noticed that they follow rules similar to those used with the integers. As in arithmetic, we extend our ideas and discuss *algebraic fractions,* or *rational expressions,* as they are usually called. In arithmetic, a **rational number** is a number that can be written in the form $\frac{a}{b}$, where a and b are integers and b is not 0. As usual, **a** is called the *numerator* and **b**, the *denominator*. A similar approach is used in algebra.

In algebra, an expression of the form $\frac{A}{B}$, where **A** and **B** are polynomials and **B** is not 0, is called an **algebraic fraction,** or a **rational expression.** Thus,

$$\frac{8}{x}, \quad \frac{x^2 + 2x + 3}{x + 5}, \quad \frac{y}{y - 1}, \quad \text{and} \quad \frac{x^2 + 3x + 9}{x^2 - 4x + 4}$$

are algebraic fractions. Of course, since we do not wish the denominators of these fractions to be 0, we must place some restrictions on these denominators. Let us see what these restrictions must be.

For $\frac{8}{x}$, x cannot be 0 ($x \neq 0$) because then we would have $\frac{8}{0}$, which is not defined. For

$$\frac{x^2 + 2x + 3}{x + 5}, \qquad x \neq -5 \quad \text{If } x \text{ is } -5, \frac{x^2 + 2x + 3}{x + 5} = \frac{25 - 10 + 3}{0}.$$

For

$$\frac{y}{y - 1}, \qquad y \neq 1 \quad \text{If } y \text{ is } 1, \frac{y}{y - 1} = \frac{1}{0}.$$

and for

$$\frac{x^2 + 3x + 9}{x^2 - 4x + 4} = \frac{x^2 + 3x + 9}{(x - 2)^2}, \quad x \neq 2 \quad \text{If } x \text{ is } 2, \frac{x^2 + 3x + 9}{x^2 - 4x + 4} = \frac{4 + 6 + 9}{4 - 8 + 4} = \frac{19}{0}.$$

To avoid stating repeatedly that the denominators of algebraic fractions must not be 0 we make the following assumption:

> The variables in an algebraic fraction may not be replaced by numbers that make the denominator 0.

A. BUILDING FRACTIONS

Now that we know that we must avoid zero denominators, recall what we did with fractions in arithmetic. First, we learned how to recognize which ones are equal, and then we used this idea to reduce or build up fractions. Let us talk about equality first.

Photo by DeAngelis.

What does the picture tell you (if you write the value of the coins using fractions)? It states that

$$\frac{1}{2} = \frac{2}{4}$$

As a matter of fact,

$$\frac{1}{2} = \frac{2}{4} = \frac{3}{6} = \frac{4}{8} = \frac{x}{2x} = \frac{x^2}{2x^2}$$

and so on.

Do you see what the pattern is? Maybe you can see it better if we write

$$\frac{1}{2} = \frac{1 \times 2}{2 \times 2} = \frac{2}{4} \quad \text{Note that } \frac{2}{2} = 1.$$

$$\frac{1}{2} = \frac{1 \times 3}{2 \times 3} = \frac{3}{6} \quad \text{Here, } \frac{3}{3} = 1.$$

$$\frac{1}{2} = \frac{1 \times 4}{2 \times 4} = \frac{4}{8}$$

$$\frac{1}{2} = \frac{1 \times x}{2 \times x} = \frac{x}{2x}$$

$$\frac{1}{2} = \frac{1 \times x^2}{2 \times x^2} = \frac{x^2}{2x^2}$$

Clearly, we can always obtain fractions that are equivalent to any given fraction by multiplying the numerator and denominator of the given fraction by the same nonzero number or expression. In symbols,

---

FUNDAMENTAL RULE OF FRACTIONS
$$\frac{A}{B} = \frac{A \cdot C}{B \cdot C}$$

Note that

$$\frac{A}{B} = \frac{A \cdot C}{B \cdot C}$$

because

$$\frac{C}{C} = 1$$

and multiplying by 1 does not change the value of the fraction.

This idea is very important when adding or subtracting fractions. For example, to add $\frac{1}{2}$ and $\frac{3}{4}$ we write

$$\frac{1}{2} = \frac{2}{4} \qquad \text{Note that } \frac{1}{2} = \frac{1 \times 2}{2 \times 2} = \frac{2}{4}.$$

$$+\frac{3}{4} = \frac{3}{4}$$

$$\overline{\qquad \frac{5}{4}}$$

Here we have used the **fundamental rule of fractions** to write the $\frac{1}{2}$ as an equivalent fraction with a denominator of 4, namely, $\frac{2}{4}$. Now, suppose we wish to write $\frac{3}{8}$ with a denominator of 16. How can we do it? First, we write the problem as

$$\frac{3}{8} = \frac{?}{16} \qquad \text{Note that } 16 = 8 \times 2.$$

$$\rule{0pt}{0pt}\llcorner \text{ Multiply by 2 } \lrcorner$$

and notice that to get 16 as the denominator, we need to multiply the 8 by 2. Of course, we must do the same to the numerator 3, obtaining

$$\underset{\displaystyle \frac{3}{8}}{\Big\downarrow} \text{ Multiply by 2 } \underset{\displaystyle = \frac{6}{16}}{\Big\downarrow} \qquad$$ By the fundamental rule of fractions, if we multiply the denominator by 2, we must multiply the numerator by 2.

$$\frac{3}{8} = \frac{6}{16} \qquad \text{Note that } \frac{3 \times 2}{8 \times 2} = \frac{6}{16}.$$

Thus,

$$\frac{3}{8} = \frac{6}{16}$$

Similarly, to write

$$\frac{5x}{3y}$$

with a denominator of $6y^3$, first write the new equivalent fraction $\dfrac{?}{6y^3}$ with the old denominator $3y$ factored out, as shown:

$$\frac{5x}{3y} = \frac{?}{6y^3} = \frac{?}{3y(2y^2)}$$

We write $6y^3$ as $3y(2y^2)$.

Multiply by $2y^2$

Since the multiplier is $2y^2$, we have

Multiply by $2y^2$

$$\frac{5x}{3y} = \frac{5x(2y^2)}{3y(2y^2)} = \frac{10xy^2}{6y^3}$$

Thus,

$$\frac{5x}{3y} = \frac{10xy^2}{6y^3}$$

We multiplied the denominator by $2y^2$, so we have to multiply the numerator by $2y^2$.

EXAMPLE 1 Write.

a. $\dfrac{5}{6}$ with a denominator of 18

b. $\dfrac{2x}{9y^2}$ with a denominator of $18y^3$

c. $\dfrac{3x}{x-1}$ with a denominator of $x^2 + 2x - 3$

Solution

a. $\dfrac{5}{6} = \dfrac{?}{18}$

Multiply by 3

Multiply by 3

$$\frac{5}{6} = \frac{15}{18}$$

Note that $\dfrac{5 \times 3}{6 \times 3} = \dfrac{15}{18}$.

Thus,

$$\frac{5}{6} = \frac{15}{18}$$

b. Since $18y^3 = 9y^2(2y)$,

$$\frac{2x}{9y^2} = \frac{?}{9y^2(2y)}$$

Multiply by $2y$

Multiply by $2y$

$$\frac{2x}{9y^2} = \frac{2x(2y)}{9y^2(2y)} = \frac{4xy}{18y^3}$$

c. We first note that $x^2 + 2x - 3 = (x-1)(x+3)$. Thus,

$$\frac{3x}{x-1} = \frac{?}{(x-1)(x+3)}$$

Multiply by $(x+3)$

Multiply by $(x+3)$

$$\frac{3x}{x-1} = \frac{3x(x+3)}{(x-1)(x+3)} = \frac{3x^2 + 9x}{x^2 + 2x - 3}$$

Problem 1 Write.

a. $\dfrac{7}{8}$ with a denominator of 16

b. $\dfrac{3x}{8y^2}$ with a denominator of $24y^3$

c. $\dfrac{4x}{x+2}$ with a denominator of $x^2 - x - 6$

B. REDUCING FRACTIONS

Now that we know how to build up fractions, we are ready for the inverse process, **reducing** fractions. In Example 1a, we wrote $\frac{5}{6}$ with a denominator of 18, that is, we found out that

$$\frac{5}{6} = \frac{5 \times 3}{6 \times 3} = \frac{15}{18}$$

Of course, you will probably agree that $\frac{5}{6}$ is written in a "simpler" form than $\frac{15}{18}$. Certainly you will agree—though it is harder to see—that

$$\frac{5}{3} = \frac{5(x+3)(x^2-4)}{3(x+2)(x^2+x-6)}$$

and that $\frac{5}{3}$ is the "simpler" of the two fractions. In algebra, the process of removing all factors common to the numerator and denominator is called **reducing to lowest terms,** or **simplifying** the fraction. How do we reduce the fraction

$$\frac{5(x+3)(x^2-4)}{3(x+2)(x^2+x-6)}$$

to lowest terms? We do it by steps.

PROCEDURE FOR REDUCING FRACTIONS TO LOWEST TERMS (SIMPLIFYING)

1. Write the numerator and denominator of the fraction in factored form.

2. Find the factors that are common to the numerator and denominator.

3. Replace the quotient of the common factors by the number 1, since $\frac{a}{a} = 1$.

4. Rewrite the fraction in simplified form.

We are now ready to reduce

$$\frac{5(x+3)(x^2-4)}{3(x+2)(x^2+x-6)}$$

to lowest terms. Here are the steps:

1. Write the numerator and denominator in factored form.

$$\frac{5(x+3)(x+2)(x-2)}{3(x+2)(x+3)(x-2)}$$

2. Find the factors that are common to the numerator and denominator (we rearranged them so the common factors are in columns).

$$\frac{5(x+2)(x+3)(x-2)}{3(x+2)(x+3)(x-2)}$$

3. Replace the quotient of the common factors by the number 1.

$$\frac{5\cancel{(x+2)}\cancel{(x+3)}\cancel{(x-2)}^{\,1}}{3\cancel{(x+2)}\cancel{(x+3)}\cancel{(x-2)}}$$

4. Rewrite the fraction in simplified form.

$$\frac{5}{3}$$

The whole procedure can be written as

$$\frac{5(x+3)(x^2-4)}{3(x+2)(x^2+x-6)} = \frac{5\cancel{(x+3)}\cancel{(x+2)}\cancel{(x-2)}}{3\cancel{(x+2)}\cancel{(x+3)}\cancel{(x-2)}} = \frac{5}{3}$$

EXAMPLE 2 Simplify.

a. $\dfrac{-5x^2y}{15xy^3}$ **b.** $\dfrac{-8(x^2-y^2)}{-4(x+y)}$

Solution

a. We use our four-step procedure:

1. Write in factored form. $\dfrac{-5x^2y}{15xy^3} = \dfrac{(-1)\cdot 5\cdot x\cdot x\cdot y}{3\cdot 5\cdot x\cdot y\cdot y\cdot y}$

2. Rearrange the common factors. $= \dfrac{5\cdot x\cdot y(-1)\cdot x}{5\cdot x\cdot y\cdot 3\cdot y\cdot y}$

3. Replace the quotient of the common factors by 1. $= \dfrac{1\cdot(-1)\cdot x}{3\cdot y\cdot y}$

4. Rewrite in simplified form. $= \dfrac{-x}{3y^2}$

Thus,

$$\frac{-5x^2y}{15xy^3} = \frac{-x}{3y^2}$$

The whole process can be indicated by writing

$$\frac{-5x^2y}{15xy^3} = \frac{\overset{-1}{\cancel{-5}}x^{\overset{x}{\cancel{2}}}y}{\underset{3}{\cancel{15}}x\underset{y^2}{\cancel{y^3}}} = \frac{-x}{3y^2}$$

b. We again use the four-step procedure given earlier:

1. Write the fractions in factored form. $\dfrac{-8(x^2-y^2)}{-4(x+y)} = \dfrac{-8(x+y)(x-y)}{-4(x+y)}$

2. Rearrange the common factors. $= \dfrac{(-1)\cdot 2\cdot 2\cdot(x+y)(x-y)\cdot 2}{(-1)\cdot 2\cdot 2\cdot(x+y)}$

3. Replace the quotient of the common factors by 1. $= 1(x-y)\cdot 2$

4. Rewrite in simplified form. $= 2(x-y)$

The abbreviated form is

$$\frac{-8(x^2-y^2)}{-4(x+y)} = \frac{\overset{2}{\cancel{-8}}\overset{1}{\cancel{(x+y)}}(x-y)}{\cancel{-4}\cancel{(x+y)}} = 2(x-y)$$ ▲

You may have noticed that in Example 2a we wrote the answer as

$$\frac{-x}{3y^2}$$

Problem 2 Simplify.

a. $\dfrac{-3xy^2}{12x^2y}$ **b.** $\dfrac{-6(x^2-y^2)}{-3(x-y)}$

It could be argued that since

$$\frac{-5}{15} = -\frac{1}{3}$$

the answer should be

$$-\frac{x}{3y^2}$$

However, to avoid confusion, we agree to write

$$-\frac{x}{3y^2} \quad \text{as} \quad \frac{-x}{3y^2}$$

with the *negative* sign in the numerator. In general, we use the following conventions:

<table>
<tr><td colspan="3">STANDARD FORM OF A FRACTION</td></tr>
</table>

$-\dfrac{a}{b}$ is written as $\dfrac{-a}{b}$ $\dfrac{-a}{-b}$ is written as $\dfrac{a}{b}$

$\dfrac{a}{-b}$ is written as $\dfrac{-a}{b}$ $-\dfrac{-a}{b}$ is written as $\dfrac{a}{b}$

$-\dfrac{-a}{-b}$ is written as $\dfrac{-a}{b}$ $-\dfrac{a}{-b}$ is written as $\dfrac{a}{b}$

The forms

$$\frac{a}{b} \quad \text{and} \quad \frac{-a}{b}$$

are called the **standard forms** of an algebraic fraction and should be used in writing the answers to problems involving these fractions.

EXAMPLE 3 Simplify.

a. $\dfrac{6x - 12y}{18x - 36y}$ **b.** $-\dfrac{y}{y + xy}$ **c.** $\dfrac{(x + 3)}{-(x^2 - 9)}$

Solution

a. $\dfrac{6x - 12y}{18x - 36y} = \dfrac{6(x - 2y)}{18(x - 2y)}$

$$= \frac{\overset{1}{\cancel{6}}(\overset{1}{\cancel{x - 2y}})}{\underset{3}{\cancel{18}}(\underset{1}{\cancel{x - 2y}})}$$

$$= \frac{1}{3}$$

b. $-\dfrac{y}{y + xy} = -\dfrac{y}{y(1 + x)}$

$$= -\frac{\overset{1}{\cancel{y}}}{\cancel{y}(1 + x)}$$

$$= \frac{-1}{1 + x}$$

Recall that if we divide two numbers with unlike signs, the result is negative.

Problem 3 Simplify.

a. $\dfrac{10x - 15y}{4x - 6y}$ **b.** $-\dfrac{x}{xy + x}$

c. $\dfrac{x + 4}{-(x^2 - 16)}$

c. $\dfrac{x + 3}{-(x^2 - 9)} = \dfrac{(x + 3)}{-(x + 3)(x - 3)}$

$$= \dfrac{\overset{1}{\cancel{(x + 3)}}}{-\cancel{(x + 3)}(x - 3)}$$

$$= \dfrac{1}{-(x - 3)}$$

$$= \dfrac{-1}{x - 3} \quad \text{in standard form} \qquad \blacktriangle$$

Is there a way in which

$$\dfrac{-1}{x - 3}$$

can be simplified further? The answer is yes. See if you can see the reasons behind each step:

$$\dfrac{-1}{x - 3} = \dfrac{-1}{-(3 - x)} \qquad -(3 - x) = -3 + x = x - 3$$

$$= \dfrac{1}{3 - x}$$

Thus,

$$\dfrac{-1}{x - 3} = \dfrac{1}{3 - x}$$

And why, you might say, is this answer simpler than the other? Because

$$\dfrac{-1}{x - 3}$$

has two negative signs, whereas

$$\dfrac{1}{3 - x}$$

has only *one*.

You should be aware of these simplifications when writing your answers! Here is a situation that occurs very frequently in algebra. Can you reduce

$$\dfrac{a - b}{b - a}$$

Look at the following steps:

$$\dfrac{a - b}{b - a} = \dfrac{\overset{1}{\cancel{(a - b)}}}{-\cancel{(a - b)}}$$

$$= \dfrac{1}{-1}$$

$$= -1$$

Thus,

$$\frac{a - b}{b - a} = -1$$

We use this idea in the next example.

EXAMPLE 4 Reduce to lowest terms.

a. $\dfrac{x^2 - 16}{4 + x}$ **b.** $-\dfrac{x^2 - 16}{x + 4}$ **c.** $\dfrac{x^2 + 2x - 15}{3 - x}$

Solution

a. $\dfrac{x^2 - 16}{4 + x} = \dfrac{(x + 4)(x - 4)}{x + 4}$ $4 + x = x + 4$ by the commutative law

$\qquad = x - 4$

b. $-\dfrac{x^2 - 16}{x + 4} = -\dfrac{(x + 4)(x - 4)}{(x + 4)}$

$\qquad = -\dfrac{\overset{1}{\cancel{(x + 4)}}(x - 4)}{\underset{.}{\cancel{(x + 4)}}}$ Note that $\dfrac{x + 4}{x + 4} = 1$.

$\qquad = -(x - 4)$

$\qquad = -x + 4$

$\qquad = 4 - x$

c. $\dfrac{x^2 + 2x - 15}{3 - x} = \dfrac{(x + 5)(x - 3)}{3 - x}$

$\qquad = \dfrac{\overset{-1}{\cancel{(x - 3)}}(x + 5)}{\cancel{(3 - x)}}$ Note that $\dfrac{x - 3}{3 - x} = -1$.

$\qquad = -(x + 5)$

We have used the fact that

$$\frac{a - b}{b - a} = -1$$

in the second step to simplify

$$\frac{(x - 3)}{(3 - x)}$$

Note that the final answer is written as $-(x + 5)$ instead of $-x - 5$, since $-(x + 5)$ is considered to be simpler.

Problem 4 Reduce to lowest terms.

a. $\dfrac{x^2 - 9}{3 + x}$ **b.** $-\dfrac{x^2 - 9}{x + 3}$

c. $\dfrac{x^2 - 3x - 10}{5 - x}$

A. Write the given fraction as an equivalent one with the indicated denominator.

ANSWERS

1. $\dfrac{3}{7}$ with a denominator of 21

1. _____

2. $\dfrac{5}{9}$ with a denominator of 36

2. _____

3. $\dfrac{-8}{11}$ with a denominator of 22

3. _____

4. $\dfrac{-5}{17}$ with a denominator of 51

4. _____

5. $\dfrac{5x}{6y^2}$ with a denominator of $24y^3$

5. _____

6. $\dfrac{7y}{5x^3}$ with a denominator of $10x^4$

6. _____

7. $\dfrac{-3x}{7y}$ with a denominator of $21y^4$

7. _____

8. $\dfrac{-4y}{7x^2}$ with a denominator of $28x^3$

8. _____

9. $\dfrac{4x}{x+1}$ with a denominator of $x^2 - x - 2$

9. _____

10. $\dfrac{5y}{y-1}$ with a denominator of $y^2 + 2y - 3$

10. _____

11. $\dfrac{-5x}{x+3}$ with a denominator of $x^2 + x - 6$

11. _____

12. $\dfrac{-3y}{y-4}$ with a denominator of $y^2 - 2y - 8$

12. _____

B. Reduce to lowest terms (simplify).

13. $\dfrac{7x^3y}{14xy^4}$ 14. $\dfrac{24xy^3}{6x^3y}$

13. _____

14. _____

15. $\dfrac{-9xy^5}{3x^2y}$ 16. $\dfrac{-24x^3y^2}{48xy^4}$

15. _____

16. _____

17. $\dfrac{-6x^2y}{-12x^3y^4}$ 18. $\dfrac{-9xy^4}{-18x^5y}$

17. _____

18. _____

19. $\dfrac{-25x^3y^2}{-5x^2y^4}$

20. $\dfrac{-30x^2y^2}{-6x^3y^5}$

21. $\dfrac{6(x^2 - y^2)}{18(x + y)}$

22. $\dfrac{12(x^2 - y^2)}{48(x + y)}$

23. $\dfrac{-9(x^2 - y^2)}{3(x + y)}$

24. $\dfrac{-12(x^2 - y^2)}{3(x - y)}$

25. $\dfrac{-6(x + y)}{24(x^2 - y^2)}$

26. $\dfrac{-8(x + 3)}{40(x^2 - 9)}$

27. $\dfrac{-5(x - 2)}{-10(x^2 - 4)}$

28. $\dfrac{-12(x - 2)}{-60(x^2 - 4)}$

29. $\dfrac{-3(x - y)}{-3(x^2 - y^2)}$

30. $\dfrac{-10(x + y)}{-10(x^2 - y^2)}$

31. $\dfrac{4x - 4y}{8x - 8y}$

32. $\dfrac{6x + 6y}{2x + 2y}$

33. $\dfrac{4x - 8y}{12x - 24y}$

34. $\dfrac{15y - 45x}{5y - 15x}$

35. $-\dfrac{6}{6 + 12y}$

36. $-\dfrac{4}{8 + 12x}$

37. $-\dfrac{x}{x + 2xy}$

38. $-\dfrac{y}{2y + 6xy}$

39. $-\dfrac{6y}{6xy + 12y}$

40. $-\dfrac{4x}{8xy + 16x}$

41. $\dfrac{3x - 2y}{2y - 3x}$

42. $\dfrac{5y - 2x}{2x - 5y}.$

43. $\dfrac{x^2 + 4x - 5}{1 - x}$

44. $\dfrac{x^2 - 2x - 15}{5 - x}$

45. $\dfrac{x^2 - 6x + 8}{4 - x}$

46. $\dfrac{x^2 - 8x + 15}{3 - x}$

47. $\dfrac{2 - x}{x^2 + 4x - 12}$

48. $\dfrac{3 - x}{x^2 + 3x - 18}$

49. $-\dfrac{3 - x}{x^2 - 5x + 6}$

50. $-\dfrac{4 - x}{x^2 - 3x - 4}$

19. ___
20. ___
21. ___
22. ___
23. ___
24. ___
25. ___
26. ___
27. ___
28. ___
29. ___
30. ___
31. ___
32. ___
33. ___
34. ___
35. ___
36. ___
37. ___
38. ___
39. ___
40. ___
41. ___
42. ___
43. ___
44. ___
45. ___
46. ___
47. ___
48. ___
49. ___
50. ___

✓ **SKILL CHECKER**

Multiply.

51. $\dfrac{3}{2} \cdot \dfrac{4}{9}$

51. ___

52. $\dfrac{3}{5} \cdot \dfrac{10}{9}$

52. _____

Factor.

53. $x^2 + 2x - 3$ **54.** $x^2 + 7x + 12$

53. _____

54. _____

55. $x^2 - 7x + 10$ **56.** $x^2 + 3x - 4$

55. _____

56. _____

8.1 USING YOUR KNOWLEDGE

There is an important relationship between fractions and *ratios*. In general, a **ratio** is a way of comparing two or more numbers. Thus, if there are 10 workers in an office, 3 women and 7 men, the ratio of women to men is

$\dfrac{3}{7} \leftarrow$ Number of women
$\phantom{\dfrac{3}{7}} \leftarrow$ Number of men

On the other hand, if there are 6 men and 4 women in the office, the **reduced ratio** of women to men is

$\dfrac{4}{6} = \dfrac{2}{3} \leftarrow$ Number of women
$\phantom{\dfrac{4}{6} = \dfrac{2}{3}} \leftarrow$ Number of men

Use your knowledge to solve the following problems.

1. A class is composed of 40 men and 60 women. Find the reduced ratio of men to women.

1. _____

2. Do you know the teacher to student ratio in your school? Suppose your school has 10,000 students and 500 teachers.
    a. Find the teacher–to–student ratio.
    b. If the school wishes to maintain a $\frac{1}{20}$ ratio and the enrollment increases to 12,000 students, how many teachers are needed?

2. **a.** _____
   b. _____

3. The transmission ratio in your automobile is defined by

$$\text{Transmission ratio} = \frac{\text{engine speed}}{\text{drive shaft speed}}$$

3. **a.** _____
   b. _____

   a. If the engine is running at 2000 revolutions per minute (rpm) and the drive shaft speed is 500 rpm, what is the transmission ratio?
   b. If the transmission ratio of a car is 5 to 1, and the drive shaft speed is 500 rpm, what is the engine speed?

By permission of Johnson, Andvis, and Bradley, *Applied Mathematics*, 4th edition, Glencoe Press.

REVIEW

Before starting this section, you should know:

1. How to multiply fractions.
2. How to reduce fractions.
3. How to factor trinomials.
4. How to factor the difference of two squares.

OBJECTIVES

You should be able to:

A. Multiply two rational expressions.
B. Divide one rational expression by another.

The photo shows a compound gear train. How fast can the last gear go? It depends on how fast the first gear goes, and the number of teeth in the gears! The formula that tells us the number of revolutions per minute (rpm) the last gear can turn is

$$\text{rpm} = \frac{T_1}{t_1} \cdot \frac{T_2}{t_2} \cdot R$$

T_1 and T_2 are the number of teeth in the driving gears; t_1 and t_2 are the number of teeth in the driven gears, and R is the number of rpm the first driving gear is turning.

A. MULTIPLICATION

Can we simplify this? Of course, if you remember how to multiply fractions in arithmetic. As you recall, in arithmetic the product of two fractions is another fraction whose numerator is the product of the original numerators and whose denominator is the product of the original denominators. In symbols,

RULE FOR MULTIPLYING FRACTIONS
$\dfrac{A}{B} \cdot \dfrac{C}{D} = \dfrac{AC}{BD}$

Thus the formula can be simplied to

$$\text{rpm} = \frac{T_1 T_2 R}{t_1 t_2}$$

If we assume that $R = 12$ and then count the teeth in the gears, we get $T_1 = 48$, $T_2 = 40$, $t_1 = 24$, and $t_2 = 24$. To find the rpm, we write

$$\text{rpm} = \frac{48}{24} \cdot \frac{40}{24} \cdot \frac{12}{1}$$

---

*Multiplication and division of arithmetic fractions is covered in Section 1.2.

Then we have the following.

1. Reduce each fraction.

$$\frac{\overset{2}{\cancel{48}}}{\underset{1}{\cancel{24}}} \cdot \frac{\overset{5}{\cancel{40}}}{\underset{3}{\cancel{24}}} \cdot \frac{12}{1} = \frac{2}{1} \cdot \frac{5}{3} \cdot \frac{12}{1}$$

2. Multiply the numerators.

$$\frac{120}{1 \cdot 3 \cdot 1}$$

3. Multiply the denominators.

$$\frac{120}{3}$$

4. Reduce the answer.

$$40$$

Thus the speed of the final gear is 40 rpm. Here is what we have done:

MULTIPLYING FRACTIONS

1. Reduce each fraction if possible.

2. Multiply the numerators to obtain the new numerator.

3. Multiply the denominators to obtain the new denominator.

4. Reduce the answer if possible.

Note that you could also write

$$\frac{2}{1} \cdot \frac{5}{\underset{1}{\cancel{3}}} \cdot \frac{\overset{4}{\cancel{12}}}{1}$$

and obtain $2 \cdot 5 \cdot 4 = 40$ as before.

EXAMPLE 1 Multiply.

a. $\dfrac{x}{6} \cdot \dfrac{7}{y}$ **b.** $\dfrac{3x^2}{2} \cdot \dfrac{4y}{9x}$

Solution

a. $\dfrac{x}{6} \cdot \dfrac{7}{y} = \dfrac{7x}{6y}$ $\leftarrow$ Multiply numerators.
$\leftarrow$ Multiply denominators.

b. $\dfrac{3x^2}{2} \cdot \dfrac{4y}{9x} = \dfrac{12x^2y}{18x}$ $\leftarrow$ Multiply numerators.
$\leftarrow$ Multiply denominators.

$$= \frac{\overset{2x}{\cancel{12x^2y}}}{\underset{3}{\cancel{18x}}} \quad \text{Reduce the answer.}$$

$$= \frac{2xy}{3} \qquad \qquad \blacktriangle$$

The procedure used to find the product in Example 1b can be shortened if we do some reduction beforehand. Thus we can write

$$\frac{\overset{1x}{\cancel{3x^2}}}{\underset{1}{\cancel{2}}} \cdot \frac{\overset{2}{\cancel{4y}}}{\underset{3}{\cancel{9x}}} = \frac{2xy}{3}$$

Use this idea in the next example.

Problem 1 Multiply.

a. $\dfrac{y}{8} \cdot \dfrac{9}{x}$ **b.** $\dfrac{7x^2}{3} \cdot \dfrac{9y}{14x}$

EXAMPLE 2   Multiply.

a. $\dfrac{-6x}{7y^2} \cdot \dfrac{14y}{12x^2}$ **b.** $8y^2 \cdot \dfrac{9x}{16y^2}$

Solution

a. Since $\dfrac{14y}{12x^2} = \dfrac{7y}{6x^2}$,

We write $\dfrac{7y}{6x^2}$

instead of

$\dfrac{14y}{12x^2}$

$\dfrac{-6x}{7y^2} \cdot \dfrac{14y}{12x^2} = \dfrac{\overset{-1}{-6x}}{\underset{1y}{7y^2}} \cdot \dfrac{\overset{1}{7y}}{\underset{1x}{6x^2}}$

$= \dfrac{-1}{xy}$

b. Since $8y^2 = \dfrac{8y^2}{1}$, we have

$8y^2 \cdot \dfrac{9x}{16y^2} = \dfrac{\overset{1}{8y^2}}{1} \cdot \dfrac{9x}{\underset{2 \; 1}{16y^2}}$

$= \dfrac{9x}{2}$

▲

Can all the problems be done as in these examples? Yes, but remember that when the numerators and denominators involved are binomials or trinomials, it is not easy to do the reductions we have just done in the examples *unless* the numerators and denominators involved are *factored*. Thus, to multiply

$$\frac{x^2 + 2x - 3}{x^2 + 7x + 12} \cdot \frac{x + 4}{x + 5}$$

we *first* factor and then multiply. Thus

Factor first.

$\dfrac{x^2 + 2x - 3}{x^2 + 7x + 12} \cdot \dfrac{x + 4}{x + 5} = \dfrac{(x - 1)\overset{1}{(x + 3)}}{(x + 4)(x + 3)} \cdot \dfrac{\overset{1}{(x + 4)}}{(x + 5)}$

$= \dfrac{x - 1}{x + 5}$

Remember, when multiplying fractions involving trinomials, factor these trinomials and reduce the answer if possible.

Problem 2   Multiply

a. $\dfrac{-3x}{4y^2} \cdot \dfrac{18y}{12x^2}$ **b.** $\dfrac{6x}{11y^2} \cdot 22y^2$

EXAMPLE 3   Multiply.

a. $(x - 3) \cdot \dfrac{x + 5}{x^2 - 9}$ **b.** $\dfrac{x^2 - x - 20}{x - 1} \cdot \dfrac{1 - x}{x + 4}$

Solution

a. Since $(x - 3) = \dfrac{x - 3}{1}$ and $x^2 - 9 = (x + 3)(x - 3)$,

$$(x - 3) \cdot \frac{x + 5}{x^2 - 9} = \frac{(x - 3)}{1} \cdot \frac{x + 5}{(x + 3)(x - 3)}$$

$$= \frac{x + 5}{x + 3}$$

b. Since $x^2 - x - 20 = (x + 4)(x - 5)$ and $\dfrac{1 - x}{x - 1} = -1$,

$$\frac{x^2 - x - 20}{x - 1} \cdot \frac{1 - x}{x + 4} = \frac{\overset{1}{\cancel{(x + 4)}}(x - 5)}{\cancel{(x - 1)}} \cdot \frac{\overset{-1}{\cancel{(1 - x)}}}{\underset{1}{\cancel{x + 4}}}$$

Remember that $\dfrac{a - b}{b - a} = -1$.

$$= -1(x - 5)$$
$$= -x + 5$$
$$= 5 - x$$

B. DIVISION

What about division? We are in luck! Division works using the same rule as in arithmetic.

RULE FOR DIVIDING FRACTIONS

$$\frac{A}{B} \div \frac{C}{D} = \frac{A}{B} \cdot \frac{D}{C} = \frac{AD}{BC}$$

Thus to divide $\frac{A}{B}$ by $\frac{C}{D}$, we simply invert $\frac{C}{D}$ (interchange the numerator and denominator) and multiply. That is, to divide $\frac{A}{B}$ by $\frac{C}{D}$, multiply $\frac{A}{B}$ by the **reciprocal** (inverse) of $\frac{C}{D}$. Thus, to divide

$$\frac{x + 4}{x - 5} \div \frac{x^2 + 3x - 4}{x^2 - 7x + 10}$$

we use the given rule and write

$$\frac{x + 4}{x - 5} \div \frac{x^2 + 3x - 4}{x^2 - 7x + 10} = \frac{x + 4}{x - 5} \cdot \frac{x^2 - 7x + 10}{x^2 + 3x - 4}$$

$$= \frac{\overset{1}{\cancel{x + 4}}}{\underset{1}{\cancel{x - 5}}} \cdot \frac{\overset{1}{\cancel{(x - 5)}}(x - 2)}{(x + 4)(x - 1)}$$

$$= \frac{x - 2}{x - 1}$$

▲

Problem 3   Multiply.

3. a. $(x - 4) \cdot \dfrac{x + 8}{x^2 - 16}$

b. $\dfrac{x^2 + x - 6}{x - 3} \cdot \dfrac{3 - x}{x + 3}$

Here is another example.

EXAMPLE 4 Divide.

a. $\dfrac{x^2 - 16}{x + 3} \div (x + 4)$ **b.** $\dfrac{x + 5}{x - 5} \div \dfrac{x^2 - 25}{5 - x}$

Solution

a. Since $(x + 4) = \dfrac{(x + 4)}{1}$,

$$\frac{x^2 - 16}{x + 3} \div \frac{(x + 4)}{1} = \frac{x^2 - 16}{x + 3} \cdot \frac{1}{(x + 4)}$$

$$= \frac{\overset{1}{\cancel{(x + 4)}}(x - 4)}{x + 3} \cdot \frac{1}{\underset{1}{\cancel{(x + 4)}}}$$

$$= \frac{x - 4}{x + 3}$$

b. $\dfrac{x + 5}{x - 5} \div \dfrac{x^2 - 25}{5 - x} = \dfrac{x + 5}{x - 5} \cdot \dfrac{5 - x}{x^2 - 25}$

$$= \frac{\cancel{x + 5}}{\cancel{x - 5}} \cdot \frac{\overset{-1}{\cancel{5 - x}}}{\underset{1}{(x + 5)(x - 5)}} \qquad \text{Note that } \frac{a - b}{b - a} = -1,$$
$$\text{so } \frac{5 - x}{x - 5} = -1.$$

$$= \frac{-1}{x - 5}$$

Of course, this answer *can* be simplified further (to show fewer negative signs), since

$$\frac{-1}{x - 5} = \frac{(-1)(-1)}{(-1)(x - 5)}$$

$$= \frac{1}{-x + 5} = \frac{1}{5 - x}$$

▲

Here is another example.

EXAMPLE 5 Divide.

a. $\dfrac{x^2 + 5x + 4}{x^2 - 2x - 3} \div \dfrac{x^2 - 4}{x^2 - 6x + 8}$ **b.** $\dfrac{x^2 - 1}{x^2 + x - 6} \div \dfrac{x^2 - 4x + 3}{x^2 - 4}$

Solution

a. $\dfrac{x^2 + 5x + 4}{x^2 - 2x - 3} \div \dfrac{x^2 - 4}{x^2 - 6x + 8} = \dfrac{x^2 + 5x + 4}{x^2 - 2x - 3} \cdot \dfrac{x^2 - 6x + 8}{x^2 - 4}$

$$= \frac{(x + 4)\overset{1}{\cancel{(x + 1)}}}{(x - 3)\underset{1}{\cancel{(x + 1)}}} \cdot \frac{(x - 4)\overset{1}{\cancel{(x - 2)}}}{(x + 2)\underset{1}{\cancel{(x - 2)}}}$$

$$= \frac{(x + 4)(x - 4)}{(x - 3)(x + 2)}$$

$$= \frac{x^2 - 16}{x^2 - x - 6}$$

Problem 4 Divide.

a. $\dfrac{x^2 - 25}{x + 1} \div (x + 5)$

b. $\dfrac{x + 6}{x - 6} \div \dfrac{x^2 - 36}{6 - x}$

Problem 5 Divide.

a. $\dfrac{x^2 - 3x + 2}{x^2 - 4x + 3} \div \dfrac{x^2 - 49}{x^2 - 5x - 14}$

b. $\dfrac{x^2 - 1}{x^2 - x - 6} \div \dfrac{x^2 - 3x + 2}{x^2 - 9}$

b. $\dfrac{x^2-1}{x^2+x-6} \div \dfrac{x^2-4x+3}{x^2-4} = \dfrac{x^2-1}{x^2+x-6} \cdot \dfrac{x^2-4}{x^2-4x+3}$

$$= \dfrac{(x+1)\overset{1}{\cancel{(x-1)}}}{(x+3)\cancel{(x-2)}} \cdot \dfrac{(x+2)\overset{1}{\cancel{(x-2)}}}{\cancel{(x-1)}(x-3)}$$

$$= \dfrac{(x+1)(x+2)}{(x+3)(x-3)}$$

$$= \dfrac{x^2+3x+2}{x^2-9}$$ ▲

A final word of warning! Be very careful when reducing fractions. Remember, you may cancel factors, but you *must not* cancel terms. Thus,

$$\dfrac{\cancel{x^2}+5x+6}{\cancel{x^2}+8x+15} = \dfrac{5x+1}{8x+15}$$

is *wrong*, because x^2 is a term, *not* a factor. The correct way is to *factor* first and then cancel. Thus,

$$\dfrac{x^2+5x+6}{x^2+8x+15} = \dfrac{\cancel{(x+3)}(x+2)}{\cancel{(x+3)}(x+5)} = \dfrac{x+2}{x+5}$$

Of course, $\dfrac{x+2}{x+5}$ *cannot* be reduced further. To write $\dfrac{x+2}{x+5} = \dfrac{2}{5}$ is wrong!

Again, x is a term, not a factor. (If you had $\dfrac{2x}{5x} = \dfrac{2}{5}$, that would be correct. In the expressions $2x$ and $5x$, x is a *factor* that may be canceled.) Why is $\dfrac{x+2}{x+5} \neq \dfrac{2}{5}$? Try it when x is 4.

$$\dfrac{x+2}{x+5} = \dfrac{4+2}{4+5} = \dfrac{6}{9} = \dfrac{2}{3}$$

Thus, the answer *cannot* be $\frac{2}{5}$!

EXERCISE 8.2

A. Multiply.

ANSWERS

1. $\dfrac{x}{3} \cdot \dfrac{8}{y}$

2. $\dfrac{-x}{4} \cdot \dfrac{7}{y}$

3. $\dfrac{-6x^2}{7} \cdot \dfrac{14y}{9x}$

4. $\dfrac{-5x^3}{6y^2} \cdot \dfrac{18y}{-10x}$

5. $7x^2 \cdot \dfrac{3y}{14x^2}$

6. $11y^2 \cdot \dfrac{4x}{33y}$

7. $\dfrac{-4y}{7x^2} \cdot 14x^3$

8. $\dfrac{-3y^3}{8x^2} \cdot -16y$

9. $(x-7) \cdot \dfrac{x+1}{x^2-49}$

10. $3(x+1) \cdot \dfrac{x+2}{x^2-1}$

11. $-2(x+2) \cdot \dfrac{x-1}{x^2-4}$

12. $-3(x-1) \cdot \dfrac{x-2}{x^2-1}$

13. $\dfrac{3}{x-5} \cdot \dfrac{x^2-25}{x+1}$

14. $\dfrac{1}{x-2} \cdot \dfrac{x^2-4}{x-1}$

15. $\dfrac{x^2-x-6}{x-2} \cdot \dfrac{2-x}{x-3}$

16. $\dfrac{x^2+3x-4}{3-x} \cdot \dfrac{x-3}{x+4}$

17. $\dfrac{x-1}{3-x} \cdot \dfrac{x+3}{1-x}$

18. $\dfrac{2x-1}{5-3x} \cdot \dfrac{3x-5}{1-2x}$

19. $\dfrac{3(x-5)}{14(4-x)} \cdot \dfrac{7(x-4)}{6(5-x)}$

20. $\dfrac{7(1-x)}{10(x-5)} \cdot \dfrac{5(5-x)}{14(x-1)}$

B. Divide.

21. $\dfrac{x^2-1}{x+2} \div (x+1)$

22. $\dfrac{x^2-4}{x-3} \div (x+2)$

23. $\dfrac{x^2-25}{x-3} \div 5(x+5)$

24. $\dfrac{x^2-16}{8(x-3)} \div 4(x+4)$

25. $(x+3) \div \dfrac{x^2-9}{x+4}$

26. $4(x-4) \div \dfrac{8(x^2-16)}{5}$

27. $\dfrac{-3}{x-4} \div \dfrac{6(x+3)}{5(x^2-16)}$

28. $\dfrac{-6}{x-2} \div \dfrac{3(x-1)}{7(x^2-4)}$

29. $\dfrac{-4(x+1)}{3(x+2)} \div \dfrac{-8(x^2-1)}{6(x^2-4)}$

30. $\dfrac{-10(x^2-1)}{6(x^2-4)} \div \dfrac{5(x+1)}{-3(x+2)}$

31. $\dfrac{x+3}{x-3} \div \dfrac{x^2-1}{3-x}$

32. $\dfrac{4-x}{x+1} \div \dfrac{x-4}{x^2-1}$

33. $\dfrac{x^2-4}{7(x^2-9)} \div \dfrac{x+2}{14(x+3)}$

34. $\dfrac{x^2-25}{3(x^2-1)} \div \dfrac{5-x}{6(x+1)}$

35. $\dfrac{3(x^2-36)}{14(5-x)} \div \dfrac{6(6-x)}{7(x^2-25)}$

36. $\dfrac{-6(x^2-1)}{35(x^2-4)} \div \dfrac{12(1-x)}{7(2-x)}$

37. $\dfrac{x+2}{x-1} \div \dfrac{x^2+5x+6}{x^2-4x+4}$

38. $\dfrac{x-3}{x+2} \div \dfrac{x^2-4x+3}{x^2-x-6}$

39. $\dfrac{x-5}{x+3} \div \dfrac{5(x-5)}{x^2+9x+18}$

40. $\dfrac{x-3}{x+4} \div \dfrac{2(x-3)}{x^2+2x-8}$

41. $\dfrac{x^2+2x-3}{x-5} \div \dfrac{x^2+6x+9}{x^2-2x-15}$

42. $\dfrac{x^2-3x+2}{x^2-5x+6} \div \dfrac{x^2-5x+4}{x^2-7x+12}$

43. $\dfrac{x^2-1}{x^2+3x-10} \div \dfrac{x^2-3x-4}{x^2-25}$

44. $\dfrac{x^2-4x-21}{x^2-10x+25} \div \dfrac{x^2+2x-3}{x^2-6x+5}$

45. $\dfrac{x^2+3x-4}{x^2+7x+12} \div \dfrac{x^2+x-2}{x^2+5x+6}$

46. $\dfrac{x^2+x-2}{x^2+6x-7} \div \dfrac{x^2-3x-10}{x^2+5x-14}$

47. $\dfrac{x^2-y^2}{x^2-2xy} \div \dfrac{x^2+xy-2y^2}{x^2-4y^2}$

48. $\dfrac{x^2+xy-2y^2}{x^2-4y^2} \div \dfrac{x^2-y^2}{x^2-2xy}$

49. $\dfrac{x^2+2xy-3y^2}{y^2-7y+10} \div \dfrac{x^2+5xy-6y^2}{y^2-3y-10}$

50. $\dfrac{x^2+2xy-8y^2}{x^2+7xy+12y^2} \div \dfrac{x^2-3xy+2y^2}{x^2+2xy-3y^2}$

29. _____
30. _____
31. _____
32. _____
33. _____
34. _____
35. _____
36. _____
37. _____
38. _____
39. _____
40. _____
41. _____
42. _____
43. _____
44. _____
45. _____
46. _____
47. _____
48. _____
49. _____
50. _____

✓ **SKILL CHECKER**

Add.

51. $\dfrac{7}{8} + \dfrac{2}{5}$

52. $\dfrac{7}{12} + \dfrac{1}{18}$

Subtract.

53. $\dfrac{7}{8} - \dfrac{2}{5}$

54. $\dfrac{7}{12} - \dfrac{1}{18}$

55. $\dfrac{5}{2} - \dfrac{1}{6}$

51. _____
52. _____
53. _____
54. _____
55. _____

8.2 USING YOUR KNOWLEDGE

1. When studying parallel resistors, the expression

$$R \cdot \frac{R_T}{R - R_T}$$

occurs, where R is a known resistance and R_T a required one. Do the multiplication.

1. _____

2. The molecular model predicts that the pressure of a gas is given by

$$\frac{2}{3} \cdot \frac{mv^2}{2} \cdot \frac{N}{v}$$

Do the multiplication.

2. _____

3. Suppose a store orders 3000 items each year. If it orders x units at a time, the number N of reorders is

$$N = \frac{3000}{x}$$

If there is a fixed \$20 reorder fee and a \$3 charge per item, the cost of each order is

$$C = 20 + 3x$$

The yearly reorder cost RC is then given by

$$RC = N \cdot C$$

Find RC.

3. _____

8.3 ADDITION AND SUBTRACTION*

Photo by Dr. Athanassio Kartsatos.

The racket in the photo is hitting the ball with such tremendous force that the ball is distorted. Can we find out how much force? The answer is

$$\frac{mv}{t} - \frac{mv_0}{t}$$

where

m = mass of ball

v = velocity of racket

v_0 = "initial" velocity of racket

t = time of contact

Since the expressions involved have the same denominator, it is very easy to subtract them. As in arithmetic, we simply subtract the numerators and keep the same denominator. Thus

$$\frac{mv}{t} - \frac{mv_0}{t} = \frac{mv - mv_0}{t} \quad \leftarrow \text{Subtract numerators}$$
$$\leftarrow \text{Keep the denominator}$$

A. ADDING AND SUBTRACTING EXPRESSIONS WITH THE SAME DENOMINATOR

As you recall from Sections 1.2 and 2.3, $\frac{1}{5} + \frac{2}{5} = \frac{3}{5}$, $\frac{1}{7} + \frac{4}{7} = \frac{5}{7}$, and $\frac{1}{11} + \frac{8}{11} = \frac{9}{11}$. The same procedure works for fractional expressions. For example,

$$\frac{3}{x} + \frac{5}{x} = \frac{3 + 5}{x} = \frac{8}{x} \quad \leftarrow \text{Add numerators}$$
$$\leftarrow \text{Keep the denominator}$$

---

*The addition and subtraction of arithmetic fractions is covered in Sections 1.2 and 2.3.

Similarly,

$$\frac{5}{x+1} + \frac{2}{x+1} = \frac{5+2}{x+1} = \frac{7}{x+1}$$

$$\frac{5}{7(x-1)} + \frac{2}{7(x-1)} = \frac{5+2}{7(x-1)} = \frac{\overset{1}{\cancel{7}}}{\underset{1}{\cancel{7}(x-1)}} = \frac{1}{x-1}$$

In these two problems we *add* the numerators and keep the same denominator.

and

$$\frac{8}{x+5} - \frac{2}{x+5} = \frac{8-2}{x+5} = \frac{6}{x+5}$$

$$\frac{8}{9(x-3)} - \frac{2}{9(x-3)} = \frac{8-2}{9(x-3)} = \frac{\overset{2}{\cancel{6}}}{\underset{3}{\cancel{9}(x-3)}} = \frac{2}{3(x-3)}$$

In these two problems we *subtract* the numerators and keep the same denominator.

EXAMPLE 1 Find.

a. $\dfrac{8}{3(x-2)} + \dfrac{1}{3(x-2)}$ **b.** $\dfrac{7}{5(x+4)} - \dfrac{2}{5(x+4)}$

Problem 1 Find.

a. $\dfrac{4}{5(x-2)} + \dfrac{6}{5(x-2)}$

b. $\dfrac{11}{3(x+2)} - \dfrac{2}{3(x+2)}$

Solution

a. $\dfrac{8}{3(x-2)} + \dfrac{1}{3(x-2)} = \dfrac{8+1}{3(x-2)} = \dfrac{\overset{3}{\cancel{9}}}{\underset{1}{\cancel{3}(x-2)}} = \dfrac{3}{x-2}$

b. $\dfrac{7}{5(x+4)} - \dfrac{2}{5(x+4)} = \dfrac{7-2}{5(x+4)} = \dfrac{\overset{1}{\cancel{5}}}{\underset{1}{\cancel{5}(x+4)}} = \dfrac{1}{x+4}$

B. ADDING AND SUBTRACTING EXPRESSIONS WITH DIFFERENT DENOMINATORS

Not all fractions have the same denominator. To add or subtract fractions with different denominators, we have to rely on our experience with arithmetic. For example, to add

$$\frac{7}{12} + \frac{5}{18}$$

we first must find a common denominator, that is, a *multiple* of 12 and 18. Of course, it is more convenient to use the smallest one available. In general, the **lowest common denominator** (LCD) of two fractions is the smallest number that is a multiple of *both* denominators. To find the LCD, we can use successive divisions as in Section 1.2.

$$\begin{array}{r|cc} 2 & 12 & 18 \\ 3 & 6 & 9 \\ \hline & 2 & 3 \end{array}$$

The LCD is $2 \times 3 \times 2 \times 3 = 36$. We can also factor both numbers and write each factor in a column, obtaining

$$
\begin{array}{rcl}
 & & \left\{\begin{array}{l}\text{Pick the number with} \\ \text{the highest exponent.}\end{array}\right. \\
12 = 2 \cdot 2 \cdot 3 & = 2^2 \cdot 3^1 & \\
18 = 2 \cdot 3 \cdot 3 = 2^1 \cdot 3^2 & & \text{Note that all the 2's and all the 3's are} \\
 & & \text{written in the } \textit{same} \text{ column.}
\end{array}
$$

Since we need the *smallest* number that is a multiple of 12 and 18, we select the factors raised to the highest power in each column, that is, 2^2 and 3^2. The product of these factors is the LCD. Thus the LCD of 12 and 18 is $2^2 \cdot 3^2 = 4 \cdot 9 = 36$, as before. We then write each fraction with a denominator of 36 and add.

$$
\frac{7}{12} = \frac{7 \cdot 3}{12 \cdot 3} = \frac{21}{36} \qquad \text{We multiply the denominator by 3 (to get}
$$
$$
36), \text{ so we do the same to the numerator.}
$$

$$
\frac{5}{18} = \frac{5 \cdot 2}{18 \cdot 2} = \frac{10}{36}
$$

$$
\frac{7}{12} + \frac{5}{18} = \frac{21}{36} + \frac{10}{36} = \frac{31}{36}
$$

The procedure can also be written as

$$
\begin{array}{r}
\dfrac{7}{12} = \dfrac{21}{36} \\[2mm]
+\dfrac{5}{18} = \dfrac{10}{36} \\[1mm]
\hline
\dfrac{31}{36}
\end{array}
$$

Similarly, to subtract

$$
\frac{11}{15} - \frac{5}{18}
$$

we can use successive divisions to find the LCD, writing

$$
3 \big|\, \underline{15 \quad 18}
$$
$$
 5 \quad\; 6
$$

The LCD is $3 \times 5 \times 6 = 90$. We can also factor the denominators and write them as shown:

$$
\begin{array}{rcl}
15 = 3 \cdot 5 & = & 3 \cdot 5 \\
18 = 2 \cdot 3 \cdot 3 & = & 2 \cdot 3^2
\end{array}
\qquad
\begin{array}{l}
\text{Note that all the 2's, all the 3's, and all the} \\
\text{5's are in the } \textit{same} \text{ column.}
\end{array}
$$

As before, the LCD is

$$
2 \cdot 3^2 \cdot 5 = 2 \cdot 9 \cdot 5 = 90
$$

Then,

$$
\frac{11}{15} = \frac{11 \cdot 6}{15 \cdot 6} = \frac{66}{90} \qquad \text{Multiply the numerator and denominator by \textbf{6}.}
$$

$$
\frac{5}{18} = \frac{5 \cdot 5}{18 \cdot 5} = \frac{25}{90} \qquad \text{Multiply the numerator and denominator by \textbf{5}.}
$$

and

$$\frac{11}{15} - \frac{5}{18} = \frac{66}{90} - \frac{25}{90}$$

$$= \frac{66 - 25}{90}$$

$$= \frac{41}{90}$$

Of course, it is possible that the denominators involved have no common factors. In this case, the LCD is the *product* of these denominators. Thus, to add $\frac{3}{5} + \frac{4}{7}$, we use $5 \cdot 7 = 35$ as the LCD and write

$$\frac{3}{5} = \frac{3 \cdot 7}{5 \cdot 7} = \frac{21}{35} \qquad \text{Multiply the numerator and denominator by \textbf{7}.}$$

$$\frac{4}{7} = \frac{4 \cdot 5}{7 \cdot 5} = \frac{20}{35} \qquad \text{Multiply the numerator and denominator by \textbf{5}.}$$

Thus,

$$\frac{3}{5} + \frac{4}{7} = \frac{21}{35} + \frac{20}{35} = \frac{41}{35}$$

Similarly, $\frac{4}{x} + \frac{5}{3}$ has $3x$ as the LCD. We then write $\frac{4}{x}$ and $\frac{5}{3}$ as equivalent fractions with $3x$ as denominator and add.

$$\frac{4}{x} = \frac{4 \cdot 3}{x \cdot 3} = \frac{12}{3x} \qquad \text{Multiply the numerator and denominator by \textbf{3}.}$$

$$\frac{5}{3} = \frac{5 \cdot x}{3 \cdot x} = \frac{5x}{3x} \qquad \text{Multiply the numerator and denominator by \textbf{x}.}$$

$$\frac{4}{x} + \frac{5}{3} = \frac{12}{3x} + \frac{5x}{3x}$$

$$= \frac{12 + 5x}{3x}$$

EXAMPLE 2   Find.

a. $\dfrac{7}{8} + \dfrac{2}{x}$ **b.** $\dfrac{2}{x - 1} - \dfrac{1}{x + 2}$

Solution

a. Since 8 and x do not have any common factors, the LCD is $8x$. We write $\frac{7}{8}$ and $\frac{2}{x}$ as equivalent fractions with $8x$ as denominator and add.

$$\frac{7}{8} = \frac{7 \cdot x}{8 \cdot x} = \frac{7x}{8x}$$

$$\frac{2}{x} = \frac{2 \cdot 8}{x \cdot 8} = \frac{16}{8x}$$

$$\frac{7}{8} + \frac{2}{x} = \frac{7x}{8x} + \frac{16}{8x}$$

$$= \frac{7x + 16}{8x}$$

Problem 2   Find.

a. $\dfrac{4}{5} + \dfrac{3}{x}$ **b.** $\dfrac{5}{x - 1} - \dfrac{3}{x + 3}$

ANSWER   (to problem on page 538)

1. a. $\dfrac{2}{x - 2}$ **b.** $\dfrac{3}{x + 2}$

b. $(x - 1)$ and $(x + 2)$ do not have any common factors. Thus the LCD of

$$\frac{2}{x - 1} - \frac{1}{x + 2}$$

is $(x - 1)(x + 2)$. We then write $\dfrac{2}{x - 1}$ and $\dfrac{1}{x + 2}$ as equivalent fractions with $(x - 1)(x + 2)$ as denominator.

$$\frac{2}{x - 1} = \frac{2 \cdot (x + 2)}{(x - 1)(x + 2)}$$

$$\frac{1}{x + 2} = \frac{1 \cdot (x - 1)}{(x + 2)(x - 1)} = \frac{(x - 1)}{(x - 1)(x + 2)}$$

Hence,

$$\frac{2}{x - 1} - \frac{1}{x + 2} = \frac{2 \cdot (x + 2)}{(x - 1)(x + 2)} - \frac{(x - 1)}{(x - 1)(x + 2)}$$

$$= \frac{2(x + 2) - (x - 1)}{(x - 1)(x + 2)}$$

$$= \frac{2x + 4 - x + 1}{(x - 1)(x + 2)}$$

$$= \frac{x + 5}{(x - 1)(x + 2)}$$ ▲

Are you ready for a harder one? We can use the sum of $\frac{7}{12}$ and $\frac{5}{18}$ as our model because the steps we used in doing that problem can be used in general. Here are the steps.

TO ADD (OR SUBTRACT) FRACTIONS WITH DIFFERENT DENOMINATORS

1. Find the LCD.

2. Write all fractions as equivalent ones with the LCD as the denominator.

3. Add (or subtract) numerators.

4. Reduce if possible.

We now use these steps to add

$$\frac{x + 1}{x^2 + x - 2} + \frac{x + 3}{x^2 - 1}$$

1. We first find the LCD of the denominators. To do this, we factor the denominators.

$$x^2 + x - 2 = (x + 2)(x - 1)$$
$$x^2 - 1 = (x - 1)(x + 1)$$

and the LCD is

$$(x + 2)(x - 1)(x + 1)$$

2. We then write $\dfrac{x+1}{x^2+2x-2}$ and $\dfrac{x+3}{x^2-1}$ as equivalent fractions with $(x+2)(x-1)(x+1)$ as denominator.

$$\frac{x+1}{x^2+x-2}=\frac{x+1}{(x+2)(x-1)}=\frac{(x+1)(\boldsymbol{x+1})}{(x+2)(x-1)(\boldsymbol{x+1})}$$

$$\frac{x+3}{x^2-1}=\frac{x+3}{(x+1)(x-1)}=\frac{(x+3)(\boldsymbol{x+2})}{(x+1)(x-1)(\boldsymbol{x+2})}$$

$$=\frac{(x+3)(x+2)}{(x+2)(x-1)(x+1)}$$

3. $\dfrac{x+1}{x^2+x-2}+\dfrac{x+3}{x^2-1}=\dfrac{(x+1)(x+1)}{(x+2)(x-1)(x+1)}+\dfrac{(x+3)(x+2)}{(x+2)(x-1)(x+1)}$

$$=\frac{(x^2+2x+1)+(x^2+5x+6)}{(x+2)(x-1)(x+1)}$$

$$=\frac{2x^2+7x+7}{(x+2)(x-1)(x+1)}$$

4. The answer is not reducible, since there are no factors common to the numerator and denominator.

We use this procedure to do the subtraction problem in the next example.

EXAMPLE 3   Subtract.

$$\frac{x-2}{x^2-x-6}-\frac{x+3}{x^2-9}$$

Solution   We use the four–step procedure given.

1. To find the LCD, we factor the denominators, obtaining

$$x^2-x-6=\quad (x-3)(x+2)$$
$$x^2-9=(x+3)(x-3)$$

Thus the LCD is $(x+3)(x-3)(x+2)$

2. We write each fraction as an equivalent one with the LCD as the denominator. Hence,

$$\frac{x-2}{x^2-x-6}=\frac{(x-2)(x+3)}{(x-3)(x+2)(x+3)}$$

$$=\frac{(x-2)(x+3)}{(x+3)(x-3)(x+2)}$$

$$\frac{x+3}{x^2-9}=\frac{(x+3)(x+2)}{(x+3)(x-3)(x+2)}$$

$$=\frac{(x+3)(x+2)}{(x+3)(x-3)(x+2)}$$

3. $\dfrac{x-2}{x^2-x-6}-\dfrac{x+3}{x^2-9}=\dfrac{(x-2)(x+3)}{(x+3)(x-3)(x+2)}-\dfrac{(x+3)(x+2)}{(x+3)(x-3)(x+2)}$

$$=\frac{(x^2+x-6)-(x^2+5x+6)}{(x+3)(x-3)(x+2)}$$

$$=\frac{x^2+x-6-x^2-5x-6}{(x+3)(x-3)(x+2)}$$

Problem 3   Subtract.

$$\frac{x-3}{(x+1)(x-2)}-\frac{x+3}{x^2-4}$$

$$= \frac{-4x - 12}{(x + 3)(x - 3)(x + 2)}$$

$$= \frac{-4(x + 3)}{(x + 3)(x - 3)(x + 2)}$$ Factor the numerator.

4. Reduce.

$$\frac{-4(x + 3)}{(x + 3)(x - 3)(x + 2)} = \frac{-4}{(x - 3)(x + 2)}$$ ▲

And now, a last word before you go on to the exercises. At this point, you should be convinced that there are great similarities between algebra and arithmetic. In fact, in this very section we use the arithmetic addition of fractions as a model to do the algebraic addition. To show that these similarities are very strong and also to give you more practice, Problems 1–20 in the exercises consist of two similar problems, an arithmetic and an algebraic one. Use the practice and experience gained in working one to do the other.

ANSWER (to problem on page 542)

3. $\dfrac{-5x - 9}{(x + 1)(x + 2)(x - 2)}$

ANSWERS

A, B. Perform the indicated operations.

1. a. $\dfrac{2}{7} + \dfrac{3}{7}$ **b.** $\dfrac{3}{x} + \dfrac{8}{x}$
1. a. _____
   b. _____

2. a. $\dfrac{5}{9} + \dfrac{2}{9}$ **b.** $\dfrac{9}{x-1} + \dfrac{2}{x-1}$
2. a. _____
   b. _____

3. a. $\dfrac{8}{9} - \dfrac{2}{9}$ **b.** $\dfrac{6}{x} - \dfrac{2}{x}$
3. a. _____
   b. _____

4. a. $\dfrac{4}{7} - \dfrac{2}{7}$ **b.** $\dfrac{6}{x+4} - \dfrac{2}{x+4}$
4. a. _____
   b. _____

5. a. $\dfrac{6}{7} + \dfrac{8}{7}$ **b.** $\dfrac{3}{2x} + \dfrac{7}{2x}$
5. a. _____
   b. _____

6. a. $\dfrac{1}{9} + \dfrac{2}{9}$ **b.** $\dfrac{3}{8(x-2)} + \dfrac{1}{8(x-2)}$
6. a. _____
   b. _____

7. a. $\dfrac{8}{3} - \dfrac{2}{3}$ **b.** $\dfrac{11}{3(x+1)} - \dfrac{9}{3(x+1)}$
7. a. _____
   b. _____

8. a. $\dfrac{3}{8} - \dfrac{1}{8}$ **b.** $\dfrac{7}{15(x-1)} - \dfrac{2}{15(x-1)}$
8. a. _____
   b. _____

9. a. $\dfrac{8}{9} + \dfrac{4}{9}$ **b.** $\dfrac{7x}{4(x+1)} + \dfrac{3x}{4(x+1)}$
9. a. _____
   b. _____

10. a. $\dfrac{15}{14} + \dfrac{3}{14}$ **b.** $\dfrac{29x}{15(x-3)} + \dfrac{4x}{15(x-3)}$
10. a. _____
    b. _____

11. a. $\dfrac{3}{4} - \dfrac{1}{3}$ **b.** $\dfrac{7}{x} - \dfrac{3}{8}$
11. a. _____
    b. _____

12. a. $\dfrac{5}{7} - \dfrac{2}{5}$ **b.** $\dfrac{x}{3} - \dfrac{7}{x}$
12. a. _____
    b. _____

13. a. $\dfrac{1}{5} + \dfrac{1}{7}$ **b.** $\dfrac{4}{x} + \dfrac{x}{9}$
13. a. _____
    b. _____

14. a. $\dfrac{1}{3} + \dfrac{1}{9}$ **b.** $\dfrac{5}{x} + \dfrac{6}{3x}$
14. a. _____
    b. _____

15. a. $\dfrac{2}{5} - \dfrac{4}{15}$ **b.** $\dfrac{4}{7(x-1)} - \dfrac{3}{14(x-1)}$
15. a. _____
    b. _____

16. a. $\dfrac{9}{2} - \dfrac{5}{8}$ **b.** $\dfrac{8}{9(x+3)} - \dfrac{5}{36(x+3)}$

16. a. _____
    b. _____

17. a. $\dfrac{4}{7} + \dfrac{3}{8}$ **b.** $\dfrac{3}{x+1} + \dfrac{5}{x-2}$

17. a. _____
    b. _____

18. a. $\dfrac{2}{9} + \dfrac{4}{5}$ **b.** $\dfrac{2x}{x+2} + \dfrac{3x}{x-4}$

18. a. _____
    b. _____

19. a. $\dfrac{7}{8} - \dfrac{1}{3}$ **b.** $\dfrac{6}{x-2} - \dfrac{3}{x+1}$

19. a. _____
    b. _____

20. a. $\dfrac{6}{7} - \dfrac{2}{3}$ **b.** $\dfrac{4x}{x+1} - \dfrac{4x}{x+2}$

20. a. _____
    b. _____

21. $\dfrac{x+1}{x^2+3x-4} + \dfrac{x+2}{x^2-16}$ **22.** $\dfrac{x-2}{x^2-9} + \dfrac{x+1}{x^2-x-12}$

21. _____

22. _____

23. $\dfrac{3x}{x^2+3x-10} + \dfrac{2x}{x^2+x-6}$ **24.** $\dfrac{x+3}{x^2-x-2} + \dfrac{x-1}{x^2+2x+1}$

23. _____

24. _____

25. $\dfrac{1}{x^2-y^2} + \dfrac{5}{(x+y)^2}$ **26.** $\dfrac{3x}{(x+y)^2} + \dfrac{5x}{(x-y)}$

25. _____

26. _____

27. $\dfrac{2}{x-5} - \dfrac{3x}{x^2-25}$ **28.** $\dfrac{x+3}{x^2-x-2} - \dfrac{x-1}{x^2+2x+1}$

27. _____

28. _____

29. $\dfrac{x-1}{x^2+3x+2} - \dfrac{x+7}{x^2+5x+6}$

29. _____

30. $\dfrac{2}{x^2+3xy+2y^2} - \dfrac{1}{x^2-xy-2y^2}$

30. _____

✓ SKILL CHECKER

Perform the indicated operations.

31. $1 \div \dfrac{20}{9}$ **32.** $1 \div \dfrac{30}{7}$

31. _____

32. _____

33. $12x\left(\dfrac{2}{x} + \dfrac{3}{2x}\right)$ **34.** $12x\left(\dfrac{4}{3x} - \dfrac{1}{4x}\right)$

33. _____

34. _____

35. $x^2\left(1 - \dfrac{1}{x^2}\right)$ **36.** $x^2\left(1 + \dfrac{1}{x}\right)$

35. _____

36. _____

8.3 USING YOUR KNOWLEDGE

In Chapter 2 we mentioned that some numbers cannot be written as the ratio of two integers. These numbers are called **irrational numbers.** Irrational numbers can be approximated by using a type of fraction called a **continued fraction.** Here is how. From a table of square roots, or a calculator,

$$\sqrt{2} \approx 1.4142 \qquad \approx \text{ means "approximately equal."}$$

Can we find some continued fraction to approximate $\sqrt{2}$?

1. Try $1 + \dfrac{1}{2}$ (write it as a decimal).

   1. _____

2. Try $1 + \dfrac{1}{2 + \frac{1}{2}}$ (write it as a decimal).

   2. _____

3. Try $1 + \dfrac{1}{2 + \dfrac{1}{2 + \frac{1}{2}}}$ (write it as a decimal).

   3. _____

4. Look at the pattern for the approximation of $\sqrt{2}$ given in Problems 1, 2, and 3. What do you think the next approximation (when written as a continued fraction) will be?

   4. _____

5. How close is the approximation for $\sqrt{2}$ given in problem 3 to the value $\sqrt{2} \approx 1.4142$?

   5. _____

You can check the addition of fractions using a calculator. For instance, we can show that:

$$\frac{x - 1}{x^2 - x - 6} - \frac{x + 4}{x^2 - 9} = \frac{-4x - 11}{(x + 3)(x - 3)(x + 2)}$$

To check this answer, you can substitute any convenient number for x and see if both sides are equal. Of course, you cannot choose numbers that will give you a 0 denominator. For this example, a simple number to use is $x = 2$. On the left you get:

$$\frac{2 - 1}{4 - 2 - 6} - \frac{2 + 4}{4 - 9} = -\frac{1}{4} + \frac{6}{5}$$

Now key in

$$\boxed{1} \boxed{\div} \boxed{4} \boxed{+/-} \boxed{+} \boxed{6} \boxed{\div} \boxed{5} \boxed{=}$$

The display shows 0.95. On the right, we have

$$\frac{-8 - 11}{(2 + 3)(2 - 3)(2 + 2)}$$

Now key in

$$\boxed{8} \boxed{+/-} \boxed{-} \boxed{1} \boxed{1} \boxed{=} \boxed{\div} \boxed{(} \boxed{2} \boxed{+} \boxed{3} \boxed{)} \boxed{=}$$
$$\boxed{\div} \boxed{(} \boxed{2} \boxed{-} \boxed{3} \boxed{)} \boxed{=} \boxed{\div} \boxed{(} \boxed{2} \boxed{+} \boxed{2} \boxed{)} \boxed{=}$$

and the display shows 0.95 again. This completes the check. If your instructor agrees, verify the answers in this section using this method. You can also verify the answers for the preceding section.

8.4 COMPLEX FRACTIONS

Courtesy Adler Planetarium, Chicago.

REVIEW

Before starting this section, you should know:

1. How to multiply fractions (Sections 1.2, 2.3).
2. How to divide fractions (Sections 1.2, 2.3).

OBJECTIVE

You should be able to:

A. Simplify a complex fraction.

This model of the planets in the solar system is similar to the model that was designed by the seventeenth-century Dutch mathematician and astronomer Christian Huygens. The gears used in the model were hard to design, since they had to make each of the planets revolve around the sun in different times. For example, Saturn goes around the sun in

$$29 + \cfrac{1}{2 + \cfrac{2}{9}} \quad \text{yr}$$

The expression

$$\cfrac{1}{2 + \cfrac{2}{9}}$$

is a *complex fraction*.

A **complex fraction** is a fraction that has one or more fractions in its numerator, denominator, or both.

Complex fractions can be simplified in either of two ways:

PROCEDURE FOR SIMPLIFYING COMPLEX FRACTIONS

1. Multiply numerator and denominator by the LCD of the fractions involved.

   or

2. Perform the operations indicated in the numerator and denominator of the complex fraction, and then divide the numerator by the denominator.

We illustrate these two methods by simplifying the complex fraction

$$\dfrac{1}{2 + \dfrac{2}{9}}$$

METHOD 1. Multiply numerator and denominator by the LCD of the fractions involved (in our case, by **9**).

$$\dfrac{1}{2 + \dfrac{2}{9}} = \dfrac{9 \cdot 1}{9\left(2 + \dfrac{2}{9}\right)} \quad \text{Note that } 9\left(2 + \dfrac{2}{9}\right) = 9 \cdot 2 + 9 \cdot \dfrac{2}{9} = 18 + 2.$$

$$= \dfrac{9}{18 + 2} = \dfrac{9}{20}$$

The answer is $\frac{9}{20}$.

METHOD 2. Perform the operations indicated in the numerator and denominator of the complex fraction and then divide the numerator by the denominator.

$$\dfrac{1}{2 + \dfrac{2}{9}} = \dfrac{1}{\dfrac{18}{9} + \dfrac{2}{9}} = \dfrac{1}{\dfrac{20}{9}}$$

$$= 1 \div \dfrac{20}{9}$$

$$= 1 \cdot \dfrac{9}{20}$$

$$= \dfrac{9}{20}$$

Either procedure also works for more complicated rational expressions, as shown in the next example, where we use Method 1.

EXAMPLE 1   Simplify.

$$\dfrac{\dfrac{1}{a} + \dfrac{2}{b}}{\dfrac{3}{a} - \dfrac{1}{b}}$$

Solution   The LCD of the fractions involved is ab, so we multiply the numerator and denominator by ab, obtaining

$$\dfrac{ab \cdot \left(\dfrac{1}{a} + \dfrac{2}{b}\right)}{ab \cdot \left(\dfrac{3}{a} - \dfrac{1}{b}\right)} = \dfrac{ab \cdot \dfrac{1}{a} + ab \cdot \dfrac{2}{b}}{ab \cdot \dfrac{3}{a} - ab \cdot \dfrac{1}{b}} \quad \begin{array}{l} \text{Note that } ab \cdot \dfrac{1}{a} = b,\ ab \cdot \dfrac{2}{b} = 2a, \\[2mm] ab \cdot \dfrac{3}{a} = 3b, \text{ and } ab \cdot \dfrac{1}{b} = a. \end{array}$$

$$= \dfrac{b + 2a}{3b - a}$$

EXAMPLE 2   Simplify.

$$\dfrac{\dfrac{2}{x} + \dfrac{3}{2x}}{\dfrac{4}{3x} - \dfrac{1}{4x}}$$

Problem 1   Simplify.

$$\dfrac{\dfrac{2}{a} - \dfrac{3}{b}}{\dfrac{1}{a} + \dfrac{2}{b}}$$

Problem 2   Simplify.

$$\dfrac{\dfrac{1}{4x} + \dfrac{2}{3x}}{\dfrac{3}{2x} - \dfrac{1}{x}}$$

Solution We must first find the LCD of x, $2x$, $3x$, and $4x$. Now,

$$x = \qquad \qquad x$$
$$2x = 2 \quad\cdot\quad x$$
$$3x = \qquad 3 \cdot x$$
$$4x = 2^2 \cdot\ x$$

We wrote the factors in columns.

The LCD is $2^2 \cdot 3 \cdot x = 12x$. Multiplying numerator and denominator by **12x,** we have

$$\frac{\dfrac{2}{x} + \dfrac{3}{2x}}{\dfrac{4}{3x} - \dfrac{1}{4x}} = \frac{12x \cdot \left(\dfrac{2}{x} + \dfrac{3}{2x}\right)}{12x \cdot \left(\dfrac{4}{3x} - \dfrac{1}{4x}\right)}$$

$$= \frac{12x \cdot \dfrac{2}{x} + 12x \cdot \dfrac{3}{2x}}{12x \cdot \dfrac{4}{3x} - 12x \cdot \dfrac{1}{4x}}$$

$$= \frac{12 \cdot 2 + 6 \cdot 3}{4 \cdot 4 - 3 \cdot 1}$$

$$= \frac{24 + 18}{16 - 3}$$

$$= \frac{42}{13}$$

EXAMPLE 3 Simplify.

$$\frac{1 - \dfrac{1}{x^2}}{1 + \dfrac{1}{x}}$$

Solution Here the LCD of the fractions involved is x^2. Thus,

$$\frac{1 - \dfrac{1}{x^2}}{1 + \dfrac{1}{x}} = \frac{x^2 \cdot \left(1 - \dfrac{1}{x^2}\right)}{x^2 \cdot \left(1 + \dfrac{1}{x}\right)}$$

$$= \frac{x^2 \cdot 1 - x^2 \cdot \dfrac{1}{x^2}}{x^2 \cdot 1 + x^2 \cdot \dfrac{1}{x}}$$

$$= \frac{x^2 - 1}{x^2 + x} \qquad \text{Factor the numerator and denominator.}$$

$$= \frac{(x + 1)(x - 1)}{x(x + 1)}$$

$$= \frac{x - 1}{x}$$

Problem 3 Simplify.

$$\frac{1 - \dfrac{1}{x^2}}{1 - \dfrac{1}{x}}$$

ANSWERS (to problems on pages 550–551)

1. $\dfrac{2b - 3a}{b + 2a}$ 2. $\dfrac{11}{6}$ 3. $\dfrac{x + 1}{x}$

A. Simplify.

ANSWERS

1. $\dfrac{\dfrac{1}{2}}{2 + \dfrac{1}{2}}$

2. $\dfrac{\dfrac{1}{4}}{3 + \dfrac{1}{4}}$

3. $\dfrac{\dfrac{1}{2}}{2 - \dfrac{1}{2}}$

4. $\dfrac{\dfrac{1}{4}}{3 - \dfrac{1}{4}}$

5. $\dfrac{a - \dfrac{a}{b}}{1 + \dfrac{a}{b}}$

6. $\dfrac{1 - \dfrac{1}{a}}{1 + \dfrac{1}{a}}$

7. $\dfrac{\dfrac{1}{a} + \dfrac{1}{b}}{\dfrac{1}{a} - \dfrac{1}{b}}$

8. $\dfrac{\dfrac{2}{a} + \dfrac{1}{b}}{\dfrac{2}{a} - \dfrac{1}{b}}$

9. $\dfrac{\dfrac{1}{2a} + \dfrac{1}{3b}}{\dfrac{4}{a} - \dfrac{3}{4b}}$

10. $\dfrac{\dfrac{1}{2a} + \dfrac{1}{4b}}{\dfrac{2}{a} - \dfrac{3}{5b}}$

11. $\dfrac{\dfrac{1}{3} + \dfrac{3}{4}}{\dfrac{3}{8} - \dfrac{1}{6}}$

12. $\dfrac{\dfrac{1}{5} + \dfrac{3}{2}}{\dfrac{5}{8} - \dfrac{3}{10}}$

13. $\dfrac{2 + \dfrac{1}{x}}{4 - \dfrac{1}{x^2}}$

14. $\dfrac{3 + \dfrac{1}{x}}{9 - \dfrac{1}{x^2}}$

15. $\dfrac{2 + \dfrac{2}{x}}{1 + \dfrac{1}{x}}$

16. $\dfrac{5 + \dfrac{5}{x^2}}{1 + \dfrac{1}{x^2}}$

17. $\dfrac{\dfrac{1}{y} + \dfrac{1}{x}}{\dfrac{x}{y} - \dfrac{y}{x}}$

18. $\dfrac{\dfrac{1}{x} + \dfrac{1}{y}}{\dfrac{y}{x} - \dfrac{x}{y}}$

1. _____

2. _____

3. _____

4. _____

5. _____

6. _____

7. _____

8. _____

9. _____

10. _____

11. _____

12. _____

13. _____

14. _____

15. _____

16. _____

17. _____

18. _____

19. $\dfrac{x - 2 - \dfrac{8}{x}}{x - 3 - \dfrac{4}{x}}$

20. $\dfrac{x - 2 - \dfrac{15}{x}}{x - 3 - \dfrac{10}{x}}$

21. $\dfrac{\dfrac{1}{x + 5}}{\dfrac{4}{x^2 - 25}}$

22. $\dfrac{\dfrac{1}{x - 3}}{\dfrac{2}{x^2 - 9}}$

23. $\dfrac{\dfrac{1}{x^2 - 16}}{\dfrac{2}{x + 4}}$

24. $\dfrac{\dfrac{3}{x^2 - 64}}{\dfrac{4}{x + 8}}$

19. _____

20. _____

21. _____

22. _____

23. _____

24. _____

✓ SKILL CHECKER

Solve.

25. $19w = 2356$

26. $18L = 2232$

27. $9x + 24 = x$

28. $10x + 36 = x$

29. $5x = 4x + 3$

30. $6x = 5x + 5$

25. _____

26. _____

27. _____

28. _____

29. _____

30. _____

8.4 USING YOUR KNOWLEDGE

At the beginning of this section, we mentioned that Saturn takes

$$29 + \dfrac{1}{2 + \dfrac{2}{9}} \quad \text{yr}$$

to go around the sun. Since we have shown that

$$\dfrac{1}{2 + \dfrac{2}{9}} = \dfrac{9}{20}$$

Saturn takes $29 + \frac{9}{20} = 29\frac{9}{20}$ yr to go around the sun. Use your knowledge to simplify the number of years it takes the following planets to go around the sun.

1. Mercury, $\dfrac{1}{4 + \frac{1}{6}}$ yr

2. Venus, $\dfrac{1}{1 + \frac{2}{3}}$ yr

3. Jupiter, $11 + \dfrac{1}{1 + \frac{7}{43}}$ yr (Write your answer as a mixed number.)

4. Mars, $1 + \dfrac{1}{1 + \frac{3}{22}}$ yr

1. _____

2. _____

3. _____

4. _____

Courtesy of J. L. Hudson.

OBJECTIVES

REVIEW

Before starting this section, you should know:

1. How to solve linear equations (Sections 3.1 through 3.3).
2. How to find the LCD of two or more rational expressions.

OBJECTIVE

A. You should be able to solve fractional equations.

The picture here shows one of the largest American flags being displayed in J. L. Hudson's Store in Detroit. By law, the ratio of length to width of a flag should be $\frac{19}{10}$. If the length of this flag is 235 ft, what should its width be to conform with the law? To solve this problem we let W be the width of the flag and set up the equation

$$\begin{array}{l} \text{Length} \to 19 \\ \text{Width} \to 10 \end{array} = \frac{235}{W}$$

This equation is an example of a *fractional equation*. A **fractional equation** is an equation containing one or more fractional expressions. To solve this equation, we must clear the denominators involved. This was accomplished in Chapter 3 (Section 3.3) by multiplying each term by the LCM (lowest common multiple) of the denominators. Since the LCM of 10 and W is **10W,** we have:

Multiply by the LCM.	$10W \cdot \dfrac{19}{10} = \dfrac{235}{W} \cdot 10W$
Simplify.	$19W = 2350$
Divide by **19.**	$W = \dfrac{2350}{19}$
Approximate the answer.	$W \approx 124$

$$\begin{array}{r} 123.6 \\ 19{\overline{\smash{\big)}\,2350.0}} \\ \underline{19} \\ 45 \\ \underline{38} \\ 70 \\ \underline{57} \\ 130 \\ \underline{114} \\ 16 \end{array}$$

(By the way, the flag is only 104 ft long and weighs 1500 lb. It is *not* an official flag.)

Of course, if there are more terms involved, multiplying each side of the equation by the LCM is equivalent to multiplying *each* term by the LCM. This is because if you have the equation

$$\frac{a}{b} + \frac{c}{d} = \frac{e}{f}$$

multiplying each side by M gives

$$M \cdot \left(\frac{a}{b} + \frac{c}{d} \right) = M \cdot \frac{e}{f}$$

or

$$M \cdot \frac{a}{b} + M \cdot \frac{c}{d} = M \cdot \frac{e}{f}$$

Thus, multiplying each side of the equation by M is equivalent to multiplying each term by M.

The procedure to solve fractional equations is very similar to the one used to solve linear equations. Because of this, you may want to review the procedure (see page 222) before going on.

EXAMPLE 1 Solve.

$$\frac{3}{4} + \frac{2}{x} = \frac{1}{12}$$

Solution The LCM of 4, x, and 12 is **12x**.

1. Multiply every term by **12x**. $12x \cdot \dfrac{3}{4} + 12x \cdot \dfrac{2}{x} = 12x \cdot \dfrac{1}{12}$

2. Simplify. $9x + 24 = x$

3. Subtract 24 from each side. $9x = x - 24$

4. Subtract x from each side. $8x = -24$

5. Divide each side by 8. $x = -3$

The answer is -3, as can easily be checked,

$$\frac{3}{4} + \frac{2}{x} \overset{?}{=} \frac{1}{12}$$

$$\begin{array}{c|c} \dfrac{3}{4} + \dfrac{2}{-3} & \dfrac{1}{12} \\ \hline \dfrac{3 \cdot 3}{4 \cdot 3} + \dfrac{2 \cdot 4}{-3 \cdot 4} & \\ \dfrac{9}{12} - \dfrac{8}{12} & \\ \dfrac{1}{12} & \end{array}$$ ▲

In some cases, the denominators involved may be more complicated. Nevertheless, the procedure used to solve the equation remains the same. Thus we use the steps on page 222 to solve

$$\frac{2x}{x-1} + 3 = \frac{4x}{x-1}$$

as shown in the next example.

EXAMPLE 2 Solve.

$$\frac{2x}{x-1} + 3 = \frac{4x}{x-1}$$

Solution Since $x - 1$ is the only denominator, it must be the LCM. We then proceed by steps.

Problem 1 Solve.

$$\frac{4}{5} + \frac{1}{x} = \frac{3}{10}$$

Problem 2 Solve.

$$\frac{8}{x-1} - 4 = \frac{2x}{x-1}$$

1. Multiply each term by the LCM, $(x - 1)$.

$$(x - 1) \cdot \frac{2x}{x - 1} + 3(x - 1) = (x - 1) \cdot \frac{4x}{x - 1}$$

2. Simplify.

$$2x + 3x - 3 = 4x$$
$$5x - 3 = 4x$$

3. Add 3.

$$5x = 4x + 3$$

4. Subtract $4x$.

$$x = 3$$

Thus the answer is $x = 3$, as can easily be verified by substituting 3 for x in the original equation. ▲

So far, the denominators used in the examples have *not* been factorable. In case they are, it is very important that we do factor them before attempting to find the LCM. For instance, to solve the equation

$$\frac{x}{x^2 - 16} + \frac{4}{x - 4} = \frac{1}{x + 4}$$

we first note that

$$x^2 - 16 = (x + 4)(x - 4)$$

We then write

The denominator has been factored.

$$\frac{x}{x^2 - 16} + \frac{4}{x - 4} = \frac{1}{x + 4}$$

$$\frac{x}{(x + 4)(x - 4)} + \frac{4}{x - 4} = \frac{1}{x + 4}$$

The solution of this equation is given in the next example.

EXAMPLE 3 Solve.

$$\frac{x}{x^2 - 16} + \frac{4}{x - 4} = \frac{1}{x + 4}$$

Solution Since $x^2 - 16 = (x + 4)(x - 4)$, we write the equation with the $x^2 - 16$ factored. Thus

$$\frac{x}{(x + 4)(x - 4)} + \frac{4}{x - 4} = \frac{1}{x + 4}$$

The LCM is $(x + 4)(x - 4)$.

1. Multiply each term by the LCM:

$$(x + 4)(x - 4) \cdot \frac{x}{(x + 4)(x - 4)} + (x + 4)(x - 4) \cdot \frac{4}{x - 4}$$

$$= (x + 4)(x - 4) \cdot \frac{1}{x + 4}$$

2. Simplify.

$$x + 4(x + 4) = x - 4$$
$$x + 4x + 16 = x - 4$$
$$5x + 16 = x - 4$$

3. Subtract 16.

$$5x = x - 20$$

4. Subtract x.

$$4x = -20$$

5. Divide by 4.

$$x = -5$$

Problem 3 Solve.

$$\frac{x}{x^2 - 9} + \frac{3}{x - 3} = \frac{1}{x + 3}$$

Thus the solution is $x = -5$. You should check this by substituting -5 for x in the original equation. ▲

By now, you have probably noticed that we always recommend checking the solution obtained when solving an equation by *direct substitution*. The importance of this step will be made obvious in the next example.

EXAMPLE 4 Solve.

$$\frac{x}{x+4} - \frac{2}{5} = \frac{-4}{x+4}$$

Solution Here the denominators are 5 and $(x + 4)$. Thus the LCM is **5(x + 4).**

1. Multiply each term by the LCM.

$$5(x+4) \cdot \frac{x}{x+4} - \frac{2}{5} \cdot 5(x+4) = \frac{-4}{x+4} \cdot 5(x+4)$$

2. Simplify. $5x - 2x - 8 = -20$

$$3x - 8 = -20$$

3. Add 8. $3x = -12$

4. Divide by 3. $x = -4$

Thus the solution seems to be $x = -4$. However, if we substitute -4 for x in the original equation, we have

$$\frac{-4}{-4+4} - \frac{2}{5} = \frac{-4}{-4+4}$$

or

$$\frac{-4}{0} - \frac{2}{5} = \frac{-4}{0}$$

└─Division─┘
by 0 is
not defined

Clearly, two of the terms are not defined. Thus, this equation has no solution. ▲

Finally, we must point out that the equations resulting when clearing denominators are not *always* linear equations. (As you recall, a linear equation is an equation in which the highest exponent of the variable is 1.) For example, to solve the equation

$$\frac{x^2}{x+2} = \frac{4}{x+2}$$

we first multiply by the LCM **(x + 2),** obtaining

$$(x+2) \cdot \frac{x^2}{x+2} = (x+2) \cdot \frac{4}{x+2}$$

or

$$x^2 = 4$$

Problem 4 Solve.

$$\frac{x}{x-2} + \frac{3}{4} = \frac{2}{x-2}$$

In this equation, the variable x has a 2 as an exponent; thus it is a quadratic equation and can be solved when written in standard form, that is, by writing the equation as

$x^2 - 4 = 0$ Recall that a quadratic equation is an equation in which the highest exponent of the variable is 2. A quadratic equation in standard form is written as $ax^2 + bx + c = 0$.

Factor. $(x + 2)(x - 2) = 0$

Use the principle of zero products. $x + 2 = 0$ or $x - 2 = 0$

Solve each equation. $x = -2$ or $x = 2$

Obviously, $x = 2$ is a solution since

$$\frac{2^2}{2 + 2} = \frac{4}{2 + 2}$$

However, for $x = -2$,

$$\frac{x^2}{x + 2} = \frac{2^2}{-2 + 2} = \frac{4}{0}$$

and the denominator $x + 2$ becomes 0. Thus $x = -2$ is *not* a solution. The only solution is $x = 2$.

EXAMPLE 5 Solve.

$$1 + \frac{3}{x - 2} = \frac{12}{x^2 - 4}$$

Solution Since $x^2 - 4 = (x + 2)(x - 2)$, the LCM is $(x + 2)(x - 2)$. We then write the equation with the denominator $x^2 - 4$ in factored form and multiply each term by the LCM as before. Here are the steps.

1. Multiply each term by the LCM $(x + 2)(x - 2)$.

$$(x + 2)(x - 2) \cdot 1 + (x + 2)(x - 2) \cdot \frac{3}{x - 2}$$

$$= (x + 2)(x - 2) \cdot \frac{12}{(x + 2)(x - 2)}$$

2. Simplify. $(x^2 - 4) + 3(x + 2) = 12$

$x^2 - 4 + 3x + 6 = 12$

$x^2 + 3x + 2 = 12$

3. Subtract 12 $x^2 + 3x - 10 = 0$
from both sides
to write in
standard form.

4. Factor. $(x + 5)(x - 2) = 0$

5. Use the principle $x + 5 = 0$ or $x - 2 = 0$
of zero products.

6. Solve each equation. $x = -5$ or $x = 2$

7. Since $x = 2$ makes the denominator $x - 2$ equal to 0, the only possible solution is $x = -5$. This solution can be verified in the original equation.

Problem 5 Solve.

$$1 - \frac{4}{x^2 - 1} = \frac{-2}{x - 1}$$

ANSWERS   (to problems on pages
558–559)
4. No solution   **5.** $x = -3$

EXERCISE 8.5

A. Solve (if possible).

1. $\dfrac{x}{4} = \dfrac{3}{2}$

2. $\dfrac{x}{8} = \dfrac{-7}{4}$

3. $\dfrac{3}{x} = \dfrac{3}{4}$

4. $\dfrac{6}{x} = \dfrac{-2}{7}$

5. $\dfrac{-8}{3} = \dfrac{16}{x}$

6. $\dfrac{-5}{6} = \dfrac{10}{x}$

7. $\dfrac{4}{3} = \dfrac{x}{9}$

8. $\dfrac{-3}{7} = \dfrac{x}{14}$

9. $\dfrac{2}{5} + \dfrac{3}{x} = \dfrac{23}{20}$

10. $\dfrac{6}{7} + \dfrac{2}{x} = \dfrac{3}{21}$

11. $\dfrac{3}{x} - \dfrac{2}{7} = \dfrac{11}{35}$

12. $\dfrac{4}{x} - \dfrac{2}{9} = \dfrac{22}{63}$

13. $\dfrac{3}{5} + \dfrac{7x}{10} = 2$

14. $\dfrac{2}{7} + \dfrac{4x}{21} = \dfrac{2}{3}$

15. $\dfrac{3x}{4} - \dfrac{1}{5} = \dfrac{13}{10}$

16. $\dfrac{2x}{3} - \dfrac{1}{4} = -1$

17. $\dfrac{3}{x+2} = \dfrac{4}{x-1}$

18. $\dfrac{2}{x-2} = \dfrac{5}{x+1}$

19. $\dfrac{-1}{x+1} = \dfrac{3}{x+5}$

20. $\dfrac{2}{x-1} = \dfrac{-3}{x+9}$

21. $\dfrac{3x}{x-3} + 2 = \dfrac{5x}{x-3}$

22. $\dfrac{2x}{x-2} + 18 = \dfrac{8x}{x-2}$

23. $\dfrac{5x}{x+1} - 6 = \dfrac{3x}{x+1}$

24. $\dfrac{5x}{x+1} - 2 = \dfrac{2x}{x+1}$

25. $\dfrac{x}{x^2-25} + \dfrac{5}{x-5} = \dfrac{1}{x+5}$

26. $\dfrac{x}{x^2-64} + \dfrac{8}{x-8} = \dfrac{1}{x+8}$

27. $\dfrac{x}{x^2-49} + \dfrac{7}{x-7} = \dfrac{1}{x+7}$

28. $\dfrac{x}{x^2-1} + \dfrac{1}{x-1} = \dfrac{1}{x+1}$

29. $\dfrac{x}{x+3} + \dfrac{3}{4} = \dfrac{-3}{x+3}$

30. $\dfrac{1}{5} + \dfrac{x}{x-2} = \dfrac{2}{x-2}$

1.
2.
3.
4.
5.
6.
7.
8.
9.
10.
11.
12.
13.
14.
15.
16.
17.
18.
19.
20.
21.
22.
23.
24.
25.
26.
27.
28.
29.
30.

31. $\dfrac{x}{x-4} - \dfrac{2}{7} = \dfrac{4}{x-4}$

32. $\dfrac{x}{x-8} - \dfrac{1}{5} = \dfrac{8}{x-8}$

33. $1 + \dfrac{2}{x-1} = \dfrac{4}{x^2-1}$

34. $1 + \dfrac{2}{x-3} = \dfrac{5}{x^2-9}$

35. $2 - \dfrac{6}{x^2-1} = \dfrac{-3}{x-1}$

36. $2 - \dfrac{4}{x^2-4} = \dfrac{-1}{x-2}$

37. $\dfrac{4}{x-3} - \dfrac{2}{x-1} = \dfrac{2}{x+2}$

38. $\dfrac{5}{x+1} - \dfrac{1}{x+2} = \dfrac{13}{x+5}$

39. $\dfrac{2x}{x^2-1} + \dfrac{4}{x-1} = \dfrac{1}{x-1}$

40. $\dfrac{3x-2}{x^2-4} + \dfrac{4}{x+2} = \dfrac{1}{x-2}$

31. _____
32. _____
33. _____
34. _____
35. _____
36. _____
37. _____
38. _____
39. _____
40. _____

✓ **SKILL CHECKER**

Solve.

41. $4(x+3) = 45$

42. $\dfrac{d}{3} + \dfrac{d}{4} = 1$

43. $\dfrac{h}{4} + \dfrac{h}{6} = 1$

44. $60(R-5) = 40(R+5)$

45. $30(R-5) = 10(R+15)$

41. _____
42. _____
43. _____
44. _____
45. _____

8.5 USING YOUR KNOWLEDGE

There are many instances in which a given formula must be changed to an equivalent form. For example, the formula

$$\frac{P}{R} = \frac{T}{V}$$

is frequently discussed in chemistry. Suppose you know P, R, and T. Can you find V? The idea is to solve for V. As before, we proceed by steps.

1. Since the LCD is RV, we multiply each term by **RV,** obtaining

$$RV \cdot \frac{P}{R} = \frac{T}{V} \cdot RV$$

2. Simplify. $VP = TR$

3. Divide by P. $V = \dfrac{TR}{P}$

Thus

$$V = \frac{TR}{P}$$

Use your knowledge of fractional equations to solve the given problem for the indicated variable:

1. The area A of a trapezoid is

$$A = \frac{h(b_1 + b_2)}{2}$$

Solve for h.

2. In an electric circuit, we have

$$\frac{1}{R} = \frac{1}{R_1} + \frac{1}{R_2}$$

Solve for R.

3. In refrigeration we find the formula

$$\frac{Q_1}{Q_2 - Q_1} = P$$

Solve for Q_1.

4. When studying the expansion of metals, we find the formula

$$\frac{L}{1 + at} = L_0$$

Solve for t.

5. In photography, we find the formula

$$\frac{1}{f} = \frac{1}{a} + \frac{1}{b}$$

Solve for f.

1. _____

2. _____

3. _____

4. _____

5. _____

8.6 RATIO, PROPORTIONS, AND WORD PROBLEMS

OBJECTIVES

REVIEW

Before starting this section you should know:

1. How to find the LCD of two or more fractions.
2. How to solve linear equations.

OBJECTIVES

You should be able to:

A. Solve proportions.
B. Solve word problems using ratios and proportions.

Courtesy of Bill Davis.

The jar of instant coffee claims that 4 oz of instant coffee are equivalent to 1 lb (16 oz) of regular coffee. In mathematics, we say that the ratio of instant coffee used to regular coffee used is **4 to 16**. The ratio **4 to 16** can be written as the fraction $\frac{4}{16}$. Clearly, a ratio is a quotient of two numbers. There are *three* ways in which the ratio of a number a to another number b can be written:

1. a to b
2. $a:b$
3. $\dfrac{a}{b}$

A. PROPORTIONS

Now, let us go back to the coffee jar in the picture. It claims that 4 oz will make 60 cups of coffee. Thus the ratio of ounces to cups, when written as a fraction, is

$$\frac{4}{60} = \frac{1}{15} \quad \begin{array}{l} \leftarrow \text{Ounces} \\ \leftarrow \text{Cups} \end{array}$$

The fraction $\frac{1}{15}$ is called the *reduced ratio* of ounces of coffee to cups of coffee. It tells us that 1 oz of coffee will make **15** cups of coffee. Now suppose you want to make 90 cups of coffee. How many ounces do you need? The ratio of ounces to cups is $\frac{1}{15}$ and we need to know how many ounces will make 90 cups. Let n be the number of ounces needed. Then

$$\frac{1}{15} = \frac{n}{90} \quad \begin{array}{l} \leftarrow \text{Ounces} \\ \leftarrow \text{Cups} \end{array}$$

Note that in both fractions the numerator tells us the number of ounces and the denominator indicates the number of cups. The equation $\frac{1}{15} = \frac{n}{90}$ is an equality between two ratios. In mathematics, an equality between ratios is called a **proportion.** Thus, $\frac{1}{15} = \frac{n}{90}$ is a proportion. To solve this proportion, which is simply a *fractional equation,* we proceed as before. First, since the LCM of 15 and 90 is **90:**

1. Multiply by the LCM. $90 \cdot \dfrac{1}{15} = 90 \cdot \dfrac{n}{90}$

2. Simplify. $6 = n$

Thus we need 6 oz of instant coffee to make 90 cups.
We use the same ideas in the next example.

EXAMPLE 1   A car travels 140 mi on 8 gal of gas.

a. What is the reduced ratio of miles to gallons (usually called miles per gallon)?

b. How many gallons will be needed to travel 210 mi?

Solution

a. The ratio of miles to gallons is

$$\frac{140}{8} = \frac{35}{2} \begin{matrix} \leftarrow \text{Miles} \\ \leftarrow \text{Gallons} \end{matrix}$$

b. Let g be the gallons needed. The ratio of miles to gallons is $\frac{35}{2}$; it is also $\frac{210}{g}$. Thus,

$$\frac{210}{g} = \frac{35}{2}$$

Multiplying by **2g**, the LCM, we have

$$2g \cdot \frac{210}{g} = \frac{35}{2} \cdot 2g$$

$$420 = 35g$$

$$g = \frac{420}{35} = 12$$

Hence 12 gal of gas are needed to travel 210 mi.   ▲

Problem 1   A car travels 150 mi on 9 gal of gas. How many gallons does it need to travel 300 mi?

Proportions are so common in algebra and other areas that there is a shortcut method to solve them. The method depends on the fact that if two fractions $\frac{a}{b}$ and $\frac{c}{d}$ are equivalent, their **cross products** are also equal. In symbols:

$$\text{If } \frac{a}{b} = \frac{c}{d} \quad \text{then} \quad ad = bc \quad \text{Note that if} \quad \frac{a}{b} = \frac{c}{d}$$

$$bd \cdot \frac{a}{b} = bd \cdot \frac{c}{d}$$

$$ad = bc$$

Thus, since $\frac{1}{2} = \frac{2}{4}$, $1 \cdot 4 = 2 \cdot 2$ and since $\frac{3}{9} = \frac{1}{3}$, $3 \cdot 3 = 9 \cdot 1$.

To solve the proportion

$$\frac{4}{3} = \frac{x}{15}$$

we use the equality of the cross products to write

$$4 \cdot 15 = 3x$$
$$60 = 3x$$
$$x = 20$$

Similarly, using cross products, the proportion

$$\frac{x - 3}{7} = \frac{5}{4}$$

can be written as

$$4(x - 3) = 7 \cdot 5$$

1. Simplify. $\quad 4x - 12 = 35$

2. Add 12. $\quad\quad 4x = 47$

3. Divide by 4. $\quad x = \dfrac{47}{4}$

EXAMPLE 2 Solve.

$$\frac{x + 3}{5} = \frac{9}{4}$$

Solution

$$\frac{x + 3}{5} = \frac{9}{4}$$

can be written as

$$4(x + 3) = 5 \cdot 9$$

1. Simplify. $\quad 4x + 12 = 45$

2. Subtract 12. $\quad\quad 4x = 33$

3. Divide by 4. $\quad x = \dfrac{33}{4}$

Problem 2 Solve.

$$\frac{x + 4}{3} = \frac{7}{2}$$

B. WORD PROBLEMS

Ratios and proportions are also used to solve word problems. For example, suppose a worker can finish a certain job in 3 days, whereas another worker can do it in 4 days. How long would it take both workers working together to complete the job? Before solving the problem, let us see how ratios play a part in the problem itself.

Since the first worker can do the job in **3** days, she does

$\dfrac{1}{3}$ of the job in 1 day

$\dfrac{2}{3}$ of the job in 2 days

and

$\dfrac{d}{3}$ of the job in d days

The second worker can finish in 4 days. He does

$\dfrac{1}{4}$ of the job in 1 day

$\dfrac{2}{4}$ of the job in 2 days

and

$\dfrac{d}{4}$ of the job in d days

If both workers together can do the job in d days, the first worker does $\frac{d}{3}$ of the job and the second one, $\frac{d}{4}$ of the job. Then

$$\dfrac{d}{3} + \dfrac{d}{4} = 1 \quad \leftarrow \begin{array}{l}\text{They must finish 1}\\ \text{whole job}\end{array}$$

Fraction done   Fraction done
by one   by the other

To solve the equation

$$\dfrac{d}{3} + \dfrac{d}{4} = 1$$

we first multiply by the LCM, **12.**

1. Multiply by the LCM. $12 \cdot \dfrac{d}{3} + 12 \cdot \dfrac{d}{4} = 12 \cdot 1$

2. Simplify. $4d + 3d = 12$
                              $7d = 12$

3. Divide by 7. $d = \dfrac{12}{7} = 1\dfrac{5}{7}$

Thus it takes $1\frac{5}{7}$ days for the workers working together to finish the job.

EXAMPLE 3 A computer can do a job in 4 hr. Computer sharing is arranged with another computer that can finish the job in 6 hr. How long would it take for both computers working together to finish the job?

Solution Let h be the number of hours it takes to complete the job when both computers are working together. Then:

1. The first computer does $\frac{h}{4}$ of the job.
2. The second computer does $\frac{h}{6}$ of the job.
3. Both working together do $\frac{h}{4} + \frac{h}{6}$ of the job.
4. Since we wish to finish one (1) complete job,

$$\dfrac{h}{4} + \dfrac{h}{6} = 1$$

Problem 3 A typist can finish a report in 5 hr. Another typist can do it in 8 hr. How many hours will it take to finish the report if they both work on it?

To solve this equation, we proceed as usual.

1. Multiply by the LCM, **12.** $12 \cdot \dfrac{h}{4} + 12 \cdot \dfrac{h}{6} = 12 \cdot 1$

2. Simplify. $3h + 2h = 12$

$5h = 12$

3. Divide by 5. $h = \dfrac{12}{5} = 2\dfrac{2}{5}$

$= 2.4 \text{ hr}$

Thus, both computers working together take 2.4 hr to do the job.   ▲

There is one more type of problem that may require ratios and proportions for its solution—the distance, rate, and time problem. We have already mentioned that the formula relating these three variables is

$$D = RT$$

Distance — — Rate, Time

Now suppose you are boating down a river on a sunny afternoon. Before you know it you have completed the 60-mi trip. Now, it is time to get back. Perhaps you can make it back in the same time! Wrong! This time you only cover 40 mi in the same time. Can you imagine, you traveled 60 mi downstream in the same time it took to travel 40 mi upstream? It must have been the current; it was flowing at 5 mph. What was the speed of the boat in still water? The problem is restated and solved in the next example.

EXAMPLE 4   A boat travels 60 mi downstream in the same time it takes to travel 40 mi upstream. If the current is flowing at 5 mi/h, what is the speed of the boat in still water?

Problem 4   A freight train travels 90 mi in the same time a passenger train covers 105 mi. If the passenger train is 5 mph faster, what is the speed of the freight train?

Solution   To solve this problem, we follow the five-step procedure given on page 247.

1. Read the problem. It asks for the speed R of the boat in still water.
2. Select R to be the speed of the boat in still water.

(Speed downstream) $R + 5 \leftarrow$ Current helps
(Speed upstream) $R - 5 \leftarrow$ Current hinders

3. Translate the problem. The time taken downstream and upstream is the same:

$T\,(\text{up}) = T\,(\text{down})$

Since $D = RT$,

$T = \dfrac{D}{R}$

$T\,(\text{up}) = T\,(\text{down})$

$\dfrac{60}{R + 5} = \dfrac{40}{R - 5}$

4. Using the cross-product rule, we have

$$60(R - 5) = 40(R + 5)$$

Simplify. $60R - 300 = 40R + 200$

Add 300. $60R = 40R + 500$

Subtract $40R$. $20R = 500$

Divide by 20. $R = 25$

Thus, the speed of the boat in still water is 25 mph.

5. Verify this answer.

A. Solve.

1. $\dfrac{x}{10} = \dfrac{5}{6}$

2. $\dfrac{x}{6} = \dfrac{-4}{3}$

3. $\dfrac{-7}{3} = \dfrac{x}{6}$

4. $\dfrac{-8}{9} = \dfrac{x}{5}$

5. $\dfrac{3}{x} = \dfrac{-7}{8}$

6. $\dfrac{4}{-x} = \dfrac{6}{11}$

7. $\dfrac{x-2}{3} = \dfrac{5}{7}$

8. $\dfrac{x-3}{4} = \dfrac{6}{5}$

9. $\dfrac{x+1}{2} = \dfrac{3}{5}$

10. $\dfrac{x+4}{5} = \dfrac{2}{7}$

11. $\dfrac{3}{x+2} = \dfrac{5}{4}$

12. $\dfrac{-2}{x+3} = \dfrac{7}{8}$

13. $\dfrac{-3}{7} = \dfrac{2}{x-1}$

14. $\dfrac{-5}{3} = \dfrac{4}{x-2}$

15. $\dfrac{4}{x+1} = \dfrac{3}{x}$

16. $\dfrac{2}{x+2} = \dfrac{5}{x}$

17. $\dfrac{3}{x-1} = \dfrac{7}{x}$

18. $\dfrac{3}{x+1} = \dfrac{4}{x+2}$

19. $\dfrac{8}{5} = \dfrac{x+2}{x+3}$

20. $\dfrac{7}{9} = \dfrac{x+1}{x-2}$

B. Solve the given word problem.

21. Do you think your gas station sells a lot of gas? The greatest number of gallons sold through a single pump is claimed by the Downtown Service Station, in New Zealand. They sold 7400 Imperial gallons in 24 hr. At that rate, how many Imperial gallons would they sell in 30 hr?

22. Do you like lemons? Bob Blackmore ate 3 whole lemons in 24 sec. At that rate, how many lemons could he eat in 1 min (60 sec)?

23. Michael Cisneros ate 25 tortillas in 15 min. How many could he eat in 1 hr at that rate?

24. Mr. Gerry Harley, of England, shaved 130 men in 60 min. If another barber takes 5 hr to shave the 130 men, how long would it take both of them working together to shave the 130 men?

ANSWERS

1. _____
2. _____
3. _____
4. _____
5. _____
6. _____
7. _____
8. _____
9. _____
10. _____
11. _____
12. _____
13. _____
14. _____
15. _____
16. _____
17. _____
18. _____
19. _____
20. _____

21. _____

22. _____

23. _____

24. _____

25. Mr. J. Moir riveted 11,209 rivets in 9 hr (a world record). If it takes another man 12 hr to do this job, how long would it take both of them to rivet the 11,209 rivets?

26. Leslie Carter ate 3 lb of potatoes in 105 sec. At that rate, how long would it take her to eat 5 lb of potatoes?

27. The fastest point-to-point train in the world is the New Tokaido, from Osaka to Okayama. This train covers 450 mi in the same time a regular train covers 180 mi. If the Tokaido is 60 mi/hr faster than the regular train, how fast is it?

28. The world's strongest current is the Saltstraumen in Norway, reaching 18 mi/hr. A motor boat can go 48 mi downstream in the same time it takes it to go 12 mi upstream. What is the speed of the boat in still water?

29. An 80-ft tree casts a 25-ft shadow. How long is the shadow of a 30-ft tree if it is measured at the same time of the day?

30. A 5-ft pole casts a 15-ft shadow. If a tree casts a 60-ft shadow at the same time of the day, how tall is the tree?

31. A map has a scale in which 1 in. represents 50 mi. One point in the map is $2\frac{1}{2}$ in. from another point in the same map. What is the actual distance between the two points?

32. A map has a scale in which 1 in. represents 10 mi. If one point on the map is $3\frac{1}{2}$ in. from another point on the map, what is the actual distance between the two points?

25. _____

26. _____

27. _____

28. _____

29. _____

30. _____

31. _____

32. _____

✓ **SKILL CHECKER**

Factor.

33. $x^2 - 16$

34. $x^2 - 36$

35. $16x^2 - 81$

36. $36x^2 - 25$

33. _____

34. _____

35. _____

36. _____

8.6   USING YOUR KNOWLEDGE

Proportions are used in many areas other than mathematics. The following problems come from the indicated areas.

1. *Physics*   A certain spring stretches 3 in. when a 7-lb force is applied to it. How many pounds of force would be required to make the spring stretch 8 in.?

2. *Photography*   Suppose you wish to enlarge a 2 × 3 in. picture so that the longer side is 10 in. How wide do you have to make it?

3. *Carpentry*   The pitch of a rafter is the ratio between the rise and the run of the rafter. What is the rise of a rafter having a pitch of $\frac{2}{5}$ if the run is 15 ft?

1. _____

2. _____

3. _____

4. *Auto Mechanics* The rear axle ratio is the ratio of the number of teeth in the ring gear to the number of teeth in the pinion gear. A car has a 3-to-1 rear axle ratio and the ring gear has 60 teeth. How many teeth does the pinion gear have?

4. _____

5. *Zoology* A zoologist took 250 fish from a lake, tagged them, and released them. A few days later 53 fish were taken from the lake and 5 of them were found to be tagged. Approximately how many fish were in the lake originally?

5. _____

SUMMARY

SECTION	ITEM	MEANING	EXAMPLE
8.1	Rational number	A number that can be written as $\frac{a}{b}$, a and b integers and b not 0	$\frac{3}{4}$, $-\frac{6}{5}$, 0, -8, and $1\frac{1}{3}$ are rational numbers.
8.1	Algebraic fraction	An expression of the form $\frac{A}{B}$, where A and B are polynomials	$\frac{x^2 + 3}{x - 1}$ and $\frac{y^2 + y - 1}{3y^3 + 7y + 8}$ are algebraic fractions.
8.1A	Fundamental rule of fractions	$\frac{A}{B} = \frac{A \cdot C}{B \cdot C}$	$\frac{3}{4} = \frac{3 \cdot 5}{4 \cdot 5}$ and $\frac{x}{x^2 + 2} = \frac{5 \cdot x}{5 \cdot (x^2 + 2)}$
8.1B	Reducing a fraction	The process of removing a common factor from the numerator and denominator of a fraction	$\frac{8}{6} = \frac{2 \cdot 4}{2 \cdot 3} = \frac{4}{3}$ is reduced.
8.1B	Standard form of a fraction	$\frac{a}{b}$ and $\frac{-a}{b}$ are the standard forms of a fraction.	$-\frac{-a}{-b}$ is written as $\frac{-a}{b}$ and $\frac{a}{-b}$ is written as $\frac{-a}{b}$.
8.2A	Multiplication of fractions	$\frac{A}{B} \cdot \frac{C}{D} = \frac{A \cdot C}{B \cdot D}$	$\frac{3}{4} \cdot \frac{7}{5} = \frac{3 \cdot 7}{4 \cdot 5} = \frac{21}{20}$
8.2C	Division of fractions	$\frac{A}{B} \div \frac{C}{D} = \frac{A \cdot D}{B \cdot C}$	$\frac{3}{7} \div \frac{5}{4} = \frac{3 \cdot 4}{7 \cdot 5} = \frac{12}{35}$
8.3A	Addition of fractions	$\frac{A}{B} + \frac{C}{D} = \frac{AD + BC}{BD}$	$\frac{1}{6} + \frac{1}{5} = \frac{5 + 6}{6 \cdot 5} = \frac{11}{30}$
8.3A	Subtraction of fractions	$\frac{A}{B} - \frac{C}{D} = \frac{AD - BC}{BD}$	$\frac{1}{5} - \frac{1}{6} = \frac{6 - 5}{5 \cdot 6} = \frac{1}{30}$
8.4	Complex fraction	A fraction that has other fractions in the numerator, denominator, or both	$\frac{\frac{x}{x + 2}}{x^2 + x + 1}$ is a complex fraction.
8.5	Fractional equation	An equation containing one or more fractional expressions	$\frac{x}{2} + \frac{x}{3} = 1$ is a fractional equation.
8.6	Ratio	A quotient of two numbers	3 to 4, 3:4, and $\frac{3}{4}$ are ratios.
8.6A	Proportion	An equality between ratios	3:4 as x:6, or $\frac{3}{4} = \frac{x}{6}$
8.6A	Cross products	If $\frac{a}{b} = \frac{c}{d}$, then $ad = bc$. ad and bc are the cross products	If $\frac{3}{4} = \frac{x}{6}$, then $3 \cdot 6 = 4 \cdot x$. $3 \cdot 6$ and $4 \cdot x$ are the cross products.

(If you need help with these exercises, look in the section indicated in brackets.)

ANSWERS

1. [8.1A] Write the given fraction with the indicated denominator.

   a. $\dfrac{5x}{8y}$ with a denominator of $16y^2$

   b. $\dfrac{3x}{4y^2}$ with a denominator of $16y^3$

   c. $\dfrac{2y}{3x^3}$ with a denominator of $15x^5$

1. a. _____
   b. _____
   c. _____

2. [8.1B] Reduce the given fraction to lowest terms.

   a. $\dfrac{-9(x^2 - y^2)}{3(x + y)}$

   b. $\dfrac{-10(x^2 - y^2)}{5(x + y)}$

   c. $\dfrac{-16(x^2 - y^2)}{-4(x + y)}$

2. a. _____
   b. _____
   c. _____

3. [8.1B] Simplify the given fraction.

   a. $\dfrac{-x}{x^2 + x}$

   b. $\dfrac{-x}{x^2 - x}$

   c. $\dfrac{-x}{x - x^2}$

3. a. _____
   b. _____
   c. _____

4. [8.1B] Reduce to lowest terms.

   a. $\dfrac{x^2 - 3x - 18}{6 - x}$

   b. $\dfrac{x^2 - 2x - 8}{4 - x}$

   c. $\dfrac{x^2 - 2x - 15}{5 - x}$

4. a. _____
   b. _____
   c. _____

5. [8.2A] Multiply.

   a. $\dfrac{3y^2}{7} \cdot \dfrac{14x}{9y}$

   b. $\dfrac{7y^2}{5} \cdot \dfrac{15x}{14y}$

   c. $\dfrac{6y^3}{7} \cdot \dfrac{28x}{3y}$

5. a. _____
  b. _____
  c. _____

6. [8.2A] Multiply.

   a. $(x - 3) \cdot \dfrac{x + 2}{x^2 - 9}$

   b. $(x - 5) \cdot \dfrac{x + 1}{x^2 - 25}$

   c. $(x - 4) \cdot \dfrac{x + 5}{x^2 - 16}$

6. a. _____
  b. _____
  c. _____

7. [8.2B] Divide.

   a. $\dfrac{x^2 - 9}{x + 2} \div (x + 3)$

   b. $\dfrac{x^2 - 16}{x + 1} \div (x + 4)$

   c. $\dfrac{x^2 - 25}{x + 4} \div (x + 5)$

7. a. _____
  b. _____
  c. _____

8. [8.2B] Divide.

   a. $\dfrac{x + 5}{x - 5} \div \dfrac{x^2 - 25}{5 - x}$

   b. $\dfrac{x + 1}{x - 1} \div \dfrac{x^2 - 1}{1 - x}$

   c. $\dfrac{x + 2}{x - 2} \div \dfrac{x^2 - 4}{2 - x}$

8. a. _____
  b. _____
  c. _____

9. [8.3A] Add.

   a. $\dfrac{3}{2(x - 1)} + \dfrac{1}{2(x - 1)}$

   b. $\dfrac{5}{6(x - 2)} + \dfrac{7}{6(x - 2)}$

   c. $\dfrac{3}{4(x + 1)} + \dfrac{1}{4(x + 1)}$

9. a. _____
  b. _____
  c. _____

10. [8.3A] Subtract.

a. $\dfrac{7}{2(x+1)} - \dfrac{3}{2(x+1)}$

b. $\dfrac{11}{5(x+2)} - \dfrac{1}{5(x+2)}$

c. $\dfrac{17}{7(x+3)} - \dfrac{3}{7(x+3)}$

10. a. _____
  b. _____
  c. _____

11. [8.3B] Add.

a. $\dfrac{2}{x+2} + \dfrac{1}{x-2}$

b. $\dfrac{3}{x+1} + \dfrac{1}{x-1}$

c. $\dfrac{4}{x+3} + \dfrac{1}{x-3}$

11. a. _____
  b. _____
  c. _____

12. [8.3B] Subtract.

a. $\dfrac{x-1}{x^2+3x+2} - \dfrac{x+7}{x^2+5x+6}$

b. $\dfrac{x+3}{x^2-x-2} - \dfrac{x-1}{x^2+2x+1}$

c. $\dfrac{x-1}{x^2+3x+2} - \dfrac{x+1}{x^2+x-2}$

12. a. _____
  b. _____
  c. _____

13. [8.4] Simplify.

a. $\dfrac{\dfrac{3}{2x} - \dfrac{1}{x}}{\dfrac{2}{3x} + \dfrac{3}{4x}}$

b. $\dfrac{\dfrac{3}{2x} - \dfrac{1}{3x}}{\dfrac{2}{x} + \dfrac{1}{4x}}$

c. $\dfrac{\dfrac{3}{2x} - \dfrac{1}{x}}{\dfrac{3}{4x} + \dfrac{4}{3x}}$

13. a. _____
  b. _____
  c. _____

14. [8.5] Solve.

a. $\dfrac{2x}{x-1} + 3 = \dfrac{4x}{x-1}$

b. $\dfrac{6x}{x-5} + 7 = \dfrac{8x}{x-5}$

c. $\dfrac{5x}{x-4} + 6 = \dfrac{7x}{x-4}$

15. [8.5] Solve.

a. $\dfrac{x}{x^2-4} + \dfrac{2}{x-2} = \dfrac{1}{x+2}$

b. $\dfrac{x}{x^2-16} + \dfrac{4}{x-4} = \dfrac{1}{x+4}$

c. $\dfrac{x}{x^2-25} + \dfrac{5}{x-5} = \dfrac{1}{x+5}$

16. [8.5] Solve.

a. $\dfrac{x}{x+6} - \dfrac{1}{7} = \dfrac{-6}{x+6}$

b. $\dfrac{x}{x+7} - \dfrac{1}{8} = \dfrac{-7}{x+7}$

c. $\dfrac{x}{x+8} - \dfrac{1}{9} = \dfrac{-8}{x+8}$

17. [8.5] Solve.

a. $3 + \dfrac{5}{x-4} = \dfrac{50}{x^2-16}$

b. $4 + \dfrac{6}{x-5} = \dfrac{84}{x^2-25}$

c. $5 + \dfrac{7}{x-6} = \dfrac{126}{x^2-36}$

18. [8.6A] A car travels 160 mi on 7 gal of gas.
  a. How many gallons will it need to travel 240 mi?
  b. Repeat the problem if the car travels 180 mi on 9 gal of gas and we wish to go 270 mi.
  c. Repeat the problem if the car travels 200 mi on 12 gal and we wish to go 300 mi.

19. [8.6A] Solve.

a. $\dfrac{x+3}{6} = \dfrac{7}{2}$

b. $\dfrac{x+4}{8} = \dfrac{9}{5}$

c. $\dfrac{x+5}{2} = \dfrac{6}{5}$

14. a. _____
    b. _____
    c. _____

15. a. _____
    b. _____
    c. _____

16. a. _____
    b. _____
    c. _____

17. a. _____
    b. _____
    c. _____

18. a. _____
    b. _____
    c. _____

19. a. _____
    b. _____
    c. _____

20. [8.6B] A person can do a job in 6 hr. Another person can do it in 8 hr.

    a. How long would it take to do the job if both of them work together?

    b. Repeat the problem if the first person takes 10 hr and the second one takes 8 hr.

    c. Repeat the problem if the first person takes 9 hr and the second one takes 6 hr.

20. a. _____

    b. _____

    c. _____

NAME

CLASS

SECTION

ANSWERS

(Answers on page 582)

1. Write $\dfrac{3x}{7y}$ with a denominator of $21y^3$.

2. Reduce $\dfrac{-6(x^2 - y^2)}{3(x - y)}$ to lowest terms.

3. Simplify $\dfrac{-x}{x + x^2}$.

4. Reduce to lowest terms $\dfrac{x^2 + 2x - 8}{2 - x}$.

In Problems 5 through 12 perform the indicated operations and simplify.

5. Multiply $\dfrac{2y^2}{7} \cdot \dfrac{21x}{4y}$.

6. Multiply $(x - 2) \cdot \dfrac{x + 3}{x^2 - 4}$.

7. Divide $\dfrac{x^2 - 4}{x + 5} \div (x - 2)$.

8. Divide $\dfrac{x + 3}{x - 3} \div \dfrac{x^2 - 9}{3 - x}$.

9. Add $\dfrac{5}{2(x - 2)} + \dfrac{1}{2(x - 2)}$.

10. Subtract $\dfrac{7}{3(x + 1)} - \dfrac{1}{3(x + 1)}$.

11. Add $\dfrac{2}{x + 1} + \dfrac{1}{x - 1}$.

12. Subtract $\dfrac{x + 1}{x^2 + x - 2} - \dfrac{x + 2}{x^2 - 1}$.

13. Simplify $\dfrac{\dfrac{1}{x} - \dfrac{2}{3x}}{\dfrac{3}{4x} + \dfrac{1}{2x}}$.

14. Solve $\dfrac{3x}{x - 2} + 4 = \dfrac{5x}{x - 2}$.

15. Solve $\dfrac{x}{x^2 - 9} + \dfrac{3}{x - 3} = \dfrac{1}{x + 3}$.

16. Solve $\dfrac{x}{x + 5} - \dfrac{1}{6} = \dfrac{-5}{x + 5}$.

17. Solve $2 + \dfrac{4}{x - 3} = \dfrac{24}{x^2 - 9}$.

18. A car travels 150 mi on 9 gal of gas. How many gallons will it need to travel 400 mi?

19. Solve $\dfrac{x + 5}{7} = \dfrac{11}{6}$.

20. A woman can paint a house in 5 hr. Another one can do it in 8 hr. How long would it take for both women working together to paint the house?

ANSWERS

1. _____

2. _____

3. _____

4. _____

5. _____

6. _____

7. _____

8. _____

9. _____

10. _____

11. _____

12. _____

13. _____

14. _____

15. _____

16. _____

17. _____

18. _____

19. _____

20. _____

IF YOU MISSED QUESTION	SECTION	EXAMPLES	PAGE	ANSWERS
1	8.1	1	516	1. $\dfrac{9xy^2}{21y^3}$
2	8.1	2	518	2. $-2(x + y)$
3	8.1	3	519	3. $\dfrac{-1}{1 + x}$
4	8.1	4	521	4. $-(x + 4)$
5	8.2	1, 2	528, 529	5. $\dfrac{3xy}{2}$
6	8.2	3	530	6. $\dfrac{x + 3}{x + 2}$
7	8.2	4	531	7. $\dfrac{x + 2}{x + 5}$
8	8.2	4, 5	531	8. $\dfrac{-1}{x - 3} = \dfrac{1}{3 - x}$
9	8.3	1a	538	9. $\dfrac{3}{x - 2}$
10	8.3	1b	538	10. $\dfrac{2}{x + 1}$
11	8.3	2	540	11. $\dfrac{3x - 1}{x^2 - 1}$
12	8.3	3	542	12. $\dfrac{-2x - 3}{(x + 2)(x + 1)(x - 1)}$
13	8.4	1, 2	550	13. $\dfrac{4}{15}$
14	8.5	1, 2	556	14. $x = 4$
15	8.5	3	557	15. $x = -4$
16	8.5	4	558	16. No solution
17	8.5	5	559	17. $x = -5$
18	8.6	1	566	18. 24
19	8.6	2	567	19. $x = \dfrac{47}{6}$
20	8.6	3	568	20. $3\dfrac{1}{13}$ hr

CHAPTER 9

RADICALS AND QUADRATIC EQUATIONS

In this chapter we learn three different methods of solving quadratic equations: taking square roots, completing the square, and using the quadratic formula. (Solution by factoring was studied in Chapter 5.) We introduce radicals (such as $\sqrt{2}$) and perform addition, subtraction, multiplication, and division (including rationalizing the denominator) in problems involving these radicals. We end the chapter with sections on the graphing of quadratic polynomials and applications.

NAME

CLASS

SECTION

(*Answers on page* 586)

ANSWERS

1. Solve $x^2 = 49$.

   1. _____

2. Solve $49x^2 - 81 = 0$.

   2. _____

3. Solve $8x^2 + 49 = 0$.

   3. _____

4. Solve $36(x + 2)^2 - 5 = 0$.

   4. _____

5. Solve $x^2 - 48 = 0$.

   5. _____

6. Solve $5x^2 = 49$.

   6. _____

7. Solve $7x^2 - 50 = 0$.

   7. _____

8. **a.** Simplify $8\sqrt{3} + 2\sqrt{3} - 4\sqrt{3}$.
   b. Simplify $\sqrt{75} + \sqrt{48}$.

   8. a. _____
       b. _____

9. Multiply $\sqrt{6}(\sqrt{50} - \sqrt{6})$.

   9. _____

10. Find the missing term in the expression $(x + 2)^2 = x^2 + 4x + \boxed{}$.

  10. _____

11. The missing terms in the expression $x^2 - 6x + \boxed{} = ()^2$ are _____ and _____, respectively.

  11. _____

12. To solve the equation $6x^2 - 32x = -5$ by completing the square, the first step will be to divide each term by _____.

  12. _____

13. To solve the equation $x^2 - 6x = -9$ by completing the square, add _____ to both sides of the equation.

  13. _____

14. The solutions of $ax^2 + bx + c = 0$ are _____.

  14. _____

15. Solve $2x^2 - 5x - 3 = 0$.

  15. _____

16. Solve $x^2 = 4x - 3$.

  16. _____

17. Solve $6x = x^2$.

  17. _____

18. Solve $\dfrac{x^2}{2} + \dfrac{5}{4}x = -\dfrac{1}{2}$

  18. _____

19. Graph the equation $y = -(x - 2)^2 + 4$.

  19. _____

20. The formula $d = 5t^2 + v_0 t$ gives the distance d (in meters, m) an object thrown downward with an initial velocity v_0 will have gone after t sec. How long would it take an object dropped from a distance of 180 m to hit the ground?

  20. _____

IF YOU MISSED QUESTION	SECTION	EXAMPLES	PAGE	ANSWERS
1	9.1	1	589	1. $x = \pm 7$
2	9.1	1, 2	589, 591	2. $x = \pm \dfrac{9}{7}$
3	9.1	3	592	3. No solution
4	9.1	4	593	4. $x = -2 \pm \dfrac{\sqrt{5}}{6}$
5	9.2	1	600	5. $x = \pm 4\sqrt{3}$
6	9.2	2	602	6. $x = \pm \dfrac{7\sqrt{5}}{5}$
7	9.2	3	603	7. $x = \pm \dfrac{5\sqrt{14}}{7}$
8	9.2	4, 5	603, 604	8. a. $6\sqrt{3}$ b. $9\sqrt{3}$
9	9.2	6	604	9. $10\sqrt{3} - 6$
10	9.3	1	611	10. 4
11	9.3	2	612	11. $9, (x-3)^2$
12	9.3	3, 4	613, 614	12. 6
13	9.3	3, 4	613, 614	13. 9
14	9.4	1	621	14. $x = \dfrac{-b \pm \sqrt{b^2 - 4ac}}{2a}$
15	9.4	1	621	15. $x = 3, x = -\dfrac{1}{2}$
16	9.4	2	621	16. $x = 1, x = 3$
17	9.4	3	622	17. $x = 0, x = 6$
18	9.4	4	623	18. $x = -2, x = -\dfrac{1}{2}$
19	9.5	1, 2, 3	630, 631, 632	19.
20	9.6	1, 2, 3	642, 643	20. 6 sec

9.1 THE QUADRATIC $X^2 = A$ AND DIVISION OF RADICALS

Courtesy of IBM Corporation.

OBJECTIVES

REVIEW

Before starting this section you should know:

1. The definition of a rational and an irrational number (Section 1.3).
2. How to solve linear equations.
3. How to square a number.

OBJECTIVES

You should be able to:

A. Solve a quadratic equation of the form

$$X^2 = A, \quad A \geq 0$$

B. Solve a quadratic equation of the form

$$(X \pm A)^2 = B$$

The tiny electronic chip in the photo is used in IBM computers. Because of size limitations, the chip is very small. As a matter of fact, it covers an area of 16 square millimeters (mm²). (It also holds 64,000 pieces of information.) If the chip is square, can you find out how long its side is? The area of a square is obtained by multiplying the length X of its side by itself. Thus, if we assume that the chip is X millimeters on its side, its area is X^2. The area is also 16; thus, we have the equation

$$X^2 = 16$$

The area of this square is $X \cdot X = X^2$.

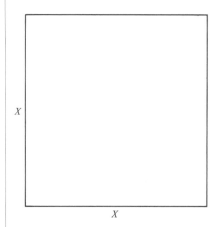

A. QUADRATICS OF THE FORM $X^2 = A$

The equation $X^2 = 16$ is a *quadratic equation*. In general, a quadratic equation is an equation that can be written in the form $ax^2 + bx + c = 0$. How can we solve $X^2 = 16$? First, the equation tells us that a certain number X multiplied by itself gives 16 as a result. Obviously, one possible answer is $X = 4$ because $4^2 = 16$. From this we conclude that the side of the chip is 4 mm long. But wait, what about $X = -4$? It is also true that $(-4)^2 = (-4)(-4) = 16$. Thus the solutions of the equation $X^2 = 16$ are $X = 4$ and $X = -4$. However, since X represents the length of the side of the chip, we must use $X = 4$ as our answer.

In mathematics, the number **4** is called the **positive square root** of 16 and -4 is called the **negative square root** of 16. These roots are usually denoted by

$$\sqrt{16} = 4 \quad \text{Read "the positive square root of 16 is 4"}$$

and

$$-\sqrt{16} = -4 \quad \text{Read "the negative square root of 16 is } -4\text{"}$$

The symbol $\sqrt{}$ is called a **radical** sign. In general, we have the following definition for the **principal** square root of a ($\sqrt{a}$):

$\sqrt{a}$ is a nonnegative number b such that $b^2 = a$ $(a \geq 0)$.

Note that $b^2 = a$ means that $\sqrt{a} \cdot \sqrt{a} = a$.

Clearly, before trying to solve quadratic equations of the form $X^2 = A$, we must know something about taking the square root of a number. Here are the first few whole numbers and their square roots:

NUMBER	0	1	2	3	4	5	6
SQUARE ROOT	$\sqrt{0} = 0$	$\sqrt{1} = 1$	$\sqrt{2}$	$\sqrt{3}$	$\sqrt{4} = 2$	$\sqrt{5}$	$\sqrt{6}$

NUMBER	7	8	9
SQUARE ROOT	$\sqrt{7}$	$\sqrt{8}$	$\sqrt{9} = 3$

The numbers with *integer* square roots are in boldface.

As you can see, not all whole numbers have integer square roots. In the table, only 0, 1, 4, and 9 do. Because of this, 0, 1, 4 and 9 are called *perfect squares*. As a matter of fact,

$\sqrt{0} = 0$ Because $0^2 = 0$
$\sqrt{1} = 1$ Because $1^2 = 1$
$\sqrt{4} = 2$ Because $2^2 = 4$
$\sqrt{9} = 3$ Because $3^2 = 9$

Can you find the next five numbers that have integer square roots? We have already given you 0, 1, 4, and 9 with roots 0, 1, 2, and 3, respectively. On the other hand, 2, 3, and 5 do not have integer roots because there are no whole numbers that give 2, 3, or 5 when squared. The numbers $\sqrt{2}$, $\sqrt{3}$, and $\sqrt{5}$ are *irrational* numbers.

An *irrational* number is a number that is *not* rational (it cannot be written in the form $\frac{a}{b}$, where a and b are integers and $b \neq 0$).

As you recall from Section 1.3, the *irrational* numbers are numbers that can be represented by decimals that *do not terminate* and *do not repeat*. The irrational numbers are part of a larger set of numbers called the *real numbers*. As a matter of fact, all the rational and the irrational numbers put together form the set of real numbers, that is,

---

The **real numbers** consist of the rational and the irrational numbers.

Many of the equations we are about to study have irrational numbers for their solution. For example, the equation $x^2 = 3$ has two *irrational* solutions. How do we obtain these solutions? By taking the square root of both sides of the equation, as shown:

Given

$$x^2 = 3$$

Then

$$x = \pm\sqrt{3} \quad \text{Note that } (\sqrt{3})^2 = 3 \text{ and } (-\sqrt{3})^2 = 3.$$

The notation $\pm\sqrt{3}$ is a shortcut to indicate that x can be $\sqrt{3}$ or $-\sqrt{3}$. On the other hand, the equation $x^2 = 16$ has *rational* roots. To solve this equation we proceed as follows:

Given

$$x^2 = 16$$

Then

$$x = \pm\sqrt{16}$$
$$x = \pm 4$$

Thus, the solutions are 4 and -4.

EXAMPLE 1 Solve.

a. $x^2 = 36$ **b.** $x^2 - 49 = 0$ **c.** $x^2 = 10$

Solution

a. Given:

$$x^2 = 36$$

Then

$$x = \pm\sqrt{36}$$
$$x = \pm 6$$

Thus, the solutions of the equation $x^2 = 36$ are 6 and -6, since $6^2 = 36$ and $(-6)^2 = 36$. (Both solutions are rational numbers.)

b. Given:

$$x^2 - 49 = 0.$$

Unfortunately, this equation is not of the form $X^2 = A$. But, by adding 49 to both sides of the equation we can remedy this situation.

Given.	$x^2 - 49 = 0$
Add 49.	$x^2 = 49$
Take the square roots.	$x = \pm\sqrt{49}$
	$x = \pm 7$

Thus, the solutions of $x^2 - 49 = 0$ are 7 and -7. (Both solutions are rational numbers.)

Problem 1 Solve.
a. $x^2 = 81$ **b.** $x^2 - 1 = 0$
c. $x^2 = 13$

Note that the equation $x^2 - 49 = 0$ can also be solved by factoring (Section 5.5) by writing $x^2 - 49 = 0$ as $(x + 7)(x - 7) = 0$. Thus, $x + 7 = 0$ or $x - 7 = 0$; that is, $x = -7$ or $x = 7$, as before.

c. Given:

$$x^2 = 10$$

Then

$$x = \pm\sqrt{10}$$

Since 10 does not have an integer root, the solutions of the equation $x^2 = 10$ are $\sqrt{10}$ and $-\sqrt{10}$, since $(\sqrt{10})^2 = \sqrt{10} \cdot \sqrt{10} = 10$ and $(-\sqrt{10})^2 = (-\sqrt{10}) \cdot (-\sqrt{10}) = 10$. (Here, both solutions are irrational.) ▲

In solving Example 1b we had to add 49 to both sides of the equation to obtain an equivalent equation of the form $X^2 = A$. Does this method work for more complicated examples? The answer is yes. In fact, when trying to solve any quadratic equation in which the *only* exponent of the variable is 2, always transform the equation into an equivalent one of the form $X^2 = A$. Here is the idea.

To solve any equation of the form

$$AX^2 - B = 0$$

write

$$AX^2 = B$$

and

$$X^2 = \frac{B}{A}$$

so that

$$X = \pm\sqrt{\frac{B}{A}}, \qquad \frac{B}{A} \geq 0$$

Thus, to solve the equation

$$16x^2 - 81 = 0$$

we proceed as follows:

Given.	$16x^2 - 81 = 0$
Add 81.	$16x^2 = 81$
Divide by 16.	$x^2 = \dfrac{81}{16}$
Then take the square roots.	$x = \pm\sqrt{\dfrac{81}{16}}$
	$x = \pm\dfrac{9}{4}$

The solutions are $\frac{9}{4}$ and $-\frac{9}{4}$. Since $(\frac{9}{4})^2 = (\frac{81}{16})$ and $(-\frac{9}{4})^2 = \frac{81}{16}$, our result is correct. Just remember to rewrite the equation in the form $X^2 = A$.

EXAMPLE 2   Solve $36x^2 - 25 = 0$.

Problem 2   Solve $9x^2 - 16 = 0$.

Solution

Given.	$36x^2 - 25 = 0$
Add 25.	$36x^2 = 25$
Divide by 36.	$x^2 = \dfrac{25}{36}$
Then take square roots.	$x = \pm\sqrt{\dfrac{25}{36}}$
	$x = \pm\dfrac{5}{6}$

The solutions are $\frac{5}{6}$ and $-\frac{5}{6}$. (Here, both solutions are rational numbers.) ▲

Of course, not all equations of the form $X^2 = A$ have solutions that are real numbers. For example, to solve the equation $x^2 + 64 = 0$, we write:

Given.	$x^2 + 64 = 0$
Subtract 64.	$x^2 = -64$

But there is no *real* number whose square is -64. If you square a non-zero real number, the answer is always positive. Thus, x^2 is positive and can never equal -64. The equation $x^2 + 64 = 0$ has *no real-number* solution.

As we have seen, not all equations have *rational-number* solutions, that is, solutions of the form $\frac{a}{b}$, where a and b are integers, $b \neq 0$. For example, the equation

$$16x^2 - 5 = 0$$

is solved as follows:

Given.	$16x^2 - 5 = 0$
Add 5.	$16x^2 = 5$
Divide by 16.	$x^2 = \dfrac{5}{16}$
Then take square roots.	$x = \pm\sqrt{\dfrac{5}{16}}$

However, $\sqrt{\frac{5}{16}}$ is *not* a rational number even though the denominator, 16, is the square of 4. As it turns out,

$$\pm\sqrt{\frac{5}{16}} = \pm\frac{\sqrt{5}}{\sqrt{16}} = \pm\frac{\sqrt{5}}{4}$$

Here is the general rule that tells us how to simplify $\pm\sqrt{\frac{5}{16}}$.

$$\sqrt{\frac{a}{b}} = \frac{\sqrt{a}}{\sqrt{b}} \qquad a \text{ and } b \text{ positive numbers}$$

This rule is discussed in 9.1 *Using Your Knowledge*.

With this in mind, the solutions of the equation $16x^2 - 5 = 0$ are

$$\frac{\sqrt{5}}{4} \quad \text{and} \quad -\frac{\sqrt{5}}{4}$$

Both solutions are irrational numbers.

EXAMPLE 3 Solve.

a. $4x^2 - 7 = 0$ **b.** $8x^2 + 49 = 0$

Solution

a. Given.

$$4x^2 - 7 = 0$$

Add 7.

$$4x^2 = 7$$

Divide by 4.

$$x^2 = \frac{7}{4}$$

Then take square roots.

$$x = \pm\sqrt{\frac{7}{4}} = \pm\frac{\sqrt{7}}{\sqrt{4}} = \pm\frac{\sqrt{7}}{2}$$

Thus, the solutions of $4x^2 - 7 = 0$ are $\dfrac{\sqrt{7}}{2}$ and $-\dfrac{\sqrt{7}}{2}$. They are both irrational numbers.

b. Given.

$$8x^2 + 49 = 0$$

Subtract 49.

$$8x^2 = -49$$

Divide by 8.

$$x^2 = -\frac{49}{8}$$

But since the square of x cannot be negative and $-\frac{49}{8}$ is, this equation has *no* real-number solution.

B. EQUATIONS OF THE FORM $(x \pm a)^2 = b$

We already mentioned that to solve an equation in which the only exponent of the variable is 2, we must transform the equation into an equivalent one of the form $X^2 = A$. Now, consider the equation

$$(x - 2)^2 = 9$$

If we think of $(x - 2)$ as X, we have the following.

Given.

$$(x - 2)^2 = 9$$

Write $x - 2$ as X.

$$X^2 = 9 \quad \text{Here we are substituting } X \text{ for } x - 2.$$

Then take square roots.

$$X = \pm\sqrt{9}$$
$$X = \pm 3$$

Write X as $x - 2$.

$$x - 2 = \pm 3$$

Add 2.

$$x = 2 \pm 3$$

Hence

$$x = 2 + 3 \quad \text{or} \quad x = 2 - 3$$
$$x = 5 \quad \text{or} \quad x = -1$$

Clearly, by thinking of $(x - 2)$ as X, we have been able to solve a more complicated equation. In the same manner, we can solve

$$9(x - 2)^2 - 5 = 0$$

Given. $\qquad\qquad 9(x-2)^2 - 5 = 0$

Add 5. $\qquad\qquad\qquad 9(x-2)^2 = 5$

Divide by 9. $\qquad\qquad (x-2)^2 = \dfrac{5}{9}$

Think of $x - 2$ as X. $\qquad x - 2 = \pm\sqrt{\dfrac{5}{9}}$

$$x - 2 = \pm\dfrac{\sqrt{5}}{\sqrt{9}} = \pm\dfrac{\sqrt{5}}{3}$$

Add 2. $\qquad\qquad\qquad x = 2 \pm \dfrac{\sqrt{5}}{3}$

$$x = 2 + \dfrac{\sqrt{5}}{3} \quad \text{or} \quad x = 2 - \dfrac{\sqrt{5}}{3}$$

EXAMPLE 4 Solve.

a. $(x+3)^2 = 9$ **b.** $(x+1)^2 - 4 = 0$ **c.** $25(x+2)^2 - 3 = 0$

Solution

a. Given. $\qquad\qquad\qquad (x+3)^2 = 9$

    Think of $(x+3)$ as X. $\qquad x + 3 = \pm\sqrt{9}$

$$x + 3 = \pm 3$$

    Subtract 3. $\qquad\qquad\qquad x = -3 \pm 3$

    Hence $\qquad\qquad x = -3 + 3 \quad \text{or} \quad x = -3 - 3$

$$x = 0 \qquad\quad \text{or} \quad x = -6$$

b. Given. $\qquad\qquad\qquad (x+1)^2 - 4 = 0$

    Add 4 (to have an equation $\qquad (x+1)^2 = 4$
of the form $X^2 = A$).

    Think of $(x+1)$ as X. $\qquad\qquad x + 1 = \pm\sqrt{4}$

$$x + 1 = \pm 2$$

    Subtract 1. $\qquad\qquad\qquad\qquad x = -1 \pm 2$

$$x = -1 + 2 \quad \text{or} \quad x = -1 - 2$$

$$x = 1 \qquad\quad \text{or} \quad x = -3$$

c. Given. $\qquad\qquad 25(x+2)^2 - 3 = 0$

    Add 3. $\qquad\qquad\quad 25(x+2)^2 = 3$

    Divide by 25. $\qquad\qquad (x+2)^2 = \dfrac{3}{25}$

    Think of $(x+2)$ as X. $\qquad x + 2 = \pm\sqrt{\dfrac{3}{25}}$

    Since $\sqrt{\dfrac{3}{25}} = \dfrac{\sqrt{3}}{\sqrt{25}}$. $\qquad x + 2 = \pm\dfrac{\sqrt{3}}{\sqrt{25}} = \pm\dfrac{\sqrt{3}}{5}$

    Subtract 2. $\qquad\qquad\qquad x = -2 \pm \dfrac{\sqrt{3}}{5}$

$$x = -2 + \dfrac{\sqrt{3}}{5} \quad \text{or} \quad x = -2 - \dfrac{\sqrt{3}}{5} \qquad \blacktriangle$$

Problem 4 Solve.
a. $(x+5)^2 = 25$
b. $(x+2)^2 - 9 = 0$
c. $16(x+1)^2 - 5 = 0$

As we mentioned before, if we square any real number, the result is not negative. For this reason, an equation such as

$(x - 4)^2 = -5$ Since $(x - 4)$ represents a real number, $(x - 4)^2$ *cannot* be negative. But -5 is, so $(x - 4)^2$ and -5 can *never* be equal.

has *no* real-number solution. Similarly,

$(x - 3)^2 + 8 = 0$

has no solution, since

$(x - 3)^2 + 8 = 0$

is equivalent to

$(x - 3)^2 = -8$

by subtracting 8.

Use this idea in the next example.

EXAMPLE 5   Solve $9(x - 5)^2 + 1 = 0$.

Solution

Given. $\qquad 9(x - 5)^2 + 1 = 0$

Subtract 1. $\qquad 9(x - 5)^2 = -1$

Divide by 9. $\qquad (x - 5)^2 = -\dfrac{1}{9}$

Since $(x - 5)$ is to be a real number, $(x - 5)^2$ *can never* be negative. But $-\frac{1}{9}$ is negative. Thus, the equation $(x - 5)^2 = -\frac{1}{9}$ (which is equivalent to $9(x - 5)^2 + 1 = 0$) has *no* real-number solution.

Problem 5   Solve $16(x - 3)^2 + 7 = 0$.

NAME

CLASS

SECTION

A. Solve.

1. $x^2 = 100$	**2.** $x^2 = 1$
3. $x^2 = 0$	**4.** $x^2 = 121$
5. $x^2 = -4$	**6.** $x^2 = -16$
7. $x^2 = 7$	**8.** $x^2 = 3$
9. $x^2 - 9 = 0$	**10.** $x^2 - 64 = 0$
11. $x^2 - 3 = 0$	**12.** $x^2 - 5 = 0$
13. $25x^2 - 1 = 0$	**14.** $36x^2 - 49 = 0$
15. $100x^2 - 49 = 0$	**16.** $81x^2 - 36 = 0$
17. $25x^2 - 17 = 0$	**18.** $9x^2 - 11 = 0$
19. $25x^2 + 3 = 0$	**20.** $49x^2 + 1 = 0$
21. $(x + 1)^2 = 81$	**22.** $(x + 3)^2 = 25$
23. $(x - 2)^2 = 36$	**24.** $(x - 3)^2 = 16$
25. $(x - 4)^2 = -25$	**26.** $(x + 2)^2 = -16$
27. $(x - 9)^2 = 81$	**28.** $(x - 6)^2 = 36$
29. $(x + 4)^2 = 16$	**30.** $(x + 7)^2 = 49$

ANSWERS

1. _____
2. _____
3. _____
4. _____
5. _____
6. _____
7. _____
8. _____
9. _____
10. _____
11. _____
12. _____
13. _____
14. _____
15. _____
16. _____
17. _____
18. _____
19. _____
20. _____
21. _____
22. _____
23. _____
24. _____
25. _____
26. _____
27. _____
28. _____
29. _____
30. _____

31. $25(x + 1)^2 - 1 = 0$ **32.** $16(x + 2)^2 - 1 = 0$

33. $36(x - 3)^2 - 49 = 0$ **34.** $9(x - 1)^2 - 25 = 0$

35. $4(x + 1)^2 - 25 = 0$ **36.** $49(x + 2)^2 - 16 = 0$

37. $9(x - 1)^2 - 5 = 0$ **38.** $4(x - 2)^2 - 3 = 0$

39. $16(x + 1)^2 + 1 = 0$ **40.** $25(x + 2)^2 + 16 = 0$

41. $x^2 = \dfrac{1}{81}$ **42.** $x^2 = \dfrac{1}{9}$

43. $x^2 - \dfrac{1}{16} = 0$ **44.** $x^2 - \dfrac{1}{36} = 0$

45. $6x^2 - 24 = 0$ **46.** $3x^2 - 75 = 0$

47. $2(x + 1)^2 - 18 = 0$ **48.** $3(x - 2)^2 - 48 = 0$

49. $8(x - 1)^2 - 18 = 0$ **50.** $50(x + 3)^2 - 72 = 0$

31. _____
32. _____
33. _____
34. _____
35. _____
36. _____
37. _____
38. _____
39. _____
40. _____
41. _____
42. _____
43. _____
44. _____
45. _____
46. _____
47. _____
48. _____
49. _____
50. _____

✓ SKILL CHECKER

Write the given fraction with the denominator indicated.

51. $\dfrac{9}{5}$ with a denominator of 25

52. $\dfrac{16}{7}$ with a denominator of 49

53. $\dfrac{25}{3}$ with a denominator of 9

54. $\dfrac{32}{3}$ with a denominator of 9

55. $\dfrac{7}{4}$ with a denominator of 16

51. _____

52. _____

53. _____

54. _____

55. _____

9.1   USING YOUR KNOWLEDGE

We have stated that

$$\sqrt{\frac{a}{b}} = \frac{\sqrt{a}}{\sqrt{b}} \quad a \text{ and } b \text{ positive numbers}$$

ANSWER   (to problem on page 594)
5. No real-number solution

This rule can be illustrated by using examples. For instance

$$\sqrt{\frac{16}{9}} = \frac{4}{3} \quad \text{because} \quad \left(\frac{4}{3}\right)^2 = \frac{16}{9}$$

Also,

$$\frac{\sqrt{16}}{\sqrt{9}} = \frac{4}{3}$$

Thus,

$$\sqrt{\frac{16}{9}} = \frac{\sqrt{16}}{\sqrt{9}}$$

Do the next few examples for practice. Find:

1. **a.** $\sqrt{\dfrac{25}{9}}$ **b.** $\dfrac{\sqrt{25}}{\sqrt{9}}$

2. **a.** $\sqrt{\dfrac{36}{49}}$ **b.** $\dfrac{\sqrt{36}}{\sqrt{49}}$

3. **a.** $\dfrac{\sqrt{25}}{\sqrt{81}}$ **b.** $\sqrt{\dfrac{25}{81}}$

These three exercises should help convince you that

$$\sqrt{\frac{a}{b}} = \frac{\sqrt{a}}{\sqrt{b}}$$

But can we prove it? Of couse we can! This proof will depend on the definition that states:

> $\sqrt{a}$ is a nonnegative number b such that $b^2 = a$ $(a \geq 0)$

Now, let us keep in mind that $\sqrt{a}$ is *nonnegative*. We can then rewrite the definition as

$$\sqrt{a} = b \quad \text{is equivalent to} \quad b^2 = a$$

Squaring on the left

$$(\sqrt{a})^2 = b^2 = a$$

Thus,

$$(\sqrt{a})^2 = a$$
$$\sqrt{a}\,\sqrt{a} = a$$

4. $\left(\dfrac{\sqrt{a}}{\sqrt{b}}\right)^2 = \left(\dfrac{\sqrt{a}}{\sqrt{b}}\right)\left(\dfrac{\sqrt{a}}{\sqrt{b}}\right)$

     Find $\dfrac{\sqrt{a} \cdot \sqrt{a}}{\sqrt{b} \cdot \sqrt{b}}$.

1. a. _____
   b. _____

2. a. _____
   b. _____

3. a. _____
   b. _____

4. _____

5. From Problem 4,

$$\left(\frac{\sqrt{a}}{\sqrt{b}}\right)^2 = \frac{a}{b}$$

Take the positive square root of both sides of this equation. What can you say about

$$\frac{\sqrt{a}}{\sqrt{b}} \quad \text{and} \quad \sqrt{\frac{a}{b}}$$

5. _____

CALCULATOR CORNER

In Example 2b, we stated that the equation

$$8x^2 + 49 = 0$$

had no real-number solution. You can verify this using your calculator. In the example, we had:

$$x^2 = -\frac{49}{8}$$

$$x = \pm\sqrt{-\frac{49}{8}}$$

Now, enter

49 $+/-$ 8 $\div$ $\sqrt{}$

You should get an ERROR message in your calculator, indicating that this equation has no real-number solution. If your instructor allows you, use your calculator to verify that the equations in Problems 5, 6, 19, 20, 25, 26, 39, and 40 do not have real-number solutions.

9.2 ADDITION, SUBTRACTION, AND MULTIPLICATION OF RADICALS

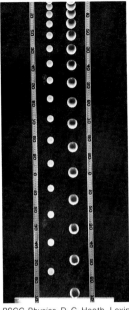

PSCC *Physics*, D. C. Heath, Lexington, Mass., © 1971, p. 247. Copied with the permission of the publisher.

OBJECTIVES

REVIEW

Before starting this section, you should know:

1. How to collect like terms.
2. How to use the Distributive Law to simplify expressions.

OBJECTIVES

You should be able to:

A. Simplify expressions involving radicals.
B. Rationalize the denominator of a given expression.
C. Add or subtract expressions involving radicals.

Suppose one ball in the picture has traveled 2 ft after being dropped. Can we find its velocity? The formula for the velocity V is given by

$$V = \sqrt{2gh} \quad g = 32, \; h \text{ the distance traveled (in feet)}$$

Thus,

$$V = \sqrt{2 \cdot 32 \cdot 2} = \sqrt{64 \cdot 2}$$

A. SIMPLIFYING RADICALS

Can we simplify $\sqrt{64 \cdot 2}$? We know that for a and b positive,

$$\sqrt{\frac{a}{b}} = \frac{\sqrt{a}}{\sqrt{b}}$$

Is there a similar rule for products, that is:

$$\sqrt{64 \cdot 2} \overset{?}{=} \sqrt{64} \cdot \sqrt{2} \overset{?}{=} 8\sqrt{2}$$

The answer is yes! For example,

$$\sqrt{4 \cdot 9} = \sqrt{4} \cdot \sqrt{9}$$

because

$$\sqrt{4 \cdot 9} = \sqrt{36} = 6$$

and

$$\sqrt{4} \cdot \sqrt{9} = 2 \cdot 3 = 6$$

Similarly,

$$\sqrt{9 \cdot 16} = \sqrt{9} \cdot \sqrt{16}$$

because

$$\sqrt{9 \cdot 16} = \sqrt{144} = 12$$

and
$$\sqrt{9} \cdot \sqrt{16} = 3 \cdot 4 = 12$$

In general, we have

$$\sqrt{a \cdot b} = \sqrt{a} \cdot \sqrt{b} \qquad a \text{ and } b \text{ positive numbers}$$

This rule is discussed in 9.2 *Using Your Knowledge.*

Of course, to use this rule when simplifying the square roots of certain numbers, you should keep in mind those numbers that have integer roots and use these numbers as *factors* of the number whose root is being sought. Here are the first few perfect squares (numbers whose square roots are integers):

1   4   9   16   25   36   49   64   81

Thus, to find $\sqrt{24}$ using this rule, we write

$$\sqrt{24} = \sqrt{4 \cdot 6} = \sqrt{4} \cdot \sqrt{6} = 2\sqrt{6}$$

Here we use **4** as a factor because $\sqrt{4} = 2$.

Similarly,

$$\sqrt{27} = \sqrt{9 \cdot 3} = \sqrt{9} \cdot \sqrt{3} = 3\sqrt{3}$$

We use **9** as a factor because $\sqrt{9} = 3$.

In general, we say that $\sqrt{A}$ is in simplified form if A *does not* contain a perfect square factor. How can we use this idea in simplifying the solution of an equation such as $x^2 = 32$? Here is how.

Given. $\qquad\qquad\qquad x^2 = 32$

Then take square roots. $\qquad x = \pm\sqrt{32}$ Note that 32 contains the perfect square factor 16.

$$x = \pm\sqrt{16 \cdot 2} = \pm\sqrt{16} \cdot \sqrt{2}$$
$$= \pm 4\sqrt{2}$$

A similar problem is given next.

EXAMPLE 1 Solve and write in simplified form.

a. $x^2 = 48$ **b.** $x^2 - 50 = 0$

Solution

a. Given. $\qquad x^2 = 48$

Then $\qquad x = \pm\sqrt{48}$
$$x = \pm\sqrt{16 \cdot 3} = \pm\sqrt{16} \cdot \sqrt{3}$$
$$= \pm 4\sqrt{3}$$

Thus the equation $x^2 = 48$ has two solutions, $4\sqrt{3}$ and $-4\sqrt{3}$.

b. Given. $\qquad\qquad\qquad x^2 - 50 = 0$

Add 50. $\qquad\qquad\qquad x^2 = 50$

Then take square roots. $\qquad x = \pm\sqrt{50}$
$$x = \pm\sqrt{25 \cdot 2} = \pm\sqrt{25} \cdot \sqrt{2}$$
$$= \pm 5\sqrt{2}$$

Thus the solutions of $x^2 - 50 = 0$ are

$5\sqrt{2}$ and $-5\sqrt{2}$

Problem 1 Solve.
a. $x^2 = 72$ **b.** $x^2 - 54 = 0$

B. RATIONALIZING DENOMINATORS

The fact that $\sqrt{ab} = \sqrt{a}\sqrt{b}$ can be used in *rationalizing* the denominators in the solutions of certain quadratic equations. For example, consider the following:

$$5x^2 = 9$$

$$x^2 = \frac{9}{5} \quad \text{Dividing by 5}$$

$$x = \pm\sqrt{\frac{9}{5}}$$

As you can see, the denominator of the answer will not be an integer. To rationalize the denominator of a fraction we write the fraction as an equivalent one with no radical in the denominator. Note that it would be easier to add

$$\frac{\sqrt{2}}{2} + \frac{\sqrt{3}}{3}$$

than

$$\frac{1}{\sqrt{2}} + \frac{1}{\sqrt{3}}$$

Thus, to **rationalize the denominator** in $\sqrt{\frac{9}{5}}$, we recall that by the fundamental rule of fractions,

$$\frac{a}{b} = \frac{a \cdot c}{b \cdot c}$$

Thus

$$\sqrt{\frac{9}{5}} = \sqrt{\frac{9 \cdot 5}{5 \cdot 5}}$$

$$= \frac{\sqrt{9 \cdot 5}}{\sqrt{5 \cdot 5}} \quad \text{Since } \sqrt{\frac{a}{b}} = \frac{\sqrt{a}}{\sqrt{b}}$$

$$= \frac{\sqrt{9} \cdot \sqrt{5}}{\sqrt{25}} \quad \text{Since } \sqrt{a \cdot b} = \sqrt{a} \cdot \sqrt{b}$$

$$= \frac{3\sqrt{5}}{5} \quad \text{Since } \sqrt{9} = 3 \text{ and } \sqrt{25} = 5$$

Hence, the solutions of $5x^2 = 9$,

$$x = \pm\sqrt{\frac{9}{5}}$$

can be written as

$$\pm\sqrt{\frac{9 \cdot 5}{5 \cdot 5}} = \pm\frac{3\sqrt{5}}{5}$$

Note that the idea in rationalizing $\sqrt{\frac{a}{b}}$ is to make the denominator a perfect square. Multiplying numerator and denominator by b *always* works, but you can save some time if you find a factor smaller than b

that will make the denominator a perfect square. Thus, to rationalize $\sqrt{\frac{3}{8}}$, you can multiply numerator and denominator by 8, obtaining:

$$\sqrt{\frac{3}{8}} = \sqrt{\frac{3 \cdot 8}{8 \cdot 8}} = \frac{\sqrt{24}}{8} = \frac{\sqrt{4 \cdot 6}}{8} = \frac{2\sqrt{6}}{8} = \frac{\sqrt{6}}{4}$$

But if we notice that $8 \cdot 2$ is a perfect square, we have

$$\sqrt{\frac{3}{8}} = \sqrt{\frac{3 \cdot 2}{8 \cdot 2}} = \sqrt{\frac{6}{16}} = \frac{\sqrt{6}}{4}$$

Save time by looking for factors that make the denominator a perfect square.

EXAMPLE 2   Solve and rationalize the denominator.

a. $7x^2 = 16$ **b.** $3x^2 - 25 = 0$

Solution

a. Given. $7x^2 = 16$

    Divide by 7. $x^2 = \dfrac{16}{7}$

    Then take square roots. $x = \pm\sqrt{\dfrac{16}{7}}$

    Rationalize the denominator. $x = \pm\sqrt{\dfrac{16 \cdot 7}{7 \cdot 7}} = \pm\dfrac{4\sqrt{7}}{7}$

    Thus, the solutions of $7x^2 = 16$ are

$$\frac{4\sqrt{7}}{7} \quad \text{and} \quad -\frac{4\sqrt{7}}{7}$$

b. Given. $3x^2 - 25 = 0$

    Add 25. $3x^2 = 25$

    Divide by 3. $x^2 = \dfrac{25}{3}$

    Take square roots. $x = \pm\sqrt{\dfrac{25}{3}}$

$$x = \pm\sqrt{\frac{25 \cdot 3}{3 \cdot 3}}$$

$$= \pm\frac{\sqrt{25 \cdot 3}}{\sqrt{3 \cdot 3}}$$

$$= \pm\frac{5\sqrt{3}}{3}$$

    Thus we have two solutions,

$$\frac{5\sqrt{3}}{3} \quad \text{and} \quad \frac{-5\sqrt{3}}{3}$$ ▲

Of course, there are equations that require the use of most of these rules. One such equation is shown next. Make sure you understand each of the steps and rules being used.

Problem 2   Solve.
a. $5x^2 = 36$ **b.** $7x^2 - 81 = 0$

EXAMPLE 3 Solve.

$3x^2 - 32 = 0$

Solution

Given.

$$3x^2 - 32 = 0$$

Add 32.

$$3x^2 = 32$$

Divide by 3.

$$x^2 = \frac{32}{3}$$

Take square roots.

$$x = \pm\sqrt{\frac{32}{3}}$$

Use the fundamental rule of fractions.

$$x = \pm\sqrt{\frac{32 \cdot 3}{3 \cdot 3}}$$

$$= \pm\frac{\sqrt{32 \cdot 3}}{\sqrt{3 \cdot 3}} = \pm\frac{\sqrt{96}}{\sqrt{9}}$$

$$= \pm\frac{\sqrt{16 \cdot 6}}{\sqrt{9}} \qquad \text{Since } \sqrt{96} = \sqrt{16 \cdot 6}$$

$$= \pm\frac{\sqrt{16} \cdot \sqrt{6}}{\sqrt{9}} \qquad \text{Since } \sqrt{A \cdot B} = \sqrt{A} \cdot \sqrt{B}$$

$$= \pm\frac{4\sqrt{6}}{3} \qquad \text{Since } \sqrt{16} = 4 \text{ and } \sqrt{9} = 3$$

Problem 3 Solve.
$5x^2 - 36 = 0$

C. OPERATIONS WITH RADICALS

Finally, we observe that expressions involving radicals can be handled using simple arithmetic rules. For example, since like terms can be combined,

$$3x + 7x = 10x$$
$$3\sqrt{6} + 7\sqrt{6} = 10\sqrt{6}$$

and

$$9x - 2x = 7x$$
$$9\sqrt{3} - 2\sqrt{3} = 7\sqrt{3}$$

Test your understanding in the next example.

EXAMPLE 4 Simplify.

a. $6\sqrt{7} + 9\sqrt{7}$ **b.** $8\sqrt{5} - 2\sqrt{5}$

Solution

a. $6\sqrt{7} + 9\sqrt{7} = 15\sqrt{7}$
(Just like $6x + 9x = 15x$)

b. $8\sqrt{5} - 2\sqrt{5} = 6\sqrt{5}$
(Just like $8x - 2x = 6x$)

Problem 4 Simplify.
a. $7\sqrt{11} + 6\sqrt{11}$ **b.** $9\sqrt{10} - 4\sqrt{10}$

▲

Of course, you may have to simplify things before combining **like radical terms**—that is, terms in which the *radical factors* are *exactly* the same. (For example, $4\sqrt{3}$ and $5\sqrt{3}$ are like radical terms.) Here is a problem that seems difficult:

$$\sqrt{48} + \sqrt{75}$$

In this case, $\sqrt{48}$ and $\sqrt{75}$ are *not* like terms. But wait!

$$\sqrt{48} = \sqrt{16 \cdot 3} = \sqrt{16} \cdot \sqrt{3} = 4\sqrt{3}$$
$$\sqrt{75} = \sqrt{25 \cdot 3} = \sqrt{25} \cdot \sqrt{3} = 5\sqrt{3}$$

Now,

$$\sqrt{48} + \sqrt{75} = 4\sqrt{3} + 5\sqrt{3}$$
$$= 9\sqrt{3}$$

Some more complicated problems follow.

EXAMPLE 5   Simplify.

a. $\sqrt{80} + \sqrt{20}$ **b.** $\sqrt{75} + \sqrt{12} - \sqrt{147}$

Solution

a.
$$\sqrt{80} = \sqrt{16 \cdot 5} = \sqrt{16} \cdot \sqrt{5} = 4\sqrt{5}$$
$$\sqrt{20} = \sqrt{4 \cdot 5} = \sqrt{4} \cdot \sqrt{5} = 2\sqrt{5}$$
$$\sqrt{80} + \sqrt{20} = 4\sqrt{5} + 2\sqrt{5} = 6\sqrt{5}$$
b.
$$\sqrt{75} = \sqrt{25 \cdot 3} = \sqrt{25} \cdot \sqrt{3}$$
$$= 5\sqrt{3}$$
$$\sqrt{12} = \sqrt{4 \cdot 3} = \sqrt{4} \cdot \sqrt{3} = 2\sqrt{3}$$
$$\sqrt{147} = \sqrt{49 \cdot 3} = \sqrt{49} \cdot \sqrt{3}$$
$$= 7\sqrt{3}$$
$$\sqrt{75} + \sqrt{12} - \sqrt{147} = 5\sqrt{3} + 2\sqrt{3} - 7\sqrt{3}$$
$$= 0$$

EXAMPLE 6   Simplify.

a. $\sqrt{5}(\sqrt{40} - \sqrt{2})$ **b.** $\sqrt{2}(\sqrt{2} - \sqrt{3})$

Solution

a. Using the distributive law,
$$\sqrt{5}(\sqrt{40} - \sqrt{2}) = \sqrt{5}\sqrt{40} - \sqrt{5}\sqrt{2}$$
$$= \sqrt{200} - \sqrt{10}$$
$$= \sqrt{100 \cdot 2} - \sqrt{10}$$
$$= 10\sqrt{2} - \sqrt{10}$$
b. $\sqrt{2}(\sqrt{2} - \sqrt{3}) = \sqrt{2}\sqrt{2} - \sqrt{2}\sqrt{3}$
$$= \sqrt{4} - \sqrt{6}$$
$$= 2 - \sqrt{6}$$

Problem 5   Simplify.
a. $\sqrt{98} + \sqrt{50}$
b. $\sqrt{20} + \sqrt{80} - \sqrt{180}$

Problem 6   Simplify.
a. $\sqrt{5}(\sqrt{60} - \sqrt{3})$ **b.** $\sqrt{7}(\sqrt{7} - \sqrt{2})$

NAME

CLASS

SECTION

A. Simplify.

1. $\sqrt{45}$

2. $\sqrt{72}$

3. $\sqrt{125}$

4. $\sqrt{175}$

5. $\sqrt{180}$

6. $\sqrt{162}$

7. $\sqrt{200}$

8. $\sqrt{245}$

9. $\sqrt{384}$

10. $\sqrt{486}$

11. $\sqrt{75}$

12. $\sqrt{80}$

13. $\sqrt{200}$

14. $\sqrt{324}$

15. $\sqrt{361}$

16. $\sqrt{648}$

17. $\sqrt{700}$

18. $\sqrt{726}$

19. $\sqrt{432}$

20. $\sqrt{507}$

B. Rationalize the denominator.

21. $\dfrac{3}{\sqrt{6}}$

22. $\dfrac{6}{\sqrt{7}}$

23. $\dfrac{-10}{\sqrt{5}}$

24. $\dfrac{-9}{\sqrt{3}}$

25. $\dfrac{\sqrt{8}}{\sqrt{2}}$

26. $\dfrac{\sqrt{48}}{\sqrt{3}}$

27. $\dfrac{-\sqrt{2}}{\sqrt{5}}$

28. $\dfrac{-\sqrt{3}}{\sqrt{7}}$

ANSWERS

1. _____
2. _____
3. _____
4. _____
5. _____
6. _____
7. _____
8. _____
9. _____
10. _____
11. _____
12. _____
13. _____
14. _____
15. _____
16. _____
17. _____
18. _____
19. _____
20. _____

21. _____
22. _____
23. _____
24. _____
25. _____
26. _____
27. _____
28. _____

29. $\dfrac{\sqrt{3}}{\sqrt{6}}$

30. $\dfrac{\sqrt{6}}{\sqrt{8}}$

31. $\dfrac{2}{\sqrt{18}}$

32. $\dfrac{3}{\sqrt{12}}$

33. $\dfrac{4}{\sqrt{12}}$

34. $\dfrac{6}{\sqrt{18}}$

35. $-\dfrac{3}{\sqrt{18}}$

36. $-\dfrac{5}{\sqrt{10}}$

37. $-\dfrac{6}{\sqrt{12}}$

38. $-\dfrac{7}{\sqrt{14}}$

39. $-\dfrac{3}{\sqrt{27}}$

40. $-\dfrac{4}{\sqrt{32}}$

29. _____
30. _____
31. _____
32. _____
33. _____
34. _____
35. _____
36. _____
37. _____
38. _____
39. _____
40. _____

Solve (write the answers in simplified form and with a rationalized denominator).

41. $x^2 = 98$

42. $x^2 = 128$

43. $x^2 - 162 = 0$

44. $x^2 - 108 = 0$

45. $6x^2 = 49$

46. $7x^2 = 36$

47. $5x^2 - 16 = 0$

48. $5x^2 - 49 = 0$

49. $6x^2 - 25 = 0$

50. $11x^2 - 49 = 0$

51. $5x^2 - 18 = 0$

52. $7x^2 - 8 = 0$

53. $3x^2 - 49 = 0$

54. $7x^2 - 25 = 0$

41. _____
42. _____
43. _____
44. _____
45. _____
46. _____
47. _____
48. _____
49. _____
50. _____
51. _____
52. _____
53. _____
54. _____

C. Perform the indicated operations.

55. $6\sqrt{7} + 4\sqrt{7}$

56. $4\sqrt{11} + 9\sqrt{11}$

57. $9\sqrt{13} - 4\sqrt{13}$

58. $6\sqrt{10} - 2\sqrt{10}$

55. _____
56. _____
57. _____
58. _____

ANSWERS (to problems on page 604)
5. a. $12\sqrt{2}$ b. 0
6. a. $10\sqrt{3} - \sqrt{15}$ b. $7 - \sqrt{14}$

59. $\sqrt{32} + \sqrt{50} - \sqrt{72}$

60. $\sqrt{12} + \sqrt{27} - \sqrt{75}$

61. $\sqrt{162} + \sqrt{50} - \sqrt{200}$

62. $\sqrt{48} + \sqrt{75} - \sqrt{363}$

63. $9\sqrt{48} - 5\sqrt{27} + 3\sqrt{12}$

64. $3\sqrt{32} - 5\sqrt{8} + 4\sqrt{50}$

65. $5\sqrt{7} - 3\sqrt{28} - 2\sqrt{63}$

66. $3\sqrt{28} - 6\sqrt{7} - 2\sqrt{175}$

67. $-5\sqrt{3} + 8\sqrt{75} - 2\sqrt{27}$

68. $-6\sqrt{99} + 6\sqrt{44} - \sqrt{176}$

69. $-3\sqrt{45} + \sqrt{20} - \sqrt{5}$

70. $-5\sqrt{27} + \sqrt{12} - 5\sqrt{48}$

71. $\sqrt{10}(\sqrt{20} - \sqrt{3})$

72. $\sqrt{10}(\sqrt{30} - \sqrt{2})$

73. $\sqrt{6}(\sqrt{14} + \sqrt{5})$

74. $\sqrt{14}(\sqrt{18} + \sqrt{3})$

75. $\sqrt{3}(\sqrt{3} - \sqrt{2})$

76. $\sqrt{6}(\sqrt{6} - \sqrt{5})$

77. $\sqrt{5}(\sqrt{2} + \sqrt{5})$

78. $\sqrt{3}(\sqrt{2} + \sqrt{3})$

79. $\sqrt{6}(\sqrt{2} - \sqrt{3})$

80. $\sqrt{5}(\sqrt{15} - \sqrt{27})$

59. _____	
60. _____	
61. _____	
62. _____	
63. _____	
64. _____	
65. _____	
66. _____	
67. _____	
68. _____	
69. _____	
70. _____	
71. _____	
72. _____	
73. _____	
74. _____	
75. _____	
76. _____	
77. _____	
78. _____	
79. _____	
80. _____	

✓ **SKILL CHECKER**

Expand.

81. $(x + 7)^2$

82. $(x + 5)^2$

83. $(x - 3)^2$

84. $(x - 5)^2$

85. $(x - 7)^2$

81. _____
82. _____
83. _____
84. _____
85. _____

9.2 USING YOUR KNOWLEDGE

We have already mentioned that a consequence of the definition of square root is that

$$\sqrt{a} \cdot \sqrt{a} = a$$

1. $\sqrt{a} \cdot \sqrt{a} = a$; what is $\sqrt{b} \cdot \sqrt{b}$?

2. What is $(\sqrt{a} \cdot \sqrt{a})(\sqrt{b} \cdot \sqrt{b})$?

1. _____
2. _____

3. From Problem 2,

$$(\sqrt{a} \cdot \sqrt{a})(\sqrt{b} \cdot \sqrt{b}) = ab$$

But

$$(\sqrt{a} \cdot \sqrt{a})(\sqrt{b} \cdot \sqrt{b}) = (\sqrt{a} \cdot \sqrt{b})(\sqrt{a} \cdot \sqrt{b})$$
$$= (\sqrt{a} \cdot \sqrt{b})^2$$

What can you say about $a \cdot b$ and $(\sqrt{a} \cdot \sqrt{b})^2$?

4. From Problem 3, $(\sqrt{a} \cdot \sqrt{b})^2 = a \cdot b$. What can you say about $\sqrt{a} \cdot \sqrt{b}$ and $\sqrt{a \cdot b}$?

3. _____

4. _____

CALCULATOR CORNER

You can use your calculator to verify the result of the operations with radicals done in this section. For instance, in Example 5a, we added $\sqrt{80} + \sqrt{20}$ to obtain $6\sqrt{5}$. This can be verified by using the following keystroke sequences:

$$\boxed{8}\,\boxed{0}\,\boxed{\sqrt{}}\,\boxed{+}\,\boxed{2}\,\boxed{0}\,\boxed{\sqrt{}}\,\boxed{=}\quad\boxed{6}\,\boxed{\times}\,\boxed{5}\,\boxed{\sqrt{}}\,\boxed{=}$$

In both instances you should get the same result, 13.416408. If your instructor allows you, verify the results of Problems 35 through 50 using your calculator.

9.3 COMPLETING THE SQUARE

Photo by Bill Davis.

OBJECTIVES

REVIEW

Before starting this section you should know:

1. The definition of a quadratic equation.
2. How to expand $(x \pm a)^2$ (Section 4.4).

OBJECTIVE

A. You should be able to solve a quadratric equation by completing the square.

The man in the picture has just batted the ball straight up at 96 ft/sec. At the end of t sec, the height h of the ball will be

$$h = -16t^2 + 96t$$

How long would it be before the ball reaches 44 ft? To solve this problem, we let $h = 44$, obtaining

$$-16t^2 + 96t = 44$$

$$t^2 - 6t = -\frac{44}{16} = -\frac{11}{4} \quad \text{Dividing by } -16$$

This equation is a **quadratic equation** (an equation in which the largest exponent of the variable is 2). Can we use the techniques we studied in Section 9.1 to solve it? We could, if we could write the equation in the form

$$(t - N)^2 = A \quad \begin{array}{l}N \text{ and } A \text{ are the numbers we need to find} \\ \text{to solve the problem.}\end{array}$$

To do this, however, we should know a little bit more about a technique used in algebra called **completing the square.** We will do this by presenting several examples and then trying to generalize the results obtained. You probably recall that

$$(X + A)^2 = X^2 + \underbrace{2AX}_{} + A^2$$

First term · Second term · First term squared · Coefficient of X · Second term squared

If you do not remember this, go back to Chapter 4 and practice with your special products.

Thus,

$$(x + 7)^2 = x^2 + 14x + 7^2$$
$$(x + 2)^2 = x^2 + 4x + 2^2$$
$$(x + 5)^2 = x^2 + 10x + 5^2$$

Do you see any relationship between the coefficient of x (14, 4, and 10, respectively) and the last term? Perhaps you will see it better if we write it in a table:

COEFFICIENT OF X	LAST TERM SQUARED
14	7^2
4	2^2
10	5^2

It seems that half the coefficient of x gives the number to be squared for the last term. Thus,

$$\frac{14}{2} = 7$$

$$\frac{4}{2} = 2$$

$$\frac{10}{2} = 5$$

Now, what numbers would you add to complete the given squares?

$(x + 3)^2 = x^2 + 6x + \boxed{}$

$(x + 4)^2 = x^2 + 8x + \boxed{}$

$(x + 6)^2 = x^2 + 12x + \boxed{}$

The correct answers are $(\frac{6}{2})^2 = 3^2 = 9$, $(\frac{8}{2})^2 = 4^2 = 16$, and $(\frac{12}{2})^2 = 6^2 = 36$. We then have:

$(x + 3)^2 = x^2 + 6x + 3^2$ The last term is always the square of half
$$6 \div 2 = 3$$ of the coefficient of the middle term.

$(x + 4)^2 = x^2 + 8x + 4^2$
$$8 \div 2 = 4$$

$(x + 6)^2 = x^2 + 12x + 6^2$
$$12 \div 2 = 6$$

Here is the procedure we have used to fill in the blanks:

RULE FOR COMPLETING THE SQUARE
1. Find the coefficient of the x term.
2. Divide the coefficient by 2.
3. Square this number to obtain the last term.

Thus, the following steps are used to complete the square in

$$(x + 8)^2 = x^2 + 16x + \boxed{}$$

1. Find the coefficient of the x term $\longrightarrow$ 16.
2. Divide the coefficient by 2 $\longrightarrow$ 8.
3. Square this number to obtain the last term $\longrightarrow 8^2$.

Hence,

$$(x + 8)^2 = x^2 + 16x + \boxed{8^2}$$

Now consider

$$(x - 9)^2 = x^2 - 18x + \square$$

Our steps to fill in the blank are as before:

1. Find the coefficient of the x term $\longrightarrow$ -18.
2. Divide the coefficient by 2 $\longrightarrow$ -9.
3. Square this number to obtain the last term $\longrightarrow$ $(-9)^2$.

Hence,

$$(x - 9)^2 = x^2 - 18x + (-9)^2$$
$$= x^2 - 18x + 9^2 \quad \text{Recall that } (-9)^2 = (9)^2; \text{ they are both 81.}$$

EXAMPLE 1 Find the missing term.

a. $(x + 10)^2 = x^2 + 20x + \square$ **b.** $\left(x - \dfrac{1}{2}\right)^2 = x^2 - x + \square$

Solution

a. 1. The coefficient of x is 20.

2. $\dfrac{20}{2} = 10$

3. The missing term is $\boxed{10^2} = \boxed{100}$.

b. 1. The coefficient of x is -1.

2. $\dfrac{-1}{2} = -\dfrac{1}{2}$

3. The missing term is $\boxed{\left(-\dfrac{1}{2}\right)^2} = \boxed{\dfrac{1}{4}}$. ▲

Can we use the patterns we have studied to look for further patterns? Of course! For example, how would you fill in the blanks in

$$x^2 + 16x + \square = (\quad)^2$$

Here the coefficient of x is 16, so $\left(\frac{16}{2}\right)^2 = 8^2$ goes in the box. Since

$$X^2 + 2AX + A^2 = (X + A)^2$$
$$x^2 + 16x + 8^2 = (x + 8)^2$$

Similarly,

$$x^2 - 6x + \square = (\quad)^2$$

is completed by reasoning that the coefficient of x is -6, so

$$\left(\dfrac{-6}{2}\right)^2 = (-3)^2 = (3)^2$$

goes in the box. Now,

Problem 1 Find the missing term.
a. $(x + 11)^2 = x^2 + 22x + \square$

b. $\left(x - \dfrac{1}{4}\right)^2 = x^2 - \dfrac{1}{2}x + \square$

$$\overset{\text{Same}}{\overbrace{X^2 - 2AX + A^2}} = (X - A)^2$$

$X^2 - 2AX + A^2 = (X - A)^2$ Since the middle term on the left has a negative sign, the sign inside the parentheses must be negative.

Same (lower bracket under $X^2 \ldots A^2$)

Hence

$$x^2 - 6x + \square \;\; = (\quad)^2$$

becomes

$$x^2 - 6x + 3^2 \;\; = (x - 3)^2$$

EXAMPLE 2   Find the missing terms.

a. $x^2 - 10x + \square = (\quad)^2$ **b.** $x^2 + 3x + \square = (\quad)^2$

Solution

a. The coefficient of x is -10; thus the number in the box should be $\left(-\frac{10}{2}\right)^2 = 5^2$, and we should have

$$\overset{\text{Same}}{x^2 - 10x + 5^2} = (x - 5)^2$$

b. The coefficient of x is 3; thus we have to add $\left(\frac{3}{2}\right)^2$. Then

$$x^2 + 3x + \left(\frac{3}{2}\right)^2 = \left(x + \frac{3}{2}\right)^2 \qquad \blacktriangle$$

We are finally ready to solve the equation involving the time it takes the ball to reach 44ft, that is,

$$t^2 - 6t = -\frac{11}{4} \qquad \text{As you recall, } t \text{ represents the time it takes the ball to reach 44 ft. (See page 609.)}$$

Since the coefficient of t is -6, we must add $\left(-\frac{6}{2}\right)^2 = (-3)^2 = \mathbf{3^2}$ to both sides of the equation. We then have

$$t^2 - 6t + 3^2 = -\frac{11}{4} + 3^2$$

$$(t - 3)^2 = -\frac{11}{4} + \frac{36}{4} \qquad \text{Note that } 3^2 = 9 = \frac{36}{4}.$$

$$(t - 3)^2 = \frac{25}{4}$$

Then

$$(t - 3) = \pm\sqrt{\frac{25}{4}} = \pm\frac{5}{2}$$

$$t = 3 \pm \frac{5}{2}$$

$$t = 3 + \frac{5}{2} = \frac{11}{2} \quad \text{or} \quad t = 3 - \frac{5}{2} = \frac{1}{2}$$

This means that the ball reaches 44 ft after $\frac{1}{2}$ sec (on the way up) and after $\frac{11}{2} = 5\frac{1}{2}$ sec (on the way down).

Problem 2   Find the missing terms.
a. $x^2 - 12x + \square = (\quad)^2$
b. $x^2 + 5x + \square = (\quad)^2$

Here is a summary of the steps needed to solve a quadratic equation by completing the square. The solution of our original equation is given in the margin so you can see how all the steps are carried out.

SOLVING A QUADRATIC EQUATION BY COMPLETING THE SQUARE

1. Write the equation with the variables in descending order on the left and the numbers on the right.

2. If the coefficient of the square term is not 1, divide each term by this coefficient.

3. Add the square of one-half of the coefficient of the first degree term to both sides.

4. Rewrite the left-hand side as a perfect square.

5. Solve the resulting equation.

We use this procedure in solving the next example.

EXAMPLE 3 Solve $4x^2 - 16x + 7 = 0$.

Solution We use the five-step procedure just given.

Given: $\qquad\qquad\qquad\qquad 4x^2 - 16x + 7 = 0$

1. Subtract 7. $\qquad\qquad\qquad 4x^2 - 16x = -7$

2. Divide by 4 so the coefficient of x^2 is 1. $\qquad x^2 - 4x = -\dfrac{7}{4}$

3. Add $\left(-\dfrac{4}{2}\right)^2 = (-2)^2 = 2^2$. $\qquad x^2 - 4x + 2^2 = -\dfrac{7}{4} + 2^2$

4. Rewrite. $\qquad\qquad\qquad (x-2)^2 = -\dfrac{7}{4} + \dfrac{16}{4} = \dfrac{9}{4}$

$$(x-2)^2 = \dfrac{9}{4}$$

5. Solve. $\qquad\qquad\qquad (x-2) = \pm\sqrt{\dfrac{9}{4}} = \pm\dfrac{3}{2}$

$$x = 2 \pm \dfrac{3}{2}$$

Thus,

$$x = \dfrac{4}{2} + \dfrac{3}{2} = \dfrac{7}{2} \quad \text{or} \quad x = \dfrac{4}{2} - \dfrac{3}{2} = \dfrac{1}{2} \qquad\qquad \blacktriangle$$

Finally, we must point out that in many cases the answers obtained when solving quadratics by completing the square are *not* rational numbers. (A rational number can be written as $\frac{a}{b}$, a and b integers, $b \neq 0$.) As a matter of fact, quadratics with rational solutions may be solved by factoring, a simpler method.

Given:

$-16t^2 + 96t = 44$

2. $\dfrac{-16t^2}{-16} + \dfrac{96t}{-16} = \dfrac{44}{-16}$

$$t^2 - 6t = -\dfrac{11}{4}$$

3. $t^2 - 6t + \left(-\dfrac{6}{2}\right)^2 = -\dfrac{11}{4} + \left(-\dfrac{6}{2}\right)^2$

$$t^2 - 6t + 3^2 = -\dfrac{11}{4} + 3^2$$

4. $(t-3)^2 = \dfrac{25}{4}$

$$(t-3) = \pm\sqrt{\dfrac{25}{4}} = \pm\dfrac{5}{2}$$

5. $t = 3 \pm \dfrac{5}{2}$

Problem 3 Solve $4x^2 - 24x + 27 = 0$.

ANSWER (to problem on page 611)

1. **a.** $\left(\dfrac{22}{2}\right)^2 = 11^2$ or 121

b. $\left(\dfrac{-\dfrac{1}{2}}{2}\right)^2 = \left(-\dfrac{1}{4}\right)^2 = \dfrac{1}{16}$

EXAMPLE 4   Solve $36x + 9x^2 + 31 = 0$.

Solution   We proceed using the five given steps.

Given. $\qquad\qquad\qquad$ $36x + 9x^2 + 31 = 0$

1. Subtract 31 and write $\qquad$ $9x^2 + 36x = -31$
in descending order.

2. Divide by 9. $\qquad\qquad$ $x^2 + 4x = -\dfrac{31}{9}$

3. Add $\left(\dfrac{4}{2}\right)^2 = 2^2$. $\qquad$ $x^2 + 4x + 2^2 = \dfrac{-31}{9} + 2^2 = -\dfrac{31}{9} + \dfrac{36}{9}$

4. Rewrite as a perfect $\qquad$ $(x + 2)^2 = \dfrac{5}{9}$
square.

5. Solve. $\qquad\qquad\qquad$ $(x + 2) = \pm\sqrt{\dfrac{5}{9}} = \pm\dfrac{\sqrt{5}}{3}$

$$x = -2 \pm \dfrac{\sqrt{5}}{3}$$

Problem 4   Solve $4x^2 + 24x + 31 = 0$.

NAME

CLASS

SECTION

A. Find the missing term(s) to make the expression a perfect square.

ANSWERS

1. $x^2 + 18x +$ ☐   **2.** $x^2 + 2x +$ ☐

3. $x^2 - 16x +$ ☐   **4.** $x^2 - 4x +$ ☐

5. $x^2 + 7x +$ ☐   **6.** $x^2 + 9x +$ ☐

7. $x^2 - 3x +$ ☐   **8.** $x^2 - 7x +$ ☐

9. $x^2 + x +$ ☐   **10.** $x^2 - x +$ ☐

11. $x^2 + 4x +$ ☐ $= ($ $)^2$ **12.** $x^2 + 6x +$ ☐ $= ($ $)^2$

13. $x^2 + 3x +$ ☐ $= ($ $)^2$ **14.** $x^2 + 9x +$ ☐ $= ($ $)^2$

15. $x^2 - 6x +$ ☐ $= ($ $)^2$ **16.** $x^2 - 24x +$ ☐ $= ($ $)^2$

17. $x^2 - 5x +$ ☐ $= ($ $)^2$ **18.** $x^2 - 11x +$ ☐ $= ($ $)^2$

19. $x^2 - \dfrac{3}{2}x +$ ☐ $= ($ $)^2$ **20.** $x^2 - \dfrac{5}{2}x +$ ☐ $= ($ $)^2$

1. _____
2. _____
3. _____
4. _____
5. _____
6. _____
7. _____
8. _____
9. _____
10. _____
11. _____
12. _____
13. _____
14. _____
15. _____
16. _____
17. _____
18. _____
19. _____
20. _____

Solve.

21. $x^2 + 2x + 7 = 0$ **22.** $x^2 + 4x + 1 = 0$

23. $x^2 + x - 1 = 0$ **24.** $x^2 + 2x - 1 = 0$

25. $x^2 + 3x - 1 = 0$ **26.** $x^2 - 3x - 4 = 0$

27. $x^2 - 3x - 3 = 0$ **28.** $x^2 - 3x - 1 = 0$

21. _____
22. _____
23. _____
24. _____
25. _____
26. _____
27. _____
28. _____

29. $4x^2 + 4x - 3 = 0$ 30. $2x^2 + 10x - 1 = 0$

31. $4x^2 - 16x = 15$ 32. $25x^2 - 25x = -6$

33. $4x^2 - 7 = 4x$ 34. $2x^2 - 18 = -9x$

35. $2x^2 + 1 = 4x$ 36. $2x^2 + 3 = 6x$

37. $(x + 3)(x - 2) = -4$ 38. $(x + 4)(x - 1) = -6$

39. $2x(x + 5) - 1 = 0$ 40. $2x(x - 4) = 2(9 - 8x) - x$

29. _____

30. _____

31. _____

32. _____

33. _____

34. _____

35. _____

36. _____

37. _____

38. _____

39. _____

40. _____

✓ **SKILL CHECKER**

Write each equation in the standard form $ax^2 + bx + c = 0$, where a, b, and c are integers.

41. $10x^2 + 5x = 12$

42. $x^2 = 2x + 2$

43. $9x = x^2$

44. $\dfrac{x^2}{2} + \dfrac{3}{5} = \dfrac{x}{4}$ (*Hint:* First multiply each term by the LCD.)

45. $\dfrac{x}{4} = \dfrac{3}{5} - \dfrac{x^2}{2}$

41. _____

42. _____

43. _____

44. _____

45. _____

9.3 USING YOUR KNOWLEDGE

Many applications of mathematics require finding the maximum or the minimum of certain algebra expressions. Thus a certain business may wish to find the price at which a product will bring *maximum* profits, whereas engineers may be interested in *minimizing* the amount of carbon monoxide produced by automobiles. Now, suppose you are the manufacturer of a certain product whose average manufacturing cost $\bar{C}$ (in dollars), based on producing x (thousand) units, is given by the expression

$$\bar{C} = x^2 - 8x + 18$$

How many units should be produced to *minimize* the cost per unit? If we consider the right-hand side of the equation, we can complete the square and leave the equation unchanged by adding and

ANSWER (to problem on page 614)

4. $x = -3 \pm \dfrac{\sqrt{5}}{2}$

subtracting the appropriate number. Thus,

$\bar{C} = x^2 - 8x + 18$

$\quad = (x^2 - 8x + \quad) + 18$

$\quad = (x^2 - 8x + 4^2) + 18 - 4^2$ Note that we have added and subtracted the square of one-half of the coefficient of x.

Then

$\bar{C} = (x - 4)^2 + 2$

Now, for $\bar{C}$ to be as small as possible (minimizing the cost), simply let x be 4, thus making $(x - 4)^2 = 0$ and $\bar{C} = 2$. This tells us that when 4 (thousand) units are produced, the minimum cost is 2. That is, the minimum average cost per unit is \$2. Use your knowledge about completing the square to solve the following problems.

$\bar{C} = (4 - 4)^2 + 2 = 2$

1. A manufacturer's average cost $\bar{C}$ (in dollars), based on manufacturing x (thousand) items, is given by

$\bar{C} = x^2 - 4x + 6$

a. How many units should be produced to minimize the cost per unit?
b. What is the minimum average cost per unit?

1. a. _____
   b. _____

2. The demand D for a certain product depends on the number x (in thousands) of units produced and is given by

$D = x^2 - 2x + 3$

What is the number of units that have to be produced so that the demand is at its lowest?

2. _____

3. Have you seen people adding chlorine to their pools? This is done to reduce the number of bacteria present in the water. Suppose that after t days, the number of bacteria per cubic centimeter is given by the expression

$B = 20t^2 - 120t + 200$

In how many days will the number of bacteria be at its lowest?

3. _____

"My final recommendation is that we reject all these proposals and continue to call it the quadratic equation."

Courtesy of Ralph Castellano.

REVIEW

Before starting this section you should know:

1. How to solve a quadratic equation by completing the square.
2. How to write a quadratic equation in standard form.

OBJECTIVES

You should be able to:

A. Write a given quadratic equation in the standard form

$$ax^2 + bx + c = 0$$

and identify the values of the real numbers a, b, and c.

B. Solve a quadratic equation using the quadratic formula.

As you were going through the preceding sections of this chapter, you probably wondered if there was a sure-fire method of solving any quadratic equation. Fortunately, there is! As a matter of fact, the derivation of this new technique uses the method of completing the square. As you recall, in each of the problems we followed the same procedure. Why not use the method of completing the square and solve the equation $ax^2 + bx + c = 0$ once and for all? We shall do that now. (You can refer to the procedure we gave to solve equations by completing the square, page 613.)

Given:

$$ax^2 + bx + c = 0$$

1. Rewrite with the numbers on the right.

$$ax^2 + bx = -c$$

2. Divide each term by a.

$$x^2 + \frac{b}{a}x = -\frac{c}{a}$$

3. Add the square of one-half the coefficient of x.

$$x^2 + \frac{b}{a}x + \left(\frac{b}{2a}\right)^2 = \left(\frac{b}{2a}\right)^2 - \frac{c}{a}$$

4. Rewrite the left side as a perfect square.

$$\left(x + \frac{b}{2a}\right)^2 = \frac{b^2}{4a^2} - \frac{c}{a}$$

$$\left(x + \frac{b}{2a}\right)^2 = \frac{b^2}{4a^2} - \frac{4ac}{4a^2}$$

5. Write the right side as a single fraction.

$$\left(x + \frac{b}{2a}\right)^2 = \frac{b^2 - 4ac}{4a^2}$$

6. Now, take the square root of both sides.

$$x + \frac{b}{2a} = \frac{\pm\sqrt{b^2 - 4ac}}{2a}$$

7. Subtract $\frac{b}{2a}$ from both sides.

$$x = -\frac{b}{2a} \pm \frac{\sqrt{b^2 - 4ac}}{2a}$$

8. Combine fractions.

$$x = \frac{-b \pm \sqrt{b^2 - 4ac}}{2a}$$

Do you realize what we have done? Any time you have a quadratic equation in the standard form $ax^2 + bx + c = 0$, you can solve the equation by simply substituting a, b, and c in the formula

$$x = \frac{-b \pm \sqrt{b^2 - 4ac}}{2a}$$

A derivation of this formula using a different method is given in 9.4 *Using Your Knowledge.*

This formula is so important that you should memorize it right now. Before using it, however, you must remember two things:

1. Write the given equation in standard form.
2. Determine the values of a, b, and c.

A. WRITING QUADRATICS IN STANDARD FORM

If we are given the equation $x^2 = 5x - 2$:

1. We must *first* write it in the standard form $ax^2 + bx + c = 0$ by subtracting $5x$ and adding 2; we have

$x^2 - 5x + 2 = 0$ This equation is in standard form.

2. When the equation is in standard form it is very easy to find the values of a, b, and c:

$$\underset{\substack{\uparrow \\ a = 1}}{x^2} \quad \underset{\substack{\uparrow \\ b = -5}}{-\ 5x} \quad \underset{\substack{\uparrow \\ c = 2}}{+\ 2} = 0 \quad \text{Recall that } x^2 = 1x^2; \text{ thus the coefficient of } x^2 \text{ is 1.}$$

If the equation contains fractions, multiply each term by the lowest common denominator (LCD) to clear the fractions. For example, consider the equation

$$\frac{x}{4} = \frac{3}{5} - \frac{x^2}{2}$$

Since the LCM of 4, 5, and 2 is **20,** we multiply each term by 20:

$$20 \cdot \frac{x}{4} = \frac{3}{5} \cdot 20 - \frac{x^2}{2} \cdot 20$$
$$5x = 12 - 10x^2$$
$$10x^2 + 5x = 12 \quad \text{Adding } 10x^2$$
$$10x^2 + 5x - 12 = 0 \quad \text{Subtracting } 12$$

Now the equation is in standard form:

$$\underset{\substack{\uparrow \\ a = 10}}{10x^2} \quad \underset{\substack{\uparrow \\ b = 5}}{+\ 5x} \quad \underset{\substack{\uparrow \\ c = -12}}{-\ 12} = 0$$

You can get a lot of practice by completing the accompanying table:

GIVEN	STANDARD FORM	a	b	c
1. $2x^2 + 7x - 4 = 0$				
2. $x^2 = 2x + 2$				
3. $9x = x^2$				
4. $\dfrac{x^2}{4} + \dfrac{2}{3}x = -\dfrac{1}{3}$				

We now use these four equations as examples.

B. SOLVING QUADRATICS WITH THE FORMULA

EXAMPLE 1   Solve $2x^2 + 7x - 4 = 0$.

Solution

1. The equation is written in standard form:

$$2x^2 \;+\; 7x \;-\; 4 = 0$$
$$\uparrow \qquad \uparrow \qquad \uparrow$$
$$a = 2 \quad b = 7 \quad c = -4$$

2. From the diagram, it is clear that $a = 2$, $b = 7$, and $c = -4$.
3. Substituting the values of a, b, and c in the formula, we obtain

$$x = \frac{-7 \pm \sqrt{(7)^2 - 4(2)(-4)}}{2(2)}$$

$$= \frac{-7 \pm \sqrt{49 + 32}}{4}$$

$$= \frac{-7 \pm \sqrt{81}}{4}$$

$$= \frac{-7 \pm 9}{4}$$

Thus,

$$x = \frac{-7 + 9}{4} = \frac{2}{4} = \frac{1}{2} \quad \text{or} \quad x = \frac{-7 - 9}{4} = \frac{-16}{4} = -4$$

Note that you could also solve $2x^2 + 7x - 4 = 0$ by factoring $2x^2 + 7x - 4 = 0$ as $(2x - 1)(x + 4) = 0$. You would obtain the same answers.

EXAMPLE 2   Solve $x^2 = 2x + 2$.

Solution   We proceed by steps as before:

1. To write the equation in standard form, subtract $2x$ and then subtract 2 to obtain

$$x^2 \;-\; 2x \;-\; 2 = 0$$
$$\uparrow \qquad \uparrow \qquad \uparrow$$
$$a = 1 \quad b = -2 \quad c = -2$$

Problem 1   Solve $3x^2 + 2x - 5 = 0$.

Problem 2   Solve $x^2 = 4x + 4$.

2. From the diagram, $a = 1$, $b = -2$, and $c = -2$.

3. Substituting these values in the quadratic formula, we have

$$x = \frac{-(-2) \pm \sqrt{(-2)^2 - 4(1)(-2)}}{2(1)}$$

$$= \frac{2 \pm \sqrt{4 + 8}}{2}$$

$$= \frac{2 \pm \sqrt{12}}{2}$$

$$= \frac{2 \pm \sqrt{4 \cdot 3}}{2}$$

$$= \frac{2 \pm 2\sqrt{3}}{2}$$

Thus,

$$x = \frac{2 + 2\sqrt{3}}{2} = \frac{2}{2} + \frac{2\sqrt{3}}{2} = 1 + \sqrt{3}$$

or

$$x = \frac{2 - 2\sqrt{3}}{2} = \frac{2}{2} - \frac{2}{2}\sqrt{3} = 1 - \sqrt{3}$$

Note that $x^2 - 2x - 2$ is *not* factorable. You *must* know the quadratic formula or you must complete the square to solve this equation!

EXAMPLE 3 Solve $9x = x^2$.

Solution

1. Subtracting $9x$, we have

$$0 = x^2 - 9x$$

or

$$x^2 \quad - \quad 9x \quad + \quad 0 = 0$$
$$\uparrow \qquad\quad \uparrow \qquad\quad \uparrow$$
$$a = 1 \quad b = -9 \quad c = 0$$

2. From the diagram, $a = 1$, $b = -9$, and $c = 0$ (because the c term is missing).

3. Substituting these values in the formula, we obtain

$$x = \frac{-(-9) \pm \sqrt{(-9)^2 - 4(1)(0)}}{2(1)}$$

$$= \frac{9 \pm \sqrt{81 - 0}}{2}$$

$$= \frac{9 \pm \sqrt{81}}{2}$$

$$= \frac{9 \pm 9}{2}$$

Thus,

$$x = \frac{9 + 9}{2} = \frac{18}{2} = 9 \quad \text{or} \quad x = \frac{9 - 9}{2} = \frac{0}{2} = 0$$

Problem 3 Solve $6x = x^2$

EXAMPLE 4   Solve $\dfrac{x^2}{4} + \dfrac{2}{3}x = -\dfrac{1}{3}$.

Problem 4   Solve $\dfrac{x^2}{4} - \dfrac{3}{8}x = \dfrac{1}{4}$.

Solution

1. We have to write the equation in standard form, but first we clear fractions by multiplying by the LCM of 4 and 3, that is, by 12:

$$12 \cdot \frac{x^2}{4} + 12 \cdot \frac{2}{3}x = -\frac{1}{3} \cdot 12$$

$$3x^2 + 8x = -4$$

We then add 4 to obtain

$$3x^2 + 8x + 4 = 0$$
$$\uparrow \qquad \uparrow \qquad \uparrow$$
$$a = 3 \quad b = 8 \quad c = 4$$

2. From the diagram, $a = 3$, $b = 8$, and $c = 4$.

3. Substituting in the formula gives

$$x = \frac{-8 \pm \sqrt{64 - 4(3)(4)}}{2(3)}$$

$$= \frac{-8 \pm \sqrt{64 - 48}}{6}$$

$$= \frac{-8 \pm \sqrt{16}}{6}$$

$$= \frac{-8 \pm 4}{6}$$

Thus,

$$x = \frac{-8 + 4}{6} = \frac{-4}{6} = -\frac{2}{3} \quad \text{or} \quad x = \frac{-8 - 4}{-6} = \frac{-12}{6} = -2 \qquad \blacktriangle$$

Now, a final word of warning. As you recall, some quadratic equations do not have real-number solutions. This is still true, even when you use the quadratic formula. The next example shows why this happens.

EXAMPLE 5   Solve $2x^2 + 3x = -3$.

Problem 5   Solve $3x^2 + 2x = -1$.

Solution

1. We add 3 to write the equation in standard form. We then have

$$2x^2 + 3x + 3 = 0$$
$$\uparrow \qquad \uparrow \qquad \uparrow$$
$$a = 2 \quad b = 3 \quad c = 3$$

2. From the diagram, $a = 2$, $b = 3$, and $c = 3$. Now,

$$x = \frac{-3 \pm \sqrt{(3)^2 - 4(2)(3)}}{2(2)}$$

$$= \frac{-3 \pm \sqrt{9 - 24}}{4}$$

$$= \frac{-3 \pm \sqrt{-15}}{4}$$

Thus,

$$x = \frac{-3 \pm \sqrt{-15}}{4}$$

But wait! If we check the definition of square root, we can see that $\sqrt{a}$ is defined only for $a \geq 0$. Hence, $\sqrt{-15}$ is not a real number, and the equation $2x^2 + 3x = -3$ has no real-number solutions.

_____
_____
_____
_____
_____

A. Write the given equation in standard form.

B. Solve the equation using the quadratic formula.

1. $x^2 + 3x + 2 = 0$

2. $x^2 + 4x + 3 = 0$

3. $x^2 + x - 2 = 0$

4. $x^2 + x - 6 = 0$

5. $2x^2 + x - 2 = 0$

6. $2x^2 + 7x + 3 = 0$

7. $3x^2 + x = 2$

8. $3x^2 + 2x = 5$

9. $2x^2 + 7x = -6$

10. $2x^2 + 7x = -6$

11. $7x^2 = 12x - 5$

12. $-5x^2 = 16x + 8$

13. $5x^2 = 11x - 4$

14. $7x^2 = 12x - 3$

15. $\dfrac{x^2}{5} - \dfrac{x}{2} = \dfrac{-3}{10}$

16. $\dfrac{x^2}{7} + \dfrac{x}{2} = -\dfrac{3}{4}$

17. $\dfrac{x^2}{2} = \dfrac{3x}{4} - \dfrac{1}{8}$

18. $\dfrac{x^2}{10} = \dfrac{x}{5} + \dfrac{3}{2}$

19. $\dfrac{x^2}{8} = -\dfrac{x}{4} - \dfrac{1}{8}$

20. $\dfrac{x^2}{3} = \dfrac{-x}{3} - \dfrac{1}{12}$

21. $6x = 4x^2 + 1$

22. $6x = 9x^2 - 4$

23. $3x = 1 - 3x^2$

24. $3x = 2x^2 - 5$

25. $x(x + 2) = 2x(x + 1) - 4$

26. $x(4x - 7) - 10 = 6x^2 - 7x$

27. $6x(x + 5) = (x + 15)^2$

28. $6x(x + 1) = (x + 3)^2$

1. _____
2. _____
3. _____
4. _____
5. _____
6. _____
7. _____
8. _____
9. _____
10. _____
11. _____
12. _____
13. _____
14. _____
15. _____
16. _____
17. _____
18. _____
19. _____
20. _____
21. _____
22. _____
23. _____
24. _____
25. _____
26. _____
27. _____
28. _____

29. $(x - 2)^2 = 4x(x - 1)$ **30.** $(x - 4)^2 = 4x(x - 2)$

29. _____

30. _____

✓ **SKILL CHECKER**

Find.

31. $(-2)^2$ **32.** -2^2

31. _____

32. _____

33. $(-1)^2$ **34.** -1^2

33. _____

34. _____

9.4 USING YOUR KNOWLEDGE

In this section we derived the quadratic formula by completing the square. The procedure depends on making the coefficient of x^2 one. But there is another way of deriving the quadratic formula. See if you can give the reason for each step.

Given: $ax^2 + bx + c = 0$

1. $4a^2x^2 + 4abx + 4ac = 0$

1. _____

2. $4a^2x^2 + 4abx = -4ac$

2. _____

3. $4a^2x^2 + 4abx + b^2 = b^2 - 4ac$

3. _____

4. $(2ax + b)^2 = b^2 - 4ac$

4. _____

5. $2ax + b = \pm\sqrt{b^2 - 4ac}$

5. _____

6. $2ax = -b \pm \sqrt{b^2 - 4ac}$

6. _____

7. $x = \dfrac{-b \pm \sqrt{b^2 - 4ac}}{2a}$

7. _____

CALCULATOR CORNER

Your calculator can be extremely helpful in finding the roots of a quadratic equation by using the quadratic formula. Of course, the roots you obtain are being approximated by decimals. It is most convenient to start with the radical part in the solution of the quadratic equation and then store this value so you can evaluate both roots without having to backtrack or copy down any intermediate steps. Let us look at the equation of Example 1,

$2x^2 + 7x - 4 = 0$

Using the quadratic formula, the solution will be obtained by following these keystrokes:

$\boxed{7}\,\boxed{x^2}\,\boxed{-}\,\boxed{4}\,\boxed{\times}\,\boxed{2}\,\boxed{\times}\,\boxed{4}\,\boxed{+/-}\,\boxed{=}$

$\boxed{\sqrt{x}}\,\boxed{\text{STO}}\,\boxed{7}\,\boxed{+/-}\,\boxed{+}\,\boxed{\text{RCL}}\,\boxed{=}\,\boxed{\div}\,\boxed{2}\,\boxed{\div}\,\boxed{2}\,\boxed{=}$

The display will show 0.5 (which was given as $\frac{1}{2}$ in the example.) To obtain the other root, key in

$\boxed{7}\,\boxed{+/-}\,\boxed{-}\,\boxed{\text{RCL}}\,\boxed{=}\,\boxed{\div}\,\boxed{2}\,\boxed{\div}\,\boxed{2}\,\boxed{=}$

which yields -4. In general, to solve the equation $ax^2 + bx + c = 0$ using your calculator, key in the following:

$\boxed{b}\ \boxed{x^2}\ \boxed{-}\ \boxed{4}\ \boxed{\times}\ \boxed{a}\ \boxed{\times}\ \boxed{c}\ \boxed{=}\ \boxed{\sqrt{x}}\ \boxed{\text{STO}}\ \boxed{b}\ \boxed{+/-}\ \boxed{+}\ \boxed{\text{RCL}}\ \boxed{=}$
$\boxed{\div}\ \boxed{2}\ \boxed{\div}\ \boxed{a}\ \boxed{=}\ \boxed{b}\ \boxed{+/-}\ \boxed{-}\ \boxed{\text{RCL}}\ \boxed{=}\ \boxed{\div}\ \boxed{2}\ \boxed{\div}\ \boxed{a}\ \boxed{=}$

9.5 GRAPHS OF QUADRATIC EQUATIONS

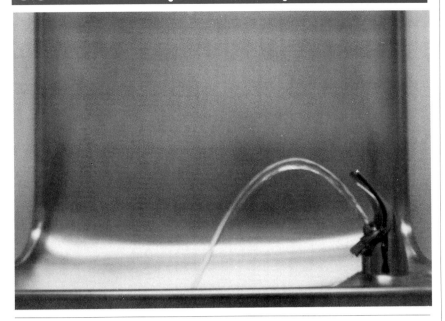

REVIEW

Before starting this section, you should know:

1. How to evaluate an expression.
2. How to graph points in the plane.
3. How to factor a quadratic expression.

OBJECTIVES

You should be able to:

A. Graph quadratic equations.
B. Find the vertex and graph factorable parabolas.

Look at the stream of water from the fountain. What shape does it have? The shape is called a **parabola.** The graph of a *quadratic equation* of the form $y = ax^2 + bx + c$ is a parabola. The simplest of these equations is $y = x^2$. This equation can be graphed in the same way as lines were graphed, that is, by selecting values for x and then finding the corresponding y-values as shown in the left-hand table. The usual shortened version is shown in the right-hand table.

x **VALUE**	y **VALUE**	x	y
$x = -2$	$y = x^2 = (-2)^2 = 4$	-2	4
$x = -1$	$y = x^2 = (-1)^2 = 1$	-1	1
$x = 0$	$y = x^2 = (0)^2 = 0$	0	0
$x = 1$	$y = x^2 = (1)^2 = 1$	1	1
$x = 2$	$y = x^2 = (2)^2 = 4$	2	4

A. GRAPHING QUADRATICS

We graph these points on a coordinate system and draw a smooth curve through them. The result is the graph of the parabola $y = x^2$. Note that the arrows at the end indicate that the curve goes on indefinitely.

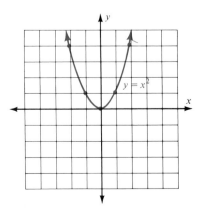

EXAMPLE 1 Graph $y = -x^2$.

Solution We could make a table of x and y values as before. However, note that for any x value, the y value will be the *negative* of the y value on the parabola $y = x^2$. (If you don't believe this, go ahead and make the table and check it.) Thus, the parabola $y = -x^2$ has the same shape as $y = x^2$, but it is turned in the *opposite* direction (opens *downward*). The graph of $y = -x^2$ is shown.

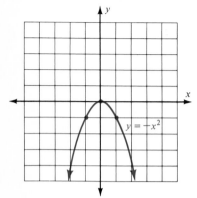

Problem 1 Graph $y = -2x^2$.

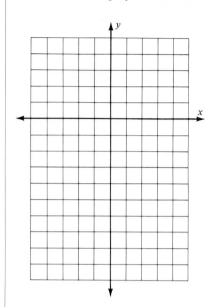

As you can see from the two preceding examples, when the coefficient of x^2 is positive (as in $y = x^2 = \mathbf{1}x^2$), the parabola opens *upward*, but when the coefficient of x^2 is negative (as in $y = -x^2 = -\mathbf{1}x^2$) the parabola opens *downward*. In general, we have the following:

Recall that it is understood that $x^2 = 1x^2$.

The graph of a quadratic equation of the form $y = ax^2 + bx + c$ is a parabola that

1. Opens upward if $a > 0$
2. Opens downward if $a < 0$

Now, what do you think will happen if we graph the parabola $y = x^2 + 1$? Two things. First, the parabola opens upward, since the coefficient of x^2 is understood to be 1. Second, all of the points will be one unit higher than those for the same value of x on the parabola $y = x^2$. Thus, we can make the graph of $y = x^2 + 1$ by following the pattern of $y = x^2$. The graphs of $y = x^2 + 1$ and $y = x^2 + 2$ are shown.

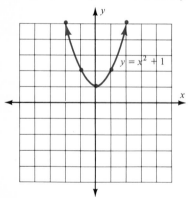

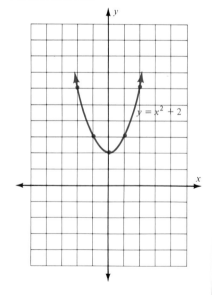

EXAMPLE 2   Graph $y = -x^2 - 2$.

Solution   Since the coefficient of x^2 (which is understood to be -1) is *negative,* the parabola opens downward. It is also 2 units *lower* than the graph of $y = -x^2$. Thus, the graph of $y = -x^2 - 2$ is as shown.

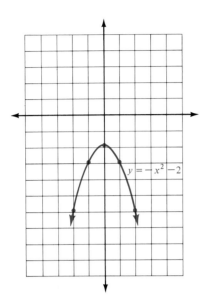

$y = -x^2 - 2$

Problem 2   Graph $y = -x^2 - 1$.

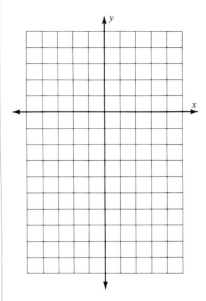

▲

So far, we have graphed only parabolas of the form $y = ax^2 + b$. How do you think the graph of $y = (x - 1)^2$ looks? As before, we make a table of values. For example,

For $x = -1,$ $y = (-1 - 1)^2 = (-2)^2 = 4$

For $x = 0,$ $y = (0 - 1)^2$ $= (-1)^2 = 1$

For $x = 1,$ $y = (1 - 1)^2$ $= (0)^2$ $= 0$

For $x = 2,$ $y = (2 - 1)^2$ $= 1^2$ $= 1$

This table looks like this.

x	y
-1	4
0	1
1	0
2	1

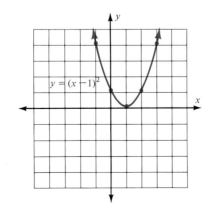

$y = (x - 1)^2$

Note that the shape of the graph is identical to that of $y = x^2$, but it is shifted 1 unit to the *right*. Similarly, the graph of $y = -(x + 1)^2$ is identical to that of $y = -x^2$ but shifted 1 unit to the *left*, as shown.

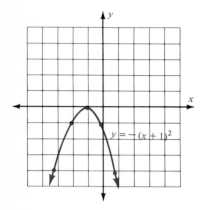

EXAMPLE 3   Graph $y = -(x - 1)^2 + 2$.

Solution   The graph of this equation is identical to the graph of $y = -x^2$ except for its position. The new parabola opens downward and it is shifted one unit to the right (because of the -1) and 2 units up (because of the $+2$). The diagram indicates these two facts, and the figure shows the finished graph of $y = -(x - 1)^2 + 2$.

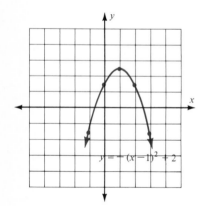

$$y = -(x - 1)^2 \quad + \quad 2$$

Opens downward ——— Shifted 1   Shifted 2
(negative)   unit right   units up

In conclusion, to graph an equation of the form

$$y = a(x \pm b)^2 \quad \pm \quad c$$

Opens upward for $a > 0$, ——   Shifts the   Moves the
downward for $a < 0$ graph right   graph up
                                               or left   or down

follow the given directions for changing the graph of $y = ax^2$.

Problem 3   Graph $y = -(x - 2)^2 - 1$.

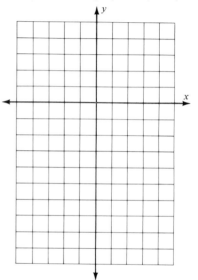

▲

ANSWER   (to problem on page 630)

1.

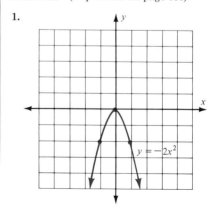

B. GRAPHING QUADRATICS USING INTERCEPTS

You have probably noticed that the graph of a parabola is **symmetric;** that is, if you draw a vertical line through the **vertex** (the high or low points on the parabola) and fold the graph along this line, the two halves of the parabola coincide. If a parabola crosses the x-axis, we can use the x-intercepts to find the vertex. For example, to graph $y = x^2 + 2x - 8$, we start by finding the x- and y-intercepts.

For $x = 0$, $y = 0^2 + 2 \cdot 0 - 8 = -8$

For $y = 0$, $0 = x^2 + 2x - 8$

$$0 = (x + 4)(x - 2)$$

Thus,

$$x = -4 \quad \text{or} \quad x = 2$$

We enter the points $(0, -8)$, $(-4, 0)$ and $(2, 0)$ in a table like this:

x	y	
0	-8	y-intercept
-4	0	x-intercepts
2	0	
?	?	Vertex

How can we find the vertex? Since the parabola is symmetric, the x-coordinate of the vertex is **exactly** halfway between -4 and 2, so

$$x = \frac{-4 + 2}{2} = \frac{-2}{2} = -1$$

We then find the y-coordinate by letting $x = -1$ in $y = x^2 + x - 8$, obtaining $y = (-1)^2 + 2 \cdot (-1) - 8 = -9$. Thus, the vertex is at $(-1, -9)$. The table now looks like this:

x	y	
0	-8	y-intercept
-4	0	x-intercepts
2	0	
-1	-9	Vertex
1	-5	
-3	-5	

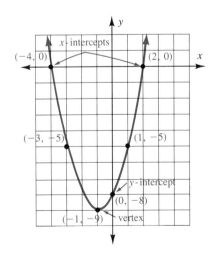

ANSWER (to problem on page 631)

2.

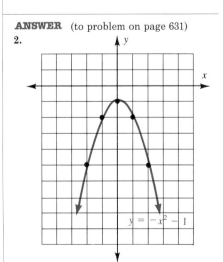

We can draw the graph using these four points or plot one or two more points. We plotted $(1, -5)$ and $(-3, -5)$. The graph is shown.

Here is the complete procedure so you can use it next.

GRAPHING A FACTORABLE PARABOLA

1. Find the y-intercept by letting $x = 0$ and then finding y.

2. Find the x-intercepts by letting $y = 0$, factoring the equation, and solving for x.

3. Find the vertex by averaging the solutions of the equation found in 2 (this is the x-coordinate of the vertex) and substituting in the equation to find the y-coordinate of the vertex.

4. Plot the points found in Steps 1 through 3 and one or two more points, if desired. The curve drawn through the points found in Steps 1 through 4 is the graph.

EXAMPLE 4   Graph $y = -x^2 + 2x + 8$.

Solution   We use the four steps discussed.

1. Since $-x^2 = -1 \cdot x^2$ and -1 is negative, the parabola opens downward. We then let $x = \mathbf{0}$, obtaining $y = -(\mathbf{0})^2 + 2 \cdot \mathbf{0} + 8 = 8$. Thus, $(0, 8)$ is on the graph.

2. We find the x-intercepts by letting $y = \mathbf{0}$ and solving for x. We have:

$0 = -x^2 + 2x + 8$

$0 = x^2 - 2x - 8$ Multiplying both sides by -1

$0 = (x - 4)(x + 2)$ to make the factorization easier

$x = 4$ or $x = -2$

We now have the x-intercepts: $(4, 0)$ and $(-2, 0)$.

3. The x-coordinate of the vertex is found by averaging 4 and -2, obtaining

$$x = \frac{4 + (-2)}{2} = \frac{2}{2} = 1$$

Letting $x = \mathbf{1}$ in $y = -x^2 + 2x + 8$, we find $y = -(\mathbf{1})^2 + 2 \cdot (\mathbf{1}) + 8 = 9$, so the vertex is at $(1, 9)$.

4. We plot these points and the two additional solutions $(2, 8)$ and $(3, 5)$. The completed graph is shown.

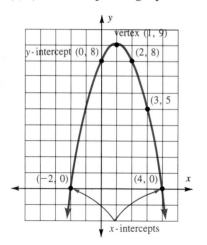

Problem 4   Graph $y = -x^2 - 2x + 8$.

ANSWER   (to problem on page 632)

3.

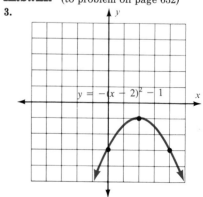

$y = -(x - 2)^2 - 1$

NAME

CLASS

SECTION

A. Graph.

1. $y = 2x^2$ (Make a table.)

2. $y = 2x^2 + 1$ (Use the results of Problem 1.)

ANSWERS

1. _____

2. _____

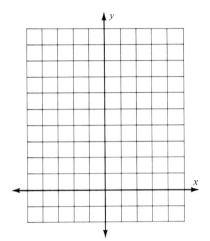

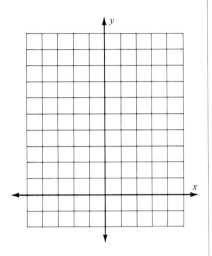

3. $y = 2x^2 - 1$ (Use the results of Problem 1.)

4. $y = 2x^2 - 2$ (Use the results of Problem 1.)

3. _____

4. _____

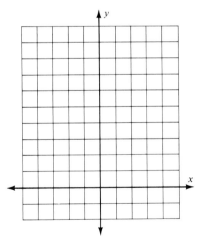

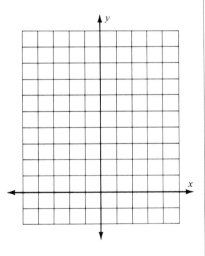

5. $y = -2x^2 + 2$ (Use the results of Problem 1.)

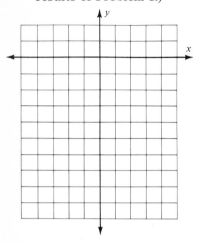

6. $y = -2x^2 - 1$ (Use the results of Problem 1.)

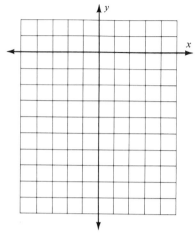

7. $y = -2x^2 - 2$ (Use the results of Problem 1.)

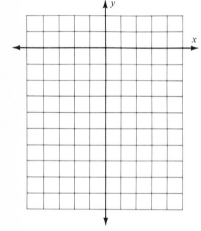

8. $y = -2x^2 + 1$ (Use the results of Problem 1.)

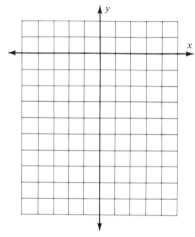

9. $y = (x - 2)^2$

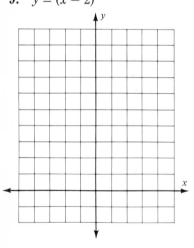

10. $y = (x - 2)^2 + 2$

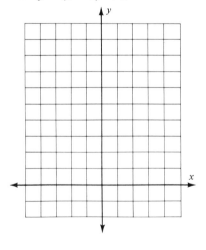

5. _____

6. _____

7. _____

8. _____

9. _____

10. _____

ANSWER   (to problem on page 634)

4.

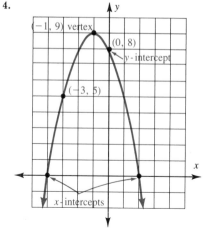

11. $y = (x - 2)^2 - 2$

12. $y = (x - 2)^2 - 1$

11. _____

12. _____

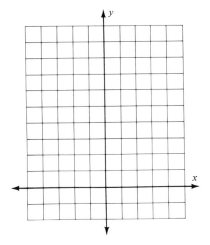

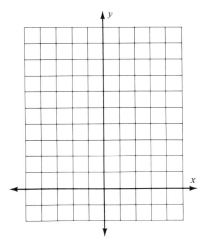

13. $y = -(x - 2)^2$

14. $y = -(x - 2)^2 + 2$

13. _____

14. _____

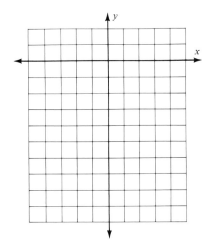

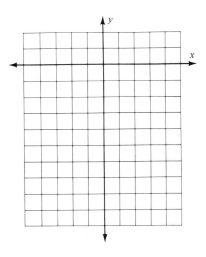

15. $y = -(x - 2)^2 - 2$

16. $y = -(x - 2)^2 - 1$

15. _____

16. _____

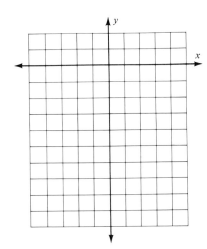

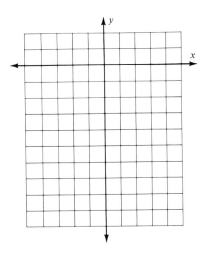

EXERCISE 9.5

In Problems 17 through 22 find the x-intercepts, the y-intercept and the vertex; then draw the graph.

17. $y = x^2 + 4x + 3$

18. $y = x^2 - 4x + 4$

19. $y = x^2 + 2x - 3$

20. $y = -x^2 - 4x - 3$

21. $y = -x^2 + 4x - 3$

22. $y = -x^2 - 2x + 3$

17. _____

18. _____

19. _____

20. _____

21. _____

22. _____

The graph shown in the margin looks somewhat like one-half of a parabola. It indicates the time it takes to travel from the earth to the moon at different speeds and will be used in Problems 23 through 26.

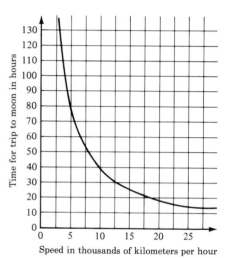

Speed in thousands of kilometers per hour

23. How long does the trip take if you travel at 10,000 kilometers per hour (km/hr)?

24. How long does the trip take if you travel at 25,000 km/hr?

25. Apollo 11 averaged about 6400 km/hr when returning from the moon. How long was the trip?

26. If you wish to make the trip in 50 hr, what should the speed be?

23. _____

24. _____

25. _____

26. _____

✓ SKILL CHECKER

Find.

27. $3^2 + 4^2$

28. $5^2 + 12^2$

29. $\dfrac{120}{0.003}$

30. $\dfrac{150}{0.005}$

27. _____

28. _____

29. _____

30. _____

9.5 USING YOUR KNOWLEDGE

The ideas studied in this section can be used in business to find ways of maximizing profits. For example, if a manufacturer can produce a certain item for $10 each, and then sell them for x dollars, the profit per item will be $x - 10$ dollars. If it is then estimated that consumers will buy $60 - x$ items per month, the total profit will be

$$\text{Total profit} = \begin{pmatrix}\text{number of} \\ \text{item sold}\end{pmatrix}\begin{pmatrix}\text{profit per} \\ \text{item}\end{pmatrix}$$

$$= (60 - x)(x - 10)$$

The graph for the total profit is the parabola shown.

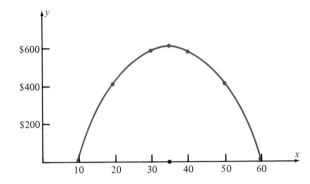

When will the profits be at a maximum? When the manufacturer produces 35 items. Note that 35 is exactly halfway between

$$10 \quad \text{and} \quad 60 \qquad \left(\frac{10 + 60}{2} = 35\right)$$

At this price, the total profits will be

$$TP = (60 - 35)(35 - 10)$$
$$= (25)(25)$$
$$= \$625$$

1. What will be the price that will maximize the profits of a certain item costing \$20 each if $60 - x$ items are sold each month (x the selling price of each item)?

2. Sketch the graph of the resulting parabola.

3. What will the maximum profits be?

1.

2.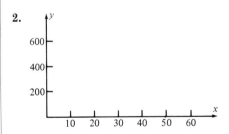

3. _____

9.6 APPLICATIONS

B.C.   **by johnny hart**

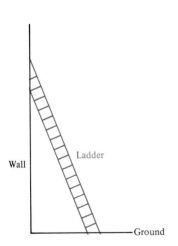

B.C. by permission of Johnny Hart and Field Enterprises, Inc.

REVIEW

Before starting this section you should know:

1. How to find the square of a number.
2. How to divide by a decimal.
3. How to solve a quadratic equation by factoring (Section 5.3).

OBJECTIVES

You should be able to:

A. Use the Pythagorean Theorem to solve right triangles.

B. Solve word problems involving quadratics.

In this very last section, we discuss several applications involving quadratic equations. One of them pertains to the theorem mentioned in the cartoon. This theorem, first proved by Pythagoras, will be stated next.

A. PYTHAGOREAN THEOREM

PYTHAGOREAN THEOREM

The square of the hypotenuse of a right triangle* equals the sum of the squares of the other two sides.

The relationship is illustrated in the margin, where a, b, and h represent the lengths of the sides of the given triangle. Thus if the lengths of the sides of a triangle are 3 and 4 units long, then the hypotenuse h is such that

$$3^2 + 4^2 = h^2$$
$$9 + 16 = h^2$$
$$h^2 = 25 = \pm\sqrt{25} = \pm 5$$

Since h represents length, h cannot be negative, so the length of the hypotenuse is 5 units.

Pythagorean Theorem

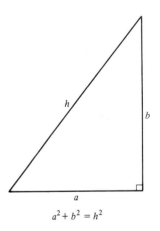

$$a^2 + b^2 = h^2$$

The answer $h = -5$ does not fit the problem (distances are always positive), so it is discarded.

---

*A right triangle is a triangle containing one right angle (90°) and the hypotenuse is the side opposite the 90° angle.

EXAMPLE 1 In the diagram on page 641 the distance from the wall to the base of the ladder is 5 ft and the distance from the floor to the top of the ladder is 12 ft. Can you find the length of the ladder?

In the diagram on page 641

Problem 1 Find the length of the ladder if the distance from the wall to the base of the ladder is 5 ft, but the height to the top of the ladder is 13 ft.

Solution We use the RSTUV procedure given earlier.

1. Read the problem carefully and decide what is asked for.
2. Select a variable to represent the length of the ladder; we choose L to be this length.
3. Translate the problem (by using the diagram below, which gives the appropriate dimensions):

$$5^2 + 12^2 = L^2$$

4. Use algebra to solve this equation. Since

$$5^2 + 12^2 = L^2$$
$$25 + 144 = L^2$$
$$L^2 = 169$$
$$L = \pm 13$$

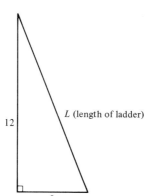

L (length of ladder)

12

5

Since L is the length of the ladder, L is positive, so $L = 13$.

5. Verify that this solution is correct ($5^2 + 12^2 = 13^2$ because $25 + 144 = 169$).

B. WORD PROBLEMS

Many applications of quadratic equations come from the field of engineering. For example, it is known that the pressure p (in pounds per square foot, lb/ft^2) exerted by a wind blowing at v mph is given by the equation

$$p = 0.003v^2 \quad \text{Note that if } p \text{ is a number the highest exponent of the variable is 2, making the equation a quadratic.}$$

If the pressure p is known, the equation $p = 0.003v^2$ is an example of a *quadratic equation*. We show how this information is used in the next example.

EXAMPLE 2 A wind pressure gauge at Commonwealth Bay registered 120 lb/ft^2 during a gale. What was the wind speed at that time?

Problem 2 Find the wind speed at the Golden Gate Bridge when a pressure gauge is registering 7.5 lb/ft^2.

Solution

1. Read the problem carefully and decide what is asked for.
2. Select a variable to represent the unknown. Since we know that $p = 120$, and that

$$p = 0.003v^2$$

we let v be the wind speed.

3. Now, substitute for p to obtain

$$0.003v^2 = 120$$

4. Use algebra to solve this equation. (The easiest way is to divide by 0.003 first.) We then have

$$v^2 = \frac{120}{0.003}$$

$$v^2 = \frac{120 \cdot 1000}{0.003 \cdot 1000} = \frac{120{,}000}{3} = 40{,}000$$

$$v = \pm\sqrt{40{,}000} = +200$$

You can also divide 120 by 0.003 obtaining

$$\begin{array}{r} 40{,}000 \\ 0.003\overline{)120.000} \\ 12 \\ \hline \end{array}$$

That is, the wind speed v was 200 mi/h. (We discard the negative answer as not suitable.) ▲

It is shown in physics that when an object is dropped or thrown downward with an initial velocity v_0, the distance d (in meters, m) traveled by the object in t seconds is given by the formula

$$d = 5t^2 + v_0 t$$ This is an approximate formula. A more exact formula is $d = 4.9t^2 + v_0 t$.

Suppose an object is dropped from a height of 125 m. How long would it be before the object hits the ground? The solution of this problem is given in the next example.

EXAMPLE 3 Use the formula $d = 5t^2 + v_0 t$ to find the time it takes an object dropped from a height of 125 m to hit the ground.

Solution Since the object is dropped, $v_0 = 0$ and $d = 125$; we then have

$$125 = 5t^2$$
$$5t^2 = 125$$
$$t^2 = 25$$ Dividing by 5
$$t = \pm\sqrt{25} = \pm 5$$

Since the time must be positive, the correct answer is $t = 5$; that is, it takes 5 sec for the object to hit the ground. ▲

Finally, quadratic equations are also used in business. Perhaps you have heard of the term to "break even." In business, the break-even point is the point at which the revenue R equals the cost C of the manufactured goods. In symbols,

$$R = C$$

Now, suppose a company produces x (thousand) items. If each item sells for \$3, the revenue R (in thousands of dollars) can be expressed by

$$R = 3x$$

If the manufacturing cost C (in thousands of dollars) is

$$C = x^2 - 3x + 5$$

and it is known that the company always produces more than 1000

Problem 3 Find the time it takes an object dropped from a height of 245 m to hit the ground.

items, the break-even point occurs when

$$R = C$$
$$3x = x^2 - 3x + 5$$
$$x^2 - 6x + 5 = 0$$
$$(x - 5)(x - 1) = 0 \qquad \text{Factoring}$$
$$x - 5 = 0 \quad \text{or} \quad x - 1 = 0 \quad \text{Using the principle of zero products}$$
$$x = 5 \quad \text{or} \quad x = 1$$

But since it is known that the company produces more than 1 (thousand) items, the break-even point occurs when the company manufactures 5 (thousand) items.

EXAMPLE 4   Find the number of items that have to be produced to break even when the revenue R (in thousands of dollars) of a company is given by $R = 3x$ and its costs (also in thousands of dollars) are $C = 2x^2 - 3x + 4$.

Solution   In order to break even,

$$R = C$$
$$3x = 2x^2 - 3x + 4$$
$$2x^2 - 6x + 4 = 0$$
$$x^2 - 3x + 2 = 0 \qquad \text{Dividing by 2}$$
$$(x - 1)(x - 2) = 0 \qquad \text{Factoring}$$
$$x - 1 = 0 \quad \text{or} \quad x - 2 = 0 \quad \text{Using the principle of zero products}$$
$$x = 1 \quad \text{or} \quad x = 2$$

This means that the company breaks even when they produce either 1 (thousand) or 2 (thousand) items.

Problem 4   Find the number of items that have to be produced to break even when the revenue R of a company (in thousands of dollars) is given by $R = 4x$ and the cost C (also in thousands of dollars) is

$$C = x^2 - 3x + 6$$

NAME

CLASS

SECTION

ANSWERS

A. If a and b represent the lengths of the sides of a right triangle, and h represents the length of the hypotenuse, find the missing side.

1. $a = 12$, $h = 13$

2. $a = 4$, $h = 5$

3. $a = 5$, $h = 15$

4. $a = b$, $h = 4$

5. $b = 3$, $a = \sqrt{6}$

6. $b = 7$, $a = 9$

7. $a = \sqrt{5}$, $b = 2$

8. $a = 3$, $b = \sqrt{7}$

9. $h = \sqrt{13}$, $a = 3$

10. $h = \sqrt{52}$, $a = 4$

1. _____

2. _____

3. _____

4. _____

5. _____

6. _____

7. _____

8. _____

9. _____

10. _____

B. Solve.

11. How long is a wire extending from the top of a 40-ft telephone pole to a point on the ground 30 ft from the foot of the pole?

12. Repeat Problem 11 if the point on the ground is 16 ft away from the pole.

13. The pressure p (in pounds per square foot) exerted on a surface by a wind blowing at v miles per hour is given by

$p = 0.003v^2$

Find the wind speed when a wind pressure gauge recorded a pressure of 30 lb/ft^2.

14. Repeat Problem 13 if the pressure is 2.7 lb/ft^2.

15. An object is dropped from a height of 320 m. How many seconds does it take for this object to hit the ground? (See Example 3.)

16. Repeat Problem 15 if the object is dropped from a height of 45 m.

17. An object is thrown downward with an initial velocity of 3 m/sec. How long does it take for the object to travel 8 m? (See Example 3.)

18. The revenue of a company is given by $R = 2x$, where x is the number of units produced (in thousands). If the cost $C = 4x^2 - 2x + 1$, how many units have to be produced before the company breaks even?

19. Repeat Problem 18 if the revenue $R = 5x$ and the cost $C = x^2 - x + 9$.

11. _____

12. _____

13. _____

14. _____

15. _____

16. _____

17. _____

18. _____

19. _____

20. If P dollars are invested at r percent compounded annually, then at the end of 2 yr the amount will have grown to $A = P(1 + r)^2$. At what rate of interest r will $1000 grow to $1210 in 2 yr? (*Hint:* $A = 1210$ and $P = 1000$.)

20. _____

21. A rectangle is 2 ft wide and 3 ft long. Each side is increased by the same amount to give a rectangle with twice the area of the original one. Find the dimensions of the new rectangle. *Hint:* Let x feet be the amount by which each side is increased.

21. _____

22. The hypotenuse of a right triangle is 4 cm longer than the shortest side and 2 cm longer than the remaining side. Find the dimensions of the triangle.

22. _____

23. Repeat Problem 22 if the hypotenuse is 16 cm longer than the shortest side and 2 cm longer than the remaining side.

23. _____

24. The square of a certain positive number is 5 more than 4 times the number itself. Find this number.

24. _____

✓ **SKILL CHECKER**

Solve.

25. $A + 20 = 32$ **26.** $A + 9 = 17$

25. _____

26. _____

27. $18 + B = 30$ **28.** $30 + B = 45$

27. _____

28. _____

9.6 USING YOUR KNOWLEDGE

There are many formulas that require the use of some of the techniques we have studied when solving for certain unknowns in these formulas. For example, consider the equation

$$v^2 = 2gh$$

How can we solve for v in this equation? In the usual way, of course! Thus we have:

$$v^2 = 2gh$$
$$v = \pm\sqrt{2gh}$$

Since the speed is positive, we simply write

$$v = \sqrt{2gh}$$

Use this idea to solve for the indicated variables in the given formulas.

1. Solve for c in the formula $E = mc^2$.

1. _____

2. Solve for r in the formula $A = \pi r^2$.

2. _____

3. Solve for r in the formula $F = \dfrac{GMm}{r^2}$.

3. _____

4. Solve for v in the formula $KE = \dfrac{1}{2}mv^2$.

4. _____

5. Solve for P in the formula $I = kP^2$.

5. _____

SUMMARY

SECTION	ITEM	MEANING	EXAMPLE
9.1	Quadratic equation	An equation that can be written in the form $ax^2 + bx + c = 0$	$x^2 = 16$
9.1A	$\sqrt{a}$, the principal square root of a, $a \geq 0$	A nonnegative number b such that $b^2 = a$	$\sqrt{16} = 4$
9.1A	Irrational number	A number that cannot be written in the form $\dfrac{a}{b}$, where a and b are integers and b is not 0	$\sqrt{2}$, $\sqrt{5}$, and $3\sqrt{7}$
9.1A	Real numbers	The rational and the irrational numbers	-7, $-\dfrac{3}{2}$, 0, $\sqrt{5}$, and 17 are real numbers.
9.1A	Quotient of radicals	$\sqrt{\dfrac{a}{b}} = \dfrac{\sqrt{a}}{\sqrt{b}}$, a and b positive	$\sqrt{\dfrac{3}{4}} = \dfrac{\sqrt{3}}{\sqrt{4}} = \dfrac{\sqrt{3}}{2}$
9.2A	Product of radicals	$\sqrt{a \cdot b} = \sqrt{a} \cdot \sqrt{b}$	$\sqrt{4 \cdot 3} = 2 \cdot \sqrt{3}$
9.2B	Rationalizing the denominator	Writing a fraction as an equivalent one with no radicals in the denominator	$\dfrac{2}{\sqrt{3}} = \dfrac{2 \cdot \sqrt{3}}{3}$
9.4	The quadratic formula	The solutions of $ax^2 + bx + c = 0$ are $x = \dfrac{-b \pm \sqrt{b^2 - 4ac}}{2a}$	The solutions of $2x^2 + 3x + 1 = 0$ are $x = \dfrac{-3 \pm \sqrt{3^2 - 4 \cdot 2 \cdot 1}}{2 \cdot 2}$, that is, -1 or $-\dfrac{1}{2}$.
9.4A	Standard form of a quadratic equation	$ax^2 + bx + c = 0$ is the standard form of a quadratic equation.	The standard form of $x^2 + 2x = -11$ is $x^2 + 2x - 11 = 0$.
9.5	Graph of a quadratic equation	The graph of a quadratic equation is a parabola.	
9.6	Pythagorean Theorem	The square of the hypotenuse of a right triangle equals the sum of the squares of the other two sides.	If a and b are the lengths of the sides and h is the length of the hypotenuse, $a^2 + b^2 = h^2$.

ANSWER (to problem on page 644)
4. 1 (thousand) or 6 (thousand)

ANSWERS

(If you need help with these exercises, look at the section indicated in brackets.)

1. [9.1A] Solve.
   a. $x^2 = 1$
   b. $x^2 = 100$
   c. $x^2 = 81$

2. [9.1A] Solve.
   a. $16x^2 - 25 = 0$
   b. $25x^2 - 9 = 0$
   c. $64x^2 - 25 = 0$

3. [9.1A] Solve.
   a. $7x^2 + 36 = 0$
   b. $8x^2 + 49 = 0$
   c. $3x^2 + 81 = 0$

4. [9.1B] Solve.
   a. $49(x + 1)^2 - 3 = 0$
   b. $25(x + 2)^2 - 2 = 0$
   c. $16(x + 1)^2 - 5 = 0$

5. [9.2A] Solve.
   a. $x^2 - 32 = 0$
   b. $x^2 - 50 = 0$
   c. $x^2 - 48 = 0$

6. [9.2B] Solve.
   a. $5x^2 = 36$
   b. $3x^2 = 49$
   c. $7x^2 = 16$

7. [9.2B] Solve.
   a. $5x^2 - 32 = 0$
   b. $7x^2 - 8 = 0$
   c. $5x^2 - 48 = 0$

8. [9.2C] Simplify.
   a. $4\sqrt{3} + 5\sqrt{3} - 5\sqrt{3}$ and $\sqrt{24} + \sqrt{150}$
   b. $7\sqrt{5} + 2\sqrt{5} - 3\sqrt{5}$ and $\sqrt{18} + \sqrt{50}$
   c. $9\sqrt{6} + 3\sqrt{6} - 2\sqrt{6}$ and $\sqrt{48} + \sqrt{75}$

9. [9.2C] Simplify
   a. $\sqrt{10}(\sqrt{50} - \sqrt{10})$
   b. $\sqrt{5}(\sqrt{60} - \sqrt{5})$
   c. $\sqrt{2}(\sqrt{300} - \sqrt{2})$

1. a. _____
   b. _____
   c. _____

2. a. _____
   b. _____
   c. _____

3. a. _____
   b. _____
   c. _____

4. a. _____
   b. _____
   c. _____

5. a. _____
   b. _____
   c. _____

6. a. _____
   b. _____
   c. _____

7. a. _____
   b. _____
   c. _____

8. a. _____
   b. _____
   c. _____

9. a. _____
   b. _____
   c. _____

10. [9.3] Find the missing term in the given expression.

    a. $(x + 3)^2 = x^2 + 6x + \square$

    b. $(x + 7)^2 = x^2 + 14x + \square$

    c. $(x + 6)^2 = x^2 + 12x + \square$

10. **a.** _____

    b. _____

    c. _____

11. [9.3] Find the missing terms in the given expression.

    a. $x^2 - 6x + \square = (\quad)^2$

    b. $x^2 - 10x + \square = (\quad)^2$

    c. $x^2 - 12x + \square = (\quad)^2$

11. **a.** _____

    b. _____

    c. _____

12. [9.3] Find the number that must divide each term in the given equation so that the equation can be solved by the method of completing the square.

    a. $7x^2 - 14x = -4$

    b. $6x^2 - 18x = -2$

    c. $5x^2 - 15x = -3$

12. **a.** _____

    b. _____

    c. _____

13. [9.3] Find the term that must be added to both sides of the given equation so that the equation can be solved by the method of completing the square.

    a. $x^2 - 4x = -4$

    b. $x^2 - 6x = -9$

    c. $x^2 - 12x = -3$

13. **a.** _____

    b. _____

    c. _____

14. [9.4B] Solve.

    a. $dx^2 + ex + f = 0$

    b. $gx^2 + hx + i = 0$

    c. $jx^2 + kx + m = 0$

14. **a.** _____

    b. _____

    c. _____

15. [9.4B] Solve.

    a. $2x^2 - x - 1 = 0$

    b. $2x^2 - 2x - 5 = 0$

    c. $2x^2 - 3x - 3 = 0$

15. **a.** _____

    b. _____

    c. _____

16. [9.4B] Solve.

    a. $3x^2 - x = 1$

    b. $3x^2 - 2x = 2$

    c. $3x^2 - 3x = 2$

16. **a.** _____

    b. _____

    c. _____

17. [9.4B] Solve.

    a. $9x = x^2$

    b. $4x = x^2$

    c. $25x = x^2$

17. **a.** _____

    b. _____

    c. _____

18. [9.4B] Solve.

    a. $\dfrac{x^2}{9} - x = -\dfrac{4}{9}$

    b. $\dfrac{x^2}{5} - \dfrac{x}{2} = \dfrac{3}{10}$

    c. $\dfrac{x^2}{3} + \dfrac{x}{6} = -\dfrac{1}{2}$

18. **a.** _____

    b. _____

    c. _____

19. [9.5] Graph.
    a. $y = -(x - 2)^2 + 1$
    b. $y = -(x - 1)^2 + 2$
    c. $y = -(x - 2)^2 + 2$

19.

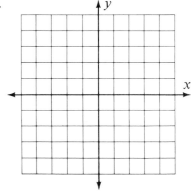

20. [9.5] The distance d (in meters, m) traveled by an object thrown downward with an initial velocity v_0 after t seconds is given by the formula

$$d = 5t^2 + v_0 t$$

Find the number of seconds it takes an object to hit the ground if the object is dropped from each height.
    a. 125 m
    b. 245 m
    c. 320 m

20. **a.** _____
    b. _____
    c. _____

ANSWERS

(*Answers on page* 654)

1. Solve $x^2 = 64$.

2. Solve $49x^2 - 25 = 0$.

3. Solve $6x^2 + 49 = 0$.

4. Solve $36(x + 1)^2 - 7 = 0$.

5. Solve $x^2 - 75 = 0$.

6. Solve $6x^2 = 49$.

7. Solve $3x^2 - 50 = 0$.

8. Simplify.
   a. $8\sqrt{7} + 2\sqrt{7} - 4\sqrt{7}$
   b. $\sqrt{12} + \sqrt{48}$

9. Multiply $\sqrt{5}(\sqrt{40} - \sqrt{5})$.

10. Find the missing term in the expression $(x + 4)^2 = x^2 + 8x + \boxed{}$.

11. The missing terms in the expressions $x^2 - 8x + \boxed{} = ()^2$ are _____ and _____, respectively.

12. To solve the equation $8x^2 - 32x = -5$ by completing the square, the first step will be to divide each term by _____.

13. To solve the equation $x^2 - 8x = -16$ by completing the square, one has to add _____ to both sides of the equation.

14. The solution of $ax^2 + bx + c = 0$ is _____.

15. Solve $2x^2 - 3x - 2 = 0$.

16. Solve $x^2 = 3x - 2$.

17. Solve $16x = x^2$.

18. Solve $\dfrac{x^2}{2} + \dfrac{5}{4}x = -\dfrac{1}{2}$.

19. Graph the equation $y = -(x - 1)^2 + 1$.

20. The formula $d = 5t^2 + v_0 t$ gives the distance d (in meters) an object thrown downward with an initial velocity v_0 will have gone after t seconds. How long would it take an object dropped from a distance of 180 m to hit the ground?

1. _____
2. _____
3. _____
4. _____
5. _____
6. _____
7. _____
8. a. _____
   b. _____
9. _____
10. _____
11. _____
12. _____
13. _____
14. _____
15. _____
16. _____
17. _____
18. _____
19. _____
20. _____

IF YOU MISSED QUESTION	SECTION	EXAMPLES	PAGE	ANSWERS
1	9.1	1	589	1. $x = \pm 8$
2	9.1	1, 2	589, 591	2. $x = \pm\dfrac{5}{7}$
3	9.1	3	592	3. No real-number solution
4	9.1	4	593	4. $x = -1 \pm \dfrac{\sqrt{7}}{6}$
5	9.2	1	600	5. $x = \pm 5\sqrt{3}$
6	9.2	2	602	6. $x = \pm\dfrac{7\sqrt{6}}{6}$
7	9.2	3	603	7. $x = \pm\dfrac{5\sqrt{6}}{3}$
8	9.2	4, 5	603, 604	8. a. $6\sqrt{7}$ b. $6\sqrt{3}$
9	9.2	6	604	9. $10\sqrt{2} - 5$
10	9.3	1	611	10. 16
11	9.3	2	612	11. $16, (x-4)^2$
12	9.3	3, 4	613, 614	12. 8
13	9.3	3, 4	613, 614	13. 16
14	9.4	1	621	14. $x = \dfrac{-b \pm \sqrt{b^2 - 4ac}}{2a}$
15	9.4	1	621	15. $x = 2,\ x = -\dfrac{1}{2}$
16	9.4	2	621	16. $x = 1,\ x = 2$
17	9.4	3	622	17. $x = 0,\ x = 16$
18	9.4	4	623	18. $x = -2,\ x = -\dfrac{1}{2}$
19	9.5	1, 2, 3	630, 631, 632	19.
20	9.6	1, 2, 3	642, 643	20. 6 sec

CHAPTER 10

APPLIED GEOMETRY

In this chapter we introduce the basic concepts of geometry as they relate to the study of algebra. The concept of lines and angles and the use of algebra in finding the length of a line segment, the measure of an unknown angle, and the solution of a triangle are discussed. We also study the perimeter, area, and volume of several common geometric figures.

NAME

CLASS

SECTION

(*Answers on page* 660)

ANSWERS

1. If $AB = 10$ and $AC = 28$, find the length of BC.

1. _____

2. If $\angle A$ and $\angle B$ are complementary angles and $m\angle A = 8\,m\angle B$, find $m\angle A$ and $m\angle B$.

2. _____

3. If $\angle A$ and $\angle B$ are supplementary angles and $m\angle A = 3\,m\angle B$, find $m\angle A$ and $m\angle B$.

3. _____

Use the figure in Problems 4 through 6.

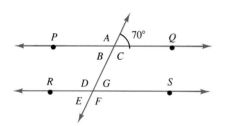

Lines PQ and RS are parallel.

4. Find $m\angle C$.

4. _____

5. Find $m\angle E$.

5. _____

6. Find $m\angle D$.

6. _____

7. In a triangle RST, $m\angle R = 53°$ and $m\angle T = 50°$. Find $m\angle S$.

7. _____

8. Find the perimeter of a rectangle 30.7 cm long and 20.8 cm wide.

8. _____

9. The perimeter of a rectangle is 228 in. If the rectangle is 80 in. long, how wide is it?

9. _____

10. The circumference of a circle is 21.98 ft. Find its diameter. (Use $\pi = 3.14$.)

10. _____

11. The area of a triangle is 14 in.^{2}. If the height h is 3.5 in., find the base b.

11. _____

12. The area of a circle is 113.04 in.^{2}. Find its diameter. ($\pi \approx 3.14$)

12. _____

13. The length of one of the legs in a right triangle is 7 in. If the hypotenuse is 9 in. long, find the length of the other leg.

13. _____

14. One leg of an isosceles right triangle is 30 in. long. Find the perimeter of the triangle.

14. _____

15. The shorter leg of a 30°-60°-90° triangle is 4.4 in. long. Find the perimeter.

15. _____

16. Triangles ABC and XYZ are similar with $\angle A = \angle X$ and $\angle B = \angle Y$. If AB, BC, and AC are 5 in., 7 in., and 8 in. long, respectively, and XY is 13 in. long, find the length of XZ. (Answer to the nearest tenth.)

16. _____

17. Find the volume of a cube 4 in. on each side.

17. _____

18. A water tank is in the shape of a 10-ft-high cone (vertex down) surmounted by a 40-ft-high cylinder. If the diameter of the cylinder and the cone is 16 ft, what is the capacity of the tank in gallons? (Use $\pi = 3.14$; 1 ft^3 = 7.5 gal.)

18. _____

19. Find the volume of a sphere 6 ft in diameter. (Use $\pi = 3.14$; answer to the nearest tenth.)

19. _____

20. The volume of a cylindrical can is 50.24 in.^{3} If the diameter is 4 in., find the height. (Use $\pi = 3.14$.)

20. _____

IF YOU MISSED QUESTION	SECTION	EXAMPLES	PAGE	ANSWERS
1	10.1	1	662	1. 18
2	10.1	2	664	2. 80°, 10°
3	10.1	3	664	3. 135°, 45°
4	10.1	4	665	4. 110°
5	10.1	4	665	5. 70°
6	10.1	4	665	6. 110°
7	10.1	5	666	7. 77°
8	10.2	1	672	8. 103 cm
9	10.2	2	673	9. 34 in.
10	10.2	3	673	10. 7 ft
11	10.2	4	675	11. 8 in.
12	10.2	5	675	12. 12 in.
13	10.3	1	682	13. $4\sqrt{2}$ in.
14	10.3	2	682	14. $(60 + 30\sqrt{2})$ in.
15	10.3	3	683	15. $(13.2 + 4.4\sqrt{3})$ in.
16	10.3	4	684	16. 20.8 in.
17	10.4	1	689	17. 64 in.^{3}
18	10.4	2	690	18. 65,312 gal
19	10.4	3	690	19. 113 ft^3
20	10.4	4	691	20. 4 in.

10.1 LINES AND ANGLES

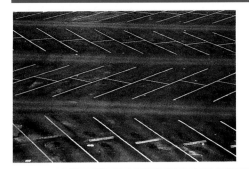

REVIEW

Before starting this section you should know how to solve linear equations.

OBJECTIVES

You should be able to:

A. Find the length of a line segment.
B. Find the measure of an angle formed by intersecting lines.
C. Find the angles of a triangle.

Do you know what the word geometry means? In Greek it means earth (*geo*) measurements (*metry*). Why do we study geometry? A student of Euclid, a Greek mathematician, asked, "What shall I gain from learning these things?" when he mastered his first geometry theorem. Euclid called a slave and said, "Give him a penny, since he must make gain from what he learns." However, Euclid aside, geometry occurs and is used in many areas such as architecture, design, geology, physics, and chemistry.

A. LINES

The basic elements in geometry are **points, lines,** and **planes.** We will not state formal definitions for these terms but will simply give you an idea of their meaning.

> A *point* is a location in space. A point has no breadth, no width, and no length.

We can picture a point A as a small dot, as in Fig. 1.

> A *line* is a set of points, and each point of the set is said to be on the line. A line extends without end in both directions.

A parking lot, for example, shows parallel lines, perpendicular lines, and intersecting lines.

Lines in a plane can be either *parallel* or *intersecting*. **Parallel** lines never meet; the distance between them is always the same. On the other hand, **intersecting** lines cross at some point in the plane.

Here are some properties that will help clarify what is meant by the word *line*:

1. Two distinct points, A and B, determine a line AB (Fig. 2). In other words, *one and only one line can be drawn through the two points*. If P and Q are points on the line AB, then the line PQ is the same as the line AB. Thus, a line may be named by any two of its points. We designate the line AB by the symbol $\overleftrightarrow{AB}$ (read, "the line AB"). Points on the same line are said to be **collinear.** If A, B, P, and Q are collinear, then $\overleftrightarrow{AB} = \overleftrightarrow{PQ}$.

FIG. 1

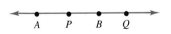

FIG. 2
The line AB

2. *Any point A on a line separates the line into three sets: the point A itself and two* **half-lines,** *one on each side of A. The half-lines do not include the point A, although A is regarded as an endpoint of both (see Fig. 3). The two half-lines may be termed half-line AB and half-line AC, respectively, and are designated by $\overset{\circ}{\overrightarrow{AB}}$ and $\overset{\circ}{\overrightarrow{AC}}$. The open circle at the end of the arrow in the symbol $\overset{\circ}{\overrightarrow{AB}}$ indicates that the half-line does* not *include the point A.*

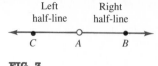

FIG. 3

It is sometimes convenient to consider the set of points consisting of a half-line and its endpoint. Such a set is called a **ray.** The ray consisting of $\overset{\circ}{\overrightarrow{AB}}$ and the point A will be designated by $\overset{\bullet}{\overrightarrow{AB}}$. The ray consisting of the half-line $\overset{\circ}{\overrightarrow{BA}}$ and the point B will be denoted by $\overset{\bullet}{\overrightarrow{BA}}$. Note that $\overset{\bullet}{\overrightarrow{AB}} \neq \overset{\bullet}{\overrightarrow{BA}}$ because rays are named using the endpoint first.

A **line segment AB** consists of the points A and B and that portion of the line AB that lies between A and B. We designate this segment by $\overset{\bullet\bullet}{AB}$. Note that $\overset{\bullet\bullet}{AB} = \overset{\bullet\bullet}{BA}$.

Figure 4 shows the figures and notations we have described.

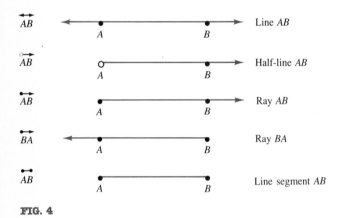

FIG. 4

The **length** of a line segment is the distance between the ends of the line segment.

For the line segment shown $AB = 6$, $BC = 2$, and $AC = AB + BC = 6 + 2 = 8$. When we give no units of measure for the lengths, the distances are assumed to be in the same units.

EXAMPLE 1 If $AB = 20$ and $AC = 32$, find the length of BC.

Problem 1 If $AB = 25$ and $AC = 32$, find the length of BC.

Solution The equation for the lengths of the line segments is:

$$AB + BC = AC$$
$$20 + BC = 32 \quad \text{Substituting } \textbf{20} \text{ for } AB, \textbf{32} \text{ for } AC$$
$$BC = 12 \quad \text{Subtracting 20 on both sides}$$

B. ANGLES FORMED BY INTERSECTING LINES

> An **angle** is the figure formed by two rays with a common end-point called the **vertex.**

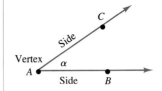

The two rays $\overrightarrow{AB}$ and $\overrightarrow{AC}$ are called the **sides** of the angle. We use the symbol $\angle$ (read "angle") in naming angles. Thus, the angle in the figure can be named $\angle \alpha$ (angle alpha), $\angle BAC$, or $\angle CAB$. (Note that the middle letter designates the vertex.) Angles are measured in degrees (°) and the measure of an angle is denoted by $m \angle A$. One complete revolution is 360° (360 degrees). One-half revolution is 180°.

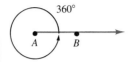

A complete revolution

> A 180° angle is called a **straight** angle. (See margin.)

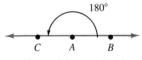

The straight angle *BAC*

One-quarter of a revolution is 90°.

> A 90° angle is called a **right** angle and is represented by the symbol ∟. (See margin.)

The right angle *XYZ*

> **Complementary angles** are two angles whose sum is 90°.

In the figure, $m \angle A = 60°$ and $m \angle B = 30°$, so that $m \angle A + m \angle B = 30° + 60° = 90°$. Thus, $\angle A$ and $\angle B$ are *complementary* angles.

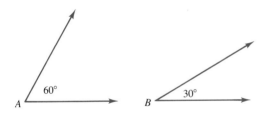

> **Supplementary angles** are two angles whose sum is 180°.

In the figure $m \angle A = 130°$ and $m \angle B = 50°$, so that $m \angle A + m \angle B = 180°$. Thus, $\angle A$ and $\angle B$ are *supplementary* angles.

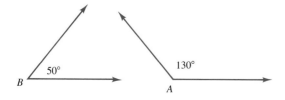

EXAMPLE 2 If $\angle A$ and $\angle B$ are complementary angles and $m \angle A = 2m \angle B$—that is, the measure of $\angle A$ is twice that of $\angle B$—find $m \angle A$ and $m \angle B$.

Solution Since $\angle A$ and $\angle B$ are complementary, their sum is 90°. Thus,

$$m \angle A + m \angle B = 90°$$
$$2m \angle B + m \angle B = 90° \quad \text{Substituting } \textbf{2m} \angle \textbf{B} \text{ for } m \angle A$$
$$3m \angle B = 90°$$
$$m \angle B = 30° \quad \text{Dividing both sides by 3}$$

Since $m \angle A = 2m \angle B, m \angle A = 60°$. Note that you can check the answer by making sure that the sum of $m \angle A$ and $m \angle B$ is 90° (which it is).

EXAMPLE 3 If $\angle A$ and $\angle B$ are supplementary angles and $m \angle A = 3m \angle B$, find $m \angle A$ and $m \angle B$.

Solution $\angle A$ and $\angle B$ are supplementary, so their sum is 180°. Thus,

$$m \angle A + m \angle B = 180°$$
$$3m \angle B + m \angle B = 180° \quad \text{Substituting } \textbf{3m} \angle \textbf{B} \text{ for } m \angle A$$
$$4m \angle B = 180°$$
$$m \angle B = 45° \quad \text{Dividing both sides by 4}$$

Since $m \angle A = 3m \angle B, m \angle A = 135°$. You can check the answer by making sure that the sum of $m \angle A$ and $m \angle B$ is 180° (which it is). ▲

Now, let us talk about railroads and railroad crossings. The sign used to denote a railroad crossing is shown in the margin; the marked angles are called **vertical angles.** In the figure $m \angle B = m \angle D$.

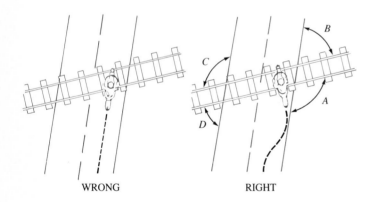

WRONG   RIGHT

If you drive a motorcycle, you should look closely at the figure; it tells you to cross the railroad tracks at right angles (because there is less danger of a wheel catching in the tracks). Now, look at the angles A, B, C, and D that we have marked on the right side of the figure. What can you say about the angles B and D? They are of equal measure, of course (and so are angles A and C). As you can see, the railroad track crosses the two parallel black lines in the figure. In geometry, a line that crosses two or more other lines is called a **transversal.** Thus, each railroad track is a transversal of the pair of parallel black lines. If a transversal crosses a pair of parallel lines, some of the resulting angles are of equal measure. The exact relationships are as follows.

Problem 2 If $\angle A$ and $\angle B$ are complementary angles and $m \angle A = 4m \angle B$, find $m \angle A$ and $m \angle B$.

Problem 3 If $\angle A$ and $\angle B$ are supplementary angles and $m \angle A = 5m \angle B$, find $m \angle A$ and $m \angle B$.

ANSWER (to problem on page 662)
1. $BC = 7$

CORRESPONDING ANGLES ARE OF EQUAL MEASURE

$$m \angle A = m \angle E \qquad m \angle B = m \angle F$$
$$m \angle C = m \angle G \qquad m \angle D = m \angle H$$

ALTERNATE INTERIOR ANGLES ARE OF EQUAL MEASURE

$$m \angle A = m \angle G \qquad m \angle D = m \angle F$$

ALTERNATE EXTERIOR ANGLES ARE OF EQUAL MEASURE

$$m \angle B = m \angle H \qquad m \angle C = m \angle E$$

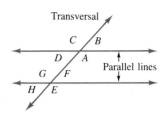

In the margin, angles A and B form a straight angle and are thus supplementary. Because $m \angle B = m \angle F$, it follows that angles A and F are also supplementary. The same idea applies to angles D and G as well as to angles B and E and angles C and H. We can summarize these facts by saying: *Interior angles on the same side of the transversal are supplementary, and exterior angles on the same side of the transversal are also supplementary.*

The next example will help to clarify and fix this information in your mind.

EXAMPLE 4 In the following figure, find the measure of the angle:

a. Y **b.** Z **c.** X **d.** R **e.** S **f.** T **g.** U

Problem 4 Find the measure of the angles in the figure if $m \angle A = 45°$.

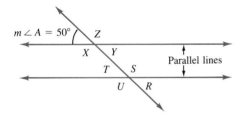

Solution

a. Since Y and A are vertical angles, $m \angle Y = m \angle A = 50°$.

b. Angles A and Z are supplementary, so $m \angle Z + m \angle A = 180°$. Substituting $m \angle A = 50°$, we have $m \angle Z + 50° = 180°$. Thus, $m \angle Z = 130°$.

c. X and Z are vertical angles. Thus, $m \angle X = m \angle Z = 130°$.

d. R and A are alternate exterior angles. Hence, $m \angle R = m \angle A = 50°$.

e. S and Y are interior angles on the same side of the transversal and so are supplementary. Therefore, $m \angle S + m \angle Y = 180°$. Since $m \angle Y = 50$, $m \angle S = 130°$.

f. T and A are corresponding angles. Thus, $m \angle T = m \angle A = 50°$.

g. U and A are exterior angles on the same side of the transversal. Thus, $m \angle U + m \angle A = 180°$. Since $m \angle A = 50°$, $m \angle U = 130°$.

C. ANGLES OF A TRIANGLE

Parallel lines and the associated angles allow us to obtain quite easily one of the most important results in the geometry of triangles. (Of course, you know what a triangle is and we give no formal definition here.) In the following figure, ABC represents any triangle. The line

XY has been drawn through the point C parallel to the side AB of the triangle. Note that $\angle 1 = \angle 2$ and $\angle 3 = \angle 4$ because they are respective pairs of alternate interior angles.

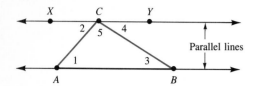

Furthermore, angles 2, 5, and 4 form a straight angle, so

$$m\angle 2 + m\angle 5 + m\angle 4 = 180°.$$

By substituting $\angle 1$ for $\angle 2$ and $\angle 3$ for $\angle 4$, we obtain

$$m\angle 1 + m\angle 5 + m\angle 3 = 180°.$$

Thus, we have shown that the *the sum of the measures of the angles of any triangle is 180°.*

EXAMPLE 5

a. In a triangle ABC, $m\angle A = 47°$ and $m\angle B = 59°$. Find the measure of $\angle C$.

b. Is it possible for a triangle ABC to be such that $\angle A$ is twice the size of $\angle B$ and $\angle C$ is three times the size of $\angle B$?

Solution

a. Because $m\angle A + m\angle B + m\angle C = 180°$, we have

$$47° + 59° + m\angle C = 180°$$
$$m\angle C = 180° - 47° - 59°$$
$$= 180° - 106° = 74°$$

b. To answer this question, let $m\angle B = x°$, so $m\angle A = 2x°$ and $m\angle C = 3x°$. Then, since the sum of the angles is 180°,

$$x + 2x + 3x = 180$$
$$6x = 180$$
$$x = 30, \qquad 2x = 60, \qquad 3x = 90$$

This means that there is such a triangle, and $m\angle A = 60°$, $m\angle B = 30°$, and $m\angle C = 90°$. (Note that this is a right triangle because one of the angles is a right angle.)

Problem 5 In a triangle ABC, $m\angle A = 42°$ and $m\angle B = 53°$.
a. Find the measure of C.
b. Is it possible for a triangle ABC to be such that $\angle A$ is three times the size of $\angle B$ and $\angle C$ is four times the size of B?

NAME

CLASS

SECTION

ANSWERS

A. In Problems 1 through 6, find the length of the indicated line segment.

1. If $AB = 20$ and $AC = 25$, find BC.

1. _____

2. If $AB = 10$ and $AC = 15$, find BC.

2. _____

3. If $BC = 12$ and $AC = 27$, find AB.

3. _____

4. If $BC = 13$ and $AC = 29$, find AB.

4. _____

5. If $AC = 15$ and $BC = 5$, find AB.

5. _____

6. If $AC = 30$ and $BC = 16$, find AB.

6. _____

B. Problems 7 through 14 refer to the following figure:

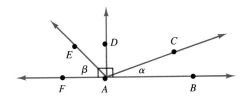

7. Name an angle that is the complement of:
   a. $\angle EAF$
   b. $\angle BAC$

7. a. _____
   b. _____

8. Name an angle that is supplementary to:
   a. $\angle \alpha$
   b. $\angle \beta$

8. a. _____
   b. _____

9. Name an angle that is supplementary to:
   a. $\angle BAD$
   b. $\angle EAF$

9. a. _____
   b. _____

10. If $m\angle\alpha = 15°$, find $m\angle CAD$.

10. _____

11. If $m\angle\beta = 55°$, find $m\angle DAE$.

11. _____

12. If $m\angle DAE = 35°$, find $m\angle\beta$.

12. _____

13. If $m\angle CAD = 75°$, find $m\angle\alpha$.

13. _____

14. If $m\angle\alpha = 15°$ and $m\angle\beta = 55°$, find $m\angle CAE$.

14. _____

Problems 15 through 22 refer to the two intersecting lines shown in the figure in the margin.

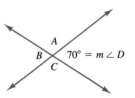

15. Name the angle that is vertical to the 70° angle.

15. _____

16. Name the two angles that are each supplementary to the 70° angle.

16. _____

17. Find $m\angle A$.

17. _____

18. Find $m\angle B$.

18. _____

19. What is the measure of an angle complementary to the 70° angle?

19. _____

20. Find the sum of the measures of angles A, B, and C.

20. _____

21. Find the sum of the measures of angles A and C.

21. _____

22. If $m\angle D = x°$ (instead of 70°), write an expression for the measure of $\angle A$.

22. _____

Problems 23 through 25 refer to the two parallel lines and the transversal shown in the figure in the margin.

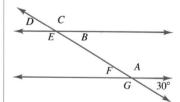

23. Find.
   a. $m\angle A$
   b. $m\angle B$
   c. $m\angle C$

23. a. _____
   b. _____
   c. _____

24. Find.
   a. $m\angle D$
   b. $m\angle E$
   c. $m\angle F$

24. a. _____
   b. _____
   c. _____

25. Name all the angles that are supplementary to $\angle B$.

25. _____

ANSWER (to problem on page 666)
5. a. 85°
   b. Yes, $m\angle A = 67.5°$; $m\angle B = 22.5°$;
     $m\angle C = 90°$

26. Refer to the angles shown in the figure in the margin.
  a. If $m \angle AOB = 30°$ and $m \angle AOC = 70°$, find $m \angle BOC$.
  b. If $m \angle AOB = m \angle COD$, $m \angle AOD = 100°$, and $m \angle BOC = 2x°$, find $m \angle COD$ in terms of x.

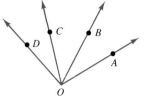

26. a. _____
    b. _____

27. If $m \angle A = 41°$, find $m \angle B$ in each case.
  a. The two angles are complementary
  b. The two angles are supplementary

27. a. _____
    b. _____

28. If $m \angle A = 19°$, find $m \angle B$ in each case.
  a. The two angles are complementary
  b. The two angles are supplementary

28. a. _____
    b. _____

29. Given that $m \angle A = (3x + 15)°$, $m \angle B = (2x - 5)°$, and the two angles are complementary, find x.

29. _____

30. Rework Problem 29 if the two angles are supplementary.

30. _____

In Problems 31 through 34, the figures show the number of degrees in each angle in terms of x. Use algebra to find x and the measure of each angle.

31.

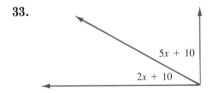

31. _____

32.

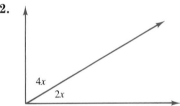

32. _____

33.

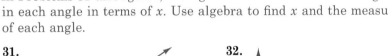

33. _____

34.

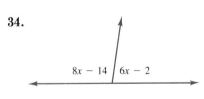

34. _____

The hour hand of a clock moves 360° (one revolution) in 12 hr.

35. Through how many degrees does the hour hand of a clock move in going from:
  a. 11 o'clock to 12 o'clock?
  b. 11 o'clock to 5 o'clock?

35. a. _____
    b. _____

36. Through how many degrees does the hour hand of a clock move in going from:
  a. 12 o'clock to 7 o'clock?
  b. 12 o'clock back to 12 o'clock?

36. _____

C. Solve the following problems.

37. In a triangle ABC, $m \angle A = 37°$ and $m \angle C = 53°$. Find $m \angle B$.

37. _____

38. In a triangle ABC, $m \angle B = 67°$ and $m \angle C = 105°$. Find $m \angle A$.

38. _____

39. In a triangle ABC, $m\angle A = (x + 10)°$, $m\angle B = (2x + 10)°$, and $m\angle C = (3x + 10)°$. Find x.

40. In a triangle ABC, $m\angle A$ is 10° less than $m\angle B$ and $m\angle C$ is 40° greater than $m\angle B$. Find the measure of each angle.

39. _____

40. _____

✓ **SKILL CHECKER**

Solve.

41. $332 + 2W = 458.50$ **42.** $550 + 2W = 758.50$

43. $141.30 = 3.14d$ **44.** $131.88 = 3.14d$

45. $28.8 = \dfrac{1}{2} \cdot b \cdot 3.6$ **46.** $54.6 = \dfrac{1}{2} \cdot b \cdot 4.2$

41. _____

42. _____

43. _____

44. _____

45. _____

46. _____

10.1 USING YOUR KNOWLEDGE

The ideas used to find the length of a given line segment can be used to complete the dimensions of an object. In Problems 1 through 6, find the indicated dimensions.

1. Find length A.

2. Find length B.

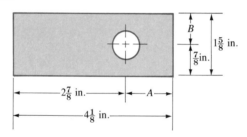

1. _____

2. _____

3. Find x for spacing the knobs in the drawer.

4. Find dimension y.

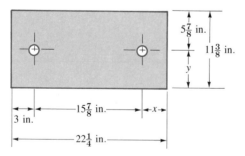

3. _____

4. _____

5. Find dimension A.

6. Find dimension B.

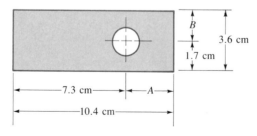

5. _____

6. _____

10.2 PERIMETER AND AREA

OBJECTIVES

REVIEW

Before starting this section you should know:

1. How to perform the fundamental operations with decimals.
2. How to solve linear equations.

OBJECTIVES

You should be able to:

A. Find the perimeter of a plane polygon or a circle.
B. Find the area of a plane region.

Courtesy of U.S. Air Force.

Do you know the name of the building or the distance around it? It is the *Pentagon*, and each of the sides is 921 ft in length. Since the building has five sides, the distance around it (the perimeter) is $5 \cdot 921 = 4605$ ft.

A pentagon is an example of a **polygon,** a figure with many sides. Polygons are named according to the *number* of sides. Here are some of the usual names.

NUMBER OF SIDES	NAME	NUMBER OF SIDES	NAME
3	Triangle	8	Octagon
4	Quadrilateral	9	Nonagon
5	Pentagon	10	Decagon
6	Hexagon	12	Dodecagon
7	Heptagon		

The **perimeter** (distance around) of a regular polygon, a polygon with equal sides and equal angles (such as a regular pentagon), is the product of the number of sides and the length of one side. Quadrilaterals (four-sided polygons) have names and definitions of their own, as follows:

1. A quadrilateral with two parallel and two nonparallel sides is called a **trapezoid.**
2. A quadrilateral with both pairs of opposite sides parallel is a **parallelogram.**
3. A parallelogram with four equal sides is a **rhombus.**
4. A parallelogram whose angles are right angles is a **rectangle.**
5. A rectangle with four equal sides is a **square.**

All these polygons are shown in the figure.

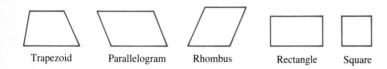

| Trapezoid | Parallelogram | Rhombus | Rectangle | Square |

A. PERIMETERS

Many of the applications of geometry involve finding the perimeter of a figure. Fencing a field, laying tile around a rectangular pool, or finding the amount of baseboard needed for a room involve **perimeters.** The table gives the formulas for the perimeters of some of the polygons we have studied.

NAME	GEOMETRIC SHAPES	PERIMETER
Triangle		$P = s_1 + s_2 + b$
Trapezoid		$P = s_1 + s_2 + b_1 + b_2$
Parallelogram		$P = 2L + 2W$
Rectangle		$P = 2L + 2W$
Square		$P = 4s$

EXAMPLE 1 The *Mona Lisa* by Leonardo da Vinci was assessed at $100 million for insurance purposes. The picture measures 30.5 by 20.9 in. Find the length of the frame around this picture.

Solution Because the picture is rectangular, its perimeter is given by

$$P = 2L + 2W$$

where $L = 30.5$ and $W = 20.9$. Thus,

$$P = 2(30.5) + 2(20.9)$$
$$= 102.8 \text{ in.} \qquad \blacktriangle$$

The ideas we have been studying here can be combined with the algebra you know to solve certain kinds of problems. Here is an interesting problem.

Problem 1 Find the length of the frame around a picture measuring 40.9 in. by 30.8 in.

EXAMPLE 2 One of the largest recorded posters was a rectangular greeting card 166 ft long and with a perimeter of 458.50 ft. How wide was this poster?

Solution The perimeter of a rectangle is $P = 2L + 2W$, and we know that $L = 166$ and $P = 458.50$. Thus, we can write

$$2(166) + 2W = 458.50$$
$$332 + 2W = 458.50$$
$$2W = 126.50$$
$$W = 63.25$$

Thus, the poster was 63.25 ft wide. ▲

What about the perimeter of a circle? First, a **circle** is a figure in which all points are the same distance (the **radius**) from a given point (the **center**). The radius r of a circle is a line segment from the center O to a point P on the circle, whereas the **diameter** d is a line segment across the circle through the center O. The length of the diameter is twice the length of the radius; that is, $d = 2r$, or $r = \frac{1}{2}d$. (See the figure.)

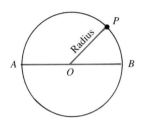

The perimeter of a circle is called the *circumference*, and the formula for it is as follows:

The **circumference** C of a circle is $C = \pi d$ or $C = 2\pi r$, where d is the diameter and r is the radius.

EXAMPLE 3 The longest ball of string on record is 40 ft in circumference. What is the diameter of this ball? Use 3.14 for π and round the answer to the nearest hundredth of a foot.

Solution The circumference is

$$C = \pi d$$
$$40 = 3.14d \quad \text{Substituting 40 for } C \text{ and 3.14 for } \pi$$
$$\frac{40}{3.14} = d \quad \text{Dividing by 3.14 on both sides}$$
$$d = 12.74 \text{ ft} \quad \text{Rounding to the nearest hundredth}$$

By the way, the ball weighs 10 tons and belongs to Francis A. Johnson of Darwin, Minnesota.

Problem 2 The students at Osaka Gakun University made a poster with a 416 ft perimeter. If the poster was 169 ft long, how wide was it?

Problem 3 Frank Stoeber's string ball is 34.54 ft in circumference. What is the diameter of this ball?

B. AREA

How do we define area? The area of a region is a measure of the amount of surface in the region. We can start by choosing the unit region to be that of a square each of whose sides is 1 *unit in length* and say that this region has an **area** of **1 square unit.** To find the area of a figure, we find the number of square units it contains. Thus, to find the area of a rectangle measuring 3 units by 4 units, we subdivide the rectangle into $3 \times 4 = 12$ unit squares and say that the area of the rectangle is $3 \times 4 = 12$ square units. In general, we have the following.

The unit of area measure

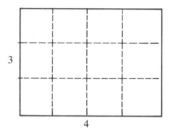

Area = 12 square units

> The **area** A of a rectangle with base b units and height h units is given by:
>
> $A = bh$

Knowing the area of a rectangle, we can find the area of a parallelogram. The idea is to construct a rectangle with the same area as the parallelogram. To do this we simply cut the right triangle ADE from one end of the parallelogram and attach it to the other end. (See the figure.) This forms a rectangle $CDEF$ with the same area as the parallelogram. Since the base and height are the same for both figures, it follows that the area of a parallelogram of base b and height h is

$A = bh$ Area of a parallelogram

(Be sure to note that h is the perpendicular height, not just a side of the parallelogram.)

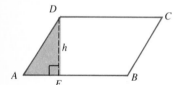
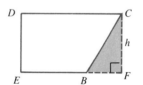

It is now an easy matter to find the area of any triangle. Suppose that triangle ABC in the margin is given. We draw a line through C parallel to $\overline{AB}$ and a line through A parallel to $\overline{BC}$. These lines meet at a point D, and the quadrilateral $ABCD$ is a parallelogram. Clearly, the line $\overline{AC}$ is a diagonal of the parallelogram and divides the parallelogram into two equal pieces. Because the area of parallelogram $ABCD$ is bh, the area of the triangle ABC is $\frac{1}{2}bh$. Thus, we have a formula for the area of any triangle with base b and height h:

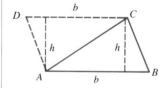

> $A = \frac{1}{2}bh$ Area of a triangle

Note: If the angle B in the figure is a right angle, the parallelogram obtained is a rectangle. This does not change the final result.

EXAMPLE 4 The area of a triangle is 3 in.^{2}. If its height h is 1.5 in., find the base.

Solution The area of a triangle is

$$A = \frac{1}{2}bh$$

$$3 = \frac{1}{2} \cdot b \cdot 1.5 \qquad \text{Letting } A = 3 \text{ and } h = 1.5$$

$$2 \cdot 3 = 2 \cdot \frac{1}{2} \cdot b \cdot 1.5 \qquad \text{Multiplying on both sides by } \mathbf{2}$$

$$6 = 1.5b \qquad \text{Simplifying}$$

$$b = 4 \qquad \text{Dividing on both sides by } 1.5$$

Thus, the base is 4 in. long. ▲

Thus far, we have been concerned entirely with the areas of polygonal regions. How about the circle? Can we find the area of a circle by using one of the preceding formulas? Interestingly enough, the answer is yes. The required area can be found by using the formula for the area of a rectangle. Here is how you can go about it. Look at the figure. Cut the lower half of the circular region into small equal slices and arrange them as shown in the figure. Then cut the remaining half of the circle into the same number of slices and arrange them as shown. The result is approximately a parallelogram whose longer side is of length πr (half the circumference of the circle) and whose shorter side is r (the radius of the circle). The more pieces you cut the circle into, the more accurate this approximation becomes. You should also observe that the more pieces you use, the more nearly the parallelogram becomes a rectangle of length πr and height r. Mathematicians have proved that the area of the circle is actually the same as the area of this rectangle, that is, $(\pi r)(r)$, or πr^2. Thus, we arrive at the formula for the area of a circle of radius r:

$$A = \pi r^2 \quad \text{Area of a circle}$$

EXAMPLE 5 The Fermi National Accelerator Laboratory has a circular atom smasher covering an area of 1.1304 mi^2. What is the diameter of this atom smasher? Use $\pi \approx 3.14$, and give the answer to the nearest hundredth of a mile.

Solution The formula for the area of a circle of radius r is

$$A = \pi r^2$$

Here, $A = 1.1304$ and $\pi \approx 3.14$. Thus,

$$1.1304 \approx 3.14 r^2$$

$$r^2 \approx \frac{1.1304}{3.14} = 0.36$$

$$r \approx \pm\sqrt{0.36}$$

$$r \approx 0.60 \text{ mi}$$

(We discard the answer $r = -0.6$ because r must be positive.) Thus, the diameter of the smasher is $2 \times 0.6 = 1.2$ mi.

Problem 4 The area of a triangle is 10 in.^{2}. If the height h is 2.5 in., find the base.

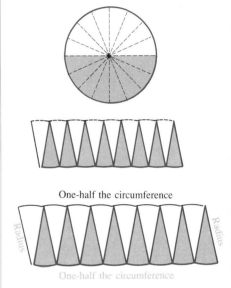

One-half the circumference

One-half the circumference

Problem 5 The Department of Energy is planning a super conductivity collider covering an area of 2122.64 mi^2. What is the diameter of this collider?

A. In Problems 1 through 8, find the perimeter of the given polygon.

ANSWERS

1.

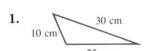

30 cm
10 cm
25 cm

1. _____

2.

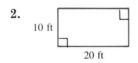

10 ft
20 ft

2. _____

3.

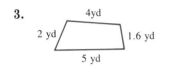

4 yd
2 yd   1.6 yd
5 yd

3. _____

4.

7.2 in.
7.2 in.

4. _____

5.

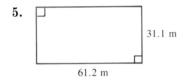

31.1 m
61.2 m

5. _____

6.

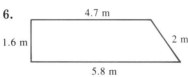

4.7 m
1.6 m   2 m
5.8 m

6. _____

7.

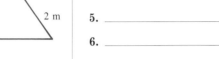

9.2 m
19.4 m

7. _____

8.

3 km
9.2 km

8. _____

9. If one side of a regular pentagon is 6 cm long, find the perimeter of the pentagon.

9. _____

10. If one side of an octagonal stop sign is 6 in. long, what is the perimeter of the stop sign?

10. _____

11. The largest rectangular omelet ever cooked was 30 ft long and had an 80-ft perimeter. How wide was it?

11. _____

12. Do you have a large pool? If you were to walk around the largest pool in the world, in Casablanca, Morocco, you would go more than 1 km. To be exact, you would go 1110 m. If the pool is 480 m long, how wide is it?

12. _____

13. A baseball diamond is actually a square. A batter who hits a home run must run 360 ft around the bases. What is the distance to first base?

13. _____

14. The playing surface of a football field is 120 yd long. A player jogging around the perimeter of this surface covers 346 yd. How wide is the playing surface?

14. _____

15. Have you seen the largest scientific building in the world? It is in Cape Canaveral, Florida. If you were to walk around the perimeter of this building, you would cover 2468 ft. If this rectangular building is 198 ft longer than it is wide, what are its dimensions?

15. _____

In Problems 16 through 20 use $\pi \approx 3.14$. Approximate your answers to the nearest hundredth.

16. One of the largest hamburgers on record had a circumference of 27.50 ft. Find the diameter of this hamburger.

16. _____

17. The largest pizza ever made had a circumference of 251.2 ft. Find its diameter.

17. _____

18. A long-playing record has a radius of 6 in. How far does a point on the rim move when the record goes around once?

18. _____

19. The circumference of the smallest functional record ever made is an amazing $4\frac{1}{8}$ in. Find the diameter of this tiny record.

19. _____

20. A strip of gold 7 cm long is used to make a wedding band (size 12). Find the diameter of this ring.

20. _____

B. In Problems 21 through 30 find the area of the given region.

21.

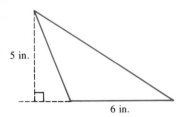

21. _____

22.

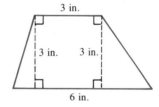

22. _____

23.

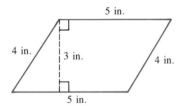

23. _____

24.

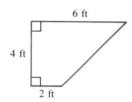

24. _____

25.

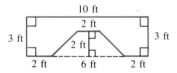

25. _____

26.

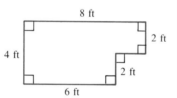

26. _____

27.

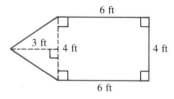

27. _____

28.

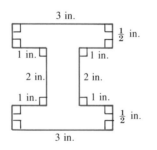

28. _____

29.

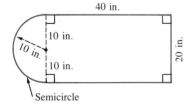

29. _____

30.

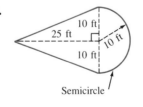

30. _____

31. The playing surface of a football field is 120 yd long and $53\frac{1}{3}$ yd wide. How many square yards of artificial turf are needed to cover this surface?

31. _____

32. The floors of three rooms in a certain house measure 9 by 10 ft, 12 by 12 ft, and 15 by 15 ft, respectively.
   a. How many square yards of carpet are needed to cover these three floors?
   b. If the price of the carpet is $14/yd^2, how much would it cost to cover these floors?

32. a. _____
    b. _____

33. The Louisiana Superdome covers an area of 363,000 ft^2. Find the diameter of this round arena to the nearest foot. Use $\pi \approx 3.14$.

33. _____

34. The largest cinema screen in the world is in the Pictorium Theater in Santa Clara, California; it covers 6720 ft^2. If this rectangular screen is 70 ft tall, how wide is it?

34. _____

35. The area of the biggest pizza was about 5024 ft^2. What was its diameter?

35. _____

✓ **SKILL CHECKER**

Solve.

36. $5^2 + b^2 = 9^2$

37. $6^2 + b^2 = 9^2$

36. _____

37. _____

38. $1^2 + b^2 = 3^2$

39. $2^2 + b^2 = 3^2$

38. _____

39. _____

40. $\dfrac{d}{4} = \dfrac{2}{3}$

41. $\dfrac{f}{3} = \dfrac{2}{3}$

40. _____

41. _____

42. $\dfrac{a}{5} = \dfrac{3}{8}$

43. $\dfrac{b}{7} = \dfrac{5}{4}$

42. _____

43. _____

44. $\dfrac{5}{12} = \dfrac{c}{6}$

45. $\dfrac{9}{22} = \dfrac{d}{11}$

44. _____

45. _____

10.2 USING YOUR KNOWLEDGE

You can use what you learned in this section to help you solve some commonly occurring problems. Do the following problems to see how.

1. A gallon of Lucite wall paint costs $14 and covers 450 ft^2. Three rooms in a house measure 10 by 12 ft, 14 by 15 ft, and 12 by 12 ft, and the ceiling is 8 ft high.
   a. How many gallons of paint are needed to cover the walls of these rooms if you make no allowance for doors and windows?
   b. What will be the cost of the paint? (The paint is sold by the gallon only.)

1. a. _____
   b. _____

2. The diagram in the margin shows the front of a house. House paint costs $17/gal and covers 400 ft².
   a. What is the minimum number of gallons of paint needed to cover the front of the house? (The paint is sold by the gallon only.)
   b. How much will the paint for the front of the house cost?

2. a. _____
   b. _____

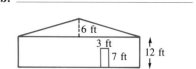

3. My house and lot are shown in the diagram below. The entire lot, except for the buildings and the drive, is lawn. A bag of lawn fertilizer costs $4 and covers 1200 ft² of grass.

3. a. _____
   b. _____

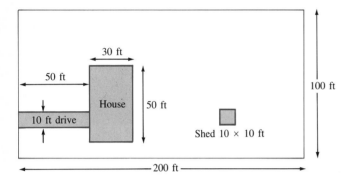

   a. What is the minimum number of bags of fertilizer needed for this lawn?
   b. What will be the cost of the fertilizer?

4. A small pizza (11-in. diameter) costs $5 and a large pizza (15-in. diameter) costs $8. Use $\pi \approx 3.14$ and find, to the nearest square inch:
   a. The area of the small pizza
   b. The area of the large pizza
   c. Which is the better deal, two small pizzas or one large pizza?

4. a. _____
   b. _____
   c. _____

5. A frozen apple pie of 8-in. diameter sells for $1.25. The 10-in. diameter size sells for $1.85.
   a. What is the unit price (price per square inch), to the nearest hundredth of a cent, of the 8-in. pie?
   b. What is the unit price of the 10-in. pie?
   c. Which pie gives you the most for your money?

5. a. _____
   b. _____
   c. _____

Courtesy of John Nye.

OBJECTIVES

REVIEW

Before starting this section you should know:

1. The Pythagorean theorem (Section 9.6).
2. How to solve quadratic equations of the form $X^2 = A$.

OBJECTIVES

You should be able to:

A. Solve right triangles.
B. Solve problems involving similar triangles.

The picture shows the triangles in the Bank of China building in Hong Kong. How many types of triangles do you see? In general, triangles are classified according to their angles.

> **1. Acute triangle:** a triangle in which the three angles are all acute (less than 90°).
>
> **2. Right triangle:** a triangle in which one of the angles is a right angle (90°).
>
> **3. Obtuse triangle:** a triangle in which one of the angles is obtuse (greater than 90°).

Triangles can also be classified according to the number of equal sides.

> **1. Scalene triangle:** no equal sides
>
> **2. Isosceles triangle:** two equal sides
>
> **3. Equilateral triangle:** three equal sides

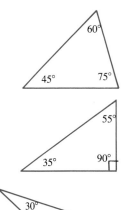

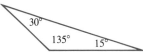

A.   SOLVING RIGHT TRIANGLES

As you recall from Section 9.6, if the lengths of two sides of a right triangle are known, the length of the third side can be found by using the *Pythagorean Theorem*. If we call the side opposite the right angle the hypotenuse and the other two sides the *legs,* the theorem is stated as follows:

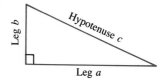

PYTHAGOREAN THEOREM

The square of the hypotenuse of a right triangle is equal to the sum of the squares of the two legs. In symbols,

$$c^2 = a^2 + b^2$$

EXAMPLE 1   The length of one of the legs of a right triangle is 5 in. If the hypotenuse is 9 in. long, find the length of the other leg.

Solution   Here $a = 5$, $c = 9$, and we wish to find b. By the Pythagorean Theorem

$$5^2 + b^2 = 9^2$$
$$25 + b^2 = 81$$

$b^2 = 56$	Subtracting 25 on both sides
$b = \sqrt{56}$	Taking the principal root
$= \sqrt{4 \cdot 14}$	Rewriting 56 as $4 \cdot 14$
$= 2\sqrt{14}$	Simplifying

Thus, $b = 2\sqrt{14}$ in.

   Note that since the answer had to be positive (b represents a length) we took the *principal* square root of both sides of the equation.   ▲

   The Pythagorean Theorem can be used to find a relationship between the sides of an isosceles right triangle (also called a **45°-45°-90° triangle**). Thus, if both legs of an isosceles right triangle are a units long, then by the Pythagorean theorem

$$a^2 + a^2 = c^2$$
$$2a^2 = c^2$$

$c = \sqrt{2}a$ Taking the principal root on both sides

We have the following result.

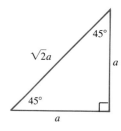

For any isosceles right triangle, the hypotenuse c equals $\sqrt{2}$ times the length a of a leg, that is,

$$c = \sqrt{2} \cdot (\text{length of a leg}) = \sqrt{2}a$$

EXAMPLE 2   A leg of an isosceles right triangle is 10 in. long. Find.
a. The length of the hypotenuse
b. The perimeter of the triangle

Solution

a. For any isosceles right triangle, the length c of the hypotenuse is

$$c = \sqrt{2} \cdot (\text{length of a leg})$$
$$c = \sqrt{2} \cdot 10 = 10\sqrt{2} \text{ in.}$$

Problem 1   The length of one of the legs of a right triangle is 6 in. If the hypotenuse is 9 in., find the length of the other leg.

Problem 2   A leg of an isosceles right triangle is 8 in. long. Find.
a. The length of the hypotenuse
b. The perimeter of the triangle

b. The perimeter P is the sum of the sides a, b, and c. Thus,

$$P = 10 + 10 + 10\sqrt{2}$$
$$= (20 + 10\sqrt{2}) \text{ in.} \qquad \blacktriangle$$

There is also a special relationship among the sides of a **30°-60°-90° triangle.** Suppose the shorter leg is a units long. If we place two of these triangles together as in the margin, we form an equilateral triangle. The sides of this triangle must be $2a$ units long. Thus, the length of the hypotenuse is twice the length of the shorter leg. We can find b using the Pythagorean theorem:

$$a^2 + b^2 = (2a)^2$$
$$a^2 + b^2 = 4a^2$$
$$b^2 = 3a^2 \qquad \text{Subtracting } a^2 \text{ on both sides}$$
$$b = \sqrt{3}a \qquad \text{Taking principal roots}$$

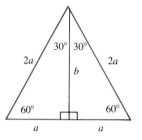

The relationship among the sides of a 30°-60°-90° triangle is given next.

For a 30°-60°-90° triangle, the hypotenuse c equals twice the length a of the shorter leg (the leg opposite the 30° angle); that is,

$$c = 2 \cdot (\text{length of shorter leg}) = 2a$$

and

$$b = \sqrt{3} \cdot a$$

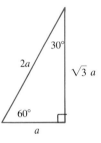

EXAMPLE 3 The shorter leg a of a 30°-60°-90° triangle is 6.8 in. long. Find the perimeter.

Solution According to the theorem,

$$c = 2 \cdot 6.8$$
$$c = 13.6$$
$$b = \sqrt{3} \cdot 6.8$$
$$= 6.8\sqrt{3}$$

The perimeter is $a + b + c = 6.8 + 13.6 + 6.8\sqrt{3} = (20.4 + 6.8\sqrt{3})$ in.

Problem 3 The shorter leg of a 30°-60°-90° triangle is 4.2 in. long. Find the perimeter.

B. SOLVING SIMILAR TRIANGLES

In everyday life many objects are similar; that is, they have the same shape. An orange, a baseball, and a basketball are similar figures. In geometry, figures that have exactly the same shape but not the same size are called **similar figures.** Thus, if you have a model car built to scale, the model and the original car are similar.

Look at the two similar triangles in the margin. Because they have the same shape, the corresponding angles are equal. In the figure, $m\angle A = m\angle D$, $m\angle B = m\angle E$, and $m\angle C = m\angle F$. The corresponding sides of the two triangles are $\overset{\leftrightarrow}{AB}$ and $\overset{\leftrightarrow}{DE}$, $\overset{\leftrightarrow}{BC}$ and $\overset{\leftrightarrow}{EF}$, and $\overset{\leftrightarrow}{AC}$ and $\overset{\leftrightarrow}{DF}$. Notice that the length of each side of the smaller triangle is one-half the length of the corresponding side of the larger triangle. The definition follows.

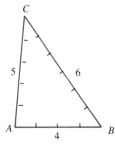
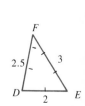

Two triangles are **similar** when they have equal corresponding angles and the corresponding sides are proportional.

Recall that the ratio of two numbers a and b is the fraction $\frac{a}{b}$. For the figure in the margin, we have

$$\frac{DE}{AB} = \frac{EF}{BC} = \frac{DF}{AC} = \frac{1}{2}$$

as the ratio of corresponding sides.

EXAMPLE 4 Two similar triangles are shown. Find f for the triangle on the right.

Problem 4 Find d for the triangle on the right.

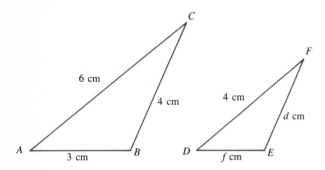

Solution Since the triangles are similar, the corresponding sides must be proportional. Thus,

$$\frac{DE}{AB} = \frac{DF}{AC}$$

so that

$$\frac{f}{3} = \frac{4}{6}$$

Since $\frac{4}{6} = \frac{2}{3}$, we have

$$\frac{f}{3} = \frac{2}{3}$$

Solving these equations for f, we get

$$f = \frac{6}{3} = 2 \text{ cm}$$

A. In Problems 1 through 9 find the unknown side of the triangle.

ANSWERS

1.

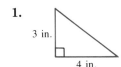
3 in.
4 in.

1. _____

2.

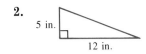
5 in.
12 in.

2. _____

3.

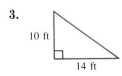
10 ft
14 ft

3. _____

4.

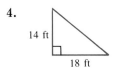
14 ft
18 ft

4. _____

5.

3 ft
2 ft

5. _____

6.

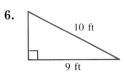
10 ft
9 ft

6. _____

7.

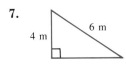
6 m
4 m

7. _____

8.

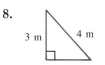
3 m
4 m

8. _____

9.

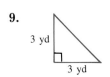
3 yd
3 yd

9. _____

In Problems 10 through 18 find (a) the hypotenuse and (b) the perimeter.

10.

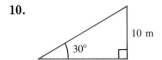
10 m
30°

10. a. _____
    b. _____

11.

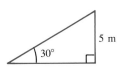
5 m
30°

11. a. _____
    b. _____

12.

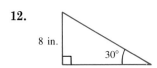
8 in.
30°

12. a. _____
    b. _____

13.

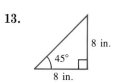
8 in.
45°
8 in.

13. a. _____
    b. _____

14.

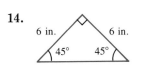
6 in.   6 in.
45°   45°

14. a. _____
    b. _____

15.

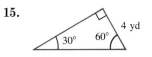
4 yd
30°   60°

15. a. _____
    b. _____

16.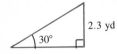

17.

16. a. _____
    b. _____

17. a. _____
    b. _____

18.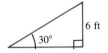

18. a. _____
    b. _____

B. The following triangles are similar. Find the lengths of the indicated unknowns.

19.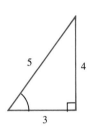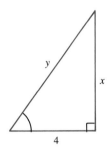

19. _____

20. Find side *DE*.

21. Find side *DE*.

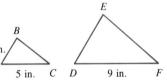

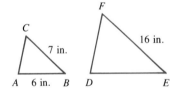

20. _____

21. _____

22.

23.

22. _____

23. _____

24.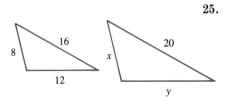

25.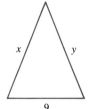

24. _____

25. _____

ANSWER (to problem on page 684)

4. $d = 2\frac{2}{3}$ cm

26.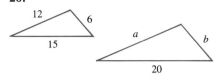

27.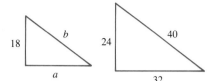

28.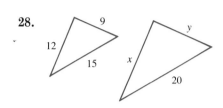

29.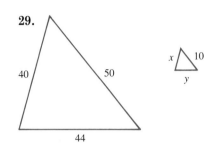

30.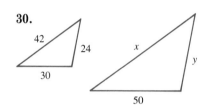

31.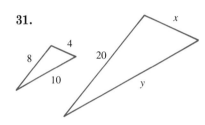

26. _____

27. _____

28. _____

29. _____

30. _____

31. _____

32. The measures of the sides of a triangle are 18 in., 20 in., and 24 in. The measure of the shortest side of a similar triangle is 27 in. Find the measures of the other two sides of the second triangle.

33. The measures of the sides of a triangle are 16 in., 16 in., and 20 in. The measure of the longest side of a similar triangle is 14 in. Find the measures of the other two sides.

34. The measure of one side of an equilateral triangle is 30 cm. Find the measure of the sides of another equilateral triangle, given that the measure of one side is 25 cm. Why are these triangles similar?

32. _____

33. _____

34. _____

✓ **SKILL CHECKER**

Solve for r.

35. $9000\pi = 40\pi r^2$

36. $6000\pi = 15\pi r^2$

37. $2250\pi = 10\pi r^2$

38. $1210\pi = 10\pi r^2$

35. _____

36. _____

37. _____

38. _____

10.3 USING YOUR KNOWLEDGE

Can we use the knowledge gained about similar triangles to perform some seemingly impossible tasks, like measuring the height of a building in your city? Of course we can! The sun's rays, the building,

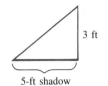

5-ft shadow

and its shadow form a triangle. How do we find a similar triangle? Suppose you wish to measure the height of a building. First measure the length of its shadow. Then take a yardstick (or any stick of known length), stand it vertically, and measure its shadow. (You must do these two things at the same time of the day.) Let us say that the yardstick (3 ft) casts a 5-ft shadow. If the building's shadow is 300 ft long, you can solve the proportion

$$\frac{h}{3} = \frac{300}{5}$$

for h, obtaining $h = 180$ ft, and you have found the height of the building!

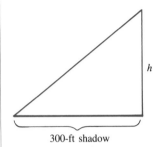

300-ft shadow

1. Find the height of a building whose shadow is 100 ft long if a yardstick casts a 4-ft shadow.

1. _____

2. Find the height of a building whose shadow is 200 ft long if a yardstick makes a 4-ft shadow.

2. _____

3. A flagpole casts a 24-ft shadow. Find its height if a yardstick casts a 4-ft shadow.

3. _____

4. A flagpole casts a 30-ft shadow. Find its height if a tree 6 ft high casts a 5-ft shadow.

4. _____

5. A building casts an 8-ft shadow. Find its height if an 8-ft flagpole casts a 4-ft shadow.

5. _____

10.4 VOLUMES

Have you seen Rubik's cube? It is a multicolored cube made up of smaller cubes that can be arranged in different patterns. Do you know how to figure the volume of a cube? In Rubik's cube, there are 3 small cubes in a row, each 1 cm long, making each edge 3 cm. Thus, the volume of this cube is $(3 \text{ cm}) \cdot (3 \text{ cm}) \cdot (3 \text{ cm}) = 27 \text{ cm}^3$ (cm^3 is read as "cubic centimeters"). The **volume** of a three-dimensional region is measured in terms of a **unit volume,** just as area is measured in terms of a unit area. For the unit volume, we choose the region enclosed by a **unit cube,** as shown in the margin. The unit volume may be the cubic inch (in.^3), cubic foot (ft^3), cubic centimeter (cm^3), cubic meter (m^3), and so on, according to the unit of length used.

Volumes may be considered in a manner similar to that used for areas. If a rectangular parallelepiped (box) is such that the lengths of its edges are all whole numbers, then the region can be cut up by planes parallel to the faces as in the margin. In general, we define the volume of a rectangular box of length a, width b, and height c to be abc.

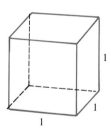

Unit cube:
Volume = 1 cubic unit

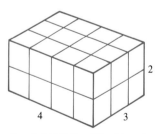

$V = 4 \times 3 \times 2 = 24$ cubic units

$V = abc = \text{length} \times \text{width} \times \text{height}$	Volume of a rectangular box

EXAMPLE 1 Find the volume of a rectangular box of length 3 in., width 5 in., and height 2 in.

Solution In this case $a = 3$, $b = 5$, and $c = 2$. The volume V is given by $V = abc$; thus $V = 3 \cdot 5 \cdot 2 = 30 \text{ in.}^3$. ▲

Problem 1 Find the volume of a rectangular box of length 4 cm, width 6 cm, and height 2 cm.

Now that we know how to find the volume of a rectangular box, can we find the volumes of other solids such as cylinders, cones, and spheres? Mathematicians have obtained the following formulas for these solids. For a circular cylinder (Fig. 5),

$V = \pi r^2 h$ Volume of a circular cylinder

Similarly, for a circular cone of radius r and height h (Fig. 6),

$V = \dfrac{1}{3}\pi r^2 h$ Volume of a circular cone

The volume of a sphere of radius r (Fig. 7) has been shown to be

$V = \dfrac{4}{3}\pi r^3$ Volume of a sphere

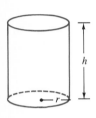

FIG. 5

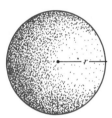

FIG. 6

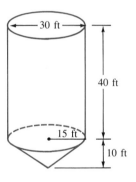

FIG. 7

EXAMPLE 2 The water tank for a small town is in the shape of a cone (vertex down) surmounted by a cylinder, as shown in the figure in the margin. If a cubic foot of water is about 7.5 gal, what is the capacity of the tank in gallons? Use $\pi \approx 3.14$.

Solution For the cylindrical portion of the tank, $r = 15$ and $h = 40$, so that

$V = \pi(15)^2(40) = 9000\pi$ ft^3

For the conical portion, $r = 15$ and $h = 10$. Thus,

$V = \dfrac{1}{3}\pi(15)^2(10) = 750\pi$ ft^3

The total volume is the sum of these—that is, 9750π ft^3. Using 7.5 gal/ft^3 and 3.14 for π, we get

9750π ft$^3 \approx (9750)(3.14)(7.5)$ gal

$\approx 229{,}612.5$ gal   (to the nearest tenth)

EXAMPLE 3 A solid metal sphere of radius 3 m just fits inside a cubical tank. If the tank is full of water and the sphere is slowly lowered into the tank until it touches bottom, how much water is left in the tank?

Solution The amount of water left in the tank is the difference between the volume of the tank and the volume of the sphere. Thus, since the edge of the cube equals the diameter of the sphere, the required volume is

$V = 6^3 - \dfrac{4}{3}\pi(3^3)$

$= 216 - 36\pi$

≈ 103 m^3 (to the nearest cubic meter)   ▲

Problem 2 If the diameter of the cylinder in the example is 20 ft, what is the capacity of the tank?

30 ft

40 ft

15 ft

10 ft

Problem 3 If the radius of the sphere in Example 3 is 4 yd, how much water is left in the tank?

ANSWER (to problem on page 689)

1. 48 cm^3

We can apply some of the algebra we have studied to problems involving volumes. For example, the inside diameter of a cylindrical soda can is about 2.5 in. If you wish to fill the can with 12 fluid ounces (fl oz) of soda, how tall should you make the can? Since the measurements of the can are in inches, first you need to know that:

$$16 \text{ fl oz} = 28.875 \text{ in.}^3$$

Thus, 12 fl oz (to the nearest ounce) represent 22 in.^{3} The volume V of the required cylinder is $V = \pi r^2 h$, where $V = 22$, $r = 1.25$, and $\pi = \dfrac{22}{7}$.

$$V = \pi r^2 h$$

$$22 = \frac{22}{7}(1.25)^2 h \quad \text{Substituting}$$

$$22 = \frac{22}{7}(1.25)^2 h$$

$$7 = (1.25)^2 h \qquad \text{Multiplying by 7 and dividing by 22}$$

$$h = 4.5 \text{ in.} \qquad \text{Dividing by } (1.25)^2 \text{ and rounding to the nearest tenth}$$

Go ahead, measure a can and see how close we came!

EXAMPLE 4 Manufacturers are now making 20 fl-oz cans (36 in.^{3}). If the diameter of the can is 3 in., how tall should it be?

Solution Here $V = 36$, $r = 1.5$, and $\pi = 3.14$.

$$V = \pi r^2 h$$

$$36 = 3.14(1.5)^2 h \quad \text{Substituting}$$

$$36 = 3.14(2.25)h \quad \text{Dividing both sides by 3.14 and 2.25 and rounding to the nearest tenth}$$

$$h = 5.1 \text{ inches}$$

Problem 4 Suppose you wish to manufacture a 50-in.^{3} can, 3 in. in diameter. How tall should you make it?

ANSWERS   (to problems on pages 690–691)
2. 102,050 gal   **3.** 244 yd^3
4. About 7.1 in.

NAME

CLASS

SECTION

In Problems 1 through 4 find the volume of the rectangular solid with the indicated dimensions.

ANSWERS

	Length	Width	Height
1.	8 cm	5 cm	6 cm
2.	15 cm	20 cm	9 cm
3.	$9\frac{1}{2}$ in.	3 in.	$4\frac{1}{2}$ in.
4.	$5\frac{1}{4}$ in.	$1\frac{3}{8}$ in.	$\frac{5}{8}$ in.

1. _____

2. _____

3. _____

4. _____

5. A swimming pool measuring 30 ft by 18 ft by 6 ft is to be excavated. How many cubic yards of dirt must be removed? (*Hint:* Change the feet to yards first.)

5. _____

6. How many cubic yards of sand are being transported in a dump truck whose loading area measures 9 ft by 4 ft by 3 ft?

6. _____

7. A rectangular tank measures 30 ft by 3 ft by 2 ft. How many gallons of water would it hold if 1 ft^3 = 7.5 gallons?

7. _____

8. A hot tub measures 4 ft by 6 ft and is 3 ft deep. How many gallons of water would it hold if 1 ft^3 = 7.5 gallons?

8. _____

9. Where do you use the most water at home? In the toilet! If the toilet tank is 18 in. long, 8 in. wide and the water is 12 in. deep, how many gallons of water does the toilet tank contain? (1 ft^3 = 7.5 gal; answer to the nearest gallon.)

9. _____

10. How much concrete is needed to build a roadway 45 ft wide, 1 ft thick, and 10 mi long (1 mile = 5280 ft)?

10. _____

In Problems 11 through 16 find the volume of a cylinder with the given radius and height. (Use $\pi \approx 3.14$ and give the answer to the nearest tenth.)

	Radius	Height
11.	10 in.	8 in.
12.	4 in.	18 in.
13.	10 cm	20 cm
14.	3.5 cm	2.5 cm
15.	1.5 m	4.5 m
16.	0.8 m	3.2 m

11. _____

12. _____

13. _____

14. _____

15. _____

16. _____

In Problems 17 through 20 use $\pi \approx 3.14$ and give the answer to the nearest tenth.

17. A cylindrical tank has a 20-ft diameter and is 40 ft high. What is its volume?

17. _____

18. A steel rod is $\frac{1}{2}$ in. in diameter and 18 ft long. What is its volume?

18. _____

19. A coffee cup has a 3-in. diameter and is 4 in. high. What is its volume? If 1 in.^{3} = 0.6 fl oz, how many fluid ounces does the cup hold?

19. _____

20. A coffee cup is 3 in. in diameter and 3.5 in. high. Find the volume and determine if a 12-fl-oz soft drink will fit in the cup (1 in.^{3} = 0.6 fl oz).

20. _____

In Problems 21 through 25 find the volume of a cone with the given radius and height. (Use $\pi \approx 3.14$ and give the answer to the nearest tenth.)

Radius	Height
21. 10 in.	6 in.
22. 18 in.	12 in.
23. 50 ft	20 ft
24. 10 ft	50 ft
25. 0.6 m	1.2 m

21. _____

22. _____

23. _____

24. _____

25. _____

In Problems 26 through 40 give the answer to the nearest tenth (if necessary, use $\pi \approx 3.14$).

26. A spherical water tank has a 24-ft radius. How many gallons of water will it hold if 1 ft^3 holds about 7.5 gal?

26. _____

27. A spherical water tank has a 7.2-m radius. How many liters of water will it hold if 1 m^3 holds 1000 L?

27. _____

28. The fuel tanks on some ships are spheres, of which only the top halves are above deck. If one of these tanks is 120 ft in diameter, how many gallons of fuel does it hold (1 ft^3 = 7.5 gal)?

28. _____

29. A popular-sized can in American supermarkets is 3 in. in diameter and 4 in. high (inside dimensions). How many fluid ounces will one of these cans hold if 1 in^3 = 0.6 fl oz?

29. _____

30. An ice cream cone is 7 cm in diameter and 10 cm deep. The inside of the cone is packed with ice cream and a hemisphere of ice cream is put on top. If ice cream weighs $\frac{1}{2}$ g/cm^3, how many grams of ice cream are there in all?

30. _____

31. A roadway 13.5 m wide and 0.3 m thick used 32,400 m^3 of concrete. How long was it?

31. _____

32. A driveway 12 ft wide and 4 in. thick used 80 ft^3 of concrete. How long was it?

32. _____

33. The volume of a cylinder with a 10-in. radius is 2512 in.^{3} How high is the cylinder?

33. _____

34. A 904-in.³ concrete cylinder is 18 in. high. What is its radius?

35. A geologist wishes to build a 6280-mm³ cylindrical soil sampling canister. If the cylinder is 20 mm high, what should its radius be?

36. The volume of a cone having a 3-in. radius is 47.10 in.³ What is the height of this cone?

37. A rocket cone has a 3-ft radius and holds 94.20 ft³ of radar equipment. How high is it?

38. An earth globe with a 10-in. radius is to be packaged in a rectangular box. Find the volume of the smallest box that can be used for packaging the globe.

39. Four cylindrical cans are to be packed in a rectangular box for shipping. If the diameter of each can is 4 in. and the height of each can is 8 in., find the volume of the smallest box that can be used for packaging the cans.

40. A regular can of dog food is 4 in. high and costs 50¢. The economy can is 6 in. high but costs 70¢. If the diameters of both cans are equal, which is the better buy?

34. _____

35. _____

36. _____

37. _____

38. _____

39. _____

40. _____

10.4 USING YOUR KNOWLEDGE

What is the surface area of a baseball? How much leather is needed to make a basketball? To solve these problems, you need to know the *surface area* of a sphere.

The **surface area** of a sphere of radius r is $4\pi r^2$.

The surface area of a sphere is 36π units.

1. What is the radius of this sphere?
2. What is the volume of this sphere?
3. If the number of units of volume ($\frac{4}{3}\pi r^3$) of a sphere equals the number of units of surface area ($4\pi r^2$), what are the radius, the volume, and the surface area of this sphere?

1. _____
2. _____
3. _____

SUMMARY

SECTION	ITEM	MEANING	EXAMPLE
10.1A	Point	A location in space with no breadth, no width, and no length	• A Point A
10.1A	Line $\overleftrightarrow{AB}$	A set of points extending without end in both directions	A B The line AB
10.1A	Parallel lines	Lines that never meet; the distance between them is always the same	
10.1A	Intersecting lines	Lines that cross at some point in the plane	

(Continued)

SECTION	ITEM	MEANING	EXAMPLE
10.1A	Ray, $\overrightarrow{AB}$	A half-line and its endpoint	The ray AB
10.1A	Line segment AB $(\overline{AB})$	The points A and B and the part of the line between A and B	The line segment AB
10.1A	Length of a line segment	The distance between the ends of the line segment	 $AB = 5$, $BC = 7$ and $AC = 12$
10.1B	Plane angle	The figure formed by two rays with a common endpoint	
10.1B	Vertex of an angle	The common point of the two rays	
10.1B	Sides of the angles	The rays forming the angle	
10.1B	Degree	$\frac{1}{360}$ of a complete revolution	
10.1B	Straight angle	One-half of a complete revolution	
10.1B	Right angle	One-quarter of a complete revolution	
10.1B	Vertical angles	The opposite angles formed by two intersecting lines	
10.1B	Supplementary angles	Angles whose measures add to 180°	
10.1B	Complementary angles	Angles whose measures add to 90°	
10.1B	Transversal	A line that crosses two or more other lines	
10.2	Perimeter	Distance around a polygon	The perimeter of a pentagon 100 ft on each side is $5 \cdot 100 = 500$ ft.
10.2	Trapezoid	A quadrilateral with two parallel and two nonparallel sides	 $P = s_1 + s_2 + b_1 + b_2$

SUMMARY *(Continued)*

SECTION	ITEM	MEANING	EXAMPLE
10.2	Parallelogram	A quadrilateral with both pairs of opposite sides parallel	$P = 2L + 2W$
10.2	Rectangle	A parallelogram whose angles are right angles	$P = 2L + 2W$
10.2	Square	A rectangle with four equal sides	$P = 4s$
10.2A	Circle	A figure in which all points are the same distance (the radius) from a given point (the center)	
10.2A	Circumference	The circumference C of a circle is $C = \pi d = 2\pi r$.	The circumference of a circle of radius 5 is $C = 10\pi$, or 31.4 units
10.2B	Area	The area of a region is a measure of the amount of surface in the region.	
10.2B	$A = bh$	The area of a rectangle of base b and height h	
10.2B	$A = bh$	The area of a parallelogram of base b and height h	
10.2B	$A = \frac{1}{2}bh$	The area of a triangle of base b and height h	
10.2B	$A = \pi r^2$	The area of a circle of radius r	
10.3	Acute triangle	All three angles are acute	
10.3	Right triangle	One of the angles is a right angle	
10.3	Obtuse triangle	One of the angles is an obtuse angle	
10.3	Scalene triangle	No equal sides	15, 9, 20

(Continued)

SECTION	ITEM	MEANING	EXAMPLE
10.3	Isosceles triangle	Two equal sides	
10.3	Equilateral triangle	A triangle with three equal sides	
10.3A	Pythagorean Theorem $c^2 = a^2 + b^2$	The square of the hypotenuse c of a right triangle equals the sum of the squares of the other two sides, a and b	
10.3B	Similar triangles	Two triangles are similar when they have equal corresponding angles and the corresponding sides are proportional.	
10.4	$V = abc$	The volume of a rectangular box of length a, width b, and height c	
10.4	$V = \pi r^2 h$	The volume of a circular cylinder of radius r and height h	
10.4	$V = \frac{1}{3}\pi r^2 h$	The volume of a circular cone of radius r and height h	
10.4	$V = \frac{4}{3}\pi r^3$	The volume of a sphere of radius r	

NAME

CLASS

SECTION

ANSWERS

(*If you need help with these exercises, look at the section indicated in brackets.*)

When necessary, use $\pi = 3.14$ and round answers to the nearest tenth.

1. [10.1A] In the figure, $AB = 10$. Find the length of BC in each case.

   a. $AC = 17$
   b. $AC = 19$
   c. $AC = 21$

   1. a. _____
      b. _____
      c. _____

2. [10.1B] If $\angle A$ and $\angle B$ are complementary angles, find $m\angle A$ and $m\angle B$ if:
   a. $m\angle A = 2\,m\angle B$
   b. $m\angle A = 4\,m\angle B$
   c. $m\angle A = 5\,m\angle B$

   2. a. _____
      b. _____
      c. _____

3. [10.1B] If $\angle A$ and $\angle B$ are supplementary angles, find $m\angle A$ and $m\angle B$ if:
   a. $m\angle A = 2\,m\angle B$
   b. $m\angle A = 4\,m\angle B$
   c. $m\angle A = 5\,m\angle B$

   3. a. _____
      b. _____
      c. _____

Use the figure in Problems 4 through 6. (Lines PQ and RS are parallel.)

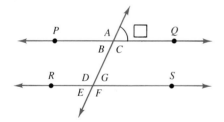

4. [10.1B]
   a. Find $m\angle C$ if the angle indicated by the box is 65°.
   b. Find $m\angle C$ if the angle indicated by the box is 67°.
   c. Find $m\angle C$ if the angle indicated by the box is 69°.

   4. a. _____
      b. _____
      c. _____

5. [10.1B]
   a. Find $m\angle E$ if the angle indicated by the box is 65°.
   b. Find $m\angle E$ if the angle indicated by the box is 67°.
   c. Find $m\angle E$ if the angle indicated by the box is 69°.

   5. a. _____
      b. _____
      c. _____

6. [10.1B]
   a. Find $m\angle D$ if the angle indicated by the box is 65°.
   b. Find $m\angle D$ if the angle indicated by the box is 67°.
   c. Find $m\angle D$ if the angle indicated by the box is 69°.

   6. a. _____
      b. _____
      c. _____

7. [10.1C] In a triangle RST, $m \angle R = 55°$ find $m \angle S$ if:
    a. $m \angle T = 50°$
    b. $m \angle T = 60°$
    c. $m \angle T = 70°$

8. [10.2A] Find the perimeter of a rectangle that is:
    a. 40.7 in. long and 20.8 in. wide
    b. 50.7 in. long and 30.8 in. wide
    c. 60.7 in. long and 40.8 in. wide

9. [10.2A] The perimeter of a rectangle is 300 in. How wide is the rectangle if it is:
    a. 100 in. long?
    b. 110 in. long?
    c. 120 in. long?

10. [10.2A] Find the diameter of a circle with circumference:
    a. 18.84 in.
    b. 12.56 in.
    c. 9.42 in.

11. [10.2B] The area of a triangle is 315 in². Find its base if:
    a. The height is 2.5 in.
    b. The height is 3.5 in.
    c. The height is 4.5 in.

12. [10.2B] Find the diameter of a circle whose area is:
    a. 12.56 cm²
    b. 28.26 cm²
    c. 50.24 cm²

13. [10.3A] The length of one leg of a right triangle is 7 in. Find the length of the other leg if:
    a. The hypotenuse is 11 in.
    b. The hypotenuse is 12 in.
    c. The hypotenuse is 13 in.

14. [10.3A] Find the perimeter of an isosceles right triangle if a leg is:
    a. 40 in. long
    b. 50 in. long
    c. 60 in. long

15. [10.3A] Find the perimeter of a 30°-60°-90° triangle if the shorter leg is:
    a. 3.2 in. long
    b. 4.2 in. long
    c. 5.2 in. long

16. [10.3B] Triangles ABC and XYZ are similar with $\angle A = \angle X$ and $\angle B = \angle Y$. If AB, BC, and AC are 8 in., 9 in., and 12 in. long, respectively, find the length of XZ if the length of XY is:
    a. 16 in.
    b. 24 in.
    c. 32 in.

17. [10.4] Find the volume of a cube that is:
    a. 3 in. on each side
    b. 4 in. on each side
    c. 5 in. on each side

7. a. _____
   b. _____
   c. _____

8. a. _____
   b. _____
   c. _____

9. a. _____
   b. _____
   c. _____

10. a. _____
    b. _____
    c. _____

11. a. _____
    b. _____
    c. _____

12. a. _____
    b. _____
    c. _____

13. a. _____
    b. _____
    c. _____

14. a. _____
    b. _____
    c. _____

15. a. _____
    b. _____
    c. _____

16. a. _____
    b. _____
    c. _____

17. a. _____
    b. _____
    c. _____

18. [10.4] A water tank is in the shape of 10-ft-high cone (vertex down) surmounted by a 40-ft-high cylinder. Find its capacity if $1 \text{ ft}^3 = 7.5$ gallons and the common diameter of the cylinder and the cone is:
   a. 12 ft
   b. 10 ft
   c. 8 ft

18. a. _____
    b. _____
    c. _____

19. [10.4] Find the volume of a sphere whose diameter is:
   a. 4 cm
   b. 6 cm
   c. 8 cm

19. a. _____
    b. _____
    c. _____

20. [10.4] Find the height of a cylindrical tank if the volume and the diameter are:
   a. 12.56 in.³ and 4 in.
   b. 28.26 in.³ and 6 in.
   c. 50.24 in.³ and 8 in.

20. a. _____
    b. _____
    c. _____

NAME

CLASS

SECTION

(Answers on page 706)

ANSWERS

1. If $AB = 10$ and $AC = 18$, find the length of BC.

1. _____

2. If $\angle A$ and $\angle B$ are complementary angles and $m \angle A = 5\, m \angle B$, find $m \angle A$ and $m \angle B$.

2. _____

3. If $\angle A$ and $\angle B$ are supplementary angles and $m \angle A = 5\, m \angle B$, find $m \angle A$ and $m \angle B$.

3. _____

Use the figure in Problems 4 through 6.

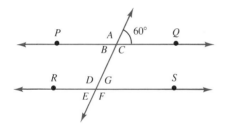

Lines PQ and RS are parallel.

4. Find $m \angle C$.

4. _____

5. Find $m \angle E$.

5. _____

6. Find $m \angle D$.

6. _____

7. In a triangle RST, $m \angle R = 53°$ and $m \angle T = 48°$. Find $m \angle S$.

7. _____

8. Find the perimeter of a rectangle 40.7 cm long and 30.8 cm wide.

8. _____

NAME _____

9. The perimeter of a rectangle is 128 in. If the rectangle is 40 in. long, how wide is it?

9. _____

10. The circumference of a circle is 37.68 ft. Find its diameter (use $\pi = 3.14$).

10. _____

11. The area of a triangle is 18 in.^{2}. If the height h is 4.5 in., find the base.

11. _____

12. The area of a circle is 78.5 in.^{2}. Find its diameter. ($\pi \approx 3.14$.)

12. _____

13. The length of one of the legs in a right triangle is 6 in. If the hypotenuse is 8 in., find the length of the other leg.

13. _____

14. One leg of an isosceles right triangle is 20 in. long. Find the perimeter of the triangle.

14. _____

15. The shorter leg of a 30°-60°-90° triangle is 3.4 in. long. Find the perimeter.

15. _____

16. Triangles ABC and XYZ are similar with $\angle A = \angle X$ and $\angle B = \angle Y$. If AB, BC, and AC are 9 in., 11 in., and 12 in. long, respectively, and XY is 10.8 in. long, find the length of XZ.

16. _____

17. Find the volume of a cube 5 in. on each side.

17. _____

18. A water tank is in the shape of a 10-ft-high cone (vertex down) surmounted by a 40-ft-high cylinder. If the diameter of the cylinder and the cone is 18 ft, what is the capacity of the tank in gallons? (Use $\pi \approx 3.14$; 1 ft^3 = 7.5 gal.)

18. _____

19. Find the volume of a sphere 10 ft in diameter. (Use $\pi \approx 3.14$; answer to the nearest tenth.)

19. _____

20. The volume of a cylindrical can is 62.80 in.^{3}. If the diameter is 4 in., find the height. (Use $\pi \approx 3.14$.)

20. _____

IF YOU MISSED QUESTION	SECTION	EXAMPLES	PAGE	ANSWERS
1	10.1	1	662	1. 8
2	10.1	2	664	2. 75°, 15°
3	10.1	3	664	3. 150°, 30°
4	10.1	4	665	4. 120°
5	10.1	4	665	5. 60°
6	10.1	4	665	6. 120°
7	10.1	5	666	7. 79°
8	10.2	1	672	8. 143 cm
9	10.2	2	673	9. 24 in.
10	10.2	3	673	10. 12 ft
11	10.2	4	675	11. 8 in.
12	10.2	5	675	12. 10 in.
13	10.3	1	682	13. $2\sqrt{7}$ in.
14	10.3	2	682	14. $(40 + 20\sqrt{2})$ in.
15	10.3	3	683	15. $(10.2 + 3.4\sqrt{3})$ in.
16	10.3	4	684	16. 14.4 in.
17	10.4	1	689	17. 125 in.^{3}
18	10.4	2	690	18. 82,660.5 gal
19	10.4	3	690	19. 523.3 in.
20	10.4	4	691	20. 5 in.

ANSWERS

EXERCISE 1.1

A 1. $\dfrac{28}{1}$

3. $\dfrac{-42}{1}$

5. $\dfrac{0}{1}$

7. $\dfrac{-1}{1}$

B 9. 3
11. 35
13. 25
15. 21
17. 32
19. 6
21. 14
23. 30
25. 3
27. 1
29. 2

C 31. $\dfrac{5}{4}$

33. $\dfrac{1}{4}$

35. $\dfrac{7}{3}$

37. $\dfrac{2}{3}$

39. 4

D 41. $\dfrac{1}{3}$

43. $\dfrac{2}{5}$

45. $\dfrac{3}{14}$

47. $\dfrac{56}{72}$

49. a. $\dfrac{1}{2}$

b. $\dfrac{24}{48}$

1.1 USING YOUR KNOWLEDGE

1. $\dfrac{5}{6}$

3. a. 6
   b. $7000
5. a. 125

b. $\dfrac{3}{10}$

EXERCISE 1.2

A 1. $\dfrac{14}{9}$

3. $\dfrac{7}{5}$

5. 2
7. 8

9. $\dfrac{39}{7}$

B 11. $\dfrac{35}{3}$

13. $\dfrac{2}{3}$

15. $\dfrac{3}{2}$

17. $\dfrac{16}{5}$

19. $\dfrac{2}{5}$

C 21. $\dfrac{3}{5}$

23. 1

25. $\dfrac{13}{8}$

27. $\dfrac{17}{15}$

29. $\dfrac{23}{6}$

31. $\dfrac{31}{10}$

33. $\dfrac{193}{28}$

D 35. $\dfrac{3}{8}$

37. $\dfrac{1}{6}$

39. $\dfrac{11}{20}$

41. $\dfrac{34}{75}$

43. $\dfrac{9}{20}$

45. $\dfrac{5}{4}$

47. $\dfrac{11}{18}$

49. $\dfrac{37}{15}$

E 51. 75 lb
53. 5600

55. $16\dfrac{5}{8}$ yd^3

57. $\dfrac{3}{10}$

59. $7\dfrac{7}{10}$

✓ **SKILL CHECKER**

61. 0.75
63. 0.66

1.2 USING YOUR KNOWLEDGE

1. 4 packs of hot dogs,
   5 packs of buns
3. $3\dfrac{1}{8}$

EXERCISE 1.3

A 1. $4 + \dfrac{7}{10}$

3. $5 + \dfrac{6}{10} + \dfrac{2}{100}$

5. $10 + 6 + \dfrac{1}{10} + \dfrac{2}{100} + \dfrac{3}{1000}$

7. $40 + 9 + \dfrac{1}{100} + \dfrac{2}{1000}$

9. $50 + 7 + \dfrac{1}{10} + \dfrac{4}{1000}$

B 11. 0.9
13. 0.09
15. 0.011
17. 0.002

19. 0.187
21. 0.2
23. 0.875
25. 0.1875
27. $0.\overline{2}$
29. $0.\overline{54}$
31. $0.\overline{27}$
33. $0.1\overline{6}$
35. $1.\overline{1}$

C 37. $\dfrac{3}{5}$

39. $\dfrac{9}{10}$

41. $\dfrac{3}{50}$

43. $\dfrac{3}{25}$

45. $\dfrac{27}{500}$

47. $\dfrac{213}{100}$

49. $\dfrac{959}{50}$

51. $\dfrac{161}{4}$

53. $\dfrac{1527}{40}$

D 55. $8 + \dfrac{9}{10}$

57. a. 0.75
   b. 0.45

59. a. $\dfrac{31}{200}$

   b. $\dfrac{39}{50}$

✓ **SKILL CHECKER**

61. $\dfrac{9}{25}$

63. $\dfrac{421}{1000}$

1.3 USING YOUR KNOWLEDGE

1. a. 0.875
   b. 0.75
3. a. 0.375
   b. 0.25
5. The prime factors of the denominators in the terminating fractions are only twos and/or fives.

EXERCISE 1.4

A 1. $\dfrac{3}{25}$

   3. $\dfrac{37}{200}$

5. $\dfrac{3}{50}$

7. $\dfrac{81}{1000}$

9. $\dfrac{21}{200}$

B 11. 0.33
13. 0.05
15. 3
17. 0.118
19. 0.005
C 21. 5%
23. 39%
25. 41.6%
27. 0.3%
29. 100%
D 31. 12.5%
33. 60%
35. $16.\overline{6}\%$
37. 15%
39. 7.5%
E 41. $1502
43. a. $15
   b. $35

45. $\dfrac{16}{25}$

47. a. 0.95
   b. 0.90
49. a. 0.477
   b. 0.241
51. a. 84%
   b. 30%
53. a. 2%

   b. $3\dfrac{19}{27}\%$

55. a. 45%
   b. 27.5%

✓ **SKILL CHECKER**

57. 10
59. 49

1.4 USING YOUR KNOWLEDGE

1. 70%
3. a. 40%

   b. $\dfrac{2}{5}$

5. a. $\dfrac{49}{100}$

   b. 0.49
7. 5%
9. 1%

EXERCISE 1.5

A 1. 23
   3. 36
   5. 57
   7. 36
   9. 78

B 11. Commutative +
13. Commutative ×
15. Distributive
17. Commutative ×
19. Associative +
C 21. $24 + 6x$
23. $8x + 8y + 8z$
25. $6x + 42$
27. $ab + 5b$
29. $30 + 6b$
D 31. $2a + 11$
33. $7a + 17$
35. $36a$
37. $20a$
39. $42a$
41. 3
43. 5
45. a
E 47. Identity for ×
49. Identity for +

✓ **SKILL CHECKER**

51. 0.0006
53. 3.22

1.5 USING YOUR KNOWLEDGE

1. 266
3. 276

EXERCISE 1.6

A 1. a. 120
      b. 275 mi
   3. 110 lb
   5. 59° F

B 7. a. $EER = \dfrac{BTU}{W}$

      b. 9
   9. a. $S = C + m$
      b. $67
C 11. 200 cm²; 60 cm
13. 13.02 m²; 14.6 m
15. 111.39 yd²; 44 yd
17. 35 cm²
19. 300 yd²
21. 314 in.²
23. 1256 ft²
25. 153.86 m²
27. 94.2 in.
29. 62.8 ft
D 31. 35
33. 10
35. 2
37. 1
39. 8
41. 2
43. 2
45. 6
47. 8
49. 1
51. 300 ft²
53. 16,328 ft
55. 300 ft²

57. $\dfrac{7}{13}$

59. $\dfrac{11}{24}$

1.6 USING YOUR KNOWLEDGE

1. 35 yd^2
3. 8
5. a. 1500 W
   b. 7.5 cents

REVIEW EXERCISES CHAPTER 1

1. a. $\dfrac{-16}{1}$

   b. $\dfrac{-14}{1}$

   c. $\dfrac{-97}{1}$

2. a. $\dfrac{6}{16}$

   b. $\dfrac{9}{24}$

   c. $\dfrac{15}{40}$

3. a. $\dfrac{1}{4}$

   b. $\dfrac{2}{8}$

   c. $\dfrac{3}{12}$

4. a. $\dfrac{2}{3}$

   b. $\dfrac{1}{2}$

   c. $\dfrac{3}{5}$

5. a. 2
   b. 4
   c. 2
6. a. 2
   b. 3
   c. 2

7. a. $\dfrac{31}{24}$ or $1\dfrac{7}{24}$

   b. $\dfrac{43}{12}$ or $3\dfrac{7}{12}$

   c. $\dfrac{129}{40}$ or $3\dfrac{9}{40}$

8. a. $\dfrac{19}{15}$

   b. $\dfrac{34}{15}$

   c. $\dfrac{49}{15}$

9. a. $60 + 8 + \dfrac{4}{10} + \dfrac{2}{100} + \dfrac{1}{1000}$

   b. $60 + 8 + \dfrac{4}{10} + \dfrac{2}{100} + \dfrac{2}{1000}$

   c. $60 + 8 + \dfrac{4}{10} + \dfrac{2}{100} + \dfrac{3}{1000}$

10. a. 0.021
    b. 0.023
    c. 0.027
11. a. $0.\overline{18}$
    b. $0.\overline{27}$
    c. $0.\overline{36}$

12. a. $\dfrac{3}{200}$

    b. $\dfrac{1}{40}$

    c. $\dfrac{9}{200}$

13. a. $\dfrac{157}{50}$

    b. $\dfrac{79}{25}$

    c. $\dfrac{159}{50}$

14. a. $\dfrac{117}{200}$

    b. $\dfrac{137}{200}$

    c. $\dfrac{157}{200}$

15. a. 0.548
    b. 0.648
    c. 0.748
16. a. 19%
    b. 29%
    c. 39%

17. a. $12\dfrac{1}{2}\%$ or 12.5%

    b. $62\dfrac{1}{2}\%$ or 62.5%

    c. $87\dfrac{1}{2}\%$ or 87.5%

18. a. Commutative +
    b. Associative ×
    c. Associative +
19. a. 28
    b. 56
20. a. $8x + 40$
    b. $7a + 7b + 7c$
21. a. $4x + 7$
    b. $5x + 9$
    c. $7x + 7$
22. a. $10x$
    b. $18x$
    c. $10x$

23. a. 4
    b. 6
    c. 6
24. a. $A = 5.89$ in.^{2},
       $P = 10$ in.
    b. $A = 5.76$ in.^{2},
       $P = 10$ in.
    c. $A = 5.61$ in.^{2},
       $P = 10$ in.
25. a. $A = 3.14$ in.^{2},
       $C = 6.28$ in.
    b. $A = 12.56$ in.^{2},
       $C = 12.56$ in.
    c. $A = 28.26$ in.^{2},
       $C = 18.84$ in.

EXERCISE 2.1

A 1. -4
   3. 49
B 5. 2
   7. 45
   9. 3
C 11. 6
   13. -4
   15. 1
   17. -7
   19. 0
   21. 3
   23. 13
   25. 2
   27. -9
   29. -12
D 31. -16
   33. -20
   35. -6
   37. 16
   39. -4
E 41. -7
   43. 9
   45. -4
   47. -23
   49. 28
F 51. 3500° C
   53. $46
   55. 14° C

✓ **SKILL CHECKER**

57. $\dfrac{7}{15}$

59. 9

2.1 USING YOUR KNOWLEDGE

1. 799
3. 284
5. 2262

EXERCISE 2.2

A 1. 36
   3. -40
   5. -81
   7. 18
   9. 18

B 11. -16
13. 25
15. -125
17. 1296
19. -32
C 21. 7
23. -5
25. -3
27. 0
29. Undefined
31. 0
33. 5
35. 5
37. -2
39. -6
D 41. $+15$ (gain)
43. -5 (loss)
45. 6 min

47. 8
49. 12

2.2 USING YOUR KNOWLEDGE

1. -3
3. 0
5. 0

EXERCISE 2.3

A 1. $-\dfrac{7}{3}$

3. 6.4

5. $-3\dfrac{1}{7}$

B 7. $\dfrac{4}{5}$

9. 3.4

11. $1\dfrac{1}{2}$

C 13. -4.7
15. -5.4
17. -8.6

19. $\dfrac{3}{7}$

21. $-\dfrac{1}{2}$

23. $-\dfrac{1}{12}$

25. $\dfrac{7}{12}$

27. $-\dfrac{13}{21}$

29. $-\dfrac{31}{18}$

D 31. -2.6
33. 5.2

35. -3.7

37. $\dfrac{4}{7}$

39. $-\dfrac{29}{12}$

E 41. -7.26
43. 2.86

45. $-\dfrac{25}{42}$

47. $\dfrac{1}{4}$

49. $-\dfrac{15}{4}$

51. $\dfrac{4}{3}$

53. $-\dfrac{1}{6}$

55. $-\dfrac{7}{2}$

57. -1

59. $\dfrac{1}{7}$

61. $-\dfrac{21}{20}$

63. $\dfrac{4}{7}$

65. $-\dfrac{5}{7}$

67. $-\dfrac{1}{2}$

69. $\dfrac{1}{6}$

71. $3x + 18$
73. $12 + 7a$

2.3 USING YOUR KNOWLEDGE

1. -1.35
3. -3.045
5. 3.875

EXERCISE 2.4

A 1. $4x - 4y$
3. $9a - 9b$
5. $12x - 6$

B 7. $-\dfrac{3}{2}a + \dfrac{6}{7}$

9. $-2x + 6y$
11. $-2.1 - 3y$
13. $-4a - 20$
15. $-6x - xy$
17. $-8x + 8y$

C 19. $-6a + 21b$
21. $0.5x + 0.5y - 1.0$

23. $\dfrac{6}{5}a - \dfrac{6}{5}b + 6$

25. $-2x + 2y - 8$
27. $-0.3x - 0.3y + 1.8$

29. $-\dfrac{5}{2}a + 5b - \dfrac{5}{2}c + \dfrac{5}{2}$

D 31. $C = \dfrac{5}{9}(F - 32)$

33. 81
35. 81
37. -27
39. 64

2.4 USING YOUR KNOWLEDGE

1. a. $v_a = \dfrac{1}{2}(v_1 + v_2)$

b. $v_a = \dfrac{1}{2}v_1 + \dfrac{1}{2}v_2$

3. a. $M = m(v_1 + v_2)$
b. $M = mv_1 + mv_2$
5. $L = \pi r_1 + \pi r_2 + 2l$

EXERCISE 2.5

A 1. $a + c$
3. $3x + y$
5. $9x + 17y$
7. $3a - 2b$
9. $-2x - 5$
B 11. $7a$

13. $\dfrac{1}{7}a$

15. bd
17. xyz
19. $-b(c + d)$
21. $(a - b)x$
23. $(x - 3y)(x + 7y)$
25. $(c - 4d)(x + y)$
C 27. $-a^2c$
29. $(a + c)^2$
31. $(2x)^2$
33. n^3
D 35. $(2x - y)^2$
37. $(yz)^4$
39. $(x + y)^4$

E 41. $\dfrac{a}{x + y}$

43. $\dfrac{a - b}{c}$

45. $\dfrac{y}{x}$

47. $\dfrac{p - q}{p + q}$

49. $\dfrac{x + 2y}{x - 2y}$

51. $H = 17 - \dfrac{1}{2}A$, 2 terms

53. $\dfrac{1}{f} = \dfrac{1}{u} + \dfrac{1}{v}$, 2 terms

55. $F = \dfrac{n}{4} + 40$, 2 terms

✓ **SKILL CHECKER**

57. -11
59. -4
61. -5
63. -13
65. 17
67. -9
69. -28

2.5 USING YOUR KNOWLEDGE

1. $V = IR$
3. $P = P_A + P_B + P_C$
5. $D = RT$
7. $E = mc^2$
9. $h^2 = a^2 + b^2$

EXERCISE 2.6

A　1. $11a$
　　3. $-5c$
　　5. $12n^2$
　　7. $-7ab^2$
　　9. $3abc$
　　11. $13ab$
　　13. $2x^2y - 9xy^2$
　　15. $3ab^2c$
　　17. $-6ab + 7xy$
　　19. $\dfrac{4}{7}a^2b + \dfrac{3}{5}a$

B　21. $11x$
　　23. $8ab$
　　25. $-7a^2b$
　　27. 0
　　29. 0

C　31. $10xy - 4$
　　33. $-2R + 6$
　　35. $-L - 2W$

D　37. $-3x - 1$
　　39. $\dfrac{x}{9} + 2$
　　41. $6a + 2b$
　　43. $3x - 4y$
　　45. $x - 5y + 36$

E　47. $-2x^3 + 9x^2 - 3x + 12$
　　49. $x^2 - \dfrac{2}{5}x + \dfrac{1}{2}$
　　51. $4a - 11$
　　53. $-7a + 10b - 3$
　　55. $-4.8x + 3.4y + 5$

✓ **SKILL CHECKER**

57. -28
59. 24

61. -2
63. 4

2.6 USING YOUR KNOWLEDGE

1. $4S$
3. 18 ft, 6 in.
5. $2h + 2w + L$
7. $7x + 2$

EXERCISE 2.7

A　1. $24x^3$
　　3. $30a^4b^3$
　　5. $3x^3y^3$
　　7. $-\dfrac{b^5c}{5}$
　　9. $\dfrac{3x^3y^3z^6}{2}$
　　11. $-30x^5y^3z^8$
　　13. $15a^4b^3c^4$
　　15. $24a^3b^5c^5$

B　17. x^4
　　19. $-\dfrac{a^2}{2}$
　　21. $2x^3y^2$
　　23. $-\dfrac{1}{2x}$
　　25. $\dfrac{2y^4}{3a^2}$
　　27. $\dfrac{3a}{4b^2c}$
　　29. $\dfrac{3a^4}{2}$
　　31. $-x^6y$
　　33. $-x^2y^2$

C　35. 367,200,000
　　37. 20 years
　　39. 80,000,000
　　41. 18×10^{16} mi or 1.8×10^{17} mi
　　43. 9.57×10^8 mi
　　45. 5×10^{10} lb

✓ **SKILL CHECKER**

47. 64
49. $4x^2$

2.7 USING YOUR KNOWLEDGE

1. a. 123,454,321
　 b. 12,345,654,321
3. a. 5^2
　 b. 7^2

EXERCISE 2.8

A　1. $\dfrac{1}{16}$
　　3. $\dfrac{1}{125}$

5. $\dfrac{1}{81}$

B　7. 2^{-3}
　　9. 4^{-5}
　　11. 3^{-5}

C　13. 3
　　15. 4
　　17. $\dfrac{1}{16}$
　　19. $\dfrac{1}{216}$
　　21. $\dfrac{1}{64}$
　　23. x^2
　　25. y^2
　　27. $\dfrac{1}{a^5}$
　　29. $\dfrac{1}{x^2}$
　　31. $\dfrac{1}{x^2}$
　　33. $\dfrac{1}{a^5}$
　　35. 1

D　37. 243
　　39. $\dfrac{1}{64}$
　　41. $\dfrac{1}{y^2}$
　　43. x^3
　　45. $\dfrac{1}{x^2}$
　　47. $\dfrac{1}{x^7}$
　　49. x^3

E　51. 81
　　53. $\dfrac{1}{9}$
　　55. 64
　　57. $\dfrac{1}{9}$
　　59. $\dfrac{1}{x^9}$
　　61. $\dfrac{1}{y^6}$
　　63. a^6
　　65. $\dfrac{8x^9}{y^6}$
　　67. $\dfrac{4y^6}{x^4}$
　　69. $-\dfrac{1}{27x^9y^6}$
　　71. $\dfrac{1}{x^{12}y^6}$

73. $x^{12}y^{12}$
F 75. $1259.71
77. $1464.10

79. 731

2.8 USING YOUR KNOWLEDGE

1. 5.335 billion
3. 4.929 billion

EXERCISE 2.9

A 1. 5.4×10^7
   3. 2.48×10^8
   5. 1.9×10^9
   7. 2.4×10^{-4}
   9. 2×10^{-9}
   11. 153
   13. 8,000,000
   15. 6,850,000,000
   17. 0.23
   19. 0.00025
B 21. 1.5×10^{10}
   23. 3.06×10^4
   25. 1.24×10^{-4}
   27. 2×10^3
   29. 2.5×10^{-3}
   31. a. 2×10^{10}
       b. 20,000,000,000
   33. 3.3
   35. 2×10^7

✓ **SKILL CHECKER**

37. $\dfrac{11}{10}$
39. -4

2.9 USING YOUR KNOWLEDGE

1. 2.99792458×10^8
3. 3.09×10^{13}
5. 3.27

REVIEW EXERCISES CHAPTER 2

1. a. 5
   b. 3
   c. 8
2. a. 7
   b. 4
   c. 9
3. a. (i)   4
      (ii) 5
      (iii) 2
   b. (i) -5
      (ii) -9
      (iii) -8
   c. (i) -6
      (ii) -5
      (iii) -3

d. (i)   3
   (ii) 7
   (iii) 4
4. a. (i) -20
      (ii) -15
      (iii) -10
   b. (i) -13
      (ii) -6
      (iii) -3
   c. (i)   1
      (ii) 2
      (iii) 1
5. a. (i)   12
      (ii) 30
      (iii) 27
   b. (i) -35
      (ii) -27
      (iii) -48
   c. (i) -21
      (ii) -36
      (iii) -54
6. a. (i)   9
      (ii) 16
      (iii) 25
   b. (i) -9
      (ii) -16
      (iii) -25
7. a. (i) -5
      (ii) -8
      (iii) -4
   b. (i)   4
      (ii) 2
      (iii) 7
   c. (i)   2
      (ii) 2
      (iii) 5
8. a. 3.4
   b. 5.6
   c. 7.2
9. a. $1\dfrac{1}{2}$

   b. $5\dfrac{1}{4}$

   c. $9\dfrac{3}{7}$

10. a. (i) -4.2
       (ii) -4.2
       (iii) -6.4

    b. (i) $-\dfrac{5}{4}$

       (ii) $-\dfrac{29}{40}$

       (iii) $-\dfrac{7}{8}$

11. a. (i) -2.4
       (ii) -3.2
       (iii) -6.2

    b. (i) $-\dfrac{17}{12}$

       (ii) $-\dfrac{11}{15}$

       (iii) $-\dfrac{19}{24}$

12. a. (i) -19.52
       (ii) -22.72
       (iii) -25.92

    b. (i) $\dfrac{6}{35}$

       (ii) $\dfrac{12}{35}$

       (iii) $\dfrac{8}{35}$

13. a. (i) $-\dfrac{4}{3}$

       (ii) $-\dfrac{3}{2}$

       (iii) $-\dfrac{2}{3}$

    b. (i) $\dfrac{5}{27}$

       (ii) $\dfrac{1}{3}$

       (iii) $\dfrac{27}{32}$

14. a. (i) $2x - 2$
       (ii) $3x - 6$
       (iii) $4x - 12$
    b. (i) $-2x + 3y$
       (ii) $-3x + 4y$
       (iii) $-4x + 5y$
    c. (i) $-2x - 4y + 2$
       (ii) $-3x - 6y + 6$
       (iii) $-4x - 8y + 12$

15. a. $\dfrac{x - y}{z}$

    b. $\dfrac{a - b}{c}$

    c. $\dfrac{p - q}{r}$

16. a. (i) $-abc$
       (ii) $-pqr$
       (iii) $-bcd$
    b. (i) 3^2x^3y
       (ii) $4^2a^3b^2$
       (iii) $2^2a^2b^4$
17. a. (i) $-7x$
       (ii) $-7x$
       (iii) $-10x$
    b. (i) $-2ab^2$
       (ii) $-2xy^2$
       (iii) $-2x^2y$
18. a. (i)   1
       (ii) $2x$
       (iii) $3x + 1$
    b. (i) $2x - 5$
       (ii) $3x - 4$
       (iii) $4x - 6$
19. a. (i) $-15a^3b^4$
       (ii) $-24a^3b^5$
       (iii) $-35a^3b^4$

b. (i) $6x^3y^3z^5$
  (ii) $12x^3y^4z^3$
  (iii) $20x^2y^3z^4$

20. a. (i) $-2x^5y^4$
  (ii) $-6x^6y^3$
  (iii) $-2x^7y^3$

  b. (i) $\dfrac{y^6}{2x^3}$

   (ii) $\dfrac{y^7}{2x^5}$

   (iii) $\dfrac{y^6}{3x^7}$

21. a. (i) 8
  (ii) 8
  (iii) 8

  b. (i) $\dfrac{1}{y^8}$

   (ii) $\dfrac{1}{y^5}$

   (iii) $\dfrac{1}{y^6}$

22. a. (i) $\dfrac{1}{x^4}$

   (ii) $\dfrac{1}{x^6}$

   (iii) $\dfrac{1}{x^8}$

  b. (i) 1
  (ii) 1
  (iii) 1

  c. (i) x
  (ii) x^3
  (iii) x^2

23. a. (i) a^6
  (ii) a^{12}
  (iii) a^{20}

  b. (i) $\dfrac{4x^4}{y^6}$

   (ii) $\dfrac{27x^6}{y^6}$

   (iii) $\dfrac{16x^8}{y^{12}}$

  c. (i) $\dfrac{x^4}{4y^8}$

   (ii) $\dfrac{x^4}{9y^8}$

   (iii) $\dfrac{x^6}{64y^9}$

24. a. (i) 4.4×10^7
  (ii) 4.5×10^7
  (iii) 4.6×10^5

  b. (i) 1.4×10^{-3}
  (ii) 1.5×10^{-4}
  (iii) 1.6×10^{-5}

25. a. (i) 2.2×10^5
  (ii) 9.3×10^6
  (iii) 1.24×10^8

b. (i) 5
  (ii) 6
  (iii) 7

EXERCISE 3.1

A 1. Yes
  3. Yes
  5. Yes
  7. No
  9. No

B 11. $x = 14$
  13. $m = 19$
  15. $y = 5$
  17. $k = 21$

  19. $z = \dfrac{11}{12}$

  21. $x = \dfrac{7}{2}$

C 23. $c = 0$
  25. $x = -3$
  27. $y = 0$
  29. $C = 3$

D 31. $p = -9$
  33. $x = -3$
  35. $m = 6$
  37. $y = 18$
  39. $a = -7$
  41. $c = -8$
  43. $x = -2$
  45. $g = -3$
  47. $x = 28$
  49. $z = 36$
  51. $b = 6$

  53. $p = \dfrac{20}{3}$

  55. $r = \dfrac{9}{8}$

  57. \$9.41
  59. 184.7

✓ **SKILL CHECKER**

61. -20

63. $-\dfrac{1}{2}$

65. $\dfrac{2}{3}$

67. 48
69. 40

3.1 USING YOUR KNOWLEDGE

1. No
3. No

EXERCISE 3.2

A 1. $x = 35$
  3. $x = -8$
  5. $b = -15$

7. $f = 6$

9. $v = \dfrac{4}{3}$

11. $x = -\dfrac{15}{4}$

B 13. $z = 11$
  15. $x = -7$
  17. $c = -7$
  19. $x = 7$

21. $y = -\dfrac{11}{3}$

23. $a = -0.6$

25. $t = \dfrac{3}{2}$

C 27. $x = -2.25$
  29. $C = -8$
  31. $a = 12$

33. $y = -\dfrac{1}{2}$ or -0.5

35. $p = 0$
37. $t = -30$
39. $x = -0.02$

D 41. $y = 12$
  43. $x = 21$
  45. $x = 20$
  47. $t = 24$
  49. $x = 1$
  51. $c = 15$
  53. $W = 10$
  55. $x = 4$
  57. $x = 10$
  59. $x = 3$
  61. 12
  63. 28
  65. 50%
  67. 150
  69. 20
  71. \$24
  73. \$12
  75. 150

✓ **SKILL CHECKER**

77. $40 - 5y$
79. $54 - 27y$
81. $-15x + 20$
83. 4
85. 3

3.2 USING YOUR KNOWLEDGE

1. \$40,000
3. 40%

EXERCISE 3.3

A 1. $x = 4$
  3. $y = 1$
  5. $z = 2$

7. $y = \dfrac{14}{5}$

9. $x = 3$
11. $x = -4$

13. $v = -8$

15. $m = -4$

17. $z = -\dfrac{2}{3}$

19. $x = 0$

21. $a = \dfrac{1}{13}$

23. $c = 35.6$

25. $x = -\dfrac{11}{80}$

27. $x = \dfrac{9}{2}$

29. $x = -20$

31. $x = -5$

33. $h = 0$

35. $w = -1$

37. $x = -3$

39. $x = \dfrac{5}{2}$

41. $x = \dfrac{1}{8}$

43. $x = 2$

45. $x = 10$

47. $x = \dfrac{27}{20}$

49. $x = \dfrac{69}{7}$

51. $r = \dfrac{C}{2\pi}$

53. $r^2 = \dfrac{A}{\pi}$

55. $s = \dfrac{A - \pi r^2}{\pi r}$

57. $V_2 = \dfrac{P_1 V_1}{P_2}$

59. $H = \dfrac{f + SH}{S}$

61. $\dfrac{a + b}{c}$

63. $a(b + c)$

65. $a - bc$

3.3 USING YOUR KNOWLEDGE

1. $C = 0.10m + 55$

3. Mileage rate

5. a. $C = 0.75S$

   b. $60

EXERCISE 3.4

A   1. $<$

3. $>$

5. $<$

7. $>$

9. $<$

B 11. $x \le 1$

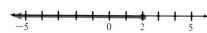

13. $y \le 2$

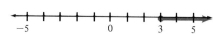

15. $x \ge 3$

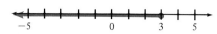

17. $a \le 3$

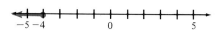

19. $z \le -4$

21. $x \le -1$

23. $x < 0$

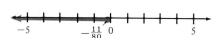

25. $x \le -\dfrac{11}{80}$

27. $x \ge -20$

29. $x \ge -2$

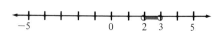

C 31. $2 < x < 3$

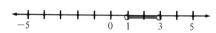

33. $1 < x < 3$

35. $-2 < x < 5$

37. $-5 < x < 1$

39. $3 < x < 5$

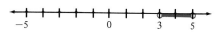

D 41. $20 < t < 40$

43. $12{,}000 < s < 13{,}000$

45. $2 \le e \le 7$

47. $\$3.50 < c < \$4.00$

49. $a < 41$

51. $w = 16$

53. $n = 40$

55. $n = 4$

3.4 USING YOUR KNOWLEDGE

1. $J = 5$ ft $= 60$ in.

3. $F = S - 3$

5. $S = 6$ ft, 5 in. $= 77$ in.

7. $B > 6$ ft, 2 in. $= 74$ in.

EXERCISE 3.5

A   1. 44, 46, 48

3. $-10, -8, -6$

5. $-13, -12$

7. 7, 8, 9

9. If the three integers are n, $n + 1$, $n + 2$, then $n + n + 2 = 2(n + 1)$ is true for any n. Thus, any three integers will do. (Try it!)

11. 20, 44

13. 39, 94

15. 43, 104

17. 37, 111, 106

19. 21, 126

B 21. 1100, 1400

23. 937

25. 222 ft

27. 3 sec

29. 304, 676

31. $T = \dfrac{20}{11}$

33. $T = 8$

35. $x = 8$

37. $P = 3500$

39. $x = 1000$

3.5 USING YOUR KNOWLEDGE

1. 84 years

A 1.

	$R \times T = D$
Bus	$R \times 8 = 400$

$R = 50$ mph

3.

	$R \times T = D$
Plane	$R \times 8 = 1632$

$R = 204$ mph

5.

	$R \times T = D$
Jet	$400 \times T = 200$

$T = \dfrac{1}{2}$ hr

7.

	$R \times T = D$
Car	$90 \times (T - 2) = 90(T - 2)$
Bus	$60 \times T = 60T$

$90(T - 2) = 60T$, $T = 6$
The car takes $6 - 2 = 4$ hr to catch the bus

9.

	$R \times T = D$
Wife	$60 \times \left(T - \dfrac{1}{2}\right) = 60\left(T - \dfrac{1}{2}\right)$
Coach	$15 \times T = 15T$

$60\left(T - \dfrac{1}{2}\right) = 15T$

$T = \dfrac{2}{3}$, $D = 15\left(\dfrac{2}{3}\right) = 10$ mi

11.

	$R \times T = D$
Car 1	$50 \times T = 50T$
Car 2	$55 \times T = 55T$

$55T + 50T = 630$, $T = 6$ hr

13.

	$R \times T = D$
Out	$480 \times T = 480T$
Return	$640 \times (7 - T) = 640(7 - T)$

$640(7 - T) = 480T$, $T = 4$
Thus, $D = 480(4) = 1920$ mi

B 15.

	%	× AMOUNT =	TOTAL
40%	0.40 ×	I	$= 0.40I$
80%	0.80 ×	10	$= 8$
65%	0.65 ×	$(I + 10)$	$= 0.65(I + 10)$

$0.40I + 8 = 0.65(I + 10)$, $I = 6$

17.

	PRICE	× AMOUNT =	TOTAL
Copper	65	× P	$= 65P$
Zinc	30	× $(70 - P)$	$= 30(70 - P)$
Brass	45	× 70	$= 3150$

$65P + 30(70 - P) = 3150$, $P = 30$
Thus, 30 lb of copper and 40 of zinc were used to make
70 lb of brass

19.

	PRICE	× AMOUNT =	TOTAL
Blue	5	× P	$= 5P$
Reg.	2	× 80	$= 160$
Mix	2.60	× $(P + 80)$	$= 2.60(P + 80)$

$5P + 160 = 2.60(P + 80)$, $P = 20$ lb

21.

	%	× AMOUNT =	TOTAL
40%	0.40 ×	$\mathcal{O}$	$= 0.40\mathcal{O}$
20%	0.20 ×	$(64 - \mathcal{O})$	$= 0.20(64 - \mathcal{O})$
30%	0.30 ×	64	$= 19.20$

$0.40\mathcal{O} + 0.20(64 - \mathcal{O}) = 19.20$, $\mathcal{O} = 32$

23.

	%	× AMOUNT =	TOTAL
50%	0.50 ×	$(30 - q)$	$= 0.50(30 - q)$
0%	0 ×	q	$= 0$
30%	0.30 ×	30	$= 9$

$0.50(30 - q) = 9$, $q = 12$

C 25.

	P	× r =	I
6%	P	× 0.06	$= 0.06P$
8%	$(20{,}000 - P)$	× 0.08	$= 0.08(20{,}000 - P)$

$0.06P + 0.08(20{,}000 - P) = 1500$
$P = \$5000$ was invested at 6% and $\$15{,}000$ at 8%

27.

P	$\times\ r\ =$	I
5%	P	$\times\ 0.05 = 0.05P$
7%	$(18{,}000 - P)$	$\times\ 0.07 = 0.07(18{,}000 - P)$

$0.05P + 0.07(18{,}000 - P) = 1100$
$P = \$8000$ was invested at 5%

29.

P	$\times\ r\ =$	I
5%	P	$\times\ 0.05 = 0.05P$
6%	$(10{,}000 - P)$	$\times\ 0.06 = 0.06(10{,}000 - P)$

$0.05P = 0.06(10{,}000 - P) + 60$
$P = \$6000$ was invested at 5% and \$4000 at 6%

✓ **SKILL CHECKER**

31. 54
33. 4

REVIEW EXERCISES CHAPTER 3

1. a. No
   b. Yes
   c. Yes
2. a. $x = \dfrac{2}{3}$
   b. $x = 1$
   c. $x = \dfrac{6}{9} = \dfrac{2}{3}$
3. a. $x = \dfrac{1}{9}$
   b. $x = \dfrac{1}{7}$
   c. $x = \dfrac{1}{18}$
4. a. $x = 5$
   b. $x = 0$
   c. $x = 3$
5. a. $x = -15$
   b. $x = -14$
   c. $x = -12$
6. a. $x = 12$
   b. $x = 15$
   c. $x = 9$
7. a. $x = 6$
   b. $x = \dfrac{24}{7}$
   c. $x = 20$
8. a. $x = 12$
   b. $x = 28$
   c. $x = 40$
9. a. $x = 17$
   b. $x = 7$
   c. $x = 9$
10. a. $x = 2$
    b. $x = 6$
    c. $x = 4$
11. a. $h = \dfrac{2A}{b}$
    b. $r = \dfrac{A}{2\pi}$
    c. $b = \dfrac{3y}{h}$
12. a.

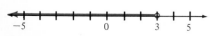

b.

c.

13. a.

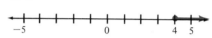

b.

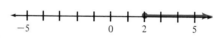

c.

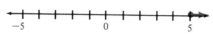

14. a.

b.

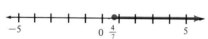

c.

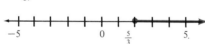

15. a. $-3 < x \le 2$

b. $-3 < x \le 2$

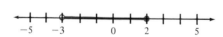

c. $-2 < x \le 1$

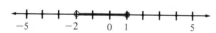

16. a. 32 and 52
    b. 14 and 33
    c. 29 and 52
17. a. \$350 and \$630
    b. \$360 and \$560
    c. \$263 and \$410
18. a. 4 hr
    b. $1\dfrac{1}{2}$ hr
    c. 2 hr
19. a. 10 lb
    b. 15 lb
    c. 25 lb
20. a. \$10,000 at 6%
       \$20,000 at 5%
    b. \$20,000 at 7%
       \$10,000 at 9%
    c. \$25,000 at 6%
       \$ 5,000 at 10%

EXERCISE 4.1

A,B 1. B, 1
    3. M, 1
    5. T, 2
    7. M, 0
    9. B, 3
B,C 11. $8x^3 - 3x$, 3
    13. $8x^2 + 4x - 7$, 2
    15. $x^2 + 5x$, 2
    17. $x^3 - x^2 + 3$, 3
    19. $4x^5 - 3x^3 + 2x^2$, 5
D 21. a. 4
      b. -8
    23. a. 7
       b. 7
    25. a. 13
       b. 9
    27. a. 9
       b. -3
    29. a. 3
       b. -1
    31. a. $(-16t^2 + 150)$ ft
       b. 134 ft
       c. 86 ft
    33. a. $(-4.9t^2 + 200)$ m
       b. 195.1 m
       c. 180.4 m

✓ **SKILL CHECKER**

35. $-7ab$
37. $3x^2y$
39. $8xy$

4.1 USING YOUR KNOWLEDGE

1. $(-32t - 10)$ ft/sec
3. a. -11.8 m/sec
   b. -21.6 m/sec

EXERCISE 4.2

A 1. $12x^2 + 5x + 6$
   3. $-2x^2 - x - 8$
   5. $-3x^2 + 7x - 5$
   7. $-x^2 - 5$
   9. $x^3 - 2x^2 - x - 2$
   11. $2x^4 - 8x^3 + 2x^2 + x + 5$
   13. $-\dfrac{3}{5}x^2 + x + 1$
   15. $0.4x^2 + 0.3x - 0.4$
   17. $-2x^2 - 2x + 3$
   19. $3x^4 + x^3 - 5x - 1$
   21. $-5x^4 + 5x^3 - 2x^2 - 2$
   23. $5x^3 - 4x^2 + 5x - 1$
   25. $-\dfrac{1}{7}x^3 - \dfrac{2}{3}x^2 + 5x$
   27. $-\dfrac{3}{7}x^3 + \dfrac{1}{9}x^2 + x - 6$
   29. $-4x^4 - 4x^3 - 3x^2 - x + 2$
B 31. $4x^2 + 7$
   33. $-x^2 - 4x - 6$
   35. $4x^2 - 7x + 6$
   37. $-3x^3 + 6x^2 - 2x$
   39. $7x^3 - x^2 + 2x - 6$
   41. $3x^2 - 7x + 7$
   43. 0
   45. $4x^3 - 3x^2 - 7x + 6$
   47. $3x^3 - 2x^2 + x - 8$
   49. $-5x^3 - 5x^2 + 4x - 9$
C 51. $2x^2 + 6x$
   53. $3x^2 + 4x$
   55. $27x^2 + 10x$

✓ SKILL CHECKER

57. $-6x^9$
59. $6y - 24$

4.2 USING YOUR KNOWLEDGE

1. $R = x - 50$
3. $R = 2x$
5. 30

EXERCISE 4.3

A 1. $45x^5$
   3. $-10x^3$
   5. $6y^3$
B 7. $3x + 3y$
   9. $10x - 5y$
   11. $8x^2 - 12x$
   13. $x^5 + 4x^4$
   15. $4x^2 - 4x^3$
   17. $3x^2 + 3xy$
   19. $8xy^2 - 12y^3$
   21. $x^2 + 3x + 2$
   23. $y^2 - 5y - 36$

25. $x^2 - 5x - 14$
27. $x^2 - 12x + 27$
29. $y^2 - 6y + 9$
31. $6x^2 + 7x + 2$
33. $6y^2 + y - 15$
35. $10z^2 + 43z^2 - 9$
37. $6x^2 - 34x + 44$
39. $16z^2 + 8z + 1$
41. $6x^2 + 11xy + 3y^2$
43. $2x^2 + xy - 3y^2$
45. $10z^2 + 13yz - 3y^2$
47. $12x^2 - 11xz + 2z^2$
49. $4x^2 - 12xy + 9y^2$
51. $x^2 + 7x + 10$
53. $96t - 16t^2$
55. $V_2CP + V_2PR$
    $- V_1CP - V_1PR$

✓ SKILL CHECKER

57. $9x^2$
59. $9A^2$

4.3 USING YOUR KNOWLEDGE

1. $-3p^2 + 65p - 100$
3. $\$18$

EXERCISE 4.4

A 1. $x^2 + 2x + 1$
   3. $x^2 + 16x + 64$
   5. $4x^2 + 4x + 1$
   7. $9x^2 + 30x + 25$
   9. $9x^2 + 12xy + 4y^2$
   11. $25x^2 + 10xy + y^2$
B 13. $x^2 - 2x + 1$
   15. $x^2 - 8x + 16$
   17. $4x^2 - 4x + 1$
   19. $25x^2 - 60x + 36$
   21. $9x^2 - 6xy + y^2$
   23. $36x^2 - 60xy + 25y^2$
   25. $4x^2 - 28xy + 49y^2$
C 27. $x^2 - 4$
   29. $x^2 - 16$
   31. $9x^2 - 4y^2$
   33. $x^2 - 36$
   35. $x^2 - 144$
   37. $9x^2 - y^2$
   39. $4x^2 - 49y^2$
   41. $x^4 + 7x^2 + 10$
   43. $x^4 + 2x^2y + y^2$
   45. $9x^4 - 12x^2y^2 + 4y^4$
   47. $x^4 - 4y^4$
   49. $4x^2 - 16y^4$

✓ SKILL CHECKER

51. $5x^2 + 5x - 10$
53. $-3x^2 + 6x + 12$

4.4 USING YOUR KNOWLEDGE

1. a. 9
   b. 5
   c. No

3. a. x^2
   b. xy
   c. y^2
   d. xy
5. $(x + y)^2 = x^2 + 2xy + y^2$

EXERCISE 4.5

A 1. $x^3 + 4x^2 + 8x + 15$
   3. $x^3 + 3x^2 - x + 12$
   5. $x^3 + 2x^2 - 5x - 6$
   7. $x^3 - 8$
   9. $x^3 - 2x^2 + 3x - 2$
   11. $x^3 - 8x^2 + 15x + 4$
B 13. $2x^3 + 6x^2 + 4x$
   15. $3x^3 + 3x^2 - 6x$
   17. $4x^3 - 12x^2 + 8x$
   19. $5x^3 - 20x^2 - 25x$
   21. $x^3 + 15x^2 + 75x + 125$
   23. $8x^3 + 36x^2 + 54x + 27$
   25. $8x^3 + 36x^2y + 54xy^2 + 27y^3$
   27. $16t^4 + 24t^2 + 9$
   29. $16t^4 + 24t^2u + 9u^2$
   31. $9t^4 - 2t^2 + \dfrac{1}{9}$
   33. $9t^4 - 2t^2u + \dfrac{1}{9}u^2$
   35. $9x^4 - 25$
   37. $9x^4 - 25y^4$
   39. $16x^6 - 25y^6$
   41. $I_D = 0.004 + 0.004V + 0.001V^2$
   43. $T_1^4 - T_2^4$
   45. $Kt_n^2 - 2Kt_nt_a + Kt_a^2$

✓ SKILL CHECKER

47. $\dfrac{1}{7}y + 2$
49. $2x + 3$
51. $4x^2$

4.5 USING YOUR KNOWLEDGE

1. $10t^3 - 159t^2 + 869t - 1308$

CALCULATOR CORNER 4.5

1. $264°$ F
3. $342°$ F
5. $588°$ F (which exceeds $500°$)

EXERCISE 4.6

A 1. $x + 3y$
   3. $2x - y$
   5. $-2y + 8$
   7. $10x + 8$
   9. $3x - 2$
B 11. $x + 2$
   13. $y - 2$
   15. $x + 6$
   17. $3x - 4$
   19. $2y - 5$ remainder -1
   21. $x^2 - x - 1$

23. $y^2 - y - 1$
25. $4x^2 + 3x + 7$ remainder 12
27. $x^2 - 2x + 4$
29. $4y^2 + 8y + 16$
31. $x^2 + x + 1$
33. $x^3 + x^2 - x + 1$
35. $m^2 - 8$ remainder 10
37. $x^2 + xy + y^2$
39. $x^2 + 2x + 4$ remainder 16

✓ **SKILL CHECKER**

41. 180
43. 120
45. 180

4.6 USING YOUR KNOWLEDGE

1. $3x + 5$
3. $50 + x - \dfrac{7000}{x}$

REVIEW EXERCISES CHAPTER 4

1. a. T
   b. M
   c. B
2. a. 4
   b. 2
   c. 2
3. a. $9x^4 + 4x^2 - 8x$
   b. $4x^2 - 3x - 3$
   c. $-4x^2 + 3x + 8$
4. a. 284
   b. 156
   c. -100
5. a. $5x^2 - x - 10$
   b. $-5x^2 + 15x + 2$
   c. $9x^2 - 7x + 1$
6. a. $-x^2 - 7x + 4$
   b. $7x^2 - 7x + 3$
   c. $-5x^2 + 4x - 11$
7. a. $-18x^7$
   b. $-40x^9$
   c. $-27x^{11}$
8. a. $-2x^3 - 4x^2y$
   b. $-6x^4 - 9x^3y$
   c. $-20x^4 - 28x^3y$
9. a. $x^2 + 15x + 54$
   b. $x^2 + 5x + 6$
   c. $x^2 + 16x + 63$
10. a. $x^2 + 4x - 21$
    b. $x^2 + 4x - 12$
    c. $x^2 + 4x - 5$
11. a. $x^2 - 4x - 21$
    b. $x^2 - 4x - 12$
    c. $x^2 - 4x - 5$
12. a. $6x^2 - 13xy + 6y^2$
    b. $20x^2 - 27xy + 9y^2$
    c. $8x^2 - 26xy + 15y^2$
13. a. $4x^2 + 12xy + 9y^2$
    b. $9x^2 + 24xy + 16y^2$
    c. $16x^2 + 40xy + 25y^2$

14. a. $4x^2 - 12xy + 9y^2$
    b. $9x^2 - 12xy + 4y^2$
    c. $25x^2 - 20xy + 4y^2$
15. a. $9x^2 - 25y^2$
    b. $9x^2 - 4y^2$
    c. $9x^2 - 16y^2$
16. a. $x^3 + 4x^2 + 5x + 2$
    b. $x^3 + 5x^2 + 8x + 4$
    c. $x^3 + 6x^2 + 11x + 6$
17. a. $3x^3 + 9x^2 + 6x$
    b. $4x^3 + 12x^2 + 8x$
    c. $5x^3 + 15x^2 + 10x$
18. a. $x^3 + 6x^2 + 12x + 8$
    b. $x^3 + 9x^2 + 27x + 27$
    c. $x^3 + 12x^2 + 48x + 64$
19. a. $25x^4 - 5x^2 + \dfrac{1}{4}$
    b. $49x^4 - 7x^2 + \dfrac{1}{4}$
    c. $81x^4 - 9x^2 + \dfrac{1}{4}$
20. a. $9x^4 - 4$
    b. $9x^4 - 16$
    c. $9x^4 - 25$
21. a. $2x^2 - x$
    b. $4x^2 - 2x$
    c. $4x^2 - 2x$
22. a. $x + 6$
    b. $x + 7$
    c. $x + 8$
23. a. $4x^2 - 4x - 4$
    b. $6x^2 - 6x - 6$
    c. $2x^2 - 2x - 2$
24. a. $2x^2 + 6x - 2$ remainder 2
    b. $2x^2 + 6x - 3$ remainder 3
    c. $3x^3 + 3x - 1$ remainder 4
25. a. $x^2 + x$ remainder $(4x + 1)$
    b. $x^2 + x$ remainder $(5x + 1)$
    c. $x^2 + x$ remainder $(6x + 1)$

EXERCISE 5.1

A 1. $3(x + 5)$
   3. $9(y - 2)$
   5. $-5(y - 4)$
   7. $-3(x + 9)$
   9. $4x(x + 8)$
  11. $6x(1 - 7x)$
  13. $-5x^2(1 + 5x^2)$
  15. $3x(x^2 + 2x + 3)$
  17. $9y(y^2 - 2y + 3)$
  19. $6x^2(x^4 + 2x^3 - 3x^2 + 5)$
  21. $8y^3(y^5 + 2y^2 - 3y + 1)$
  23. $\dfrac{1}{7}(4x^3 + 3x^2 - 9x + 3)$
  25. $\dfrac{1}{8}y^2(7y^7 + 3y^4 - 5y^2 + 5)$

B 27. $(x + 2)(x^2 + 1)$
  29. $(y - 3)(y^2 + 1)$
  31. $(2x + 3)(2x^2 + 1)$
  33. $(3x - 1)(2x^2 + 1)$

35. $(y + 2)(4y^2 + 1)$
37. $(2a + 3)(a^2 + 1)$
39. $(x^2 + 4)(3x^2 + 1)$
41. $(2y^2 + 3)(3y^2 + 1)$
43. $(y^2 + 3)(4y^2 + 1)$
45. $(a^2 - 2)(3a^2 - 2)$
47. $\alpha L(t_2 - t_1)$
49. $(R - 1)(R - 1)$

✓ **SKILL CHECKER**

51. $x^2 + 8x + 15$
53. $2x^2 - 9x + 4$
55. $4x^2 - 4x - 3$

5.1 USING YOUR KNOWLEDGE

1. $-w(l - z)$
3. $a(a + 2S)$
5. $-16(t^2 - 5t - 15)$

EXERCISE 5.2

A 1. $(y + 2)(y + 4)$
   3. $(x + 2)(x + 5)$
   5. $(y + 5)(y - 2)$
   7. $(x + 7)(x - 2)$
   9. $(x + 1)(x - 7)$
  11. $(y + 2)(y - 7)$
  13. $(y - 2)(y - 1)$
  15. $(x - 4)(x - 1)$
B,C 17. $(2x + 3)(x + 1)$
   19. $(2x + 3)(3x + 1)$
   21. $(2x + 1)(3x + 4)$
   23. $(2x - 1)(x + 2)$
   25. $(3x - 2)(x + 6)$
   27. $(4y - 3)(y - 2)$
   29. Not factorable
   31. $2(3y + 1)(y - 2)$
   33. $(3y + 2)(4y - 3)$
   35. $3(3y + 1)(2y - 3)$
   37. $(3x + 1)(x + 2)$
   39. $(5x + 1)(x + 2)$
   41. Not factorable
   43. $(3x + 1)(x - 2)$
   45. $(5x + 2)(3x - 1)$
   47. $4(2x + y)(x + 2y)$
   49. $(2x + 3y)(3x - y)$
   51. $(7x - 3y)(x - y)$
   53. $(3x + y)(5x - 2y)$
   55. Not factorable
   57. $(2g + 9)(g - 4)$
   59. $(2R - 1)(R - 1)$
   61. $x^2 + 8x + 16$
   63. $x^2 - 6x + 9$
   65. $9x^2 + 12xy + 4y^2$
   67. $25x^2 - 20xy + 4y^2$
   69. $9x^2 - 16y^2$

5.2 USING YOUR KNOWLEDGE

1. $(L - 3)(2L - 3)$
3. $(5t - 7)(t - 1)$

EXERCISE 5.3

A 1. Yes
   3. No
   5. Yes
   7. No
   9. Yes

B 11. $(x + 1)^2$
  13. $3(x + 5)^2$
  15. $(3x + 1)^2$
  17. $(3x + 2)^2$
  19. $(4x + 5y)^2$
  21. $(5x + 2y)^2$
  23. $(y - 1)^2$
  25. $3(y - 4)^2$
  27. $(3x - 1)^2$
  29. $(4x - 7)^2$
  31. $(3x - 2y)^2$
  33. $(5x - y)^2$

C 35. $(x + 7)(x - 7)$
  37. $(3x + 7)(3x - 7)$
  39. $(5x + 9y)(5x - 9y)$
  41. $(x^2 + 1)(x + 1)(x - 1)$
  43. $(4x^2 + 1)(2x + 1)(2x - 1)$

✓ SKILL CHECKER

45. $R^2 - r^2$
47. $6(x + 1)(x + 4)$
49. $2(x + 3)(x - 3)$

5.3 USING YOUR KNOWLEDGE

1. $D(x) = (x - 7)^2$
3. $C(x) = (x + 6)^2$

EXERCISE 5.4

A 1. $3(x + 2)(x - 3)$
   3. $5(x + 4)(x - 2)$
   5. $3x(x^2 + 2x + 7)$
   7. $2x^2(x^2 - 2x - 5)$
   9. $2x^2(2x^2 + 6x + 9)$
  11. $(x + 2)(3x^2 + 1)$
  13. $(x + 1)(3x^2 + 2)$
  15. $(x + 1)(2x^2 - 1)$
  17. $3(x + 4)^2$
  19. $k(x + 2)^2$
  21. $4(x - 3)^2$
  23. Not factorable
  25. $3x(x + 2)^2$
  27. $2x(3x + 1)^2$
  29. $3x^2(2x - 3)^2$
  31. $(x^2 + 1)(x + 1)(x - 1)$
  33. $(x^2 + y^2)(x + y)(x - y)$
  35. $(x^2 + 4y^2)(x + 2y)(x - 2y)$

B 37. $-(x + 3)^2$
  39. $-(x + 2)^2$
  41. $-(2x + y)^2$
  43. $-(3x + 2y)^2$
  45. $-(2x - 3y)^2$
  47. $-2x(3x + 2y)^2$
  49. $-2x(3x + 5y)^2$
  51. $-x(x + 1)(x - 1)$

53. $-x^2(x + 2)(x - 2)$
55. $-x^2(2x + 3)(2x - 3)$
57. $-2x(2x + 3)(2x - 3)$
59. $-2x^2(3x + 2)(3x - 2)$

✓ SKILL CHECKER

61. $(2x + 3)(5x - 1)$
63. $(2x + 1)(x - 3)$

5.4 USING YOUR KNOWLEDGE

1. $\dfrac{2\pi A}{360}(IR + Kt)$

3. $\dfrac{3S}{2bd^3}(d + 2z)(d - 2z)$

EXERCISE 5.5

A 1. $x = -3, x = -\dfrac{1}{2}$

   3. $x = 1, x = -\dfrac{3}{2}$

   5. $y = 3, y = \dfrac{2}{3}$

   7. $y = 1, y = -\dfrac{1}{3}$

   9. $x = -\dfrac{4}{3}, x = -\dfrac{1}{2}$

  11. $x = 2, x = -\dfrac{1}{3}$

  13. $x = -2, x = \dfrac{4}{5}$

  15. $x = 1, x = \dfrac{8}{5}$

  17. $y = 5, y = \dfrac{2}{3}$

  19. $y = -2, y = -\dfrac{1}{2}$

  21. $x = -\dfrac{1}{3}$

  23. $y = 4$

  25. $x = -\dfrac{2}{3}$

  27. $y = \dfrac{5}{2}$

  29. $x = -5$
  31. $x = 4, x = 1$

  33. $x = -2, x = -\dfrac{5}{2}$

  35. $x = 1$
  37. 1 sec
  39. 1 sec

✓ SKILL CHECKER

41. $x = 3$
43. $x = 6$

5.5 USING YOUR KNOWLEDGE

1. $x = 7$

REVIEW EXERCISES CHAPTER 5

1. a. $5x^3(2 + 3x)(2 - 3x)$
  b. $7x^4(2 - 5x^2)$
  c. $8x^7(2 - 5x^2)$

2. a. $\dfrac{1}{7}x^2(3x^4 - 5x^3 + 2x^2 - 1)$

  b. $\dfrac{1}{9}x^3(4x^4 - 2x^3 + 2x^2 - 1)$

  c. $\dfrac{1}{8}x^5(3x^4 - 7x^3 + 3x^2 - 1)$

3. a. $(x + 7)(3x^2 + 1)$
  b. $(x + 6)(3x^2 + 1)$
  c. $(x + 2)(4x^2 + 1)$
4. a. $(x + 7)(x + 1)$
  b. $(x + 9)(x + 1)$
  c. $(x + 5)(x + 1)$
5. a. $(x - 5)(x - 2)$
  b. $(x - 7)(x - 2)$
  c. $(x - 4)(x - 2)$
6. a. $(2x + 3)(3x - 2)$
  b. $(2x + 1)(3x - 1)$
  c. $(2x + 5)(3x - 1)$
7. a. $(2x - 5y)(3x - y)$
  b. $(2x - y)(3x - 2y)$
  c. $(2x - y)(3x - 4y)$
8. a. $(x + 2)^2$
  b. $(x + 3)^2$
  c. $(x + 4)^2$
9. a. $(3x + 2y)^2$
  b. $(3x + 5y)^2$
  c. $(3x + 4y)^2$
10. a. $(x - 2)^2$
  b. $(x - 3)^2$
  c. $(x - 4)^2$
11. a. $(2x - 3y)^2$
  b. $(2x - 5y)^2$
  c. $(2x - 7y)^2$
12. a. $(x + 6)(x - 6)$
  b. $(x + 7)(x - 7)$
  c. $(x + 9)(x - 9)$
13. a. $(4x + 9y)(4x - 9y)$
  b. $(5x + 8y)(5x - 8y)$
  c. $(3x + 10y)(3x - 10y)$
14. a. $3x(x^2 - 2x + 9)$
  b. $3x(x^2 - 2x + 10)$
  c. $4x(x^2 - 2x + 8)$
15. a. $2x(x - 2)(x + 1)$
  b. $3x(x - 3)(x + 1)$
  c. $4x(x - 4)(x + 1)$
16. a. $(x + 4)(2x^2 + 1)$
  b. $(x + 5)(2x^2 + 1)$
  c. $(x + 6)(2x^2 + 1)$
17. a. $k(3x + 2)^2$
  b. $k(3x + 5)^2$
  c. $k(2x + 5)^2$
18. a. $-3x^2(x + 3)(x - 3)$
  b. $-4x^2(x + 4)(x - 4)$
  c. $-5x^2(x + 2)(x - 2)$

19. a. $x = 5$ or $x = -1$
   b. $x = 6$ or $x = -1$
   c. $x = 7$ or $x = -1$

20. a. $x = -\dfrac{5}{2}$ or $x = 2$

   b. $x = -\dfrac{5}{2}$ or $x = 1$

   c. $x = -\dfrac{3}{2}$ or $x = 1$

EXERCISE 6.1

A 1.

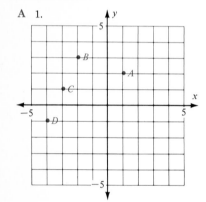

3.

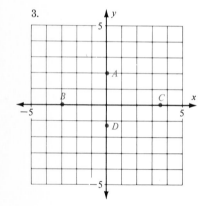

5.

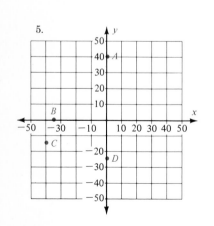

B 7. a. $A\left(0, 2\dfrac{1}{2}\right)$

   b. $B(-3, 1)$
   c. $C(-2, -2)$

   d. $D\left(0, -3\dfrac{1}{2}\right)$

   e. $E\left(1\dfrac{1}{2}, -2\right)$

   9. a. $A(10, 10)$
      b. $B(-30, 20)$
      c. $C(-25, -20)$
      d. $D(0, -40)$
      e. $E(15, -15)$
   11. $(20, 140)$
   13. $(45, 150)$
C 15. Yes
   17. Yes
   19. No
D 21. $(3, 0)$
   23. $(-2, 2)$
   25. $(0, -3)$
   27. $(3, 0)$
   29. $(-3, 2)$
E 31. $(103, 95)$
   33. $(1050, 956)$
   35. $(176, 35)$

✓ **SKILL CHECKER**

37. $y = 9$
39. $y = 3$

6.1 USING YOUR KNOWLEDGE

1. 86° F
3. Less than 10%
5. 12° F

EXERCISE 6.2

A 1. $2x + y = 4$

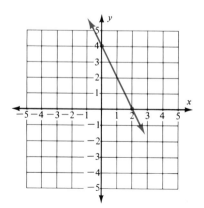

3. $-2x - 5y = -10$

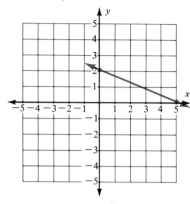

5. $y = 3x + 3$

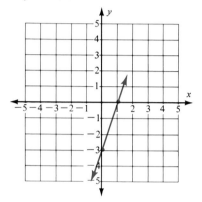

7. $6 = 3x - 6y$

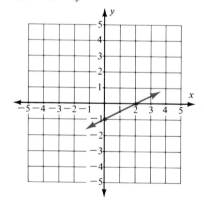

9. $-3y = 4x + 12$

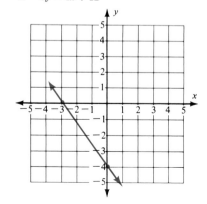

B 11. $y = -4$

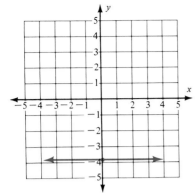

13. $2y + 6 = 0$

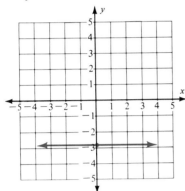

15. $x = \dfrac{-5}{2}$

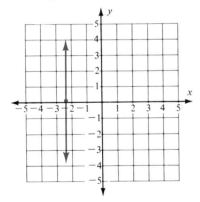

17. $2x + 4 = 0$

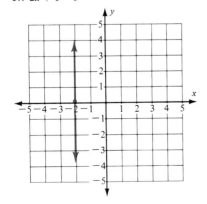

19. $2x - 9 = 0$

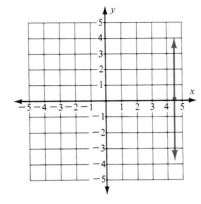

C 21. a. $y = 2x + 4$
   b. 2
   c. 4
   d.

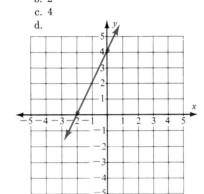

23. a. $y = \dfrac{5}{2}x + 5$

   b. $\dfrac{5}{2}$

   c. 5
   d.

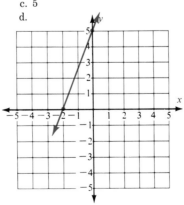

25. a. $y = -\dfrac{1}{2}x$

   b. $-\dfrac{1}{2}$

   c. 0

d.

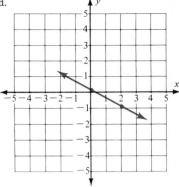

27. a. $y = -\dfrac{1}{2}x - 2$

   b. $-\dfrac{1}{2}$

   c. -2

   d.

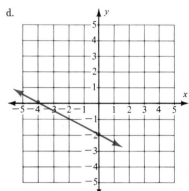

29. a. $y = \dfrac{1}{2}x + 1$

   b. $\dfrac{1}{2}$

   c. 1
   d.

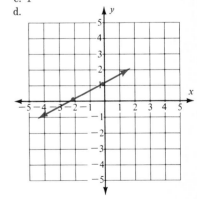

✓ **SKILL CHECKER**

31. $>$
33. $>$
35. $<$

6.2 USING YOUR KNOWLEDGE

1. $y = 2x + 50$
3. a. $2
   b. $75

EXERCISE 6.3

1. $2x + y > 4$

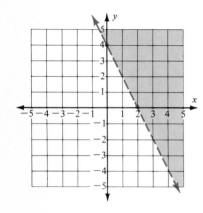

3. $-2x - 5y \leq -10$

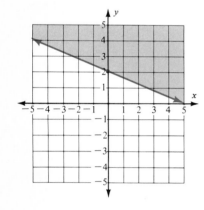

5. $y \geq 3x - 3$

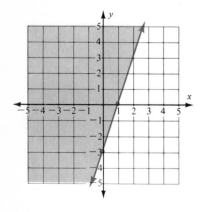

7. $6 < 3x - 6y$

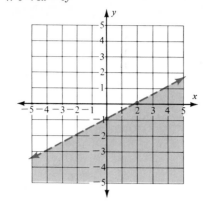

9. $3x + 4y \geq 12$

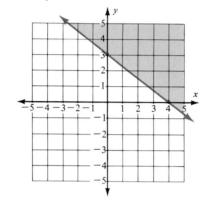

11. $10 < -2x + 5y$

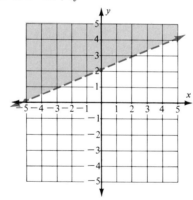

13. $x \geq 2y - 4$

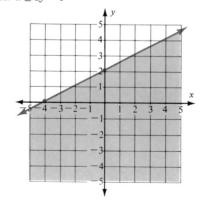

15. $y < -x + 5$

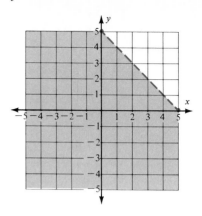

17. 9
19. -9

6.3 USING YOUR KNOWLEDGE

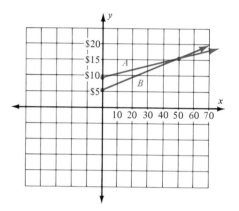

1. See the preceding graph.
3. When they travel 50 mi.
5. When they travel less than 50 mi.

EXERCISE 6.4

A 1. 1
   3. -1
   5. $-\dfrac{1}{8}$
   7. $\dfrac{1}{4}$
   9. 0

B 11. $y = \dfrac{1}{2}x + \dfrac{3}{2}$
   13. $y = -x + 6$
   15. $y = 5$
C 17. $y = 2x - 3$
   19. $y = -4x + 6$

21. $y = \dfrac{3}{4}x + \dfrac{7}{8}$

23. $y = 2.5x - 4.7$

25. $y = -3.5x + 5.9$

D 27. Parallel

29. Not parallel

31. Parallel

✓ SKILL CHECKER

33.

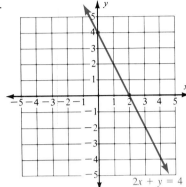

$2x + y = 4$

35.

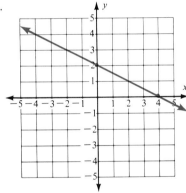

6.4 USING YOUR KNOWLEDGE

1. -500

3. $(0, 3000)$

REVIEW EXERCISES CHAPTER 6

1.

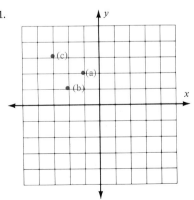

(c)
(a)
(b)

2. a. $(1, 1)$
   b. $(-1, -2)$
   c. $(3, -3)$

3. a. No
   b. No
   c. No

4. a. $x = 3$
   b. $x = 4$
   c. $x = 2$

5.

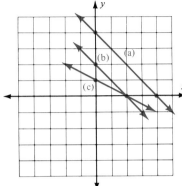

(b) (a)
(c)

6.

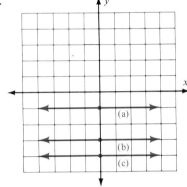

(a)
(b)
(c)

7.

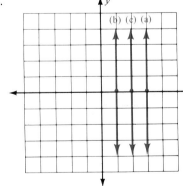

(b) (c) (a)

8. a. 3
   b. 2
   c. 4

9. a. 3
   b. 4
   c. 4

10. a.

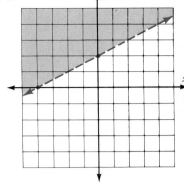

b.

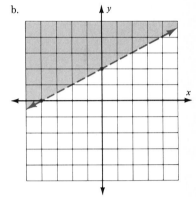

c.

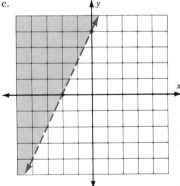

11. a.

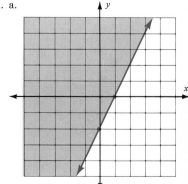

b.

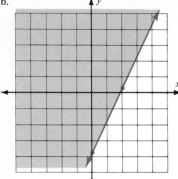

3.

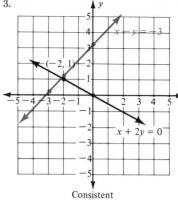

Consistent

11.

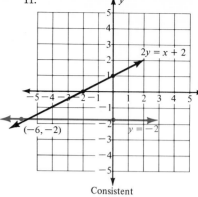

Consistent

c.

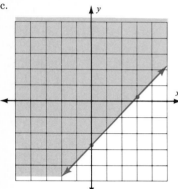

5.

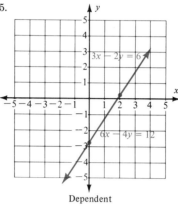

Dependent

13.

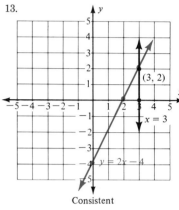

Consistent

12. a. 1
    b. -1
    c. -4
13. a. $y = -2x + 1$
    b. $y = -3x + 4$
    c. $y = -4x + 7$
14. a. $y = 5x - 2$
    b. $y = 4x - 3$
    c. $y = 6x - 4$
15. a. $2x - 4y = 8$
    b. $2x - 4y = 8$
    c. Neither

7.

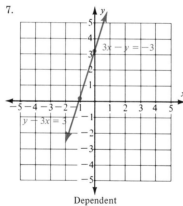

Dependent

15.

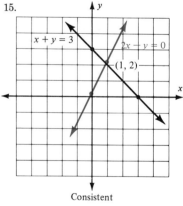

Consistent

EXERCISE 7.1

A,B 1.

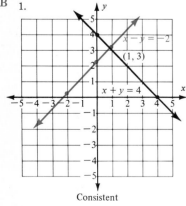

Consistent

9.

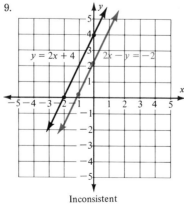

Inconsistent

17.

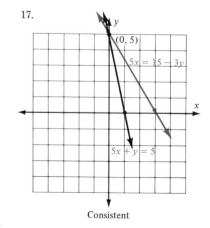

Consistent

19.

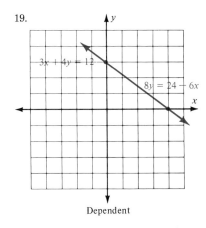

Dependent

✓ SKILL CHECKER

21. $x = 4$
23. Yes
25. No

7.1 USING YOUR KNOWLEDGE

1. $y = 0.15x + 5$
3.

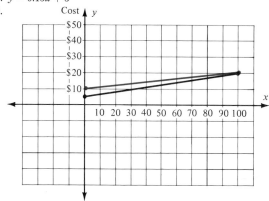

5. Hurt's

EXERCISE 7.2

A,B 1. Consistent, (2, 0)
3. Consistent, (2, 3)
5. Inconsistent
7. Inconsistent
9. Dependent
11. Consistent, $(-1, -1)$
13. Inconsistent
15. Consistent, (4, 1)
17. 4 days
19. 5 days

✓ SKILL CHECKER

21. $x = 8$
23. $x = -1$

7.2 USING YOUR KNOWLEDGE

1. (80, 320)
3. a. $5x = 4x + 500$
   b. $x = 500$

EXERCISE 7.3

A,B 1. (1, 2)
3. (0, 2)
5. Inconsistent
7. Inconsistent
9. $(10, -1)$
11. $(-26, 14)$
13. (6, 2)
15. $(2, -1)$
17. (3, 5)
19. $(-3, -2)$
21. (8, 6)
23. (1, 2)

✓ SKILL CHECKER

25. $n + d = 300$
27. $4(x - y) = 48$
29. $m = n - 3$

7.3 USING YOUR KNOWLEDGE

1. Tweedledee—$120\frac{2}{3}$

   Tweedledum—$119\frac{2}{3}$

EXERCISE 7.4

A 1. 59, 43
3. 21, 105
5. Not possible
7. Pikes: 14,110 ft
   Longs: 14,255 ft
B 9. 5 nickels
   20 dimes
11. 15 nickels
   5 dimes
13. 4 fives
   6 ones
C 15. 1050 mi
17. Boat: 15 mph
   Current: 3 mph
19. 20 mi

✓ SKILL CHECKER

21. $\frac{3}{8} = \frac{6}{16}$

23. $\frac{1}{3}$

REVIEW EXERCISES CHAPTER 7

1. a.

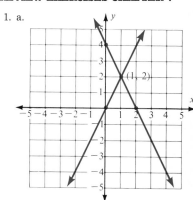

(1, 2) is the solution

b.

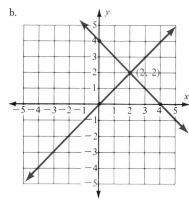

(2, 2) is the solution

c.

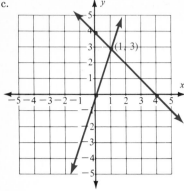

(1, 3) is the solution

2. a.

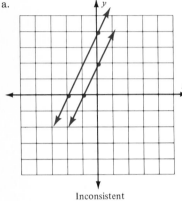

Inconsistent

b.

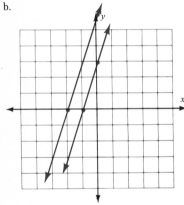

Inconsistent

c.

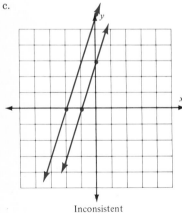

Inconsistent

3. a. Inconsistent, no solution
   b. Inconsistent, no solution
   c. $x = 1$
      $y = 1$
4. a. Dependent, infinitely many solutions
   b. Dependent, infinitely many solutions
   c. Dependent, infinitely many solutions
5. a. $x = -1$
      $y = 2$
   b. $x = 2$
      $y = -1$
   c. $x = -1$
      $y = -2$
6. a. Inconsistent, no solution
   b. Inconsistent, no solution
   c. Inconsistent, no solution
7. a. Dependent, infinitely many solutions
   b. Dependent, infinitely many solutions
   c. Dependent, infinitely many solutions
8. a. 20 nickels, 20 dimes
   b. 40 nickels, 10 dimes
   c. 50 nickels, 5 dimes
9. a. 70, 110
   b. 60, 120
   c. 50, 130
10. a. 550 mph
    b. 520 mph
    c. 500 mph

EXERCISE 8.1

A 1. $\dfrac{9}{21}$

3. $\dfrac{-16}{22}$

5. $\dfrac{20xy}{24y^3}$

7. $\dfrac{-9xy^3}{21y^4}$

9. $\dfrac{4x^2 - 8x}{x^2 - x - 2}$

11. $\dfrac{-5x^2 + 10x}{x^2 + x - 6}$

B 13. $\dfrac{x^2}{2y^3}$

15. $\dfrac{-3y^4}{x}$

17. $\dfrac{1}{2xy^3}$

19. $\dfrac{5x}{y^2}$

21. $\dfrac{x - y}{3}$

23. $-3(x - y) = 3y - 3x$

25. $\dfrac{-1}{4(x - y)} = \dfrac{1}{4(y - x)}$

27. $\dfrac{1}{2(x + 2)}$

29. $\dfrac{1}{x + y}$

31. $\dfrac{1}{2}$

33. $\dfrac{1}{3}$

35. $\dfrac{-1}{1 + 2y}$

37. $\dfrac{-1}{1 + 2y}$

39. $\dfrac{-1}{x + 2}$

41. -1

43. $-(x + 5)$

45. $2 - x$

47. $\dfrac{-1}{x + 6}$

49. $\dfrac{1}{x - 2}$

✓ **SKILL CHECKER**

51. $\dfrac{2}{3}$

53. $(x + 3)(x - 1)$

55. $(x - 2)(x - 5)$

8.1 USING YOUR KNOWLEDGE

1. $\dfrac{2}{3}$

3. a. 4 to 1
   b. 2500 rpm

EXERCISE 8.2

A 1. $\dfrac{8x}{3y}$

3. $\dfrac{-4xy}{3}$

5. $\dfrac{3y}{2}$

7. $-8xy$

9. $\dfrac{x + 1}{x + 7}$

11. $\dfrac{2 - 2x}{x - 2}$

13. $\dfrac{3x + 15}{x + 1}$

15. $-(x + 2) = -x - 2$

17. $\dfrac{x + 3}{x - 3}$

19. $\dfrac{1}{4}$

B 21. $\dfrac{x - 1}{x + 2}$

23. $\dfrac{x - 5}{5x - 15}$

25. $\dfrac{x + 4}{x - 3}$

27. $\dfrac{-5x - 20}{2x + 6}$

29. $\dfrac{x - 2}{x - 1}$

31. $\dfrac{x + 3}{1 - x^2} = \dfrac{-x - 3}{x^2 - 1}$

33. $\dfrac{2x - 4}{x - 3}$

35. $\dfrac{x^2 + 11x + 30}{4}$

37. $\dfrac{x^2 - 4x + 4}{x^2 + 2x - 3}$

39. $\dfrac{x + 6}{5}$

41. $x - 1$

43. $\dfrac{x^2 - 6x + 5}{x^2 - 6x + 8}$

45. 1

47. $\dfrac{x + y}{x}$

49. $\dfrac{xy + 2x + 3y^2 + 6y}{xy - 2x + 6y^2 - 12y}$

✓ **SKILL CHECKER**

51. $\dfrac{51}{40}$

53. $\dfrac{19}{40}$

55. $\dfrac{14}{6} = \dfrac{7}{3}$

8.2 USING YOUR KNOWLEDGE

1. $\dfrac{RR_T}{R - R_T}$

3. $\dfrac{60,000 + 9000x}{x}$

EXERCISE 8.3

A,B 1. a. $\dfrac{5}{7}$

b. $\dfrac{11}{x}$

3. a. $\dfrac{2}{3}$

b. $\dfrac{4}{x}$

5. a. 2

b. $\dfrac{5}{x}$

7. a. 2

b. $\dfrac{2}{3(x + 1)}$

9. a. $\dfrac{4}{3}$

b. $\dfrac{5x}{2(x + 1)}$

11. a. $\dfrac{5}{12}$

b. $\dfrac{56 - 3x}{8x}$

13. a. $\dfrac{12}{35}$

b. $\dfrac{x^2 + 36}{9x}$

15. a. $\dfrac{2}{15}$

b. $\dfrac{5}{14(x - 1)}$

17. a. $\dfrac{53}{56}$

b. $\dfrac{8x - 1}{(x + 1)(x - 2)}$

19. a. $\dfrac{13}{24}$

b. $\dfrac{3x + 12}{(x - 2)(x + 1)}$

21. $\dfrac{2x^2 - 2x - 6}{(x + 4)(x - 1)(x - 4)}$

23. $\dfrac{5x^2 + 19x}{(x + 5)(x - 2)(x + 3)}$

25. $\dfrac{6x - 4y}{(x - y)(x + y)^2}$

27. $\dfrac{10 - x}{x^2 - 25}$

29. $\dfrac{-2(3x + 5)}{(x + 2)(x + 3)(x + 1)}$

✓ **SKILL CHECKER**

31. $\dfrac{9}{20}$

33. $24 + 18 = 42$

35. $x^2 - 1$

8.3 USING YOUR KNOWLEDGE

1. 1.5

3. $1.41\overline{6}$

5. 0.0025

EXERCISE 8.4

A 1. $\dfrac{1}{5}$

3. $\dfrac{1}{3}$

5. $\dfrac{ab - a}{b + a}$

7. $\dfrac{b + a}{b - a}$

9. $\dfrac{6b + 4a}{48b - 9a}$

11. $\dfrac{26}{5}$

13. $\dfrac{x}{2x - 1}$

15. 2

17. $\dfrac{1}{x - y}$

19. $\dfrac{x + 2}{x + 1}$

✓ **SKILL CHECKER**

21. $\dfrac{x - 5}{4}$

23. $\dfrac{1}{2(x - 4)}$

25. $W = 124$

27. $x = -3$

29. $x = 3$

8.4 USING YOUR KNOWLEDGE

1. $\dfrac{6}{25}$ years

3. $11\dfrac{43}{50}$ years

EXERCISE 8.5

A 1. 6

3. 4

5. -6

7. 12

9. 4

11. 5

13. 2

15. 2

17. -11

19. -2

21. No solution

23. $-\dfrac{3}{2}$

25. -6

27. -8

29. No solution

31. No solution

33. -3

35. $-\dfrac{5}{2}$

37. $\dfrac{1}{7}$

39. $-\dfrac{3}{5}$

✓ **SKILL CHECKER**

41. $x = \dfrac{33}{4}$

43. $h = \dfrac{12}{5}$

45. $R = 15$

8.5 USING YOUR KNOWLEDGE

1. $h = \dfrac{2A}{b_1 + b_2}$

3. $Q_1 = \dfrac{PQ_2}{P + 1}$

5. $f = \dfrac{ab}{a + b}$

EXERCISE 8.6

A 1. $\dfrac{25}{3}$

3. -14

5. $\dfrac{-24}{7}$

7. $\dfrac{29}{7}$

9. $\dfrac{1}{5}$

11. $\dfrac{2}{5}$

13. $\dfrac{-11}{3}$

15. 3

17. $\dfrac{7}{4}$

19. $\dfrac{-14}{3}$

B 21. 9250

23. 100

25. $\dfrac{36}{7} = 5\dfrac{1}{7}$ hr

27. 100 mph

29. $9\dfrac{3}{8}$ ft

31. 125 miles

✓ **SKILL CHECKER**

33. $(x + 4)(x - 4)$

35. $(4x + 9)(4x - 9)$

8.6 USING YOUR KNOWLEDGE

1. $\dfrac{56}{3} = 18\dfrac{2}{3}$ lb

3. 6 ft

5. 2650

REVIEW EXERCISES CHAPTER 8

1. a. $\dfrac{10xy}{16y^2}$

   b. $\dfrac{12xy}{16y^3}$

   c. $\dfrac{10x^2y}{15x^5}$

2. a. $-3(x - y)$
   b. $-2(x - y)$
   c. $4(x - y)$

3. a. $\dfrac{-1}{x + 1}$

   b. $\dfrac{-1}{x - 1} = \dfrac{1}{1 - x}$

   c. $\dfrac{-1}{1 - x} = \dfrac{1}{x - 1}$

4. a. $-(x + 3) = -x - 3$
   b. $-(x + 2) = -x - 2$
   c. $-(x + 3) = -x - 3$

5. a. $\dfrac{2xy}{3}$

   b. $\dfrac{3xy}{2}$

   c. $8xy^2$

6. a. $\dfrac{x + 2}{x + 3}$

   b. $\dfrac{x + 1}{x + 5}$

   c. $\dfrac{x + 5}{x + 4}$

7. a. $\dfrac{x - 3}{x + 2}$

   b. $\dfrac{x - 4}{x + 1}$

   c. $\dfrac{x - 5}{x + 4}$

8. a. $\dfrac{-1}{x - 5} = \dfrac{1}{5 - x}$

   b. $\dfrac{-1}{x - 1} = \dfrac{1}{1 - x}$

   c. $\dfrac{-1}{x - 2} = \dfrac{1}{2 - x}$

9. a. $\dfrac{2}{x - 1}$

   b. $\dfrac{2}{x - 2}$

   c. $\dfrac{1}{x + 1}$

10. a. $\dfrac{2}{x + 1}$

    b. $\dfrac{2}{x + 2}$

    c. $\dfrac{2}{x + 3}$

11. a. $\dfrac{3x - 2}{(x + 2)(x - 2)}$

    b. $\dfrac{4x - 2}{(x + 1)(x - 1)}$

    c. $\dfrac{5x - 9}{(x + 3)(x - 3)}$

12. a. $\dfrac{-6x - 10}{(x + 1)(x + 2)(x + 3)}$

    b. $\dfrac{7x + 1}{(x + 1)^2(x - 2)}$

    c. $\dfrac{-4x}{(x + 1)(x + 2)(x - 1)}$

13. a. $\dfrac{6}{17}$

    b. $\dfrac{14}{27}$

    c. $\dfrac{6}{25}$

14. a. 3
    b. 7
    c. 6

15. a. -3
    b. -5
    c. -6

16. a. No solution
    b. No solution
    c. No solution

17. a. $-6, \dfrac{13}{3}$

    b. $-7, \dfrac{11}{2}$

    c. $-8, \dfrac{33}{5}$

18. a. $10\dfrac{1}{2}$ gal

    b. $13\dfrac{1}{2}$ gal

    c. 18 gal

19. a. 18

   b. $10\dfrac{2}{5}$

   c. $\dfrac{-13}{5}$

20. a. $3\dfrac{3}{7}$ hr

   b. $4\dfrac{4}{9}$ hr

   c. $3\dfrac{3}{5}$ hr

EXERCISE 9.1

A 1. ± 10
   3. 0
   5. No real number solution
   7. $\pm\sqrt{7}$
   9. ± 3
  11. $\pm\sqrt{3}$

  13. $\pm\dfrac{1}{5}$

  15. $\pm\dfrac{7}{10}$

  17. $\pm\dfrac{\sqrt{17}}{5}$

  19. No real number solution
  21. $8, -10$
  23. $8, -4$
  25. No real number solution
  27. $18, 0$
  29. $0, -8$

  31. $-\dfrac{4}{5}, -\dfrac{6}{5}$

  33. $\dfrac{25}{6}, \dfrac{11}{6}$

  35. $\dfrac{3}{2}, \dfrac{-7}{2}$

  37. $1 \pm \dfrac{\sqrt{5}}{3}$

  39. No real number solution

  41. $\pm\dfrac{1}{9}$

  43. $\pm\dfrac{1}{4}$

  45. ± 2
  47. $2, -4$

  49. $\dfrac{5}{2}, -\dfrac{1}{2}$

51. $\dfrac{45}{25}$

53. $\dfrac{75}{9}$

55. $\dfrac{28}{16}$

9.1 USING YOUR KNOWLEDGE

1. a. $\dfrac{5}{3}$

  b. $\dfrac{5}{3}$

3. a. $\dfrac{5}{9}$

  b. $\dfrac{5}{9}$

5. $\dfrac{\sqrt{a}}{\sqrt{b}} = \sqrt{\dfrac{a}{b}}$

EXERCISE 9.2

A 1. $3\sqrt{5}$
   3. $5\sqrt{5}$
   5. $6\sqrt{5}$
   7. $10\sqrt{2}$
   9. $8\sqrt{6}$
  11. $5\sqrt{3}$
  13. $10\sqrt{2}$
  15. 19
  17. $10\sqrt{7}$
  19. $12\sqrt{3}$

B 21. $\dfrac{\sqrt{6}}{2}$

  23. $-2\sqrt{5}$
  25. 2

  27. $\dfrac{-\sqrt{10}}{5}$

  29. $\dfrac{\sqrt{2}}{2}$

  31. $\dfrac{\sqrt{2}}{3}$

  33. $\dfrac{2\sqrt{3}}{3}$

  35. $-\dfrac{\sqrt{2}}{2}$

  37. $-\sqrt{3}$

  39. $-\dfrac{\sqrt{3}}{3}$

  41. $\pm 7\sqrt{2}$
  43. $\pm 9\sqrt{2}$

  45. $\pm\dfrac{7\sqrt{6}}{6}$

  47. $\pm\dfrac{4\sqrt{5}}{5}$

  49. $\pm\dfrac{5\sqrt{6}}{6}$

  51. $\pm\dfrac{3\sqrt{10}}{5}$

  53. $\pm\dfrac{7\sqrt{3}}{3}$

C 55. $10\sqrt{7}$
  57. $5\sqrt{13}$
  59. $3\sqrt{2}$
  61. $4\sqrt{2}$
  63. $27\sqrt{3}$
  65. $-7\sqrt{7}$
  67. $29\sqrt{3}$
  69. $-8\sqrt{5}$
  71. $10\sqrt{2} - \sqrt{30}$
  73. $2\sqrt{21} + \sqrt{30}$
  75. $3 - \sqrt{6}$
  77. $\sqrt{10} + 5$
  79. $2\sqrt{3} - 3\sqrt{2}$

81. $x^2 + 14x + 49$
83. $x^2 - 6x + 9$
85. $x^2 - 14x + 49$

9.2 USING YOUR KNOWLEDGE

1. b
3. $a \cdot b = (\sqrt{a}\,\sqrt{b})^2$

EXERCISE 9.3

A 1. 81
   3. 64

   5. $\dfrac{49}{4}$

   7. $\dfrac{9}{4}$

   9. $\dfrac{1}{4}$

  11. $4, (x + 2)^2$

  13. $\dfrac{9}{4}, \left(x + \dfrac{3}{2}\right)^2$

  15. $9, (x - 3)^2$

  17. $\dfrac{25}{4}, \left(x - \dfrac{5}{2}\right)^2$

  19. $\dfrac{9}{16}, \left(x - \dfrac{3}{4}\right)^2$

  21. No real number solution

  23. $-\dfrac{1}{2} \pm \dfrac{\sqrt{5}}{2}$

  25. $-\dfrac{3}{2} \pm \dfrac{\sqrt{13}}{2}$

27. $\dfrac{3}{2} \pm \dfrac{\sqrt{21}}{2}$

29. $\dfrac{1}{2}, -\dfrac{3}{2}$

31. $2 \pm \dfrac{\sqrt{31}}{2}$

33. $\dfrac{1}{2} \pm \sqrt{2}$

35. $1 \pm \dfrac{\sqrt{2}}{2}$

37. $-\dfrac{1}{2} \pm \dfrac{3}{2} = 1, -2$

39. $-\dfrac{5}{2} \pm \dfrac{3\sqrt{3}}{2}$

41. $10x^2 + 5x - 12 = 0$
43. $x^2 - 9x = 0$
45. $10x^2 + 5x - 12 = 0$

9.3 USING YOUR KNOWLEDGE

1. a. 2 (thousand)
   b. $2
3. 3 days

EXERCISE 9.4

A,B 1. $-1, -2$
  3. $-2, 1$

  5. $\dfrac{-1 \pm \sqrt{17}}{4}$

  7. $\dfrac{2}{3}, -1$

  9. $-\dfrac{3}{2}, -2$

  11. $\dfrac{5}{7}, 1$

  13. $\dfrac{11 \pm \sqrt{41}}{10}$

  15. $\dfrac{3}{2}, 1$

  17. $\dfrac{3 \pm \sqrt{5}}{4}$

  19. -1

  21. $\dfrac{3 \pm \sqrt{5}}{4}$

  23. $\dfrac{-3 \pm \sqrt{21}}{6}$

  25. ± 2
  27. $\pm 3\sqrt{5}$

  29. $\pm \dfrac{2\sqrt{3}}{3}$

31. 4
33. 1

9.4 USING YOUR KNOWLEDGE

1. Multiply each term by $4a$.
3. Add b^2.
5. Take the square root of each side of the equation.
7. Divide by $2a$.
9. Add the expression on the right.

EXERCISE 9.5

1.

x	y
0	0
1	2
-1	2
2	8
-2	8

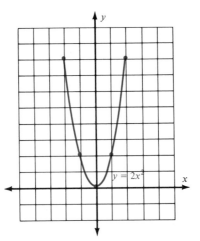

3. $y = 2x^2 - 1$ (Use the results of Problem 1.)

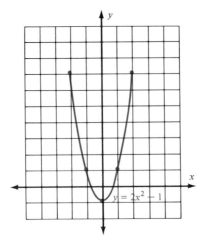

5. $y = -2x^2 + 2$ (Use the results of Problem 1.)

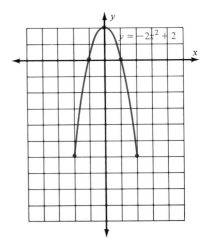

7. $y = -2x^2 - 2$ (Use the results of Problem 1.)

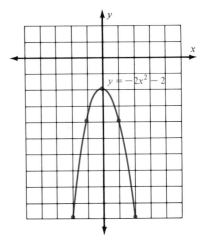

9. $y = (x - 2)^2$

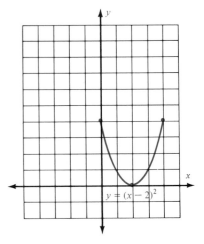

11. $y = (x - 2)^2 - 2$

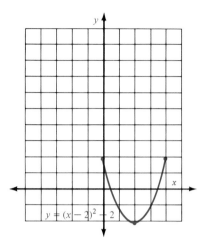

13. $y = -(x - 2)^2$

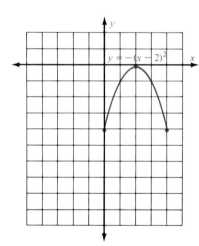

15. $y = -(x - 2)^2 - 2$

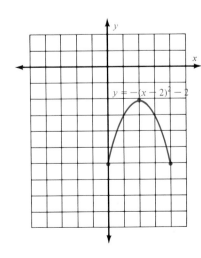

17.

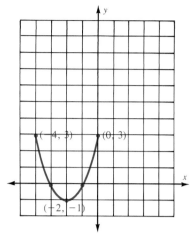

19.

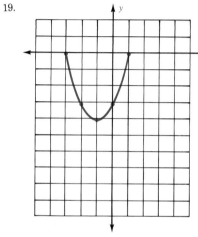

21.

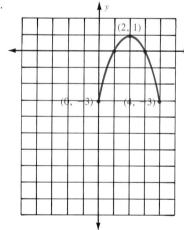

23. About 38 hr
25. About 60 hr

27. 25
29. 40,000

9.5 USING YOUR KNOWLEDGE

1. $40
3. $400

EXERCISE 9.6

A 1. 5
   3. $\sqrt{200} = 10\sqrt{2}$
   5. $\sqrt{15}$
   7. 3
   9. 2
B 11. 50 ft
   13. 100 mph
   15. 8 sec
   17. 1 sec
   19. 3 (thousand)
   21. 3 by 4
   23. 10, 24, 26

✓ **SKILL CHECKER**

25. $A = 12$
27. $B = 12$

9.6 USING YOUR KNOWLEDGE

1. $c = \sqrt{\dfrac{E}{m}} = \dfrac{\sqrt{Em}}{m}$

3. $r = \sqrt{\dfrac{GMm}{F}} = \dfrac{\sqrt{GFMm}}{F}$

5. $P = \sqrt{\dfrac{I}{k}} = \dfrac{\sqrt{Ik}}{k}$

REVIEW EXERCISES CHAPTER 9

1. a. ± 1
   b. ± 10
   c. ± 9

2. a. $\pm \dfrac{5}{4}$

   b. $\pm \dfrac{3}{5}$

   c. $\pm \dfrac{5}{8}$

3. a. No real number solution
   b. No real number solution
   c. No real number solution

4. a. $-1 \pm \dfrac{\sqrt{3}}{7}$

   b. $-2 \pm \dfrac{\sqrt{2}}{5}$

   c. $-1 \pm \dfrac{\sqrt{5}}{4}$

5. a. $\pm 4\sqrt{2}$
   b. $\pm 5\sqrt{2}$
   c. $\pm 4\sqrt{3}$

6. a. $\pm\dfrac{6\sqrt{5}}{5}$

   b. $\pm\dfrac{7\sqrt{3}}{3}$

   c. $\pm\dfrac{4\sqrt{7}}{7}$

7. a. $\pm\dfrac{4\sqrt{10}}{5}$

   b. $\pm\dfrac{2\sqrt{14}}{7}$

   c. $\pm\dfrac{4\sqrt{15}}{5}$

8. a. $4\sqrt{3};\ 7\sqrt{6}$

   b. $6\sqrt{5};\ 8\sqrt{2}$

   c. $10\sqrt{6};\ 9\sqrt{3}$

9. a. $10\sqrt{5}-10$

   b. $10\sqrt{3}-5$

   c. $10\sqrt{6}-2$

10. a. 9

    b. 49

    c. 36

11. a. $9,\ (x-3)^2$

    b. $25,\ (x-5)^2$

    c. $36,\ (x-6)^2$

12. a. 7

    b. 6

    c. 5

13. a. 4

    b. 9

    c. 36

14. a. $\dfrac{-e\pm\sqrt{e^2-4df}}{2d}$

    b. $\dfrac{-h\pm\sqrt{h^2-4gi}}{2g}$

    c. $\dfrac{-k\pm\sqrt{k^2-4jm}}{2j}$

15. a. $1,\ -\dfrac{1}{2}$

    b. $\dfrac{2\pm\sqrt{44}}{4}=\dfrac{1\pm\sqrt{11}}{2}$

    c. $\dfrac{3\pm\sqrt{33}}{4}$

16. a. $\dfrac{1\pm\sqrt{13}}{6}$

    b. $\dfrac{2\pm\sqrt{28}}{6}=\dfrac{1\pm\sqrt{7}}{3}$

    c. $\dfrac{3\pm\sqrt{33}}{6}$

17. a. 0, 9

    b. 0, 4

    c. 0, 5

18. a. $\dfrac{9\pm\sqrt{65}}{2}$

    b. $3,\ -\dfrac{1}{2}$

    c. No real number solution

19.

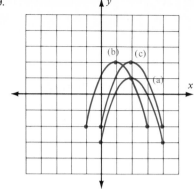

20. a. 5 sec

    b. 7 sec

    c. 8 sec

EXERCISE 10.1

A 1. 5

   3. 15

   5. 10

B 7. a. $\angle DAE$

      b. $\angle CAD$

   9. a. $\angle DAF$

      b. $\angle BAE$

   11. 35°

   13. 15°

   15. $\angle B$

   17. 110°

   19. 20°

   21. 220°

   23. a. 150°

       b. 30°

       c. 150°

   25. $\angle C,\ \angle E,\ \angle A,\ \angle G$

   27. a. 49°

       b. 139°

   29. 16

   31. $x=15;\ 35°$ and $145°$

   33. $x=10;\ 30°$ and $60°$

   35. a. 30°

       b. 180°

C 37. 90°

   39. 25°

41. $W=63.25$

43. $d=45$

45. $b=16$

10.1 USING YOUR KNOWLEDGE

1. $1\dfrac{1}{4}$ in.

3. $3\dfrac{3}{8}$ in.

5. 3.1 cm

EXERCISE 10.2

A 1. 65 cm

   3. 12.6 yd

   5. 184.6 m

   7. 57.2 m

   9. 30 cm

   11. 10 ft

   13. 90 ft

   15. 716 ft by 518 ft

   17. 80 ft

   19. About 1.31 in.

B 21. 15 in.²

   23. 15 in.²

   25. 22 ft²

   27. 30 ft²

   29. 957 in.²

   31. 6400

   33. 680 ft

   35. About 80 ft

37. $b=\pm3\sqrt{5}$

39. $b=\pm\sqrt{5}$

41. $f=2$

43. $b=\dfrac{35}{4}$

45. $d=\dfrac{9}{2}$

10.2 USING YOUR KNOWLEDGE

1. a. $2\dfrac{2}{3}$

   b. $42

3. a. 15

   b. $60

5. a. 2.49 cents per in.²

   b. 2.36 cents per in.²

   c. The 10-in. pie

EXERCISE 10.3

A 1. 5 in.

   3. $2\sqrt{74}$ ft

   5. $\sqrt{5}$ ft

   7. $2\sqrt{5}$ m

   9. $3\sqrt{2}$ yd

   11. a. 10 m

       b. $(15+5\sqrt{3})$ m

   13. a. $8\sqrt{2}$ in.

       b. $(16+8\sqrt{2})$ in.

   15. a. 8 yd

       b. $(12+4\sqrt{3})$ yd

   17. a. $6\sqrt{2}$ yd

       b. $(12+6\sqrt{2})$ yd

B 19. $x=5\dfrac{1}{3},\ y=6\dfrac{2}{3}$

   21. $13\dfrac{5}{7}$ in.

   23. $x=16,\ y=8$

25. $x = 12$, $y = 12$
27. $a = 24$, $b = 30$
29. $x = 8$, $y = 8\frac{4}{5}$
31. $x = 10$, $y = 25$
33. Both $11\frac{1}{5}$ in.

✓ **SKILL CHECKER**

35. $r = 15$
37. $r = 15$

10.3 USING YOUR KNOWLEDGE

1. 75 ft
3. 18 ft
5. 16 ft

EXERCISE 10.4

1. 240 cm^3
3. $128\frac{1}{4}$ in.^{3}
5. 120
7. 1350
9. 7.5
11. 2312 in.^{3}
13. 6280 cm^3
15. 31.8 m^3
17. 12,560 ft^3
19. 28.3 in.^{3}; 17.0 oz
21. 628 in.^{3}
23. 52,333.3 ft^3
25. 0.4 m^3
27. 1,562,665
29. 17.0

31. 8000 m
33. 8 in.
35. 10 mm
37. 10 ft
39. 512 in.^{3}

10.4 USING YOUR KNOWLEDGE

1. 3 units
3. $r = 3$ units, $V = 36\pi$ cubic units, $S = 36\pi$ square units

REVIEW EXERCISES CHAPTER 10

1. a. 7
   b. 9
   c. 11
2. a. $m\angle A = 60°$, $m\angle B = 30°$
   b. $m\angle A = 72°$, $m\angle B = 18°$
   c. $m\angle A = 75°$, $m\angle B = 15°$
3. a. $m\angle A = 120°$, $m\angle B = 60°$
   b. $m\angle A = 144°$, $m\angle B = 36°$
   c. $m\angle A = 150°$, $m\angle B = 30°$
4. a. 115°
   b. 113°
   c. 111°
5. a. 65°
   b. 67°
   c. 69°
6. a. 115°
   b. 113°
   c. 111°
7. a. 75°
   b. 65°
   c. 55°
8. a. 123 in.
   b. 163 in.
   c. 203 in.

9. a. 50 in.
   b. 40 in.
   c. 30 in.
10. a. 6 in.
    b. 4 in.
    c. 3 in.
11. a. 252 in.
    b. 180 in.
    c. 140 in.
12. a. 4 cm
    b. 6 cm
    c. 8 cm
13. a. $6\sqrt{2}$ in.
    b. $\sqrt{95}$ in.
    c. $2\sqrt{30}$ in.
14. a. $(80 + 40\sqrt{2})$ in.
    b. $(100 + 50\sqrt{2})$ in.
    c. $(120 + 60\sqrt{2})$ in.
15. a. $(9.6 + 3.2\sqrt{3})$ in.
    b. $(12.6 + 4.2\sqrt{3})$ in.
    c. $(15.6 + 5.2\sqrt{3})$ in.
16. a. 24 in.
    b. 36 in.
    c. 48 in.
17. a. 27 in.^{3}
    b. 64 in.^{3}
    c. 125 in.^{3}
18. a. 36,738 gal
    b. 25,512.5 gal
    c. 13,328.0 gal
19. a. 33.5 cm^3
    b. 133.0 cm^3
    c. 267.9 cm^3
20. a. 1 in.
    b. 1 in.
    c. 1 in.

INDEX

Equation (continued)
  solution of an, 410
  $y = mx + b$, 424
Equations
  consistent, 468, 482, 489
  dependent, 468, 482, 489, 501
  equivalent, 192, 273
  fractional, 555, 573
  inconsistent, 468, 482, 489, 501
  linear, 220, 273
  linear, procedure for solving, 222
  of lines, 445
  literal, 224
  quadratic, solving by factoring, 385
  simultaneous, 465
  solving, 219
  solving linear, 222
  solving by three-step rule, 196
  systems of, 501
Equilateral triangle, 681, 698
Equivalent equation(s), 192, 273
Equivalent fractions, 10, 73
Euclid, 661
Evaluating
  expressions, 66
  formulas, 62
  polynomials, 286
Expanded form of a decimal, 31, 73
Exponent, 147, 151, 157
Exponential notation, 130
Exponents, 102–103, 130, 174
  first law of, 159, 175
  negative, 157
  second law of, 160, 175
  third law of, 161, 175
Expression(s)
  adding, 137
  division of, 147
  evaluating, 66
  multiplication, 147, 175
  rational, 513
  subtracting, 137

F

Factor, 103, 345
  greatest common (GCF), 346
Factoring
  common factors, 346
  the difference of two squares, 371, 393
  general strategy, 377–378, 393
  by grouping, 348
  polynomials, 345
  solving quadratics by, 385
  squares of binomials, 369–370, 393
  by trial and error, 361
  trinomials, 355, 358
Factors, 62, 147
Figures, similar, 683, 698
Finding the LCD, 21–22
First law of exponents, 159, 175
FOIL method, 333
  to multiply binomials, 303
Formula, quadratic, 619–620, 647

Formulas, 61
  evaluating, 62
  geometric, 63
  using, 61
  writing, 61
Fraction
  algebraic, 513, 573
  complex, 549, 573
  continued, 547
  to percent, 45
  standard form of a, 519, 573
Fractional equations, 555, 573
Fractions
  addition and subtraction of, 537–538, 541, 573
  addition of, 20
  building, 514
  common, 9
  to decimals, 32
  division of, 19, 73, 530, 573
  equivalent, 10, 73
  fundamental rule of, 515, 573
  multiplication of, 17, 73, 527–528, 573
  to percents, 45
  reducing, 11, 517, 573
  simplifying, 517, 549
  subtraction of, 23, 73
Fundamental rule of fractions, 515, 573

G

Gear ratio, 15
General factoring, 393
General problems, 497
Geometry, 661
Geometry, analytic, 419
Gesselmann, Harrison A., 271
Graph of a linear equation, 451
Graphing horizontal lines, 423
  inequalities, 437
  linear equations, 427
  lines, 419
  ordered pairs, 408
  parabolas, 629–634
  quadratics, 629–634, 647
  solving systems by, 465–466
  vertical lines, 423
Greater than ($>$), 232
Greater than or equal to, 231
Greatest common divisor (GCD), 12
Greatest common factor (GCF), 346
Grouping symbols, 51, 130
  removing, 141
Grouping, factoring by, 348
Guess-and-correct procedure, 271

H

Half-line, 662
Horizontal line(s), 451
  graphing, 423
Huygens, Christian, 549
Hypotenuse, 682

I

Identity for addition, 54, 74
  for multiplication, 53, 74
Identity laws, 55
Inconsistent equations, 468, 482, 489, 501
Inequalities
  addition and subtraction properties, 234
  compound, 240
  graphing, 437
  multiplication and division properties of, 236, 238
  solving, 231–232
Inequality, 273
  addition property of, 273
  division property of, 273
  linear, 452
  multiplication property of, 273
  subtraction property of, 273
Integer problems, 248
Integers, 9, 73
  addition of, 93–94, 174
  addition and subtraction, 91, 96, 174
  division of, 104
  multiplication and division of, 101–102, 104
  negative, 9, 73
  positive, 9, 73
  subtraction of, 95, 174
Intercept, x, 421, 451
Intercept, y, 421, 451
Interest, compound, 162
Intersecting lines, 469, 661, 695
Inverse
  additive, 54, 74, 91–92
  multiplicative, 55, 74, 207
Inverse laws, 55
Inverses, 109
Investment problems, 262
Irrational numbers, 588, 647
Irrational numbers, 10, 35, 73, 546
Isosceles triangle, 681, 698

L

Law of exponents
  for division, 150, 175
  for multiplication, 148, 175
Least common denominator (LCD), 20, 73
Least common multiple (LCM), 20, 29
Legs of a right triangle, 682
Length, 662, 696
Less than ($<$), 232
Less than or equal to ($\leq$), 231
Light year, 173
Like
  radicals, 603
  terms, 137, 175
  terms, adding, 137
  terms, subtracting, 139
Line
  segment, 662, 696
  finding the equation of, 448
  horizontal, 451
  vertical, 451